19th Topical Conference on Refinery Processing 2016

Topical Conference at the 2016 AIChE Spring Meeting and 12th Global Congress on Process Safety

Houston, Texas, USA
10-14 April 2016

ISBN: 978-1-5108-2488-1

Printed from e-media with permission by:

Curran Associates, Inc.
57 Morehouse Lane
Red Hook, NY 12571

Some format issues inherent in the e-media version may also appear in this print version.

Printed by Curran Associates, Inc. (2016)

For permission requests, please contact AIChE
at the address below.

AIChE
120 Wall Street, FL 23
New York, NY 10005-4020

Phone: (800) 242-4363
Fax: (203) 775-5177

www.aiche.org

Additional copies of this publication are available from:

Curran Associates, Inc.
57 Morehouse Lane
Red Hook, NY 12571 USA
Phone: 845-758-0400
Fax: 845-758-2633
Email: curran@proceedings.com
Web: www.proceedings.com

TABLE OF CONTENTS

Effective Relief System Evaluation through Process Safety Information Optimization

Srihari K Maganti
PSM Consultant
15995 N Barkers Landing, Suite 145, Houston, TX 77079
srihari.maganti@provenanceconsulting.com

Sara Tzaperas
Line of Service Manager, ARTS™
815 Brazos Street, Unit 204, Austin, TX 78701
sara.tzaperas@provenanceconsulting.com

Prepared for Presentation at
American Institute of Chemical Engineers
2016 Spring Meeting
11th Global Congress on Process Safety
Houston, Texas
April 10-14, 2016

Effective Relief System Evaluation through Process Safety Information Optimization

Srihari K Maganti
PSM Consultant
15995 N Barkers Landing, Suite 145, Houston, TX 77079
srihari.maganti@provenanceconsulting.com

Sara Tzaperas
Line of Service Manager, ARTS™
815 Brazos Street, Unit 204, Austin, TX 78701
sara.tzaperas@provenanceconsulting.com

Keywords: Process Safety Information, Data Repository, Data Aggregation, Data Validation

Abstract

Process Safety Information (PSI) is a vital element of Process Safety Management (PSM) per Occupational Safety and Health Administration (OSHA) CFR 1910.119 (d). An engineer needs a variety of PSI documents for relief system evaluation, such as Process and Instrumentation Diagrams (P&IDs), Process Flow Diagrams (PFDs), Heat and Material Balances (H&MBs), relief device specification sheets, equipment specification sheets, equipment mechanical drawings, manufacturer data reports, Process Hazard Analysis (PHA) documentation, Management of Change (MOC) documentation, etc.,. Obtaining accurate PSI can be a daunting task for a Relief Systems Engineer. Process safety optimization can be achieved by an efficient and reliable data repository that can save valuable engineering time. Such a repository can be built by aggregating equipment information from multiple data repositories into a single point of access. This paper describes the use of a data aggregation and validation tool and outlines other PSI needs that can optimize a relief system evaluation.

1. Introduction

Data management plays a major role in every industry. This is particularly true for the oil and gas industry where facilities often have complex workflows, requiring significant amounts of data to perform any operational activity. PSI is the foundation of these activities and must be kept complete and accurate in order to maintain a safe and operable environment [1]. As defined by OSHA CFR 1910.119 (d), PSI consists of information pertaining to process technology (such as process chemistry, process simulation, heat and material balance, PFDs, and P&IDs), process equipment (such as U1-A form, mechanical drawings, specification sheets, repair reports, and MOC documentation), and process hazards (such as Material Safety Data Sheets and PHA documentation).

In order to perform a Relief Systems Evaluation (RSE), a Relief Systems Engineer needs a variety of PSI documentation and PSI data points, such as inlet, outlet, orifice sizes, and set pressure for a relief device; maximum allowable working pressure (MAWP) and dimensions for fixed equipment; and Heat and Material Balance and process simulation. This information often resides in multiple data repositories across many departments within a facility. Obtaining accurate and up-to-date PSI can be a daunting task, and PSI discrepancies can lead to errors or misjudgment, which and pose a significant risk to the facility [2]. Accuracy, completeness, and ultimate confidence in this data are frequent challenges to a Relief Systems Engineer, so significant time is often spent tracking down the right information.

In a typical RSE, the task of obtaining accurate and up-to-date PSI is a step embedded within the evaluation process; as discrepant or missing information is discovered by the Engineer, he or she seeks out the accurate value by reviewing the available data or document sources at the facility. However, there are pitfalls to this process that allow for significant inefficiencies and inaccuracy of data. Redefining the RSE process to focus on effective data management significantly reduces cost and increases accuracy. Effective data management must focus on integration of the multi-departmental data and document sources, traceability of the required data back to the validating source, optimization of the process to gather the data, and employing a data management tool to facilitate this process. Data management is the foundation of an RSE, and an effective RSE can only be achieved through effective data management.

2. Typical PSI validation process for RSE – Case Study

As a case study, assume an RSE is to be performed for an entire unit or facility. Assume that this facility has access to historical RSEs in an electronic relief systems design database. Additionally, assume the facility has access to various electronic and/or hard copy documentation and PSI located in multiple data repositories that are maintained by separate departments. Examples of data repositories include maintenance databases such as SAP and Maximo, and inspection databases such as PCMS, Meridium, and UltraPIPE.

2.1 Typical validation process

When performing an RSE, the steps involved in the typical validation process are as follows:

2.1.1. *Initiate relief system evaluation*
The Relief Systems Engineer starts by reviewing and evaluating the previous RSE design database on a system by system basis.

2.1.2. *Discover missing and discrepant data*
As the Engineer evaluates the system, the Engineer reviews the PSI residing in the RSE design database for that system to check for accuracy and completeness. To check for accuracy, the data in the RSE design database is typically cross-referenced with the data on the P&ID. This cross reference is performed to catch any changes made since the last evaluation. Furthermore, the Engineer checks for the accuracy of the data by performing a sanity check of the data while evaluating the system, e.g., asking *is the relief valve set pressure less than or equal to the MAWP of the vessel?* Finally, the Engineer checks for completeness by determining what information required for the evaluation is not available in the RSE design database or on the P&ID.

2.1.3. Gather PSI for missing and discrepant data
Once missing or potentially discrepant information is identified, the Engineer seeks out the accurate value. The process by which this information is sought out is typically ad hoc based on where he or she thinks the information will most likely reside. For example, if a vessel MAWP is missing or discrepant, the following steps may be taken:

a. Search electronic documentation database for the manufacturer's data report (U-1A), nameplate rubbing, vendor drawing, etc.
b. Search hard copy files for paper documentation
c. Call the Inspection Department to see what relevant information resides in their database
d. Field verify to obtain data stamped on the nameplate
e. Call the vendor

2.1.4. Populate RSE design database with new PSI
Once the correct data value is obtained, the value is populated by the Engineer into the RSE design database.

2.1.5. Complete RSE
Now that all required available data is obtained, the Engineer can finish the evaluation.

2.2 *Pitfalls associated with typical validation process*

When an RSE is performed for one or two systems, the above validation process may not be unreasonable; the engineer can take the time to ensure the accuracy and completeness of the data by manually comparing and gathering appropriate PSI. However, for larger RSE efforts or even smaller efforts repeated over time, there are a number of pitfalls in this process that can cause significant time delays and reduce confidence in the data.

2.2.1. Inefficient PSI Gathering
The process to *Gather PSI for missing and discrepant data* can lead to redundant or inefficient work over the lifetime of the RSE effort. For example, for the first system, the Relief Systems Engineer may need to call the Inspection Department to ask for assistance in verifying the MAWP of a vessel. Then for the next system a few days later, the Engineer may need to call the Inspection Department again to ask for the dimensions of a different vessel. This redundant effort is a time burden on both the Relief Systems Engineer and the Inspection Department. This same inefficiency can also be applied for any step in the PSI gathering process: going to the field to obtain nameplate information on neighboring equipment, walking to the file room to review a single hard copy file each time for different equipment, etc. Furthermore, if the RSE is performed by an offsite contractor, these inefficient requests for information system by system place an unnecessary financial or time burden on the client.

Additionally, the PSI gathering process can take time and create a bottleneck for the Engineer performing the RSE. If the Engineer has to request information from a different department or the vendor, it may take days, weeks, or months to receive a response. Once the information is provided, the Engineer has to familiarize themselves with the system again before starting the evaluation. Though this may seem to be minimal, over time the delays and re-familiarization can cause significant time delays.

2.2.2. Reduced Confidence in Data

During the *Discover missing and discrepant data* step, it is the responsibility of the Relief Systems Engineer to discover the discrepant information. However, that process allows for possible oversights. For example, in the case study outlined above, the discovery of discrepant data is dependent on up-to-date and accurate P&IDs; however, if the P&ID data is lacking or if there was a change that accidentally did not get captured on the P&ID, the Engineer may not catch the inaccuracy.

Furthermore, the data that is available on the P&ID or in the RSE design database is not always easily traced back to the validation source, such as the manufacturer's data report or nameplate. Therefore, when there is a discrepancy, it is hard for the Engineer to know what to trust or where to look for the correct value.

3. Optimized PSI validation process for RSE

The pitfalls associated with validation process can be minimized by redefining the validation process. Before the Relief Systems Engineer even begins the RSE work, the PSI data validation effort can be defined, executed, and documented. Streamlining and frontloading this effort drastically increases the efficiency and accuracy of the RSE data validation process. Additionally, utilization of a PSI Verification and Traceability Software tool can facilitate this effort, allowing for increased efficiency in the process, increased accuracy and confidence in the data, and long term reusability of the effort.

3.1. Optimized Validation Process

3.1.1. Define required PSI and establish hierarchy of validating sources

Prior to starting an RSE, the Engineer can generally define the PSI data points required for the evaluation. For example, for each vessel, the Engineer will need to know the MAWP, design code, dimensions, orientation, head types, height above grade, and vacuum rating. As the Engineer performs the RSE, other data points may ultimately be required, but a predefined list can capture a majority of the required data. Similar predefined lists can be generated for each equipment type as well as for the relief valve.

Once the required PSI data points are identified, the Engineer can establish a list of sources from which each data point can be validated. For example, if a vessel MAWP is missing or discrepant, typical validating sources are the manufacturer's data report, manufacturer's nameplate, vendor drawing, specification sheet, P&ID, inspection software, and contacting the vendor. These validating sources may vary from facility to facility but can be defined by the Engineer prior to the execution of the RSE.

Now that the required data points and the recommended validating sources for each data point have been defined, a hierarchical procedure to find this information can be established. A potential hierarchy of these sources to obtain vessel MAWP may be as follows:

1. Check the electronic documentation for the manufacturer's data report (U-1A), nameplate rubbing, vendor drawing, etc.
2. Check the hard copy files for paper documentation
3. Check the nameplate on vessel in the field
4. Check the P&ID
5. Call Inspection Department to see what resides in their database
6. Call the vendor

Again, this hierarchy may vary for different facilities; for example, for a vessel MAWP, some facilities may have high confidence in their P&IDs while others would prefer to confirm with the manufacturer's nameplate. Either way, this process can be pre-defined by the Engineer.

3.1.2. *Identify missing and discrepant PSI*

Now that the required data points have been determined, each data point can be reviewed to identify missing and discrepant values. This may seem like a daunting task; however, this task can be streamlined by leveraging the facility's electronic data repositories. Data from each departmental repository can often be extracted in mass and then loaded to a single system for easy data comparison. Leveraging other departmental data repositories increases the accuracy of the data: rather than just having the Engineer manually compare the RSE design database to the P&IDs, this process also compares it to other departmental repositories, which is likely to identify additional discrepancies.

Employing a Verification and Traceability Software tool can make this task extensively automated. For example, Figure 1 below shows an example of data from each departmental repository uploaded to a single system. Once in the same system, the data can be cross referenced to identify discrepant and missing data: the ✖ represents discrepant values and the ⦸ represents missing values.

	Inspection Database	Drawings (P&IDs)	Maintenance Database	Relief Systems Database
Unit No	Unit 101	Unit 101	Unit 101	Unit 101
MAWP	✖ 385	✖ 385	✖ 385	✖ 305
MAWT	✖ 475	✖ 475	✖ 475	✖ 500
MDMT	⦸	n/a	n/a	n/a
Diameter	⦸	1' 11"	n/a	1.9
Length	⦸	✖ 13' 2-3/16"	n/a	✖ 3.7
Shell Material	SA-516-70	SA-516-70	n/a	n/a
Head Material	SA-516-70	SA-516-70	n/a	n/a
Design Code	S8/D1	n/a	n/a	n/a
Head Type	n/a	n/a	n/a	Elliptical
Config/Section	n/a	n/a	n/a	VV
Vacuum Rating	n/a	n/a	n/a	⦸

⦸ Missing Value

✖ Discrepant Value

Figure 1: *Integrated equipment information from different data resources using ARTS™*

3.1.3. PSI Gathering and Validating

With the potential discrepant and missing values now identified, a Data Technician can follow the pre-defined hierarchy of sources to validate these data points. Since the data comparison effort was done simultaneously, the process for gathering the PSI can be streamlined. For example, using the example hierarchy above, the Data Technician can first access the electronic database to search for the manufacturer's data report, nameplate rubbing, vendor drawing, etc. for all vessels with discrepant or missing data; this search may be successful for confidently validating 40% of the discrepant and missing values. Then for the remaining 60% of the values, the Data Technician can make one trip to the hard copy file location and search for all possible files at once; this search may be successful for an additional 20% of the values. And so forth. Eventually the gathering and validation is performed for each step in the hierarchy, after which all options for tracking down available data are exhausted.

By consolidating the validation of all discrepant and missing values into one PSI gathering and validation effort, redundant steps are minimized, reducing the overall time spent to perform this work. It also minimizes the request burden on other departments; for example, once the Data Technician reaches the "contact the Inspection Department" step in the hierarchy, the Data Technician only needs to contact them one for a single all-encompassing request, rather than multiple small requests. Furthermore, this entire effort can be performed by a Data Technician rather than an Engineer; once the Engineer defines the pre-defined lists and hierarchy, the actual gathering and validation steps can be proceduralized and executed by a lower cost resource, allowing the Engineer to focus on engineering work.

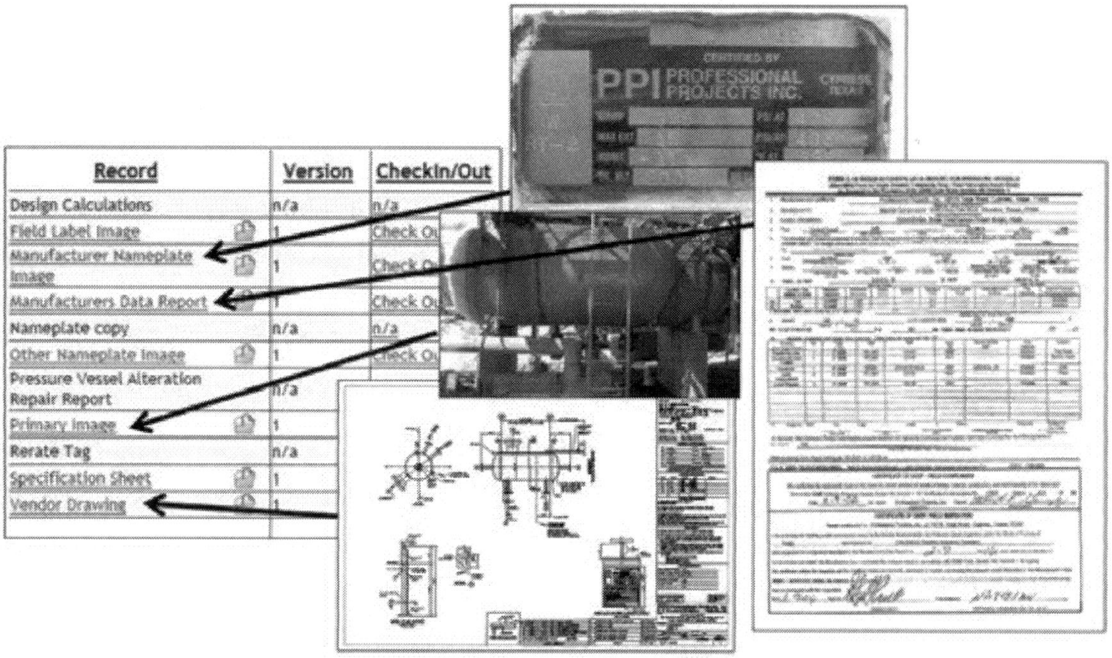

Figure 2: *Link all validating documentation with the equipment asset obtained from ARTS™*

If a Verification and Traceability Software tool is employed, this gathering and validation process can be documented for quality checking and use for future efforts. As seen in Figure 2, all of the PSI obtained during the gathering step can be linked to the equipment assets within the tool. Then, since both the data and validating documentation are all linked together into one system, the data points can be literally traced back to the validating source as seen in Figure 3. This traceability dramatically increases the Engineer's confidence in the validation process: not only can they see the value has been validated, but they can also see by which validating source it was validated. If this system is maintained over time as changes are made to the facility, then this system can be used again, *eliminating the entire data validation effort for all future RSE work.*

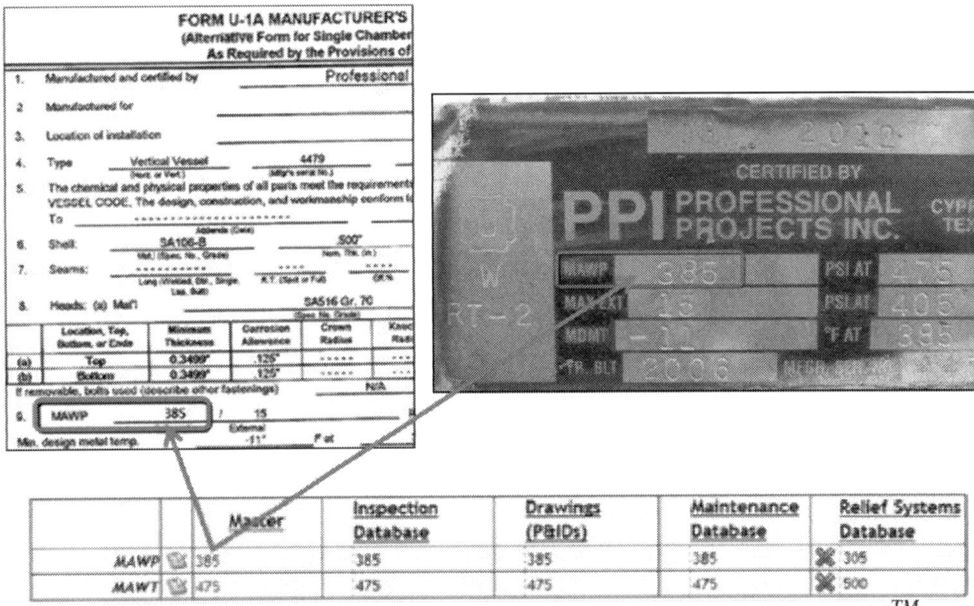

Figure 3: *PSI data points for traceability obtained from ARTS™*

3.1.4. Perform Relief Systems Evaluation

With the data validation process complete, the Relief Systems Engineer can begin the RSE. For each data point defined in the very beginning of this process, the Engineer knows what validating sources were reviewed to ensure accuracy of the data. If a Verification and Traceability Software tool was used, the Engineer can reference this tool for confirmation. Therefore, for each of these data points, the Engineer does not need to perform any additional PSI gathering or validation work.

Invariably, there are certain data points that may not be able to be determined through the validation process, for example, the documentation for a 70-year-old tank may be unavailable. If the result of the data gathering and validation process was a lack of information, then the Engineer knows exactly what validating sources were reviewed; knowing that all validating sources were exhausted; it is up to the Engineer to make an assumption based on best engineering judgement. Ultimately, the Engineer will not be required to perform any additional PSI gathering, allowing the Engineer to complete the effort more efficiently.

3.2. Benefits of the optimized process

For a large RSE or for smaller efforts that are repeated over time, reorganizing the overall RSE process as outlined can achieve increased efficiency, improved accuracy of the data, and therefore improved accuracy of the results of the RSE.

1. Once defined by the Relief Systems Engineer, the PSI validation effort can be performed by a Data Technician, allowing the Engineer to focus their time on performing the RSE rather than tracking down validating sources.
2. Front loading the PSI validation effort allows for a more streamlined effort with fewer trips to field, fewer trips to hard copy files, and fewer calls to other departments and vendors.
3. The Relief Systems Engineer does not begin RSE until the PSI validation effort is complete, allowing the Engineer to complete the effort more efficiently.
4. Consolidation of the data and documents into one location prior to starting the RSE allows the Engineer to easily navigate and reference the PSI.
5. Leveraging data from various departmental repositories allows for the data in the RSE design database to be cross checked for accuracy.
6. Data points are traceable back to the validating source, improving confidence in the data.

Furthermore, this improved process has benefits outside of the single RSE effort, particularly if a PSI Verification and Traceability Software is utilized.

1. Lowers the overall facility risk level by allowing discrepancies to be identified and corrected
2. Facilitates knowledge sharing between all departments
3. Provides accessibility of PSI for all departments
4. Provides validated and traceable data for future efforts
5. Improves efficiency and accuracy of long term PSI maintenance
6. Allows for a quality check of MOC process
7. Provides protection and confidence in PSI and RSE for future audits.

4. Conclusion

Redefining the RSE process with a focused effort to leverage effective data management and PSI optimization significantly reduces cost and increases accuracy. Pre-defining the PSI validation process and completing it prior to beginning the RSE effort increases the efficiency of both the validation and RSE efforts, shifts the burden of data gathering away from the Relief Systems Engineer, and reduces costs by allowing the Engineer to focus on engineer work – all while increasing accuracy. Furthermore, employing a PSI Verification and Traceability Software can significantly reduce the risk to the facility as it allows on-going cross checking of data sources and traceability between data points and validating documentation. Data management is the foundation of an RSE, and an effective RSE can only be achieved through effective data management. If done well, the value of that data validation effort can extend beyond the RSE, benefiting the entire facility on a daily basis.

5. References

[1] Mahmoud, Reza S., "Data Mining and Data Gathering in a Refinery", Proceedings of the 10th WSEAS international conference on Applied computer science, 2010.

[2] http://provenanceconsulting.com/software/arts-asset-and-records-traceability-system/

16AIChE Spring Meeting, Houston, TX

12TH GLOBAL CONGRESS ON PROCESS SAFETY

Paper 24b: Extracting Value From Plant Data

Doug White
Principal Consultant
Emerson Process Management
Houston, Texas

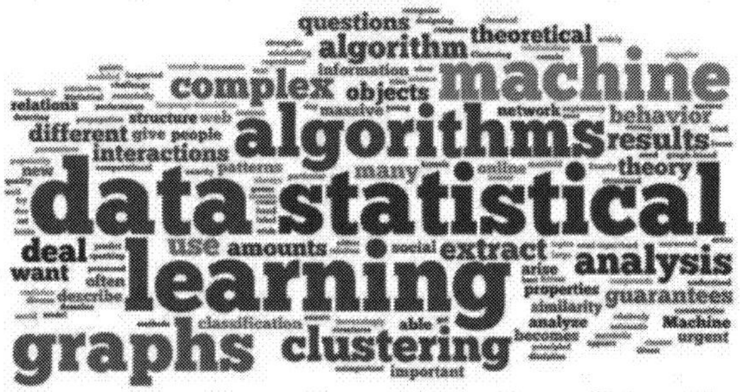

Speaker

Doug White

Principal Consultant

PlantWeb Solutions Group

Emerson Process Management

Background: Many years experience designing, justifying, installing and commissioning advanced real time computer/ automation applications in the process industries.

EMERSON. Process Management

Data is Increasing

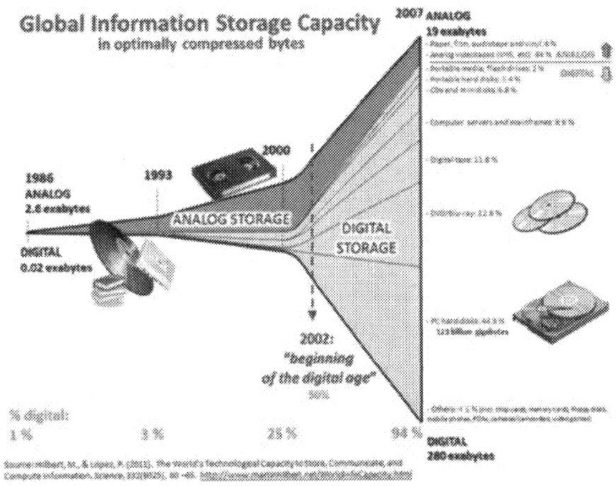

Exponential growth of global storage capacity and processing power

3/27/2016 16AIChE Spring: Paper 24b 3

Data Is Increasing

- Walmart's Online Database
  - More than 250 billion rows
  - More than 20 million updates per hour
  - 3 years history available online
  - 24,000 queries per hour from 1500 users
- Internet Devices
  - Gardner group estimates 5 billion connected devices in 2015 – 25 billion by 2020
  - Plant device connectivity is increasing similarly
- Process Company
  - 10 million tags across 15 sites
  - Implementing system to provide real time access to 3 years of 1 minute sampled historical data for tags

3/27/2016 16AIChE Spring: Paper 24b 4

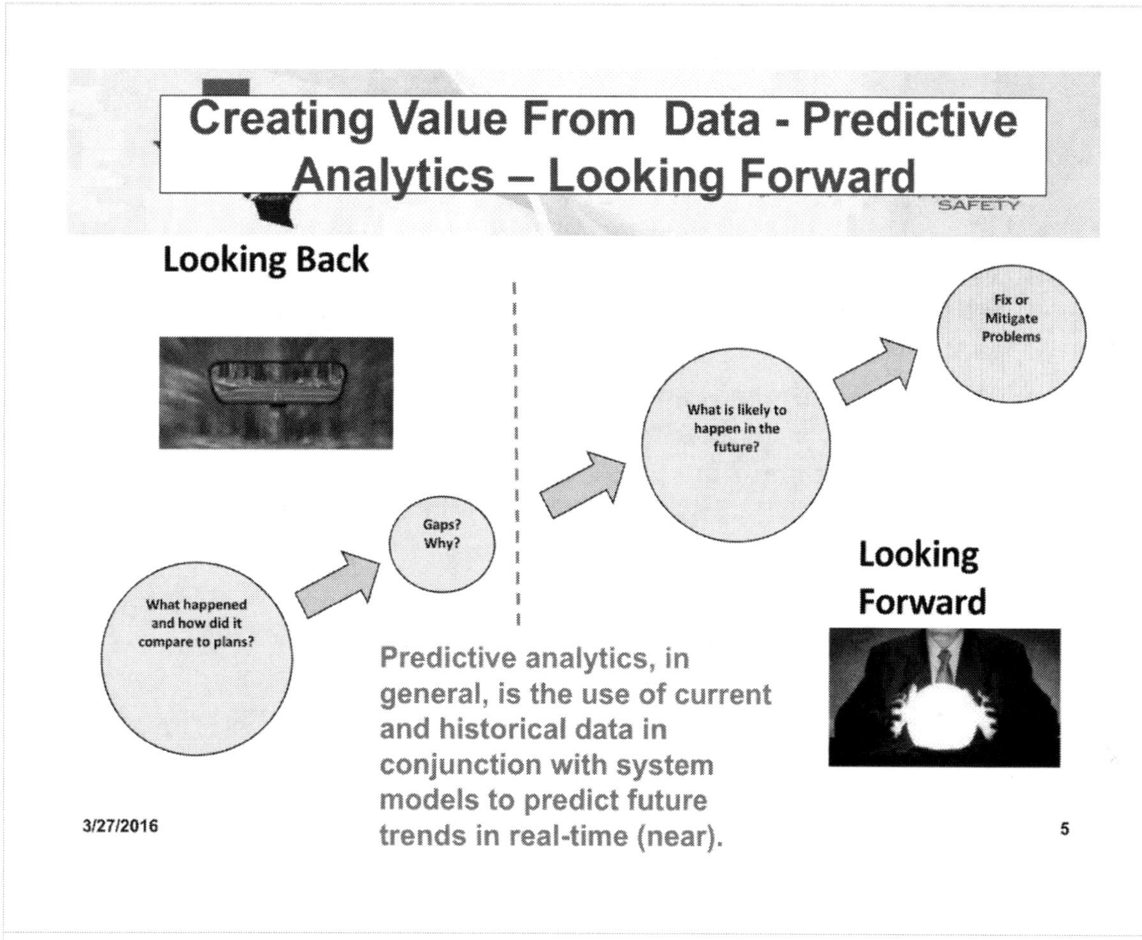
Creating Value From Data - Predictive Analytics – Looking Forward
SAFETY
Looking Back
What happened and how did it compare to plans?
Gaps? Why?
What is likely to happen in the future?
Fix or Mitigate Problems
Looking Forward
Predictive analytics, in general, is the use of current and historical data in conjunction with system models to predict future trends in real-time (near).
3/27/2016
5

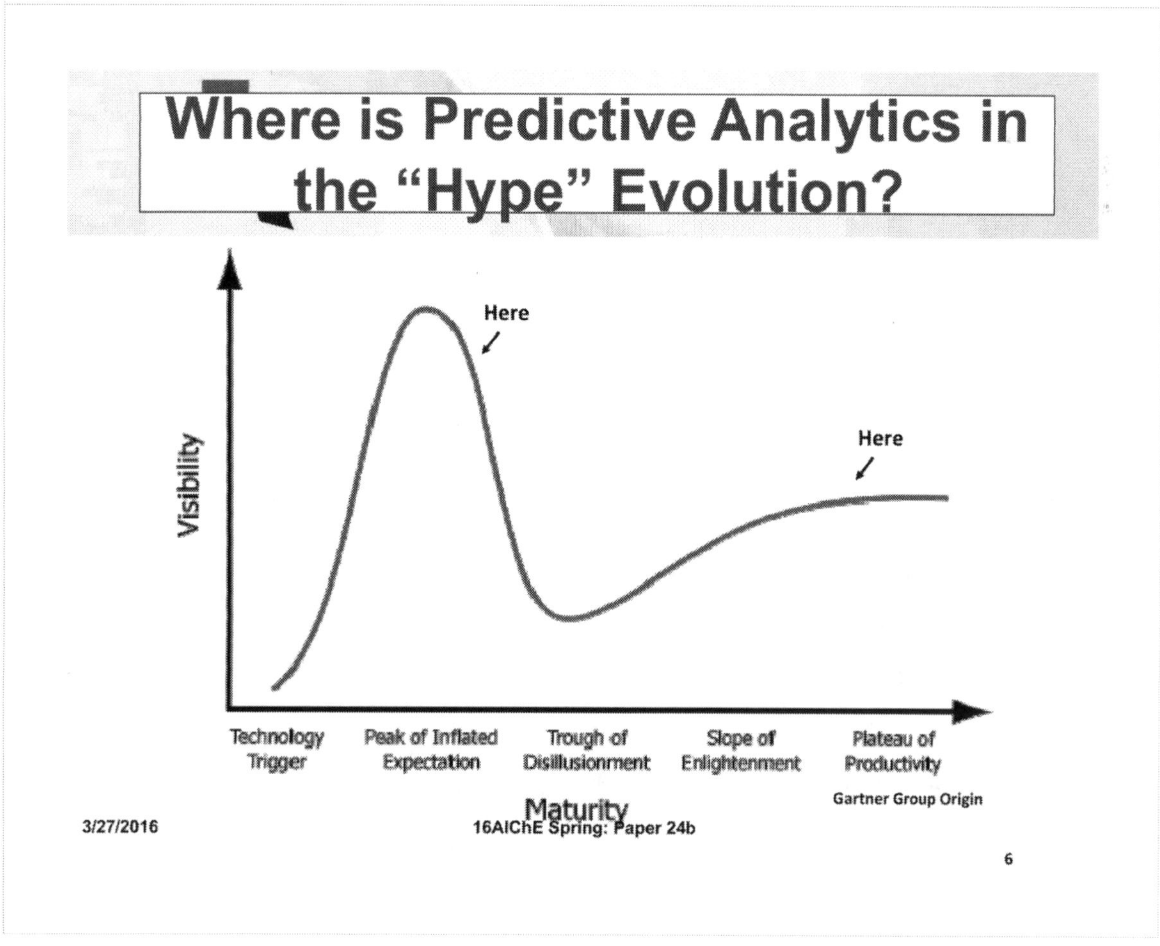
Where is Predictive Analytics in the "Hype" Evolution?
Here
Here
Visibility
Technology Trigger
Peak of Inflated Expectation
Trough of Disillusionment
Slope of Enlightenment
Plateau of Productivity
Maturity
Gartner Group Origin
3/27/2016
16AIChE Spring: Paper 24b
6

Predictive Analytics - Moneyball

- In 2001, the Oakland A's (salaries $44m) lost to the New York Yankees (salaries $125m) in the divisional playoffs and then lost 3 top players to free agency.
- Billy Beane, the manager, hires Peter Brand, a new graduate, to implement statistical analysis of previous seasons to determine what it takes to win. He develops correlations that go against some of the traditional baseball ideas. Beane supports using the approach to pick new players.
- In 2002, The A's set a league record by winning 20 games in a row on the way to winning 95 games overall and making the league playoffs – and doing this again in 2003.

3/27/2016 16AIChE Spring: Paper 24b 7

Predictive Analytics - Predicting Supreme Court Decisions

- Andrew Martin, a professor of political science at Washington University, developed a statistical model of supreme court decisions based on actual decisions from 1994 to 2001 (no change in justices during that period.
- He used the model to predict decisions for the 2002 term. Results were compared against a panel of distinguished legal individuals experienced with the supreme court.
- For the 68 cases in the 2002 term:
  - Model accuracy – 75%
  - Experts accuracy – 59%

3/27/2016 16AIChE Spring: Paper 24b 8

Predictive Analytics – Predicting Flu Trajectory

- Google researchers used 50 million health related search queries for the period 2003 to 2008 to correlate with CDC data on the number and region for physician visits for influenza like symptoms
- Using analytics techniques they found which terms correlated most strongly with the physician visit data (correlation = 0.9)
- Using a predictive model for 2008/ 2009 flu season they were able to predict the number of visits per region about two weeks prior to the CDC report – allowing earlier response

AFPM
American
Fuel & Petrochemical
Manufacturers

How does this apply (and not apply) to the process industries?

Plant ("Near") Real Time Data Includes Many Types

Numeric Data	Spectral Data	Video/ Images	Text
Process/ Offsites/ Utilities/ Terminal Data	Analyzers – NIR, etc	Cameras – Flares, Process, Security, Etc	Daily Plans, Notes, etc
Lab Data	Rotating Machinery – Vibration, etc.	Thermography	Maintenance - Inspection, Work Orders, etc
Accounting, KPI's, Planning, Scheduling, etc	Infrared/ Acoustic/ Ultrasonic – Event Detection	Misc	E-Mail

Typical Refinery DCS IO (Input / Output) Numbers – Continued Increase

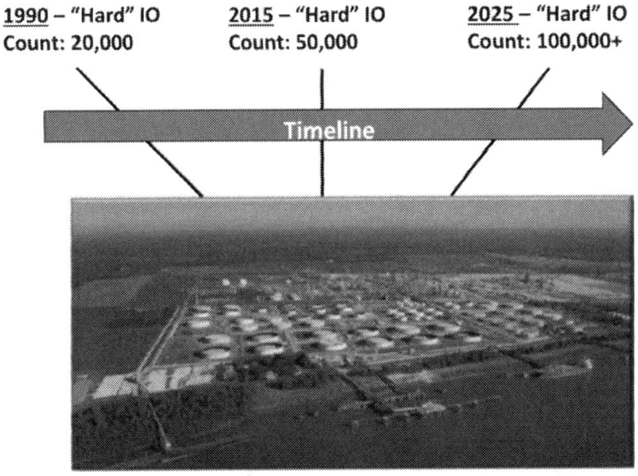

Photo Source: US DOE

Remote Access - "Internet of Things" in Plants

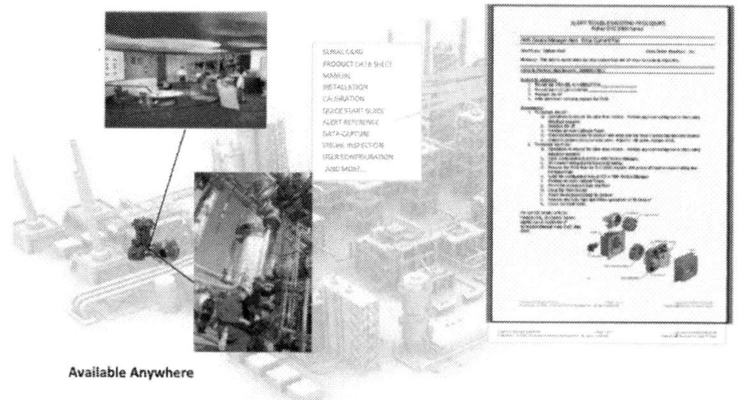

Available Anywhere

Photo Ref: Emerson Process Management

Move the data to the expert, not the expert to the data!

Plant Data Typical Challenges

- Data
  - What data should we collect?
  - How should we store it? Particularly non-numeric data?
  - How should we search and retrieve data of interest?
  - How should we get the best data in the right hands?
  - How should we then act on that data?
  - How do we do this quick and seamless with as little investment as possible?
- Platforms and Tools
  - What infrastructure should I use?
  - What tools should I use?
  - Should I store on-premise or in the cloud?
  - Should I use open source?
  - How do I scale?

What are the Components of a Predictive Analytics Implementation?

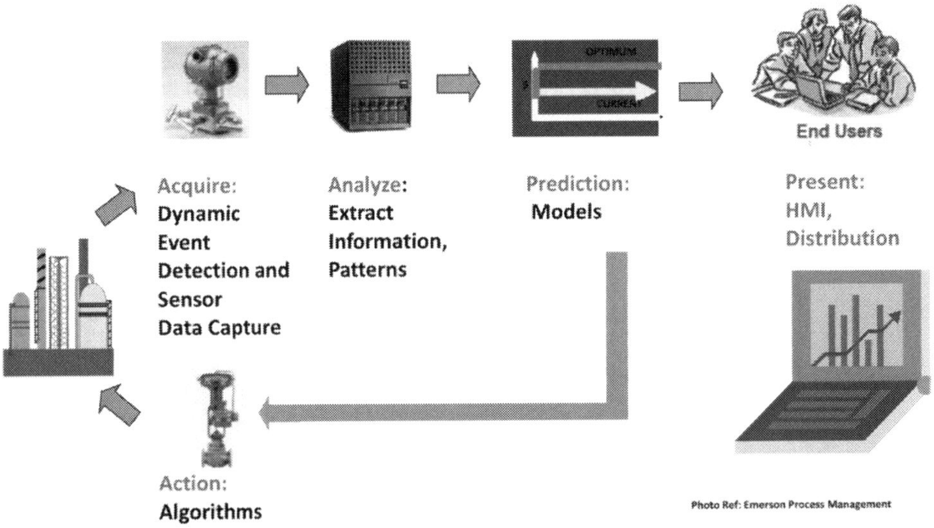

Move from reaction to events to anticipation and optimization

Major Predictive Analytics Techniques

- Numerical Data (with adjustment for time series dependence)
  - Develop Predictive Model
    - Linear Models
      - Continuous Variables – *Generalized Linear Regression*
      - Binary Outcome (Yes/No) – *Logistics Regression*
      - Discrete Outcome (1,2,3,4,..) *Classification and Reference Trees (CART + Random Forest)*
    - Non-Linear Models
      - Continuous Variables - *Neural Nets*
      - Discrete Outcome (1,2,3,4,..) - *(CART + Random Forest)*
  - Detect Anomaly
    - *Principal Component Analysis - PCA*
- Textual Data
  - *Textual Analytics*
- Spectral Data
  - *Nearest Neighbor*
  - *Regression*
- Video/ Images
  - *Neural Nets*

AFPM
American
Fuel & Petrochemical
Manufacturers

Making cents from data - where is the payback?

Plant Operating Objectives - The Four Zero's

- ***Safety*** **– the goal is <u>zero serious safety incidents</u>**
- ***Sustainability*** **– the goal is <u>zero significant environmental incidents, excess energy use and excess waste</u>**
- ***Reliability*** **– the goal is <u>zero unscheduled downtime</u>**
- ***Financial*** **- the goal is <u>zero lost profit opportunities</u>**

How can Predictive Analytics support these objectives?

Plant Decision Cycle

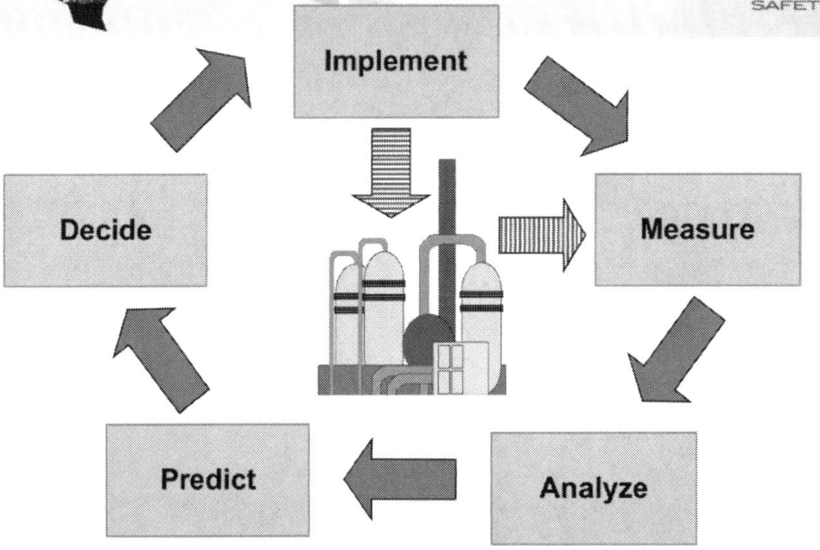

To have a financial impact predictive analytics has to improve this cycle – reduce delays, reduce uncertainty, etc.!

What impact can predictive analytics have?

- **Safety**
  - Avoiding incidents through early detection of potential hazardous situations
- **Availability/ Reliability**
  - Anomaly detection – identifying precursor events to unscheduled equipment outage or problems
  - Performance monitoring – detecting loss of process/ equipment performance before it impacts production capacity
- **Sustainability**
  - Comparing current usage of resources such as energy to its expected usage under current conditions and determining possible causes of variation
- **Financial Optimization**
  - Detecting and dissecting complex interacting constraints on production
  - Determining reasons for product quality/ yield issues
  - Understanding patterns and relationships – developing statistical models that explain them

AFPM
American
Fuel & Petrochemical
Manufacturers

Typical analytics project

Data from plants is often natively unsuitable for analysis

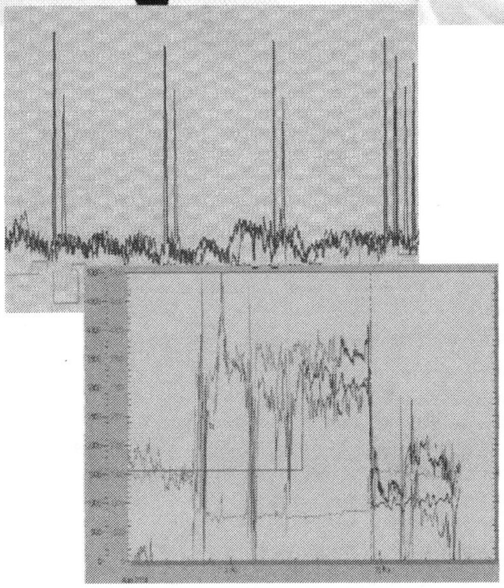

- We need to be able to preprocess and shape data before it can be used to generate models.
- These same techniques must be able to be used off-line to generate models and on-line to preprocess data before it is used to generate predictions, provide fault detections, and generate recommendations.

Cleaning Data – Outlier Detection

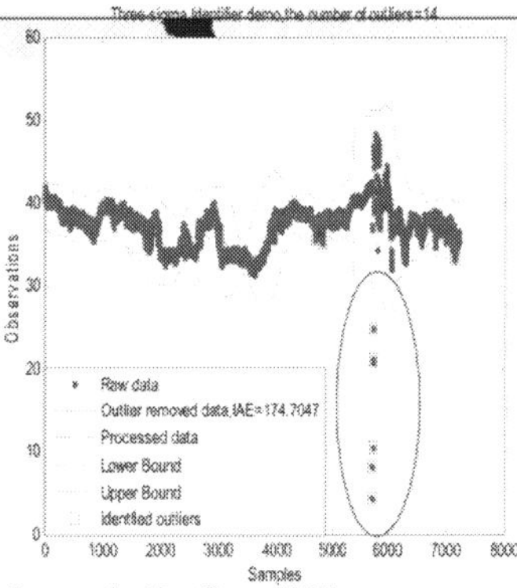

3-sigma outlier identifier on TD601 steam output data (moving window size=n/20=360)

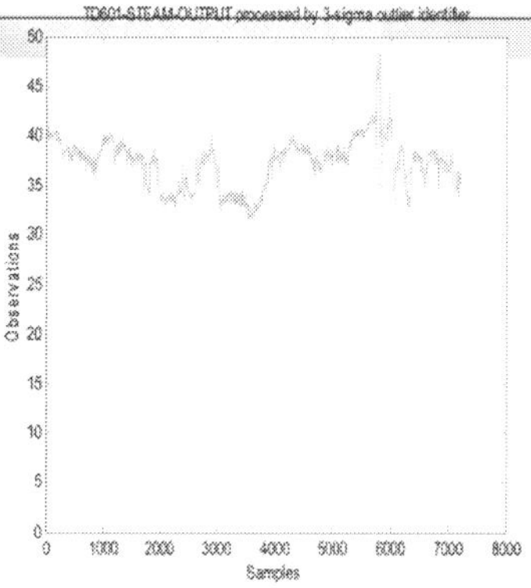

TD601 steam output data processed by 3-sigma outlier identifier(moving window size=n/20=360)

- **Compressed data**
- **Missing data**
- **Outliers**
- **Noisy data**

THE UNIVERSITY OF TEXAS AT AUSTIN
Cockrell School
McKetta Department of Chemical Engineering

Time Relationships Matter

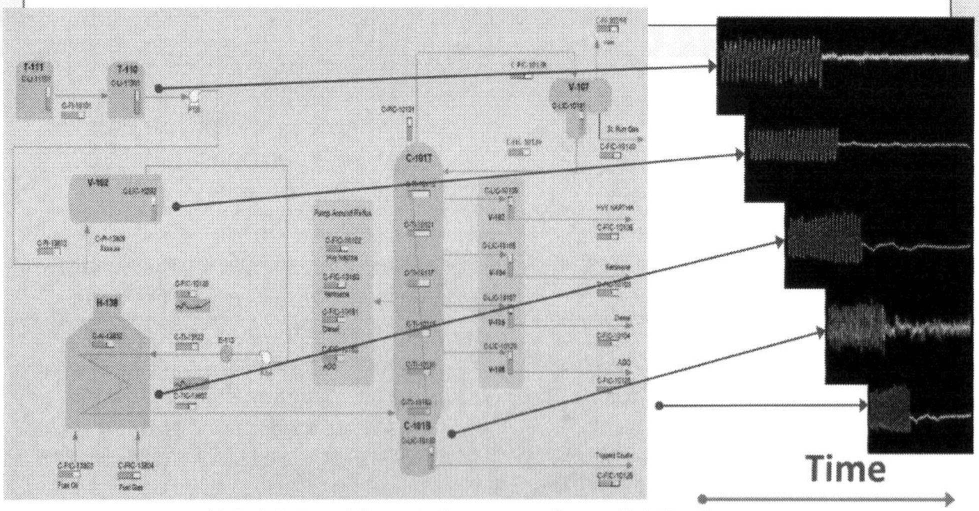

Materials travel through the process in a well defined arrangement. This flow of materials and the time delays are critical to being able to perform any kind of meaningful analysis.

By explicitly defining measurements and actuators along the flow path it is possible for an application to utilize this information to access and utilize information in context.

Finding sources of variations and oscillations

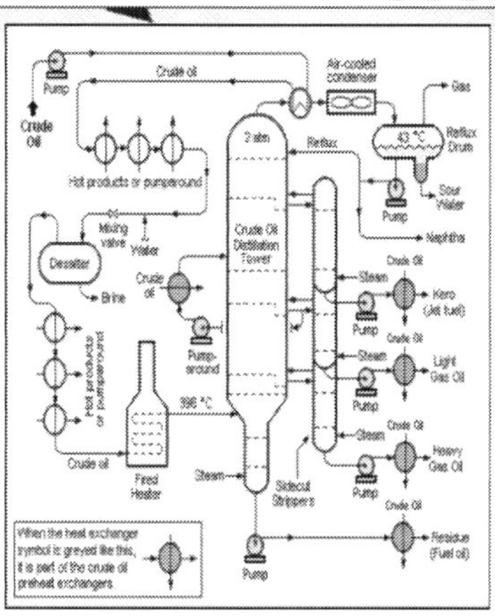

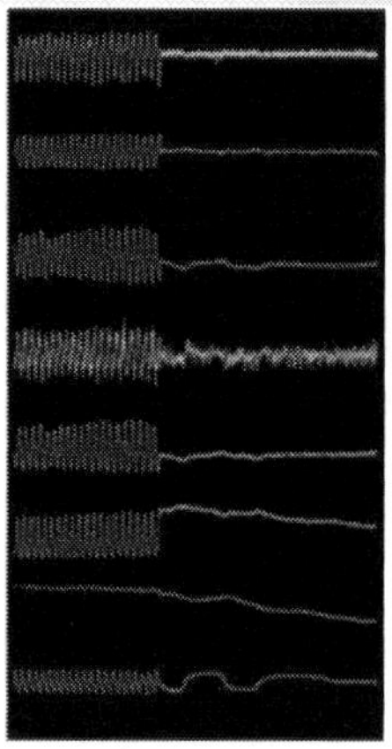

How do I know even where to look?

Parameters in processes are highly coupled. Variation in one parameter shows up in product quality and is reflected in many parameters downstream. Finding the cause can be difficult and time consuming.

Even knowing to look for it can often go undetected.

Cross Correlation and Causality

Output (columns) / Input (rows)

	X1	X2	X3	X4	X5
X1	0	2	3	2	4
X2	-2	0	1	0	2
X3	-3	-1	0	-1	1
X4	-2	0	1	0	1
X5	-4	-2	-1	-1	0

Table 1 Cross correlation time delay

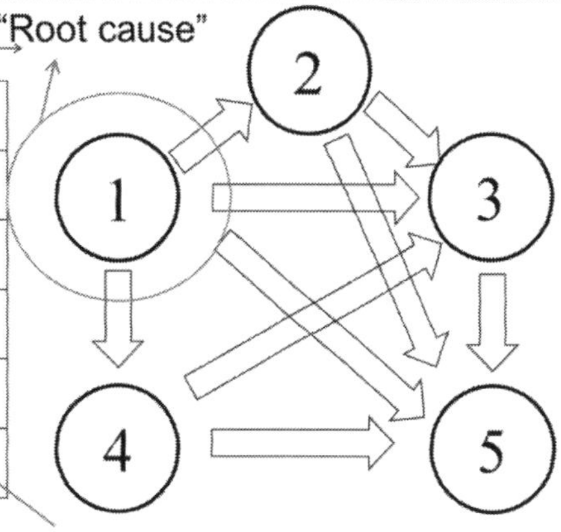

Figure 2 Process topology obtained by cross correlation

Correlation is not Causation!

Selecting parameters and time delays

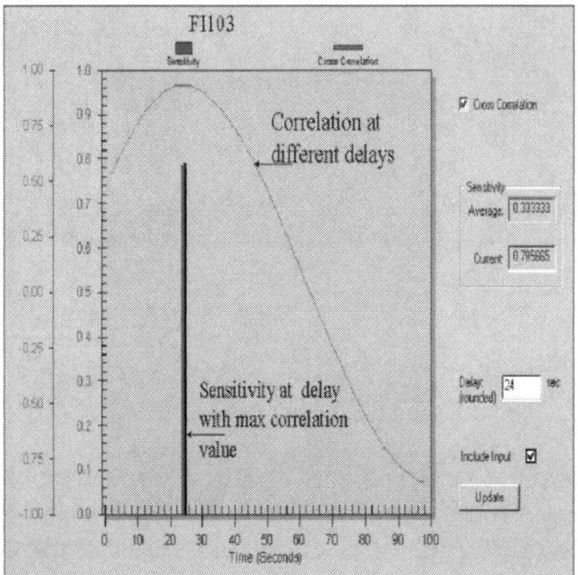

Develop and Evaluate Models

Utilize Block Diagramming Environment to develop and evaluate models

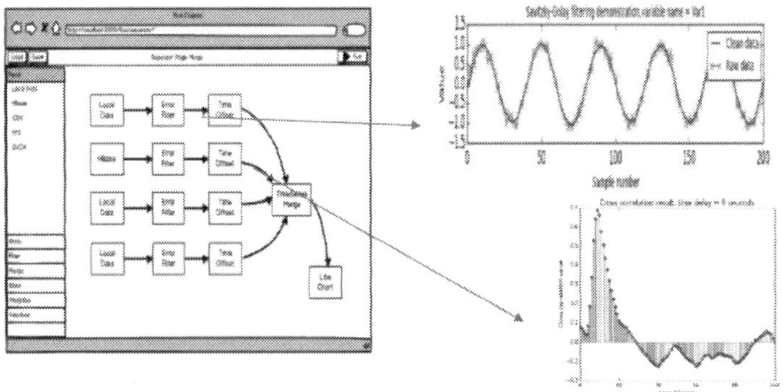

Interact with your model ***visualizing results at each stage of the data processing, when you are done,*** *deploy the model* ***for run-time execution***

Visualizing Predictions

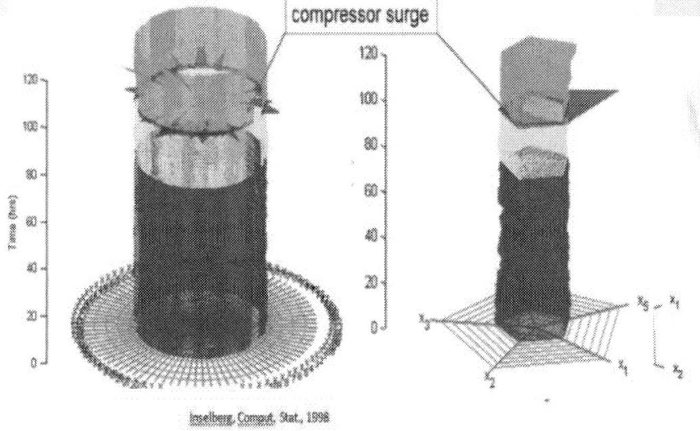

Inselberg, Comput. Stat., 1998

1 to 4 days before surge – Blue
1 day before surge – Yellow
2 hours before surge – Purple
Post surge - Green

Center for Operator Performance
An Industry/University Center for Excellence

THE UNIVERSITY OF TEXAS AT AUSTIN
McKetta Department of Chemical Engineering
Cockrell School

Reference: Dunia, R; G. Rochelle; T.F. Edgar; M. Nixon; "Multivariate Modeling of a carbon dioxide removal process"; Computers & Chem Eng;60 (2014); pp. 381-395

AFPM
American
Fuel & Petrochemical
Manufacturers

Case Studies

Case Study – Predictive Analytics – Waste Water Plant

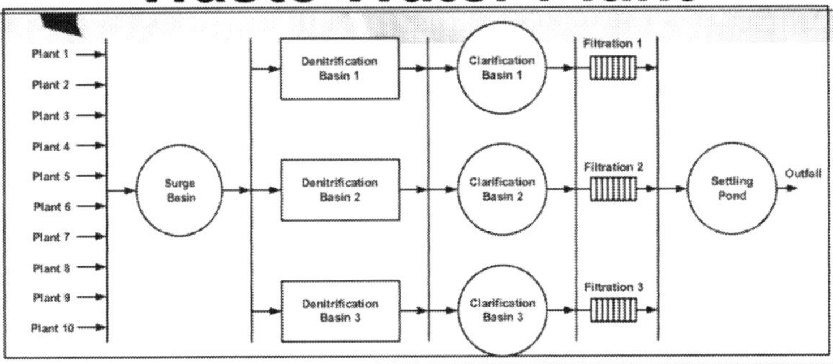

- Multiple effluent NH3 spikes in outfall violating consent decree and incurring fines
- Residence time of the treatment plant >2 weeks making issue ID difficult
- More than 200 measured variables to be considered as possible predictors

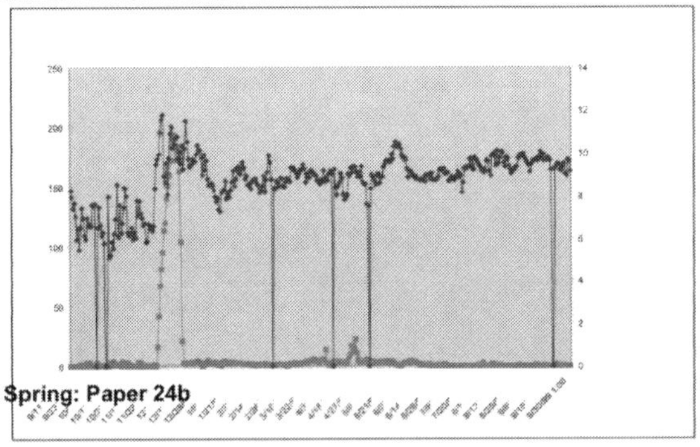

Case Study – Predictive Analytics – Waste Water Plant

- 5 years data analyzed for correlations with simultaneous identification of time lag
- Identification of strong correlation with pH of effluent from one of the plants – large drop in pH preceded excursion by one day
- Vessel cleaning correspondence with pH drop

Case Study – Predictive Analytics - Maintenance

- **Significance**
  - Sipchem Jubail site is considered a world-class complex with 4 large integrated plants
  - Successful implementation of complete integrated maintenance program
  - Built on installed base to create multi-phase asset management program
- **Project Scope**
  - Implementation of Asset Management including Reliability Centered Maintenance on 6,000 assets
  - Uses predictive analytics and diagnostics from field instrumentation
- **Audited Value**
  - 12% decrease in maintenance costs
  - 2% increase in Plant availability

Saudi International Petrochemical Company (Sipchem) actively develops and invests in petrochemical and chemical industries.

Sources: Abdullah Al Ghamde; John E. Hill,; "*Saudi International Petrochemical Company (Sipchem) Reliability Process,*" Paper 01A241; 2010 Emerson Global User Exchange; Sudheer Prabhakaran; John E. Hill,;"*Generating Actual Savings with Sipchem's Reliability Program,*" Paper 1B-2487; 2012 Emerson Global User Exchange

Summary

- Predictive analytics is an evolving technology with many potential applications in the process industries
- Implementation of Predictive Analytics in the process industries has distinctive requirements/ issues due to the time series character and data
- Current applications in process fault detection, availability, safety and optimization have proven value – there will be more in the future

Thank you for attending. Questions?

446613 ASTM D6299 and Leveraging Real-Time Analytics from the Lab

Abstract:

The EPA's new Tier 3 environmental standards require changes with employing ASTM D6299 to closely monitor emissions. This requires tighter control and deploying new analytic procedures. Frequently comparing control limits with T tests and F tests are adding steps which may be streamlined using automated procedures. We will also explore methods for automating lab instrument monitoring by streamlining the SQC duplicates and standards testing. Comparing real-time visibility of lab results with online analyzers can also provide powerful insights, but these projects face challenges with data alignment, and integrity. Issues with lead/lag relationships and data method variety such as sample vs event vs time based data, create complications for proper aggregation. These issues can be complicated with measurement errors, gaps in data, and noisy data. The session will showcase examples of how these challenges were overcome, and how additional value was achieved by being able to identify online instruments which may need attention. These methods provide the potential for refineries to run at higher levels of efficiency, reduce the need to downgrade product, and meet new regulatory standards.

Paper 27a

Vacuum Distillation – Process Configurations

Andrew W. Sloley

Advisian
Energy Center II
575 N Dairy Ashford St
Houston, TX 77079

Prepared for Presentation at the
American Institute of Chemical Engineers
2016 Spring Meeting
Houston, TX April 10-14, 2016

UNPUBLISHED

1. Summary

Vacuum tower configuration choices control product yields and energy recovery. Conventional fuels vacuum towers produce products for downstream cracking and have reasonable energy efficiency. However, less common configurations may allow for either diesel recovery, more effective energy recovery, or both.

Dieselization of the European auto fleet has shifted world-wide refining patterns. In all regions, diesel now often makes up the most valuable product from the refinery. The most profitable diesel is the diesel already in the crude oil. Adept molecule management to recover that diesel maximizes refinery profits.

Current energy prices do not favor intensive investment for direct energy savings. Nevertheless, energy savings are becoming more important. This is mainly driven by regulatory pressure from:

- Energy intensity regulations.
- Avoiding emissions (nitrous oxides and others).

Dieselization and energy recovery combine to focus attention on the crude vacuum unit. Significant diesel remains in most vacuum tower feeds. Many refiners find recovering diesel as a vacuum tower product attractive. Diesel recovery also changes temperatures in the vacuum unit, this may change heat recovery. Understanding how diesel recovery and temperature levels interact is critical to evaluating both new units and revamping existing units. Multiple vacuum tower configurations can allow the top vacuum tower product to meet a diesel boiling range specification.

Results are shown for running a typical crude oil for the most common configuration changes. These include:

- Additional pumparounds for heat recovery improvement.
- Diesel recovery with internal fractionation with net top-bottom reflux.
- Diesel recovery with internal fractionation with zero top-bottom reflux.
- Diesel recovery with no internal fractionation.

The results show that:

- Increased fractionation will increase diesel yields.
- Increased fractionation reduces pumparound temperatures and makes heat recovery more expensive (more capital) or less efficient (less heat recovered).

The more diesel the unit produces, the more difficult the heat integration. Trade-offs for adding diesel recovery to both existing plants and to new units must balance the value of diesel recovery against capital costs, yields, and energy recovery.

2. Process Configurations – Fuels and Lubricants

Refinery vacuum units fall into two distinct types: fuels units and lubricants units. Fuels units directly produce fuel streams (diesel) and feeds for downstream conversion units that make fuels. Modern downstream units typically include Fluid Catalytic Cracker (FCC) units and Hydrocracker (HC) units. Products going to FCC units may be hydrotreated before reaching the FCC. Lubricants units produce lubricant base oils for further processing into lubricants. Lubricants units often produce some fuel streams as well.

In concept, the conventional fuels unit vacuum tower operates as a series of drums that combine de-entrainment and absorption heat removal in one physical structure. No intentional reflux is present. Fuels units that recover diesel as a product often add limited internal fractionation capability.

In contrast, a lubricants unit has significant internal reflux and fractionation capability. The fractionation is required to meet viscosity or other lubricant specifications.

The focus here is to examine operation of fuels vacuum units. Table 1 shows the cases evaluated. These cases cover the range of the most economically important options.

Unit *configuration* is the mechanical layout of the unit. Six configurations will be discussed. Results will be presented for four configurations. These four have been selected to illustrate the advantages and disadvantages of the major options examined. Configuration options covered here include:

- Number of pumparounds (heat removal levels)
- Number of fractionation sections
- Location of fractionation sections
- Disposition of internal streams

Unit *operation* is the process objective in making a stream. The options examined here focus on the highest side-draw product from the vacuum column. Plants may use specific names for the top product. The results of three different types of top products are discussed. These are:

- Top product without a fixed specification.
- Top product to a diesel specification.
- Top product to a hydrocracker feed specification.

The appendix lists standard terms and acronyms used.

Table 1 Case Summary

Case	1	2	3	4	5	6	7	8	9
Pumparounds	2	3	3	3	2	2	2	2	3
Products	2	3	3	3	2	2	3	2	3
Diesel Prod					Yes	Yes	Yes	*	Yes
Diesel at specification					Too light	Yes	Yes	*	Yes
MPA duty from		TPA	BPA	Both PA					
Diesel fractionation bed					Yes	Yes	Yes	No	No
Internal flow to HVGO					Yes	Yes	No	No	No
Notes	* infeasible – zero diesel production								

2.1. Case 1 – Conventional Vacuum Unit – 2 Pumparound, 2 Product

Figure 1 shows the conventional refinery vacuum tower. This remains the most common unit configuration. Two pumparounds remove duty from the vacuum tower. Even though built in one shell, this tower operates as three independent sections:

- De-entrainment and washing above the feed entry.
- Bottom pumparound for heat removal.
- Top pumparound for heat removal.

No fractionation or reflux is present between the TPA and BPA sections. The wash oil return from the BPA and the wash bed create some fractionation between the feed and BPA. However, the composition gradient created is small. The wash bed is mainly a de-entrainment operation.

Case 1 is used as the base case for comparison of the different options.

2.2. Cases 2, 3, 4 – Addition of a Pumparound – 3 Pumparound, 3 Product

Figure 2 shows a vacuum unit with three pumparounds and three products. An additional pumparound is added. The purpose of the third pumparound is to increase the heat recovery possible. The extra MPA section adds an additional temperature level for heat integration. It also shifts duty from the TPA and the BPA into the MPA.

The individual cases vary how the heat shifts:

- Case 2 shifts duty from the TPA into the new MPA.
- Case 3 shifts duty from the BPA into the new MPA.
- Case 4 splits the duty in the MPA. Some comes from the TPA and some from the BPA.

The new pumparound operates at a temperature in-between the BPA temperature and the TPA temperature.

2.3. Cases 5, 6 – Top Product as Diesel, Internal Fractionation

Figure 3 shows a vacuum unit with two pumparounds and two products. The top product meets diesel distillation specifications. The tower includes a fractionation section between the TPA (diesel) and the BPA. The reflux from the bottom of the fractionating beds falls into the BPA section. This case is internal fractionation with net TPA-BPA reflux.

The cases vary the diesel product rate:

- Case 5 has a product rate matched to Case 7. This gives a lower diesel endpoint distillation.
- Case 6 has a product rate maximized for the configuration.

The intent of this option is to produce more diesel product and less gas oil product.

2.4. Case 7 – Top Product as Diesel, Internal Fractionation – 3 Product

Figure 4 shows a vacuum unit with two pumparounds and three products. The top product meets diesel distillation specifications. The tower includes a fractionation section between the TPA (diesel) and BPA. The reflux from the bottom of the fractionating bed is drawn as a LVGO product. The intent of this operation is to produce more diesel product and less gas oil product. Case 7 has the same number of theoretical stages as Case 5/6. This case is internal fractionation with zero TPA-BPA reflux.

2.5. Case 8 – Top Product as Diesel, No Fractionation

Figure 5 shows a vacuum unit with two pumparounds and two products. The configuration is the same as Case 1. However, in this case the attempt is made to meet diesel product specifications on the top product. No fractionating bed has been added in this case.

2.6. Case 9 – 3 Pumparound, 3 Product, Top Product as Diesel

Figure 6 shows a vacuum unit with three pumparounds and three products. This is the same configuration as Cases 2/3/4. In this case, the top product meets a diesel product specification. No fractionating bed has been added. This case is diesel production with no internal reflux.

2.7. Other Fuels Configurations

Many other unit configurations are possible and have been built. Figure 7 shows another example. This configuration adds a fractionating bed between the top pumparounds of a unit that is similar to Cases 2/3/4.

In fuels units with fractionation, the distillation specification of the top product may be either set by diesel product requirements or downstream processing requirements. Many units that have gas oil feeding a hydrocracker may have a distillation requirement to protect the reactor catalyst. Figure 8 shows a vacuum unit producing hydrocracker feed. In concept, the hydrocracker feed units are the same as the diesel production units. The major difference is in the distillation specification used.

2.8. Lubricants Vacuum Units

For comparison, Figure 9 shows a vacuum lubricants unit. Lubricants units typically have multiple fractionating beds. Product specifications are set by a combination of performance criteria (viscosity, viscosity index) and product properties (density, distillation). Compared to fuels units, vacuum units often include a large number of fractionating stages and a much more reflux.

3. Basis

The basis for comparison uses a 31.6° API (0.8676 SG) medium density, sweet, moderate sulfur crude (Tia Juana Light) [1]. Tia Juana Light has a density and yield distribution very close to the current (2016) average U.S. crude. The upstream atmospheric tower configuration is the same as in previous work [2] [3] . The upstream atmospheric tower operates at a flash zone pressure of 30 psig and 690°F (308 kPa and 366°C). This gives a 718°F (381°C) cutpoint for the atmospheric tower.

All the vacuum tower configurations use a wet vacuum tower with stripping trays in the bottom of the tower. Vacuum heater, flash zone, and stripping section conditions were kept constant for all cases. Vacuum heater outlet temperature was 775°F (413°C) and vacuum tower cutpoint was 1,046°F (563°C).

All cases kept a constant pressure profile across the wash bed, BPA section and TPA section. Cases that had a fractionation section or a MPA section added used a pressure drop of zero across that section. Of course, real towers do have pressure drop across all sections. The intent of keeping constant pressures was to simplify the analysis of potential heat shifts and fractionation. Evaluation for actual plants must include pressure drop changes when adding pumparounds or fractionation beds.

Performance limits on the vacuum tower have been imposed in actual volumes. Equipment reliability requirements are evaluated in actual volumes. Actual volumetric liquid rate at the bottom of the wash bed (overflash) is kept constant in all case. Yields are shown in standard volumes. Purchasers buy standard volumes. Hence, refinery economics analysis uses standard volumes.

Total distillate yields for all cases are very close. The largest difference is 0.13%. Yield differences arise from changes in wash oil composition and changes in HVGO (BPA) draw temperature. These factors modify wash section fractionation and change the number of standard barrels lost to the bottoms for the same actual volume of overflash at the bottom of the wash bed.

Table 2 shows the results of comparing the different operations. Cases 1-4 focus on improved heat recovery from the base operation. Cases 5-9 examine the results of converting the top product to diesel-boiling range material.

4. Heat Removal – Pumparound and Product

Heat removal in a vacuum tower covers three duties:

- Condensation of the vapor feed.
- Cooling of the vapors down to overhead temperature.
- Condensation of the liquid down to product temperatures.

Net product may leave the vacuum unit as either a hot stream or a cold stream (Figure 10). Hot stream products would leave the unit upstream of the pumparound exchangers. Cold stream products would leave the unit downstream of the pumparound exchangers. Hot-product or cold-product towers do not fundamentally differ in how the vacuum tower works. They do differ in the amount of heat recovery. As product yields shift, the amount of duty in the rundown cooling changes.

Table 2 shows duties for the following:

- PA Q: duty for the pumparound only (hot product tower)
- Prod Q to PA return temperature: extra duty for cooling the product to the pumparound return temperature. This would be the extra duty for a cold product tower.

- Prod Q to final temperature: extra duty from cooling the product from the pumparound return temperature to the final product temperature.
- PA level duty: PA Q plus the Prod Q to PA return temperature.

5. Base Case – Case 1 – Conventional Vacuum Unit – 2 Pumparound, 2 Product

The conventional vacuum unit recovers 15.9 Mbtu/hr (12.5 GJ/hr) TPA level duty and 122.4 Mbut/hr (86.4 GJ/hr) BPA level duty. This base load has more heat recovery into BPA than most typical units. Nevertheless, this heat recovery remains within the feasible range. Units with this split of heat balance between BPA (88%) and TPA (12%) exist.

Figure 11 shows the heat availability for the case. The plot shows total recovered duty versus temperature for a cold product operation. The duty from the pumparound return temperature to the final product temperature is not include ('Prod Q to final temperature' is not included). The desired direction for the heat available curves is shift them up and extend them to the right.

The crude heat required curve has not been added. Attempting to plot vacuum tower heat recovery against crude preheat without examining the atmospheric tower heat recovery would be extremely difficult. In order to focus on vacuum tower effects, the analysis here is being restricted to the generic consequences of an isolated shift in vacuum tower duty and temperature levels.

Economics of vacuum tower operations vary with each specific crude unit. Useful recovered heat generally goes into two areas:

- Preheating crude feed or,
- Steam generation (or boiler feed water preheat).

At current North American pricing (January 2016), a new unit should target a crude preheat temperature of 515°F to 565°F (268°C to 296°C) [3]. Existing units operate in the range of 400°F to 605°F (204°C to 318°C).

6. Fuels Operation and Heat Recovery (Cases 2-4)

Cases 2-4 show the addition of a third pumparound and a third product. Adding an MPA creates an extra level of heat availability. The limiting cases are: all the energy recovered from the MPA is pulled from the TPA (Case 2), or all the heat recovered from the MPA is pulled from the BPA (Case 3). A final example (Case 4) shows an intermediate case where some of the heat comes from the BPA and some from the TPA. Figure 12 plots the heat availability for Case 1 to Case 4.

Adding a pumparound increases cost in the vacuum tower, pumps, external piping, controls, and heat exchangers. Even if economic, the costs of an added pumparound must be compared to the costs of improved heat recovery in other places in the crude unit.

6.1. Case 2 Shifting Duty from TPA into the MPA

Shifting duty from the TPA to the MPA (Case 2) shows the largest gain in the lower temperature region below 300°F (149°C). Above that point, Case 2 has only minor benefits.

In general, most crude units have ample low temperature heat available. Case 2 only helps low temperature heat recovery. Unless the crude unit is running at low energy efficiency, extra low level heat recovery (Case 2) occurs 'below the pinch.' The pinch is where the temperature difference between the heat recovery streams and the feed streams is the smallest. Generally, heat recovery economics are attractive when energy recovery duty is shifted from below the pinch temperature to above the pinch temperature. Even when energy recovery at low temperatures is attractive, an added

MPA must compete with many other ways to recover heat in the crude unit. These other methods will often be more cost effective.

The improved Case 2 heat recovery is also at too low a temperature for effective steam generation or boiler feed water preheat. In addition, the total heat recovered in Case 2 drops. Case 2 operation where TPA duty shifts to an MPA is unlikely to be economic.

6.2. Case 3 Shifting Duty from the BPA into the MPA

Shifting duty from the BPA to the MPA (Case 3) shows three major differences in Figure 12:

- Higher temperatures for the same duty above 450°F (232°C). This improves the heat recovery effectiveness of the hotter heat exchangers.
- Higher total heat recovery duty (the line goes further to the right). This improves total energy recovered.
- Lower temperatures for the same duty below 400°F (204°C). This reduces the heat recovery effectiveness of the colder heat exchangers.

The first two are benefits, the third is a disadvantage. The attractiveness of Case 3 strongly depends upon the balance of possible heat recovery losses in the cold exchangers versus heat recovery gains in the hot exchangers.

For a new unit, Case 3 is not clear cut. In some units, adding a pumparound would be justified, in others it would not be justified. Detailed analysis is required to check the economic benefits of the third pumparound. In general, the larger the capacity of the unit, the more likely the third pumparound would be justified.

Factors that make adding Case 3 to an existing unit more attractive would include:

- Excess surface area in low-temperature exchangers (this reduces the possible energy loss of lower temperatures in the cold exchangers).
- Tight limitations on the fired heater after the heat exchangers (this increases the value of incremental heat recovery at high temperatures).
- Existing pumparound pump capacity limits (this reduces the cost of extra capital for the new MPA pumps).

6.3. Case 4 MPA Duty from a Split Between BPA and TPA

Case 4 takes some of the MPA duty from the TPA and some of the MPA duty from the BPA. This operation improves the temperatures in both the low temperature region and the high temperature region. Around the middle temperature range of 350°F to 400°F (177°C to 204°C) the duty available drops slightly.

The performance of heat exchangers improve for both the cold exchangers and the hot exchangers. The cost is performance of heat exchangers in the middle range degrade. For a well heat-integrated unit, the pinch temperature is likely to be in this range. Additionally, the total heat recovered drops (the end point of the line on Figure 12 shifts left).

For most new units, Case 4 is less attractive than Case 3.

For existing units, Case 4 is most likely to be attractive if the unit is less energy efficient or subject to high fouling rates. Less energy efficient units have higher temperature differences in the heat exchangers, so the small temperature drop in the mid duty section has a smaller cost. High fouling rates make extra temperature driving force in the high temperature range more valuable.

7. Fuels Operation and Diesel Recovery (Cases 5-9)

Cases 5-7 show the addition of a fractionation section to allow for controlling the endpoint of the top product. This allows the top product to easily meet diesel specifications. Two theoretical stages of fractionation were added. For most units, two theoretical stages will capture nearly all of the diesel in the vacuum tower feed [4]. Optimization may show more stages are justified. Nevertheless, two stages is close enough to the optimum answer for accurate comparisons of different vacuum tower configurations. Cases 8 and 9 show attempts to make diesel as a top product with vacuum towers without fractionation sections.

7.1. Cases 5-6 Diesel Product with 2 Pumparounds and 2 Products

Case 5 and Case 6 use a fractionating bed between the TPA and BPA (Figure 3). The bottom liquid from the fractionating bed drops into the BPA section. The TPA product is diesel boiling range material. Case 5 uses yields that match the performance of Case 7. Case 6 has maximum diesel yields at the specification limit for diesel. Case 6 recovers more diesel than Case 5.

Figure 13 compares the heat recovery from Case 5 and Case 6 with the base operation (Case 1). Forcing the TPA to meet diesel distillation specifications drops the TPA draw temperature and reduces the total amount of energy recovery. The lower the distillation temperature of the diesel product, the colder the TPA. Meeting the diesel specification on the TPA also forces lighter material into the BPA. The BPA temperature drops as well. Case 5 shows lower energy recovery and lower temperatures than Case 6.

Economics of diesel recovery depend upon plant specifics. However, for all cases, diesel recovery makes heat integration more difficult, resulting in either lower heat recovery or more capital required for the same heat recovery.

7.2. Case 7 Diesel Product with 2 Pumparounds and 3 Products

Case 7 takes the liquid from the bottom of the fractionation bed and draws it from the tower as a separate product (Figure 4). Creating an extra product stream prevents lighter liquid from dropping into the BPA section. This configuration allows for recovering some diesel without causing BPA heat-integration problems.

In this configuration, there is no fractionation between the combined diesel-LVGO and the HVGO. The maximum diesel yield for a given specification is lower. For comparison, Figure 14 shows the base case (Case 1), Case 5 and Case 7. Case 5 and Case 7 have the same diesel yield.

The diesel recovery option without reflux (Case 7) improves the temperature level of the available duty and slightly reduces the total quantity of energy available. The temperatures improve (rise) due to the production of LVGO at a higher temperature. The LVGO product is available for heat integration even thought the unit has not LVGO pumparound. In existing units with difficult heat integration, using the no reflux configuration can be an attractive option for diesel recovery.

8. Fuels Operation and Diesel Recovery (Cases 8-9) with Conventional Units

Cases 8-9 show conventional fuels vacuum units without fractionation attempting to produce diesel as the top product. Case 8 is the conventional two pumparound tower. Case 9 is the vacuum tower with three pumparounds.

For the base configuration with two pumparounds (Figure 5), converting the TPA into a diesel boiling range specification stream requires shifting duty into the BPA. The BPA condenses more material, making the TPA distillation lighter. This shift increase the BPA rate and decreases the TPA rate. For the base example shown, the TPA rate reaches zero before the TPA distillation matches the required diesel specification. The operation is infeasible.

Some conventional vacuum units can operate with the TPA at a diesel specification. These units combine:

- Low upstream diesel recovery. This increases the rate of diesel entering the vacuum tower.
- Low vacuum unit feed vaporization. This reduces the fraction of heavy material in the BPA.

Case 9 shows meeting a diesel specification on the TPA with three pumparounds (Figure 6). The MPA section creates an effective one-stage of separation between the TPA and BPA. Even one stage of fractionation significantly improves diesel recovery performance.

Figure 15 compares the heat recovery performance of:

- Case 1 (Conventional)
- Case 5 (Fractionation added, TPA-BPA reflux)
- Case 7 (Fractionation added, no TPA-BPA reflux)
- Case 9 (Three pumparound, no internal reflux)

The three pumparound configuration (Case 9) has the best heat recovery of all the diesel recovery configurations. However it has the lowest diesel yield (Table 2).

9. Hydrocracker Feed as Product versus Diesel as Product

Instead of diesel, some fuels vacuum units produce feed for hydrocracking. Hydrocracking catalyst life can vary strongly with the distillation end-point of its feed. In general, hydrocracker feed units have the same conceptual trade-offs as diesel recovery units. The difference between the units is the distillation specification placed on the products. Within capacity and heat balance constraints, many have some ability to shift specifications between diesel operation and hydrocracker feed operation.

10. Conclusions

Conventional vacuum fuels units produce feed streams for downstream conversion units. The downstream units include both FCC and hydrocracking units. Increased diesel demand makes recovering diesel from the vacuum tower more attractive. Multiple vacuum tower configurations can allow the top vacuum tower product to meet a diesel boiling range specification. The two major factors to consider are:

- Increased fractionation will increase diesel yields.
- Increased fractionation will reduce temperatures and make heat recovery more expensive (more capital) or less efficient (less heat recovered).

Trade-offs for adding diesel recovery to both existing plants and to new units must balance the value of diesel recovery against capital costs, yields, and energy recovery.

11. References

[1] R. N. Watkins, *Petroleum Refinery Distillation*. Houston, TX: Gulf Publishing Company, 1973.

[2] A. W. Sloley, "Crude Distillation Sequences," presented at the AIChE Fall National Meeting, San Francisco, CA, 2013.

[3] A. W. Sloley, "Crude Unit Heat Integration - Best Practices," presented at the AIChE Spring Meeting, Austin, TX, 2015.
[4] A. W. Sloley, "Increase diesel recovery," *Hydrocarbon Processing*, vol. 87, no. 6, pp. 85–102, Jun. 2008.

12. Tables and Figures

Table 2E Vacuum Unit Comparisons

Case		1	2	3	4	5	6	7	9
Pumparounds		2	3	3	3	2	2	2	3
Products		2	3	3	3	2	2	3	3
Top prod as diesel						Yes	Yes	Yes	Yes
Diesel at spec						Too Light	Yes	Yes	Yes
MPA Duty from			TPA	BPA	Both PA				
Internal flow to hvgo						Yes	Yes		
Vacuum heater outlet	F	775	775	775	775	775	775	775	775
Vacuum tower cutpoint	F	1,046	1,046	1,050	1,050	1,044	1,045	1,046	1,046
Top PA Duty	Mbtu/hr	11.3	2.2	14.1	2.0	17.6	13.5	9.2	2.3
Mid PA Duty	Mbtu/hr		7.6	6.1	16.7				
Bot PA Duty	Mbtu/hr	77.1	77.1	66.0	66.0	69.6	73.4	77.1	77.1
Top PA Draw Temp	F	290	214	302	209	277	275	259	215
Top PA Return Temp	F	100	100	100	100	100	100	100	100
Mid PA Draw Temp	F		369	482	402			400	371
Mid PA Return Temp	F		200	200	200			200	200
Bot PA Draw Temp	F	558	558	597	597	536	542	558	558
Bot PA Return Temp	F	300	300	300	300	300	300	300	300
Top PA Rate	BPD	9,043	3,099	10,543	2,896	15,316	11,880	9,008	3,189
Mid PA Rate	BPD		6,157	2,787	11,178				
Bot PA Rate	BPD	35,241	35,241	25,676	25,676	35,114	36,075	35,241	35,241
Top Prod	BPD	3,680	583	4,930	449	960	1,380	960	618
Mid Prod	BPD		3,070	2,459	6,904			2,701	3,035
Bot Prod	BPD	20,716	20,716	17,206	17,206	23,283	22,917	20,716	20,716
Top Prod D86 5 vol %	F	560	504	572	501	512	523	515	506
Top Prod D86 95 vol %	F	745	636	745	641	616	638	638	638
Mid Prod D86 5 vol %	F		604	693	623			620	605
Mid Prod D86 95 vol %	F		752	839	800			758	753
Bot Prod D86 5 vol %	F	733	733	771	771	699	709	733	733
Bot Prod D86 95 vol %	F	1,006	1,006	1,018	1,018	1,000	1,001	1,006	1,006
Top Prod Q to TPA ret temp	Mbtu/hr	4.6	0.4	6.6	0.3	1.1	1.6	1.0	0.4
Mid Prod Q to MPA ret temp	Mbtu/hr		3.8	5.4	10.3			4.0	3.8
Bot Prod Q to BPA ret temp	Mbtu/hr	45.3	45.3	44.2	44.2	46.1	46.6	45.3	45.3
Top Prod Q to final temp	Mbtu/hr	0.0	0.0	0.0	0.0	0.0	0.0	0.0	0.0
Mid Prod Q to final temp	Mbtu/hr		1.0	0.8	2.2			0.9	1.0
Bot Prod Q to final temp	Mbtu/hr	17.8	17.8	14.8	14.8	19.9	19.6	17.8	17.8
Top prod final temp	F	100	100	100	100	100	100	100	100
Mid prod final temp	F		150	150	150			150	150
Bot prod final temp	F	180	180	180	180	180	180	180	180
Total distillate	BPD	24,396	24,368	24,594	24,560	24,243	24,297	24,376	24,369
TPA Level Duty	Mbtu/hr	15.9	2.6	20.7	2.3	18.7	15.1	10.2	2.7
MPA Level Duty	Mbtu/hr	0.0	11.3	11.5	27.0	0.0	0.0	4.0	3.8
BPA Level Duty	Mbtu/hr	122.4	122.4	110.2	110.2	115.7	120.0	122.4	122.4
Total Duty	Mbtu/hr	138.4	136.4	142.4	139.6	134.4	135.1	136.6	129.0

Table 2M Vacuum Unit Comparisons

Case		1	2	3	4	5	6	7	9
Pumparounds		2	3	3	3	2	2	2	3
Products		2	3	3	3	2	2	3	3
Top prod as diesel						Yes	Yes	Yes	Yes
Diesel at spec						Too Light	Yes	Yes	Yes
MPA Duty from			TPA	BPA	Both PA				
Internal flow to hvgo						Yes	Yes		
Vacuum heater outlet	C	413	413	413	413	413	413	413	413
Vacuum tower cutpoint	C	563	563	565	565	562	563	563	563
Top PA Duty	GJ/hr	11.9	2.3	14.9	2.1	18.6	14.3	9.7	2.4
Mid PA Duty	GJ/hr		8.0	6.4	17.6				
Bot PA Duty	GJ/hr	81.4	81.4	69.6	69.6	73.4	77.4	81.4	81.4
Top PA Draw Temp	C	143	101	150	98	136	135	126	102
Top PA Return Temp	C	38	38	38	38	38	38	38	38
Mid PA Draw Temp	C		187	250	205			204	188
Mid PA Return Temp	C		93	93	93			93	93
Bot PA Draw Temp	C	292	292	314	314	280	283	292	292
Bot PA Return Temp	C	149	149	149	149	149	149	149	149
Top PA Rate	M3/D	1,013	347	1,181	324	1,715	1,331	1,009	357
Mid PA Rate	M3/D		690	312	1,252				
Bot PA Rate	M3/D	3,947	3,947	2,876	2,876	3,933	4,040	3,947	3,947
Top Prod	M3/D	412	65	552	50	108	155	108	69
Mid Prod	M3/D		344	275	773			302	340
Bot Prod	M3/D	2,320	2,320	1,927	1,927	2,608	2,567	2,320	2,320
Top Prod D86 5 vol %	C	293	262	300	260	267	273	268	263
Top Prod D86 95 vol %	C	396	336	396	338	325	337	337	337
Mid Prod D86 5 vol %	C		318	367	329			327	319
Mid Prod D86 95 vol %	C		400	448	427			403	400
Bot Prod D86 5 vol %	C	389	389	410	410	371	376	389	389
Bot Prod D86 95 vol %	C	541	541	548	548	538	538	541	541
Top Prod Q to TPA ret temp	GJ/hr	0.5	0.0	0.7	0.0	0.1	0.2	0.1	0.0
Mid Prod Q to MPA ret temp	GJ/hr		0.4	0.6	1.2			0.4	0.4
Bot Prod Q to BPA ret temp	GJ/hr	5.1	5.1	5.0	5.0	5.2	5.2	5.1	5.1
Top Prod Q to final temp	GJ/hr	0.0	0.0	0.0	0.0	0.0	0.0	0.0	0.0
Mid Prod Q to final temp	GJ/hr		0.1	0.1	0.3			0.1	0.1
Bot Prod Q to final temp	GJ/hr	2.0	2.0	1.7	1.7	2.2	2.2	2.0	2.0
Top prod final temp	C	38	38	38	38	38	38	38	38
Mid prod final temp	C		66	66	66			66	66
Bot prod final temp	C	82	82	82	82	82	82	82	82
Total distillate	M3/D	2,732	2,729	2,755	2,751	2,715	2,721	2,730	2,729
TPA Level Duty	GJ/hr	12.5	2.4	15.6	2.1	18.7	14.5	9.8	2.5
MPA Level Duty	GJ/hr	0.0	8.4	7.0	18.8	0.0	0.0	0.4	0.4
BPA Level Duty	GJ/hr	86.4	86.4	74.6	74.6	78.6	82.6	86.4	86.4
Total Duty	GJ/hr	98.9	97.2	97.2	95.5	97.3	97.1	96.7	89.3

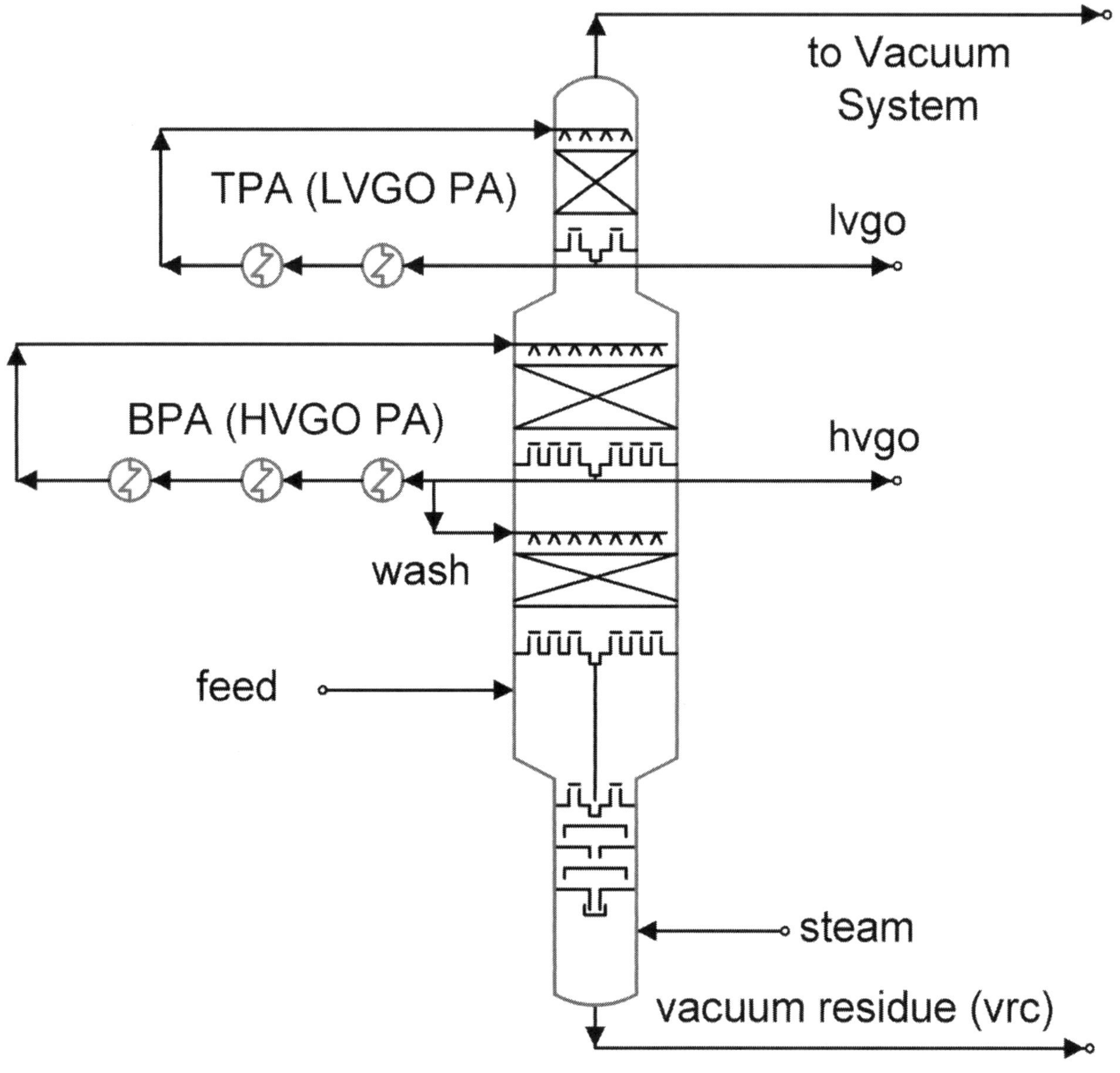

Figure 1 – Case 1: 2 Pumparound, 2 Product, Conventional

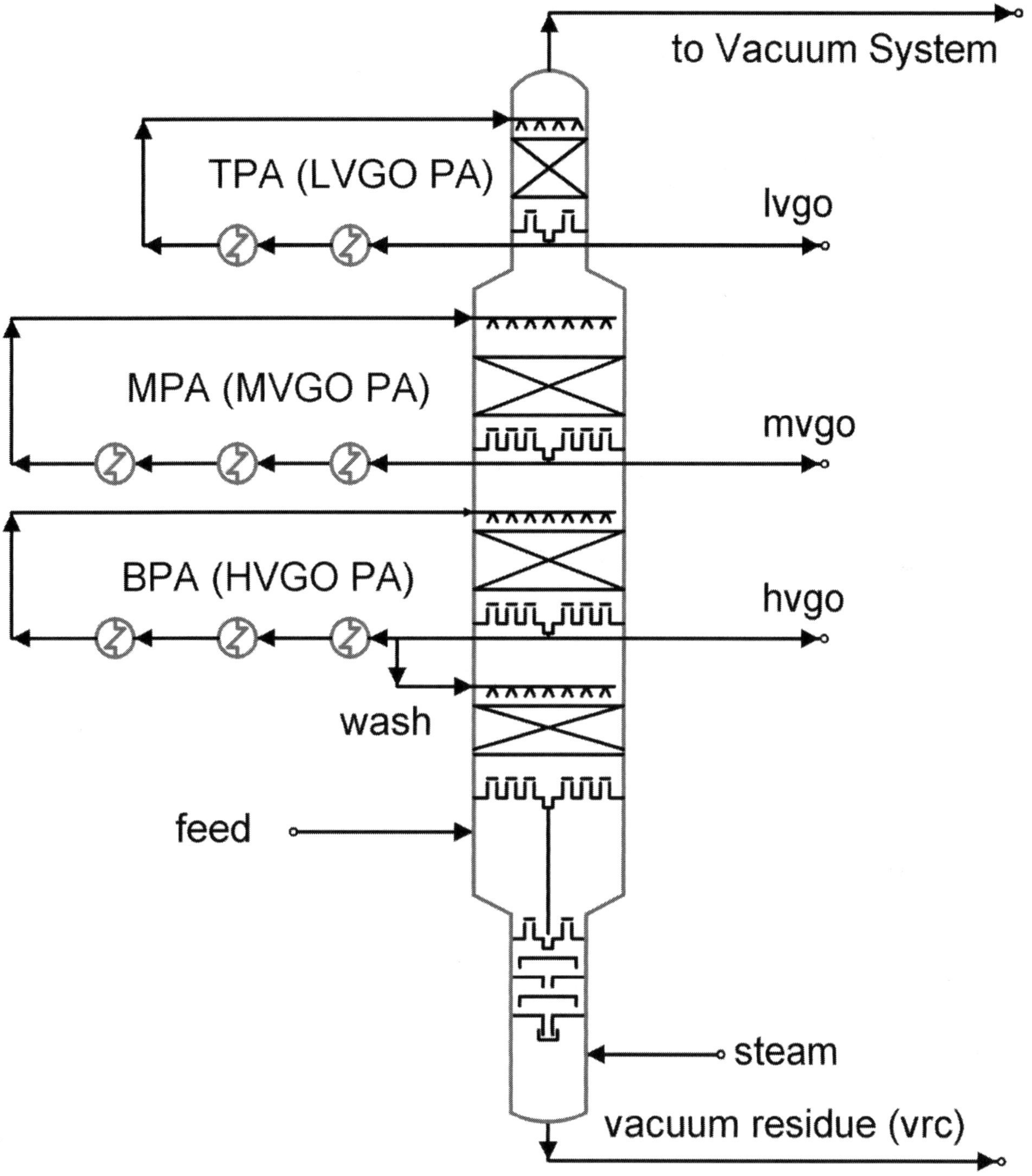

Figure 2 – Case 2, 3, 4: 3 Pumparound, 3 Product

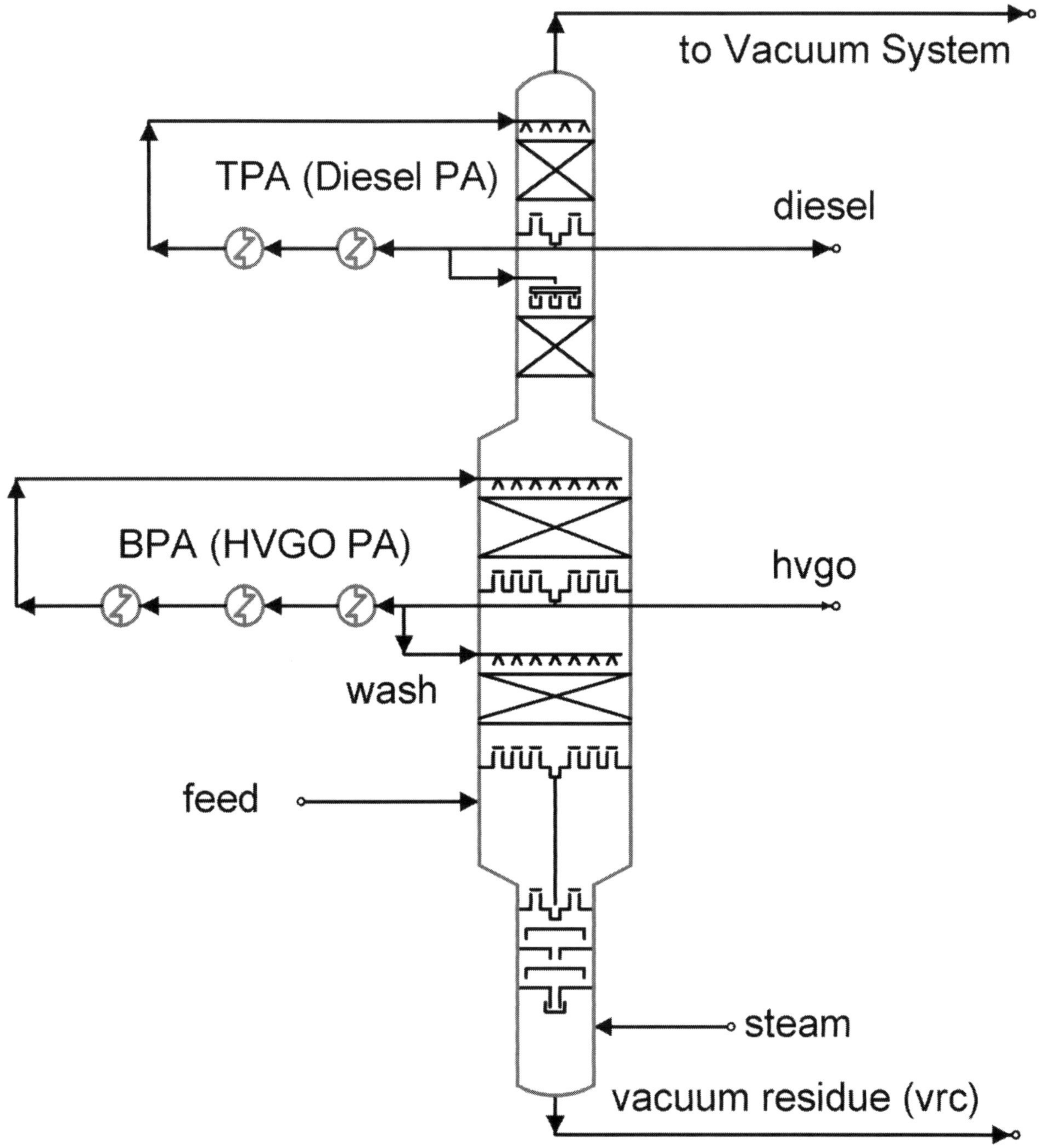

Figure 3 – Case 5, 6: 2 Pumparound, 2 Product, Diesel Operation

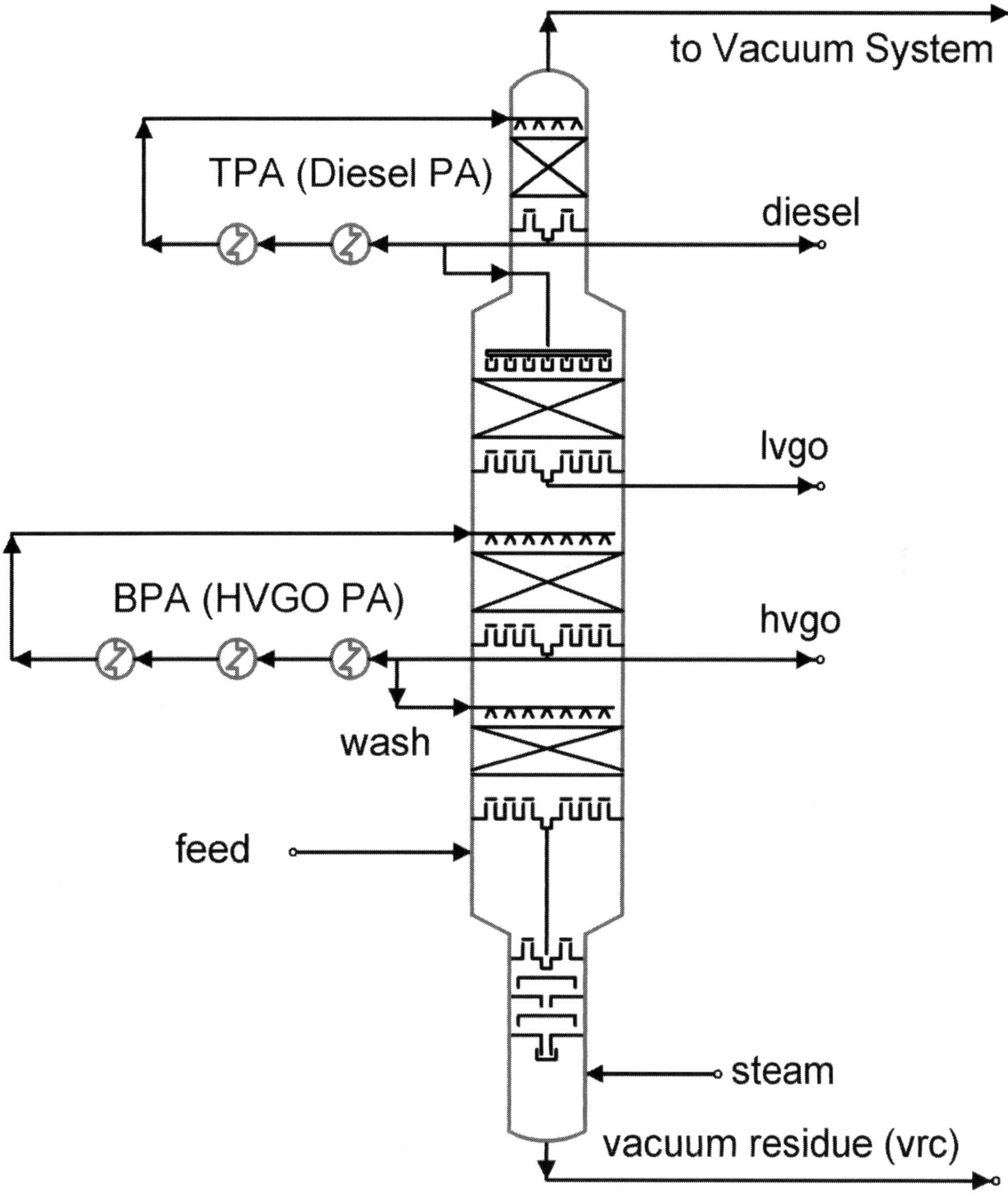

Figure 4 – Case 7: 2 Pumparound, 3 Product – Top Product as Diesel

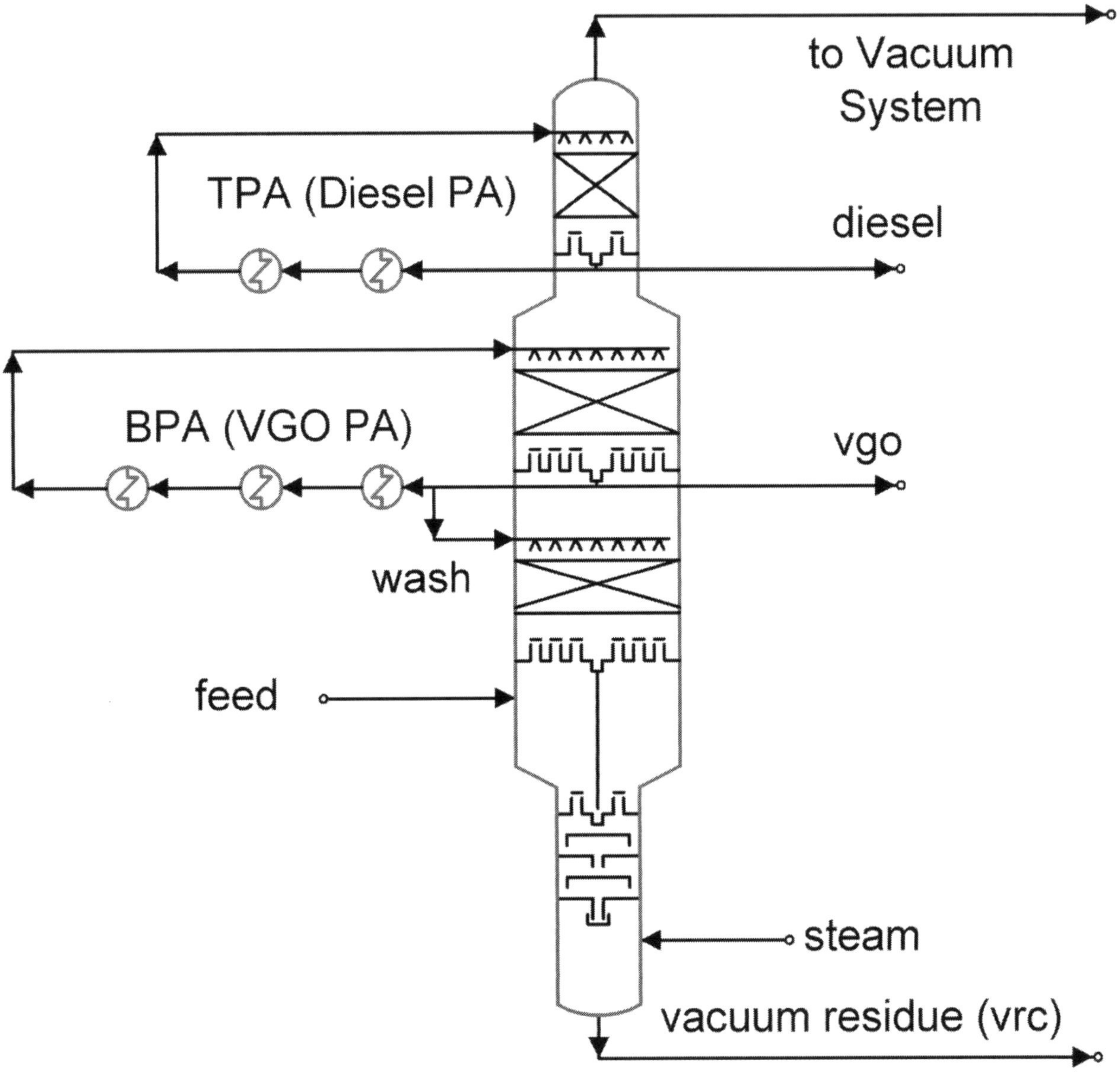

Figure 5 - Case 8: 2 Pumparound, 2 Product - Top Product as Diesel, No Internal Fractionation

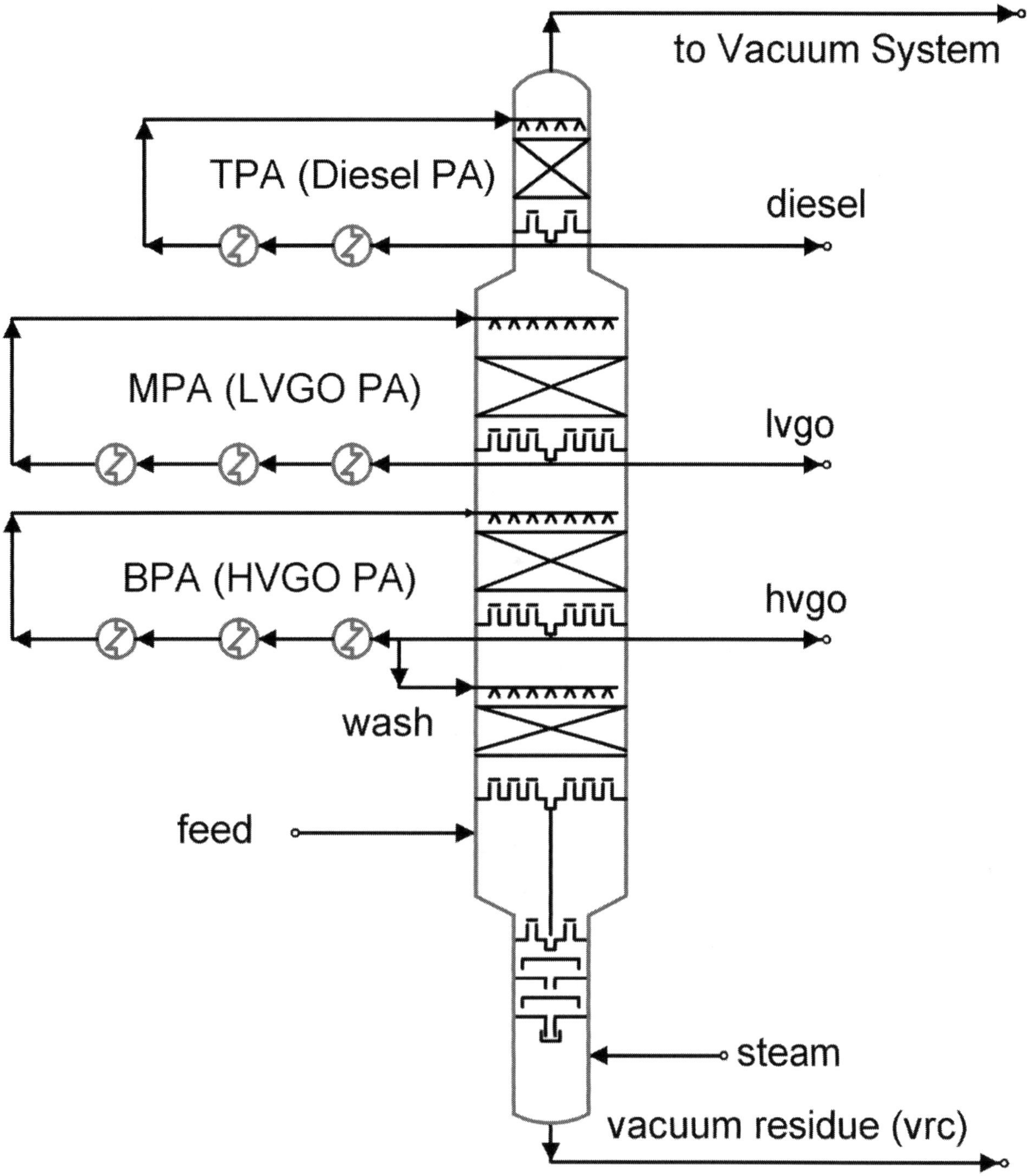

Figure 6 – Case 9: 3 Pumparound, 3 Product – Top Product as Diesel, No Internal Fractionation

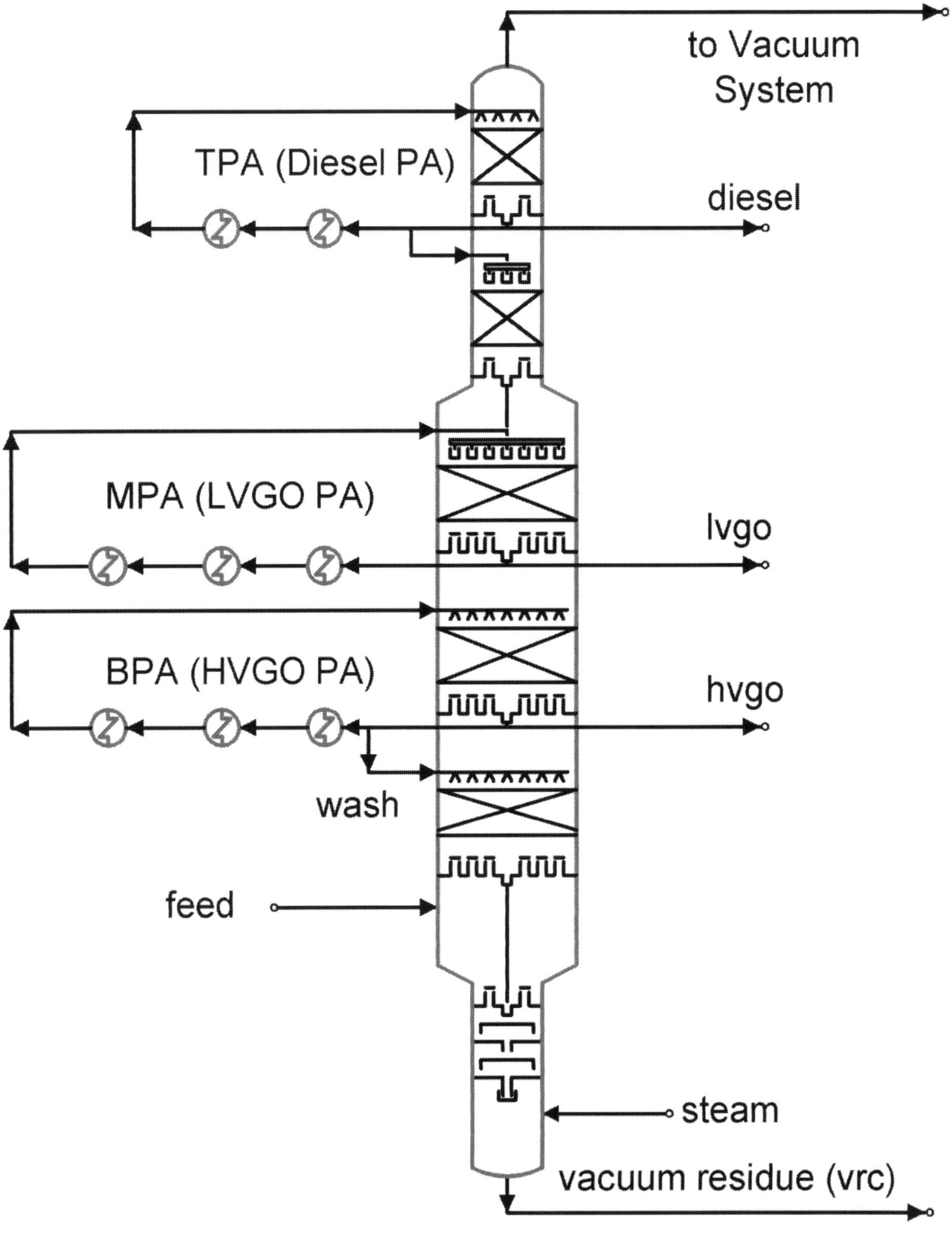

Figure 7 –3 Pumparound, 3 Product – Top Product as Diesel, Internal Fractionation

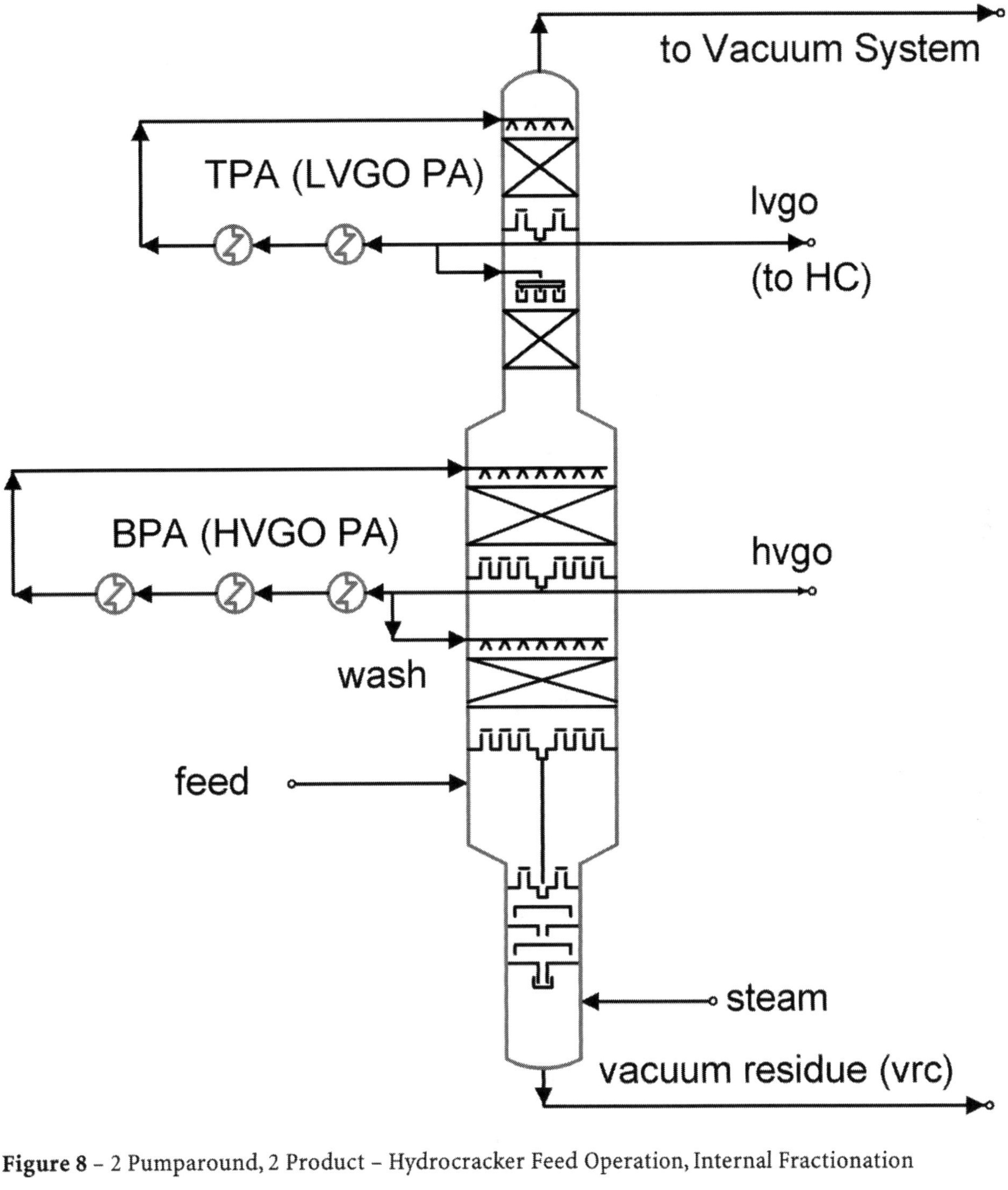

Figure 8 – 2 Pumparound, 2 Product – Hydrocracker Feed Operation, Internal Fractionation

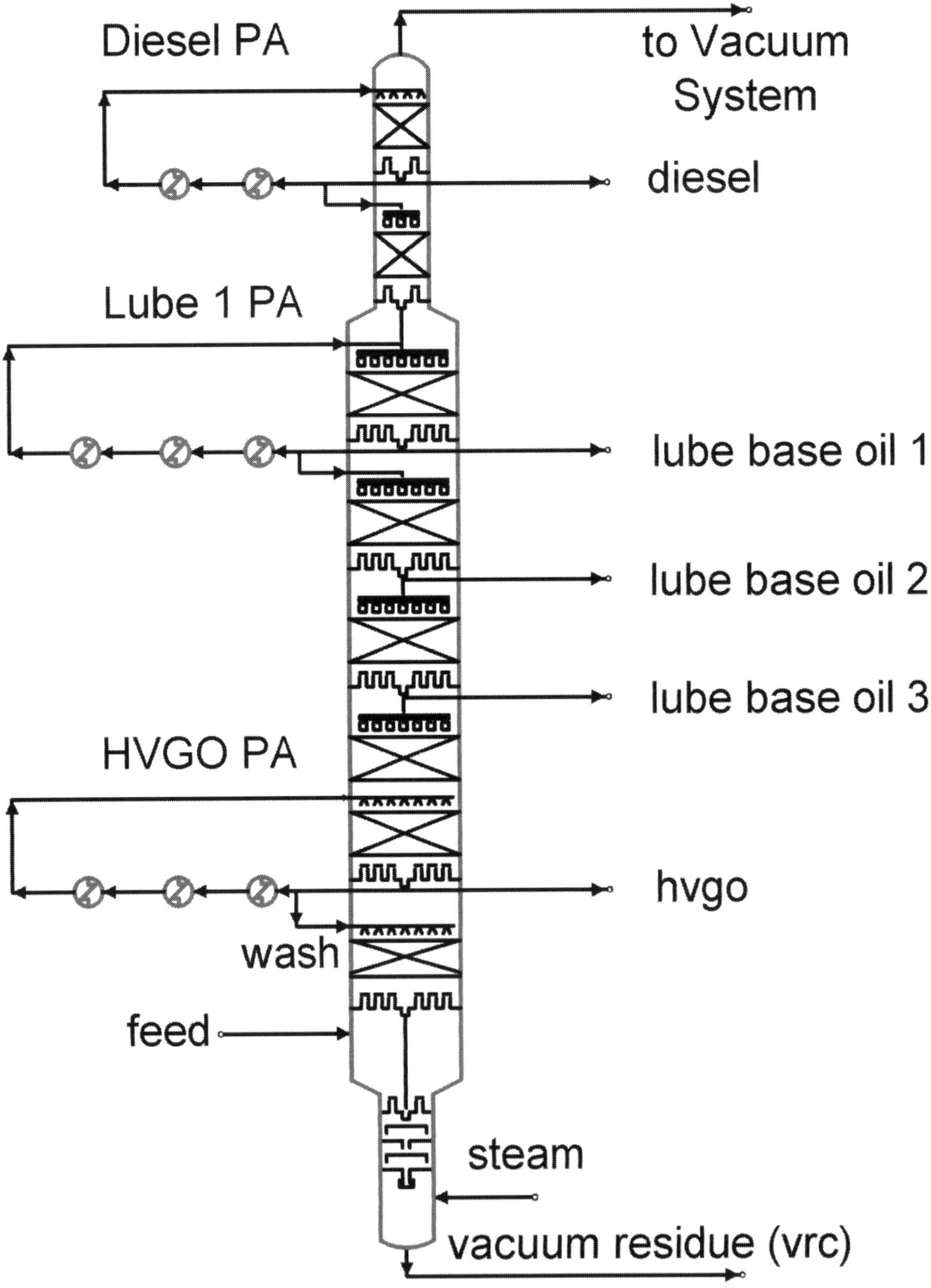

Figure 9 – Lubricants Vacuum Tower

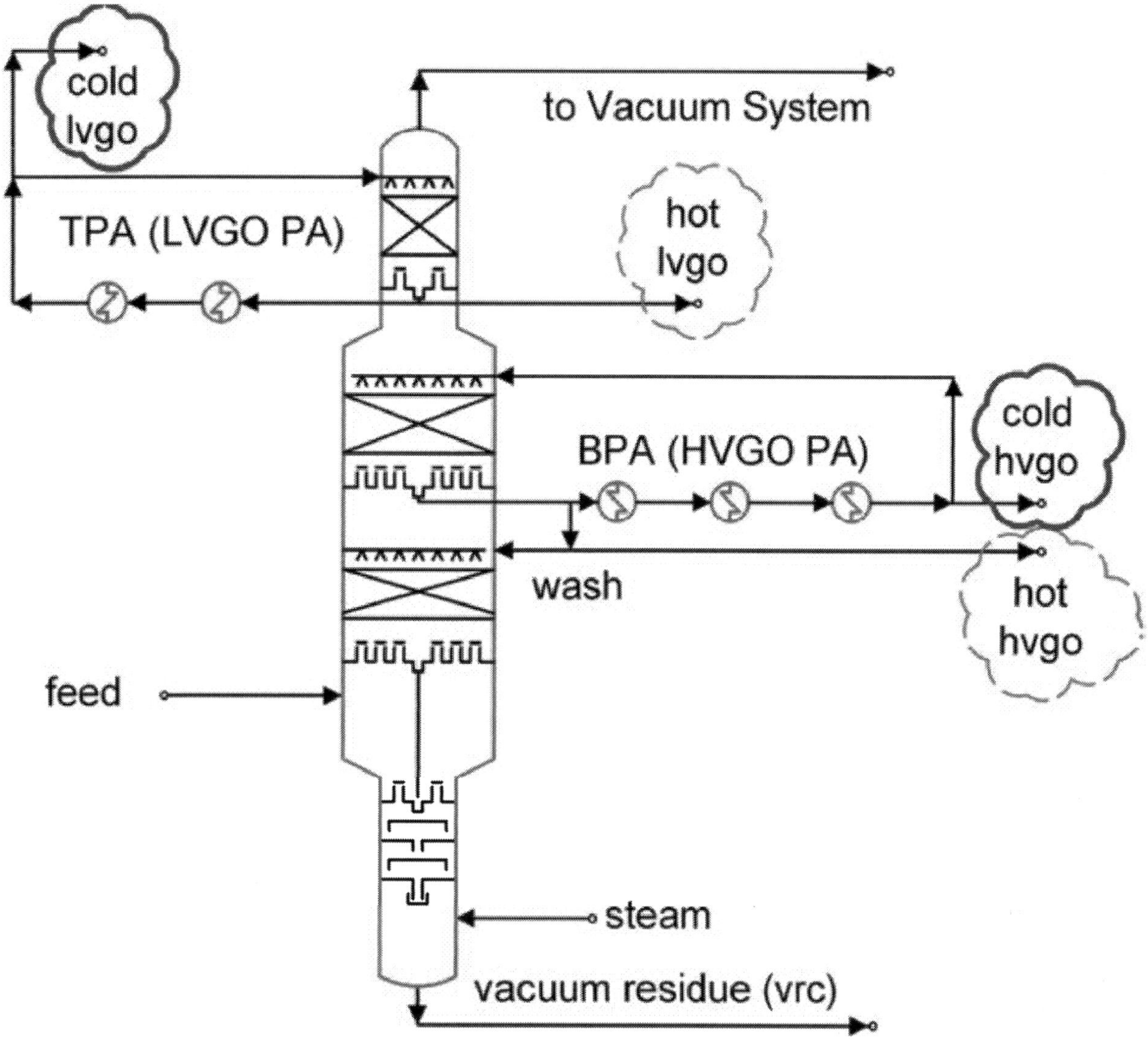

Figure 10 Hot-product Versus Cold-Product

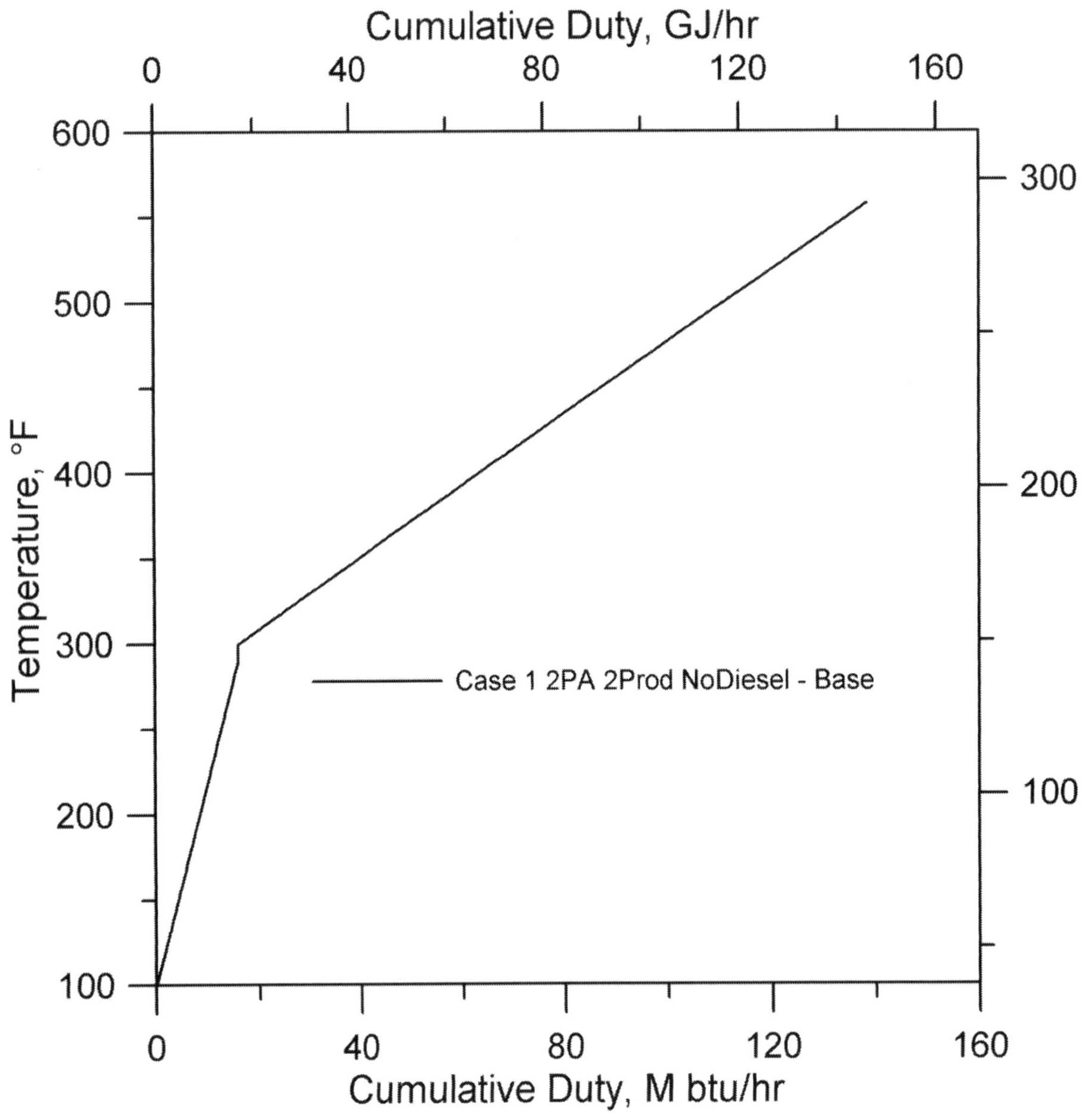

Figure 11 Case 1 (Base) Cumulative Duty Recovered Versus Temperature

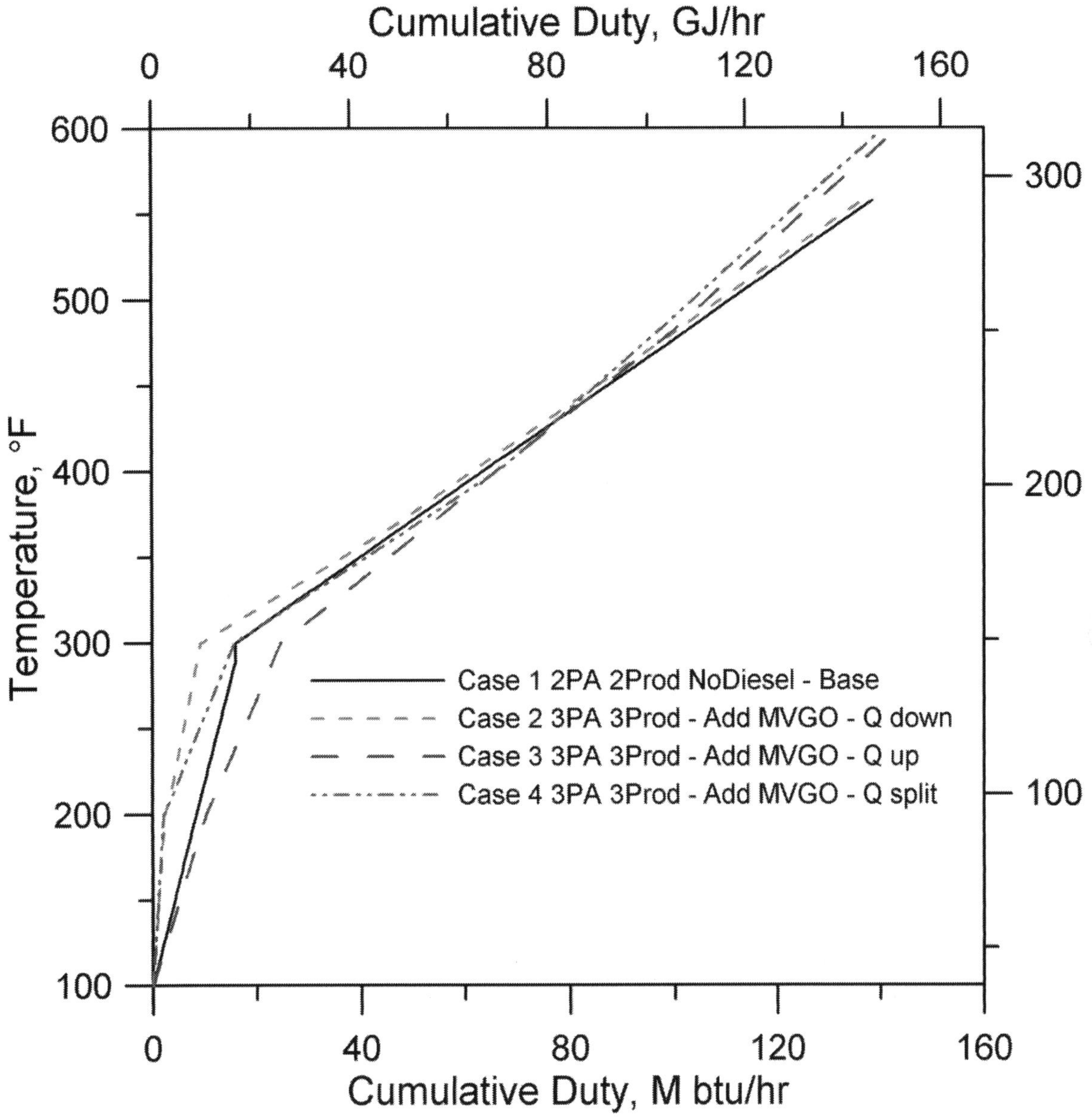

Figure 12 Cumulative Duty Recovered Versus Temperature – Adding a Pumparound

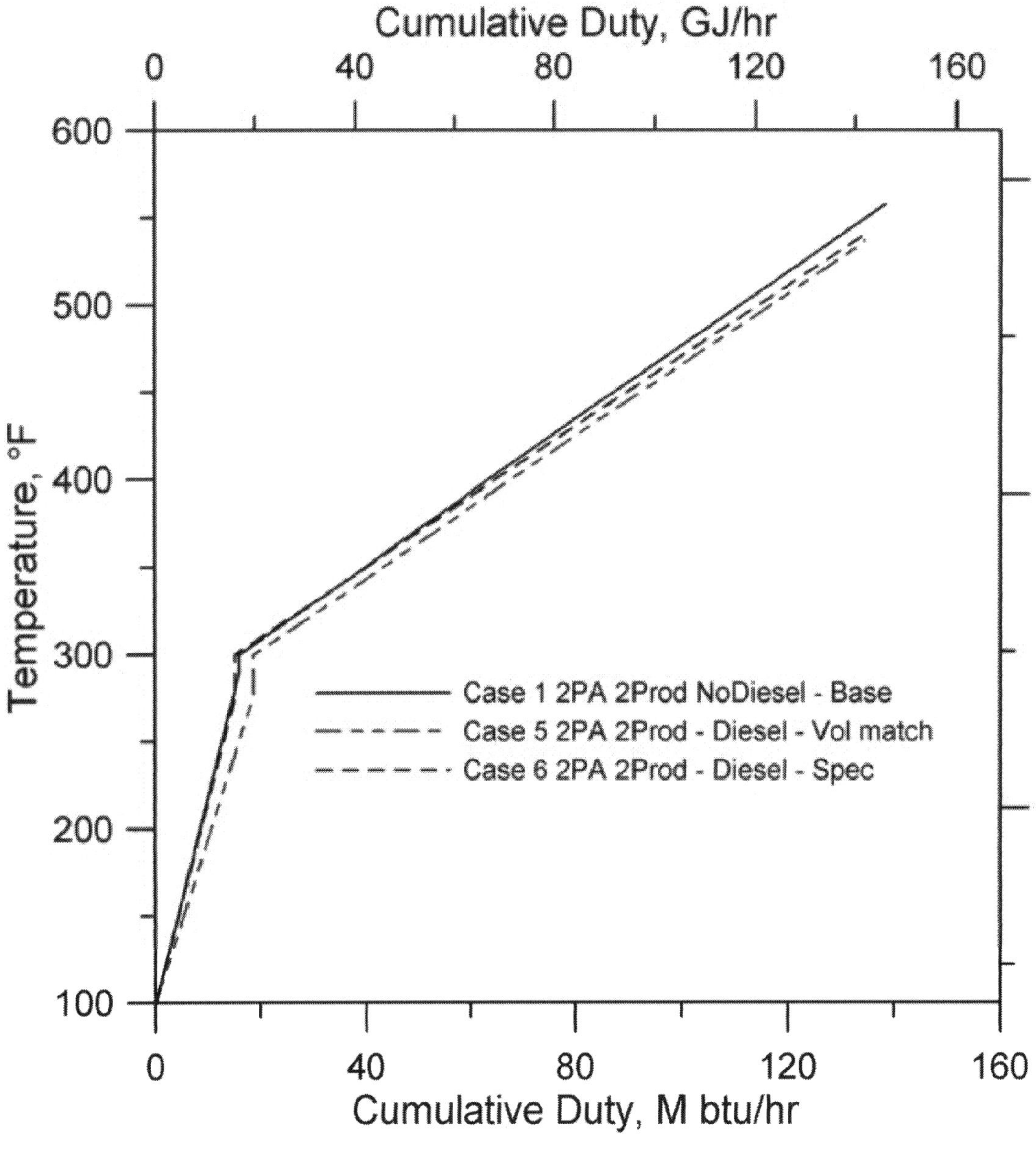

Figure 13 Cumulative Duty Recovered Versus Temperature – Diesel Recovery

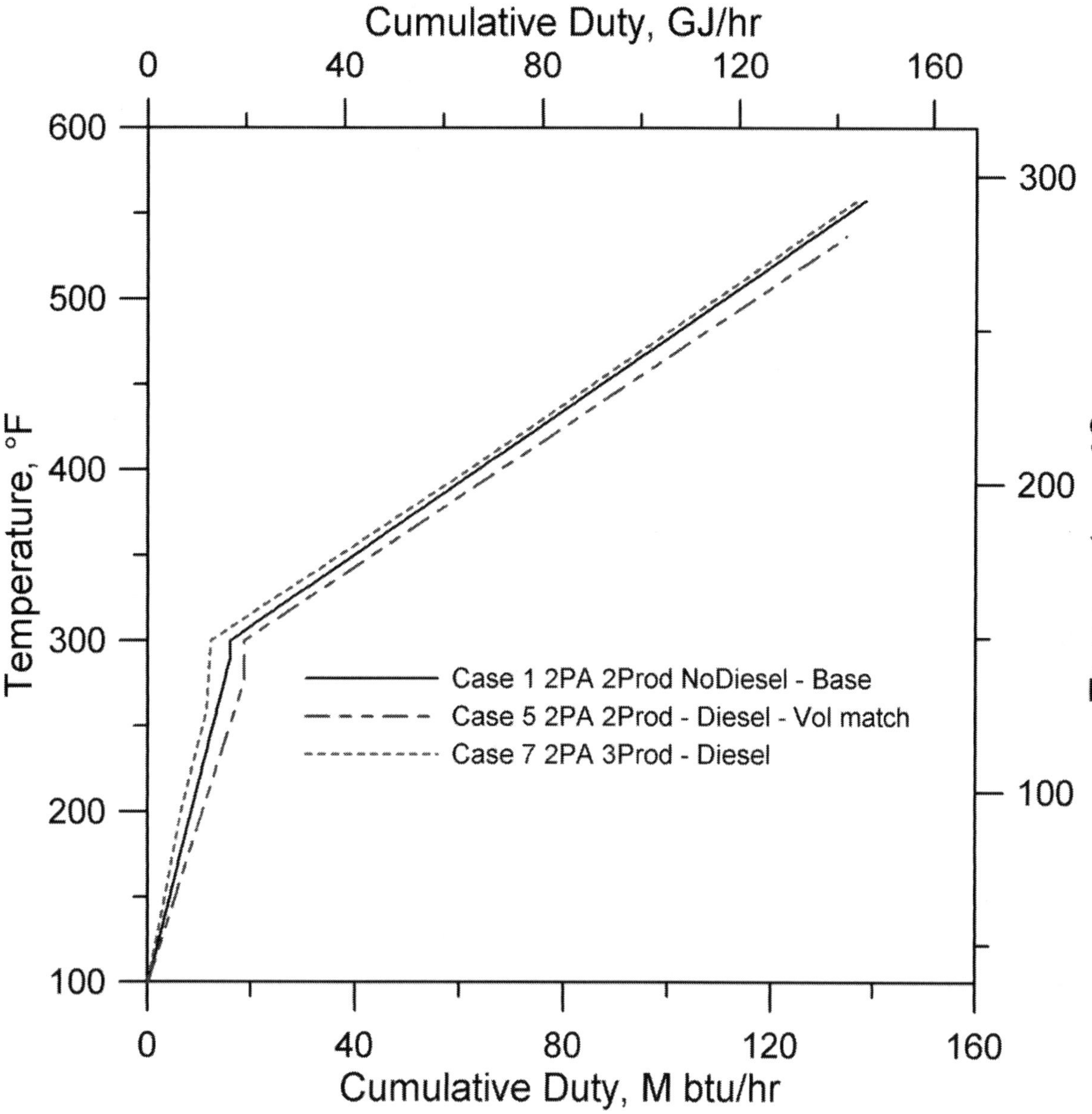

Figure 14 Cumulative Duty Recovered Versus Temperature – Diesel Reflux Options

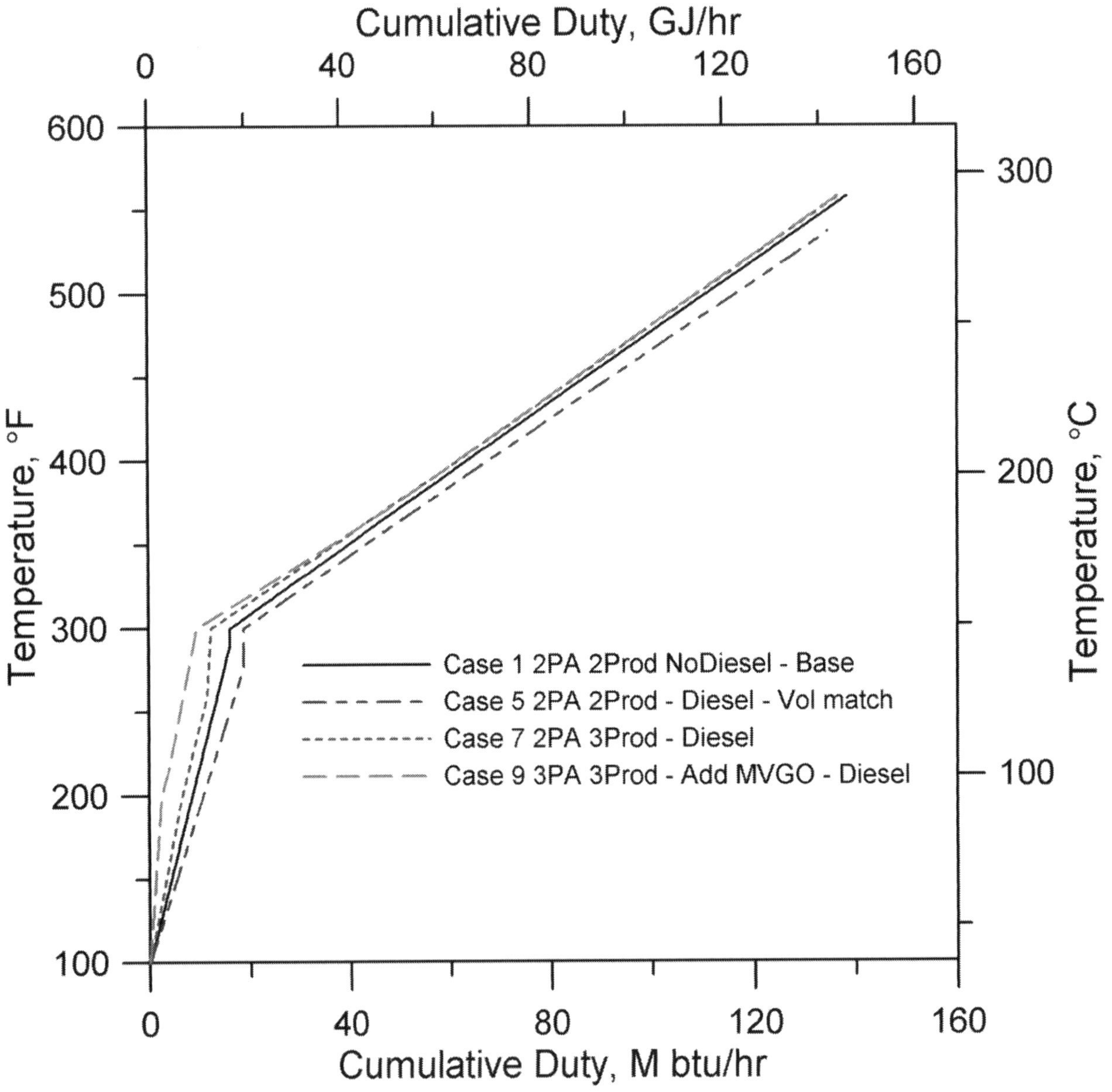

Figure 15 Cumulative Duty Recovered Versus Temperature – Diesel Options Expanded

13. Appendix – Glossary, Definitions, Abbreviations and Conventions

Abbreviations and terms in text and figures that may not have been defined or that may be non-standard usage.

These definitions are tailored to the most common industry use of the term in crude distillation and petroleum refining. Many plants use their own variations on these definitions and other terms in place of those listed. Other technical fields may have different meanings for these terms.

Term or abbreviation	Definition
API degrees, API°	See API Gravity
API Gravity	A measure of liquid density defined by the American Petroleum Institute. The scale was derived from the reading marks on a specific type of measurement instrument. $API^{O} = \frac{141.5}{SG} - 131.5$ $\quad SG = \frac{141.5}{API^{O} + 131.5}$ Where SG is specific gravity at 60°F (15.6°C). Water has an API gravity of 10. An API gravity lower than 10 is denser than water, an API gravity higher than 10 is less dense than water.
ASTM Distillation	Old usage. ASTM Distillation quoted without a descriptive qualifier refers to an ASTM D86 distillation. This is a batch distillation without reflux.
Boiling range	See distillation range.
BPA	Bottom pumparound.
bpd	Barrels per day, standard volume barrels at 60°F (15.6°C)
CP	Cutpoint
Cutpoint	The ASTM D2892 distillation temperature that would achieve the same split between overhead products and bottoms products that the unit achieves. The D2892 distillation basis is atmospheric equivalent temperature. Typical usage is volume percent.
Distillation range	Range of AEBP boiling temperatures for a stream. Typically expressed as an ASTM D-2892 equivalent (true boiling range) or an ASTM D-86 equivalent (ASTM boiling range) on a 5-95 volume percent basis. Other basis may be used.
Flash zone	Feed entry point into a main fractionator.
MB	Million btu/hr (abbreviated to fit in figures), mmbtu/hr
MBtu/hr	Million btu/hr, mmbtu/hr
mmbtu/hr	Million btu/hr, industry standard usage remains the old form using a

	lower-case m for thousand. 'mm' equals thousand X thousand to get million.
MPA	Middle pumparound.
Specific gravity	Density of a stream at 60°F (15.56°C) compared to the density of water at 60°F (15.56°C).
T_{XX}	Temperature at which XX volume (or weight) percent of liquid is distilled.
TBP	True boiling point. Distillation curve based on ASTM D2892 or an equivalent.
TPA	Top pumparound.
Transfer line	Pipe connecting the outlet of a fired heater to the downstream main fractionator.

Figure conventions;

- Lower case text and abbreviations in figures refer to a stream.
- Upper case text and abbreviations in figures refer to an equipment item.

Paper 27b

Atmospheric Crude Distillation

Equipment State of the Art

Andrew W. Sloley

Advisian
Energy Center II
575 N Dairy Ashford St
Houston, TX 77079

Prepared for Presentation at the
American Institute of Chemical Engineers
2016 Spring Meeting
Houston, TX April 10-14, 2016

UNPUBLISHED

1. Summary

Atmospheric crude distillation is the first processing step in the oil refinery. Overall distillation performance sets feed rates and properties to downstream units. Improved separation increases refinery profits and flexibility.

This paper provides an overview of the state-of-the art on atmospheric crude distillation equipment. The major focus is on towers with trays. However, some comments are included on specific situations where packing may be preferred.

Specific topics covered include:

- Service types (fractionating, heat-transfer, de-entrainment, stripping)
- Vapor and liquid rate zones
- Conventional versus high-capacity trays
- Blowing in wash trays
- Mechanical reliability
- Stripping trays and packing

In many ways refinery atmospheric crude distillation towers fit conventional mass-transfer equipment requirements. However, specific parts of the atmospheric crude distillation service merit special attention. The discussion highlights areas requiring special care, areas where conventional equipment fits, and situations well suited for modern high-capacity column internals.

Services commonly include heat transfer (pumparounds), fractionation, de-entrainment (feed entry) and stripping trays. In modern atmospheric towers the dominant choices revolve around trays. For a given capacity, a tower with trays usually has a lower investment cost than a tower with packing, even if the packed tower might have a smaller diameter. An exception is the wash zone of the atmospheric crude column. For that service, grid or structured packing often gives better overall economics.

High-capacity trays have been a major recent development. They often allow for significant capacity increases in existing towers. For new units conventional trays with sloped downcomers are usually the favored choice for minimum investment.

Process upsets often subject the stripping section of these towers to large mechanical forces. Pressure relief design (trap door) trays and mechanically reinforced trays are often favored for this service.

2. Atmospheric Tower Description

Figure 1 shows the conventional layout of an atmospheric crude tower. Two examples of major physical components present in typical towers for this service are illustrated. The first example has a packed wash section. By choice, the packing should be structured packing, grid or a combination of the two. The second example has trays throughout the column.

Historically most atmospheric crude towers use the second configuration: entirely trayed. However, the first configuration, the tower with the packed wash section, has significant technical advantages that often make it more profitable even though investment is higher.

3. Predominant Equipment

3.1. Trays

Trays are the predominant fractionation equipment in atmospheric crude towers. Economics are the driving force behind this choice. For a given capacity, even though a conventional tower with trays has a larger diameter than a tower with packing, the overall installed cost of the tower with trays (including auxiliaries and internals) is lower than the cost of a tower with packing.

Functionally, trays have three operating areas:

- Active area: the region where the vapor and liquid mix. Mass transfer occurs in this area.
- Disengagement area: the region above the tray where liquid droplets disengage from the rising vapor.
- Downcomer: the region where vapor bubbles disengage from the liquid. The downcomer also accumulates a liquid head that provides the driving force for flow down the column. The downcomer must provide enough inventory to overcome the total of:
    - Vapor pressure drop trough the tray
    - Head losses for liquid flowing down the tower
    - Hold-up provided by the outlet weir (if any)
    - Crest height over the outlet weir (if any)

Figure 2 illustrates these sections for a conventional tray.

Three other factors favor trays in atmospheric tower services:

- Atmospheric towers have multiple side draw products and multiple vapor feeds from sidestrippers. Sumps can be easily added to trays for partial liquid draws. The modest vapor rates from the side-strippers in this service usually allow for the use of tower nozzles without internals. The trays have sufficient pressure drop to adequately distribute the vapor. Even if high rates are present in the side-stripper returns, relatively simple internal pipes can distribute the vapor.
- Liquid draws for pumparounds are relatively straightforward. Again simple sump and piping arrangements are adequate.
- The trays in the top of the column often suffer from salt deposition. Water injection may clean out soluble salts. This is more effective for cleaning trays than packing.

3.2. Equipment Capacity Evaluation

Conventional trays have multiple mechanical features that determine the trays' upper capacity, lower capacity, and efficiency. For atmospheric crude columns the major mechanical choices are tower diameter, tray spacing, and number of liquid flow paths for the tray. As a rule-of-thumb, Table 1 shows the basic relations between these variables.

Table 1 Major Tray Capacity Trade-offs

Feature	Liquid Handling	Vapor Handling
Tray spacing, more distance	Increases maximum rate No change on minimum rate	Increases maximum rate No change on minimum rate
Tray spacing, less distance	Decreases capacity No change on minimum rate	Decreases capacity No change on minimum rate
Tray diameter, more diameter	Increases maximum rate Increases minimum rate	Increases maximum rate Increases minimum rate
Tray diameter, less diameter	Decreases maximum rate Decreases minimum rate	Decreases maximum rate Decreases minimum rate
Tray passes, more	Increases maximum rate	Approximately no change in rates
Tray passes, less	Decreases maximum rate	Approximately no change in rates

Tray design and performance is highly complex. Many specific situations can create exceptions to these rules. In certain cases, what look like minor modifications to a tray layout may cause large changes in tray performance. Engineers should consult detailed references to tray operation when evaluating specific circumstances [1] [2].

3.3. Active Area Devices

Current industry practice uses both sieve trays and valve trays in atmospheric tower service. Based on surveys by the author, ~75% of the trays in atmospheric columns, excluding stripping trays, are valve trays. The trend over the last 10 years has been an increasing number of fixed valves. Nevertheless, movable valves remain the most common valve-type. Sieve trays cost less but have lower vapor handling capacity. Correctly applied sieve trays and valve trays have similar capacity and efficiency for the different services.

A few units remain in service with bubble-cap trays and tunnel-cap trays. Due to higher pressure drop, lower capacity, and higher cost these trays are much less common today compared to 30 years ago. The benefit of bubble-cap and tunnel-cap trays was capability to run at very low liquid rates.

3.4. Flexibility of Multiple-Pass Trays

Multiple pass trays divide the liquid flow into different flow paths. Figure 3 shows one-pass and two-pass trays. Figure 4 shows three-pass and four-pass trays. Multiple flow paths have two significant benefits:

- Increased liquid rate per flow path decreases weir load (volumetric flow per unit length of weir). Lower weir loads reduce tray pressure drop and increase tray capacity.
- Decreased flow path length across the active area. If the flow path length becomes too long, performance may suffer from vapor cross-flow channeling [3] [4]. For most towers, flow path lengths of over 8-10 ft (2.4-3.0 m) create situations where cross-flow channeling may occur. In specific cases, lower flow path lengths may create a problem.

The author rarely uses three-pass trays. They have a non-symmetrical layout and may be difficult to design. Other options, including some of the multiple downcomer trays (§3.5) often perform more effectively than three-pass trays.

Increasing the number of flow paths increases maximum liquid handling capacity. Many towers use multiple flow paths to handle pumparound liquid rates.

Increasing the number of flow paths also increases the minimum vapor rate required. At low vapor rates, some of the tray sections may remain inactive [5]. Inactive tray sections reduce heat-transfer and mass-transfer efficiency. Preventing inactive sections requires a minimum number of valves to open [6]. Table 2 summarizes guidelines for this.

Table 2 Vapor Turn-Down for Valve Trays

Flow paths	Minimum Number of Valves Open %
2	40
3	60
4	80

The minimum number of valves open does not directly correlate with vapor rate turndown. A valve tray may have 100% of its valves open at much less than 100% of the tray's vapor handling capacity. Vapor rate flexibility can be improved by using valves with multiple weights and using caged valves. Both steps increase the tray pressure drop. The higher the tray pressure drop, the higher the flexibility.

3.5. High-Capacity Trays

Current high-capacity tray technology uses one or more of the following to increase tray capacity:

- Increased weir length to reduce tray pressure drop. This increases both vapor and liquid handling capacity (Figure 5).
- Directional valves to use the vapor to push liquid across the tray (Figure 6).
- Hanging downcomers to increase the active area on the tray. This increases vapor handling rate (Figure 7).

- Centrifugal devices to increase vapor-liquid separation.
- Other de-entrainment devices to improve vapor-liquid separation.

Techniques are listed in chronological order of development. The first three have extensive application in atmospheric crude towers. When done correctly, they may significantly improve tower capacity. The last two, centrifugal separators and use of de-entrainment devices, are relatively rare due to their expense. In general, high capacity devices have lower operating flexibility than conventional trays [7].

4. Equipment Services

Equipment inside the atmospheric tower can be classified into the unit operation function (heat transfer…) or the service (diesel pumparound…).

The basic equipment functions in the refinery atmospheric column are:

- Heat-transfer (always present)
- De-entrainment (always present)
- Mass-transfer (always present)
- Stripping (nearly always present)
- Auxiliary equipment (always present)
    - Structural components (beams, rings)
- Auxiliary equipment (often present)
    - Collector trays
    - Feed entry devices

Figure 8 illustrates the location of the functions in an atmospheric tower:

- Mass-transfer – fractionation between products
- Heat-transfer – direct contact heat exchange for the pumparounds
- De-entrainment – removal of liquid droplets from rising vapor
- Stripping – recovery of light material from tower bottoms

In a refinery atmospheric tower, most of the internals are in mass-transfer operation. The other operations, while important, use up less of the tower.

5. Mass-Transfer

Mass-transfer equipment has the intent of fractionating between adjacent product cuts. The refinery atmospheric column creates the feeds to the downstream units. Some flexibility is possible on most of the products. Nevertheless, fractionation requirements are critical for the atmospheric column. The trays must work properly and have a reliable efficiency.

5.1. Tray Efficiency – General

Tray efficiency is normally stated as a percent of the effectiveness of a theoretical plate. A tray with the same effectiveness as a single theoretical plate has 100% efficieny. Tray efficiency is determined by the effectiveness of the vapor-liquid mixing on the tray and how much liquid or vapor bypasses the mixing. A partial list of mechanical and hydraulic factors that affect tray efficiency includes:

- System physical properties
- Absolute vapor and liquid rates

- Vapor-liquid ratio on the tray
- Backmixing on the tray
- Mechanical layout including
    - Device type (sieve hole, fixed-valve, movable valve, bubble-cap, other)
    - Device (sieve hole or valve) size
    - Flow path length
    - Downcomer inlet and outlet velocities
    - Downcomer weir height
    - Weir geometry (setback, picketed, other)
- Condition (fouled, corroded, leaking, blocked)

Different methods of defining tray efficiency have been used. The most common definition is the Murphree tray efficiency [8] [9]. Murphree tray efficiency is used here.

Many correlations are available for calculating tray efficiency. The author has always taken the position that the more ways are in use to calculate something, the more likely it is that none of them work very well. The reason someone develops another correlation (or is willing to pay for it to be done) is that the previous methods do not give useful results. The author's library of papers on tray efficiency has roughly 75 different methods and variations attempting to create a unified approach to tray efficiency calculation. Most of them only apply to specific physical conditions, tray diameters, and limited equipment operating characteristics. None of them work well as a general method.

Tray efficiency is best determined from operating data. For atmospheric distillation column fractionating trays, the most important factor determining tray efficiency is the volumetric ratio of vapor and liquid on the tray. This assumes well-designed, properly installed, and maintained trays.

Discussions of tray efficiency often use detailed correlations combining physical properties such as viscosity, diffusivity, and composition gradients. However, for the mixtures of hydrocarbons present in crude oils at a vapor-liquid equilibrium (boiling-condensing), the differences due to these factors are minor. Only liquid-phase viscosity variation has any significant effect. For the heavier products in the column (going down the column), the viscosity increases, reducing tray efficiency.

5.2. Tray Efficiency – Liquid and Vapor Hydraulics

The major factor determining tray efficiency in atmospheric crude distillation is the liquid-vapor volume ratio on the tray. Figure 9 shows a conventional plot of vapor-liquid profiles in an optimized atmospheric column [10]. The conventional plot gives a rough idea of the vapor-liquid loads. Descending the column, the liquid loads in the fractionation sections generally drop. This results from two factors:

- Energy integration favors recovering heat at as high a temperature as possible. This shifts condensation into the lower parts of the column.
- Product draws take liquid from the column.

The example shown has large liquid rates between trays 21-32 and trays 3-8. This is due to large economic incentives for maximizing the middle distillate (kerosene plus diesel) yields versus other products. But for the typical tower, liquid rates will more closely align with those shown on trays 40-35 and trays 17-12.

While Figure 9 gives a general idea of the vapor and liquid loads, it does not give full insight into the tray hydraulics. Figure 10 gives a better comparison. Figure 10 modifies the presentation in four ways:

- The rates are converted to wet rates. This includes steam present in the vapor. This gives the actual vapor load required for tray hydraulics.
- The rates are converted to actual volumes.
- The vapor has been changed from net vapor (vapor leaving the stage) to entering vapor (vapor entering the stage). Fractionation calculations use net vapor and net liquid rates. This is the vapor leaving the stage and liquid leaving the stage. For proper evaluation of tray pressure balance the required rates are entering vapor and leaving liquid.
- The results are shown as a ratio of liquid-to-vapor on the tray.

All fractionation sections have less than 1% of the system volume as liquid. In fact, liquid fractions drop to as low as 0.2%. Very low liquid rates shift the tray operating regime from the froth regime into a spray regime [11] [12] [13]. Spray regimes tend to have lower efficiencies. The section on wash services (§7.2) includes a more detailed discussion of some problems with low liquid rates and operation in the spray regime.

5.3. Tray Efficiency - Recommendations

Based on plant operating data the following efficiencies are recommended for the services listed:

Table 3 Recommended Fractionation Service Efficiencies [2]

Service	Based on Tray Type	Efficiency, %
Light naphtha-heavy naphtha	Valve	75
Heavy naphtha-kerosene	Valve	70
Kerosene-Diesel	Valve	70
Diesel-Gas Oil	Valve	65
Wash	Valve	50
Stripping	Sieve	30
Side Stripping	Valve	25

Observed efficiency is typically within 10% of these values. The efficiencies quoted are for typical operating conditions and reflect conventional liquid-vapor ratios and physical properties for these services. Wash, stripping and sidestripping services are included for completeness.

Operating data, when available, should be used to calculate actual efficiencies for existing columns.

6. Heat-Transfer Services

The heat-transfer service in the refinery atmospheric tower takes sub-cooled pumparound return liquid and condenses the rising vapors. The heat-transfer returns the pumparound liquid to saturated conditions. As they lose heat, the vapors condense and create the net product liquid. Industrial units have been built with from zero (spray chamber) to 10 trays in pumparound sections.

Tray calculations for heat-transfer are complex [14] [15] [16]. The dominant factors are the relative vapor and liquid rates and mixing energy. These set the surface area and vapor bubble size. Experience has shown that 2-3 trays will meet heat-transfer requirements in nearly all pumparounds. Most units use three trays in each pumparound. Revamps that add a pumparound to an existing tower often use only two trays to minimize tower changes.

7. De-Entrainment Services

The atmospheric tower feed enters in the tower flash zone. The high-velocity feed creates small liquid droplets. The feed entry device coupled with the lower column vapor velocity allow the larger droplets to fall towards the tower bottom. However, the smaller droplets still entrain up the tower.

The section of the tower between the lowest product side draw and the feed is often called the wash zone. The name wash zone recognizes that the major purpose of this section of the tower is to de-entrain liquid droplets from the vapor leaving the flash zone.

7.1. Fractionation versus De-entrainment in the Wash Section

Either fractionation or de-entrainment performance may set the liquid rate required in this section. Figure 11 shows two atmospheric crude towers. One tower has AGO as the lowest side-draw product. The other tower has diesel as the lowest side draw product.

AGO normally has no distillation specification on the heavy end. Downstream processing economics favor making the product as heavy as possible (maximizing product yields). Entrainment of atmospheric residue contaminants limits the product quality. The trays, or packing, between the flash zone and the AGO are purely a de-entrainment device.

Diesel has a distillation specification on the heavy end [17]. Specific plants may vary from this because of blending and other constraints, but the atmospheric crude tower must produce a product stream close to the specification values. This distillation specification is in addition to the de-entrainment requirements. In a tower with diesel as the lowest side-draw, either the fractionation requirements or the de-entrainment requirements set the liquid load in the wash section. In nearly all cases, the fractionation load sets a higher liquid requirement.

7.2. Minimum Liquid Load for Trays versus Packing

One lower limit for tray capacity is normally quantified as a minimum liquid load on the outlet weir. At low liquid rates with low vapor density and high liquid density, tray operations change from continuous liquid-phase (froth) to continuous vapor-phase (spray) [18]. The lower liquid limit quoted before this transition causes efficiency to fall is 1 gpm/inch of weir length [6] (1.49 lpm/cm).

When the vapor phase is continuous, the liquid no longer flows smoothly over the outlet weir. Figure 12 illustrates this by showing three zones on a tray in spray operation. In all areas the rising vapor creates drops of liquid spray over the tray deck. In the inlet area, spray back against the downcomer is collected and falls back to the tray. In the center of the tray, the spray droplets fall back down to the tray. Close to the outlet, the vapor projects some of the spray droplets over the downcomer. The droplets over the downcomer fall to the tray below. The tray functions in a diffusion-like operation with the driving force for flow coming from a mass-density gradient across the tray. This operation is called tray blowing [19]. When trays enter blowing operation both fractionation and de-entrainment effectiveness drop.

Figure 1 shows atmospheric crude columns with both a packed wash section a trayed wash section. The tower with the trayed wash section shows one-pass trays. Using one-pass trays is an attempt to reduce the weir length and prevent blowing. However, if the diameter is too large, the flow gradient across the tray can create cross-flow channeling and other problems.

The tray with the lowest liquid load is the tray immediately above the flash zone. Table 4 illustrates the effect of some different options to resolve tray limits. The top half of Table 4 shows the effect on yields for a constant feed temperature. The bottom half of Table 4 shows the change in required feed temperature for constant yields. The results are at a constant tower pressure and stripping steam-to-bottoms ratio.

Table 4 Trays or Packing in Wash Zone Service

Packing					110% of min liquid
Tray passes		2	1	1	1
Liquid at weir	gpm/in	1.00	1.00	0.50	0.30
Weir length	Inches	188 *	110	110	110
Flow path length	Inches	76	152	152	152
Overflash	Vol % of crude	4.87	2.86	1.43	0.86
Constant heat input					
Distillate yield	Vol %	56.91	58.48	59.63	60.10
Feed Temp	°F	690.0	690.0	690.0	690.
Heat duty	M btu/hr	222.6	222.6	222.6	222.6
Stripping Stm	Lbs/hr	17,990	17,320	16,840	16,640
Bottoms Temp	°F	668.2	669.9	670.9	671.2
Constant Yield					
Distillate yield	Vol %	60.10	60.10	60.10	60.10
Feed Temp	°F	715.0	702.7	693.8	690.0
Heat duty	M btu/hr	248.4	235.7	226.5	222.6
Stripping Stm	Lbs/hr	16,640	16,640	16,640	16,640
Bottoms Temp	°F	692.8	682.5	676.7	671.2

* bottom tray has outboard downcomers.

Table 4 shows four cases;

- A two-pass tray at 1 gpm/in (1.49 lpm/cm). This tray has a high liquid flow requirement. Overflash is the amount of liquid that must flash in the tower over the rate of tower products to produce the liquid required for the wash section. Numbers for this are often quoted, but the author finds the concept of little use. Liquid rates depend highly on the tray configuration. However, the value here of 4.87% is very close to oft-quoted values of 5%. All the other trays have better performance.
- A one-pass tray at 1 gpm/in (1.49 lpm/cm).
- A one-pass tray with 0.50 gpm/in (0.75 lpm/cm) weir load. This is around the lower limit for even mediocre de-entrainment on trays.
- A one-pass tray with 0.30 gpm/in (0.60 lpm/cm) weir load. This was chosen for matching performance to a tower with a packed wash section. For a conventional tray this is likely to have ineffective de-entrainment.

Minimum liquid requirements on the bottom tray have significant yield or operating effects. For all liquid rates, the two pass tray example has a 71% higher liquid demand (188" weir length compared to 110" weir length). Tray design can optimize these rates; nevertheless, the possible improvements all have limits. Some approaches include:

- Smaller tower diameter by using a larger tray spacing, 30" instead of 24", for example.

- Parallel baffles in the wash section (see §8 on stripping trays for examples) to reduce effective weir lengths.
- Splash baffles [5]
- Picket-fence weirs [20]

Of these, picket-fence weirs are the most common. Figure 13 shows a tray with a picket-fence weir. The pickets act as deflectors that prevent too much liquid from spray above the downcomer. Effective outlet weir load increases. The liquid mass inventory on the tray goes up and both mass-transfer and de-entrainment improve.

The normal limit for structured packing or grid in wash service is a minimum liquid rate of 0.15 gpm/ft^2 (6.11 lpm/m^2) liquid load at the bottom of the bed. For a 16' (4.88 m) diameter tower this gives a liquid load of 30.1 gpm (114 lpm). The final case in Table 4 (right column) has a liquid load equal to 110% of the minimum packing load. At this liquid rate the de-entrainment effectiveness exceeds the performance of trays operating with 1 gpm/in (1.49 lpm/cm) of weir length.

Further optimization might include using wash sections of reduced diameter. This decreases the liquid load requirement and improves unit operation.

Wash services in atmospheric columns have easier service requirements than wash services in vacuum columns. A full discussion of wash section performance with packing can be found in references on vacuum towers [21].

7.3. Wash Oil Distributors

Wash oil distributors merit some additional comments. Vendor recommendations vary. The two major options are either spray nozzle distributors or trough distributors. Both present challenges. The ideal de-entrainment device would do no heat transfer and no mass transfer. Neither device comes is clearly superior.

The author typically recommends spray nozzle distributors. Spray distributors have the major benefit of low initial cost and minimal maintenance cost and time and have reasonable performance. Sprays are very effective heat-transfer devices (bad), but they give low mass-transfer efficiency in the bed (good). These characteristics compete against each other. Finally, low residence times in the distributor pipes inside the tower tend to reduce coking.

Trough distributors have the benefit of eliminating the liquid sprayed onto the vessel wall, and reduce splash back from the liquid entering the packing surface. They do no heat transfer. However, their longer residence time increases the likelihood of coking, especially at high temperature operation.

Spray distributors and trough distributors have similar risk profiles for plugging from scale, dirt, and other solids. The balance of pressure drop used across the orifice versus the number of distribution points results in distributors with roughly equal plugging risk due to scale. Even with liquid distribution baffles or drip point multipliers, hole sizes are small in trough distributors.

Systems that pump liquid back to the tower for the wash distributor should use parallel basket strainers to remove scale. Piping downstream of the basket strainers should be 5-Cr alloy or better to prevent scale created in the pipe plugging the distributor.

7.4. Liquid Collection Beneath Wash Bed

A liquid collector should be installed beneath the wash bed. The liquid collector is typically a chimney tray. It performs the following:

- Acts as a vapor distributor to the packed bed above
- Collects liquid to bypass the feed zone below, avoiding quenching the flash zone and reducing unit yield
- Reduces probability of liquid re-entrainment at high vapor rates

8. Stripping Services

Nearly all atmospheric columns have a stripping section below the feed. Liquid in the feed is collected and stripped with steam. Stripping allows for reduced flash zone operating temperature at the same yield, reduces light material entering the downstream vacuum tower, and provides flash control of the bottoms product. Trays are the most common device used.

8.1. Tray Process Design and Layout

Stripping trays in this service are normally sieve trays with large-diameter holes (>= 0.75 inches, 19 mm). Occasionally conventional 0.5 inch (12.7 mm) diameter holes are used. Some units have used fixed-valves for the stripping trays, but that has few benefits and increases the risk of plugging.

Vapor rates change dramatically across each tray. It is extremely difficult to create a useful tray layout that will work across the entire stripping section. The bottom tray has such a low vapor rate that the low hole area it requires would cause the top tray to have excessive pressure drop and flood. The top tray has such a high vapor rate that the bottom tray with that hole area would weep and have extremely low efficiency. Multiple tray layouts are required.

Various parameters need to be evaluated to set the stripping tray layout. These include minimum dry tray pressure drop, weeping, hole spacing, and total tray pressure drop. Even with customized tray designs, the trays will have low efficiency. At best, tray efficiency will be 25-30%.

Table 5 illustrates the difficulty. It includes two cases for the example unit:

- Low steam: 2. 5 lbs of steam per standard barrel of bottoms (0.75 weight percent steam to bottoms rate).
- High steam: 10 lbs of steam per standard barrel of bottoms (2.97 weight percent steam to bottoms rate).

Though these cases are labeled low-steam and high-steam, most commercial units run rates less than the low-steam rate here. Typical rates are often as low as 1.0 lbs of steam per barrel of bottoms (~0.30 weight percent steam to bottoms rate). The low-steam rate defined here is the upper end of most commercial practice. The rate of 10 lbs of steam per barrel is approaching an optimum design for a unit with 12-16 real trays (3-4 theoretical trays). The economic optimum for specific plants will vary based on the economic structure for the plant.

The basis for Table 5 is 8 trays with 25% tray efficiency. Tray design analysis showed the following zones for effective tray design;

- Trays 8-5, top 4 trays, one design.
- Trays 2-4, middle trays, one design.
- Tray 1, bottom tray, one design.

Table 5 Tray Limits for Example Stripping Section (Two-Pass Trays)

Tray	Tower diameter In (m)	Limited by	Operating Problems
Low Steam Rate	90 (2.29)		
8-5		Liquid weir load, high	
2-4		Liquid weir load, high	
1		Liquid weir load, high Hole area, low (3.3%) Hole pitch, high	Weeping
High Steam Rate	126 (3.20)		
8-5		Liquid weir load, at limit	
2-4		Liquid weir load, at limit	
1		Hole area, low (3.3%) Hole pitch, high	Weeping

Note that the diameters shown are 90" (2.29m) and 126" (3.20m). These would be reduced diameter sections below the feed zone. The diameter just above the feed zone for this tower is 192" (4.88m).

Industrial practice takes four approaches to dealing with reduced diameter sections:

- Use very large downcomers to reduce the tray active area appropriately
- Reduce the diameter of the bottom section of the overall tower
- Install a reduced diameter can section inside a full-sized tower shell
- Use channel baffles to reduce the effective active area

Channel baffles are often used in conjuction with reduced diameter sections in order to optimize active area and weir load. Figure 14 shows a channel baffle tray. The side baffles (channels) isolate a portion on the outside of the tray. All the vapor and liquid flow through the center section only.

8.2. Tray Mechanical Design

Operating conditions subject the stripping trays to severe service conditions. Tray failure occurs routinely. The two major sources of tray failure are pressure surge due to flashing of water and upsets due to high liquid level. Unit startup creates most of the problems. Some units have never started up successfully with the stripping trays in place. A systems approach is needed to keep the stripping trays intact. Steps include configuration of external piping to reduce the risk of water getting to the tower in the stripping steam, reliable level measurement, sound operating procedures and mechanical upgrades to the trays.

Conventional tray design uses two design criteria. The first is a point load of 300 lbs (136 kg) to meet installation requirements. The second is a distributed load of 60 lbs per ft^2 (293 kg/m^2) to meet operating requirements. These criteria are insufficient for severe mechanical services. A distributed load design of 288 lbs per ft^2 (1,406 kg/m^2) is required for the stripping section. A combination of strengthened construction and special features including through-bolting, shear clips, truss-lugs and extra-strength components can meet this requirement [22].

8.3. Alternative Tray Mechanical Configurations – The Trap-Door Tray

One alternative to handling pressure surges is the trap-door tray. Figure 15 and Figure 16 show the trap-door tray configuration [23]. The trap-door tray was invented and first applied in an atmospheric crude column [24] suffering from tray damage. In service, a chain with a weight attached runs between the trap-door and the tray deck. The chain prevents the panel from excessive opening and the weight pulls the panel closed after it opens.

8.4. Packed Stripping Sections

Packed stripping sections have been used in atmospheric crude towers. A smooth-surface structured packing is the best choice for this, although it is an ambitious service. The benefit is the ability of the packing to handle higher steam rates, have an increased number of stages, and to easily deal with the varying vapor load in the stripping section. The major risk is plugging of the packing or distributors due to coking or solids. Distributor plugging has been a larger source of problems than packing coking. Packed stripping sections work best with clean crudes.

Reports of using random packing in atmospheric tower stripping sections date back to 1979 at Guam Refining [25]. Documented cases of structured packing in atmospheric tower stripping sections date back to 1987 [26]. In the author's experience some units with random packing in the stripping section have operated over 20 years with minimal problems. Other units with packing in the residue stripping services have had annual coking events. Before taking this path, all benefits and costs should be weighed.

9. Auxiliary Equipment

In atmospheric crude towers auxiliary equipment may include collector trays, water draw trays, feed inlet devices, and other equipment. Appropriate references should be consulted for these items [1] [21] [27].

10. Conclusions

Distillation equipment design for atmospheric crude distillation is all about trays. With one exception trays have the appropriate capacity, flexibility, and cost performance to make them the favored distillation internal in this service. Both valve trays and sieve trays work effectively. In the wash service, just above the feed zone, structured packing or grid is often the superior technical choice.

Tray design for this service fits well into the conventional tray category. Revamping units for higher throughput favors high-capacity trays. The stripping section at the bottom of the tower requires careful attention to mechanical fabrication details to keep the trays intact and in place.

11. References

[1] H. Z. Kister, *Distillation design*. McGraw-Hill, Inc., 1992.

[2] A. W. Sloley, "Distillation design: trays," in *Encyclopedia of Chemical Processing*, vol. 2, 5 vols., S. Lee, Ed. New York, NY: Marcel Dekker, 2005, pp. 749–764.

[3] H. Z. Kister, "Can valve trays experience vapour cross flow channeling?," *The Chem Eng*, Jun. 1993.

[4] H. Z. Kister, K. F. Larson, and Madsen, P. E., "Vapor cross-flow channeling on sieve trays: fact or myth?," *Chem Eng Prog*, vol. 88, no. 11, pp. 86–93, Nov. 1992.

[5] A. W. Sloley and P. B. Fleming, "Successfully downsize trayed columns," *Chem Eng Prog*, vol. 90, no. 3, pp. 39–47, Mar. 1994.
[6] "Bulletin 681R1 44 frequently-asked questions and answers about trays and packing." Glitsch, Inc., 1983.
[7] A. W. Sloley, "Should you switch to high-capacity trays?," *Chem Eng Prog*, vol. 95, no. 1, pp. 23–35, Jan. 1999.
[8] E. V. Murphree, "Rectifying column calculations," *Ind Eng Chem*, vol. 17, no. 7, pp. 747–750, Jul. 1925.
[9] E. V. Murphree, "Graphical rectifying column calculations," *Ind Eng Chem*, vol. 17, no. 9, pp. 960–964, Sep. 1925.
[10] A. W. Sloley, "Atmospheric distillation process - fundamental concepts," presented at the AIChE Spring Meeting, New Orleans, LA, 2014.
[11] M. J. Lockett, "The froth to spray transition on sieve trays," *Trans IChemE*, vol. 59, pp. 26–34, 1981.
[12] M. Prado, "The bubble-to-spray transition on sieve trays: mechanisms of the phase inversion," University of Texas, Austin, TX, 1986.
[13] M. Prado, K. L. Johnson, and J. R. Fair, "Bubble-to-spray transition on sieve trays," *Chem Eng Prog*, vol. 83, no. 3, pp. 32–40, Mar. 1987.
[14] J. R. Fair, "Direct contact gas-liquid heat exchanger for energy recovery," *Trans ASME*, vol. 112, no. 8, pp. 216–222, Aug. 1990.
[15] J. R. Fair, "Process heat transfer by direct fluid-phase contact," presented at the Process Heat Transfer Symposium, 12th National Heat Transfer Conference, Tulsa, OK, 1971, vol. 118, pp. 1–11.
[16] T. P. Ognisty and Y. S. Kao, "Performance of fractionation devices in direct contact heat transfer," presented at the PACHEC IV Pacific Area Chemical Engineering Conference, Acapulco, Mexico, 1988.
[17] "D975-15b Standard specification for diesel fuel oils," ASTM International, West Conshohocken, PA, 2015.
[18] H. Z. Kister, "Design and layout for sieve and valve trays (Column internals 3)," *Chem Eng*, vol. 87, no. 18, pp. 119–123, Sep. 1980.
[19] A. W. Sloley, "Improve tray operations," *Hyd Proc*, vol. 80, no. 6, pp. 85–86, Jun. 2001.
[20] D. R. Summers and A. W. Sloley, "How to handle low-liquid loadings," *Hyd Proc*, vol. 86, no. 1, pp. 67–78, Jan. 2007.
[21] A. W. Sloley, "Refinery vacuum towers - state of the art," presented at the AIChE National Meeting, San Francisco, CA, 2013.
[22] G. H. Shiveler, "Use heavy-duty trays for severe services," *Chem Eng Prog*, vol. 91, no. 8, pp. 72–81, Aug. 1995.
[23] A. R. DiNicolantonio and H. R. Theysen, "Hinged pressure relief tray," 4,133,852, 09-Jan-1979.
[24] A. R. DiNicolantonio, "Private communication," 1997.
[25] L. D. Nicholson and H. A. Erskine, "Crude-unit revamp cuts resid yield," *OGJ*, vol. 78, no. 12, pp. 102–117, Mar. 1980.
[26] P. Roy and A. C. Mercer, "The use of structured packing in a crude oil atmospheric distillation column," presented at the Distillation and Absorption, Brighton, UK, 1987, vol. 104, pp. A103–A114.
[27] A. W. Sloley, "Don't get drawn into distillation difficulties," *Chem Eng Prog*, vol. 94, no. 6, pp. 63–78, Jun. 1998.

[28] G. J. Keller, "Apparatus for liquid and vapor or gas mass transfer," 3,729,179, April 24, 29173.
[29] W. Bruckert, "Vapor-liquid contact system and method," 3,410,540, 12-Nov-1968.
[30] J. Forgrieve, "Commerical jet-tray fractionators," in *Proceedings of the International Symposium on Distillation*, Brighton, UK, 1960, pp. 185–189.
[31] J. O. Nye, "Distillation tray/downcomer," 5,242,628, 07-Sep-1993.
[32] K. T. Chuang and A. E. Everatt, "Gas-liquid contacting apparatus," 4,504,426, 12-Mar-1985.

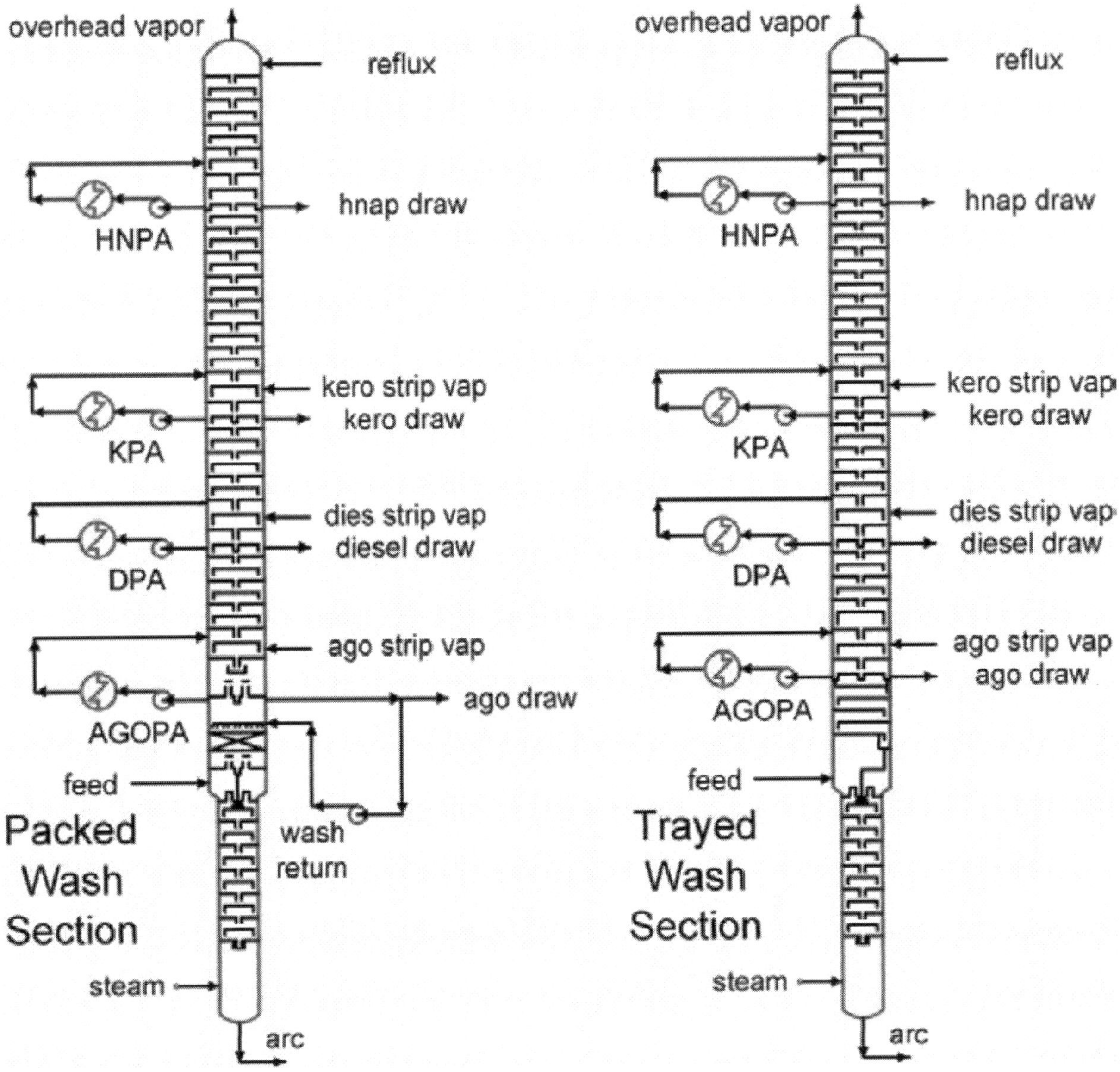

Figure 1 Atmospheric Crude Distillation Tower Configuration Examples

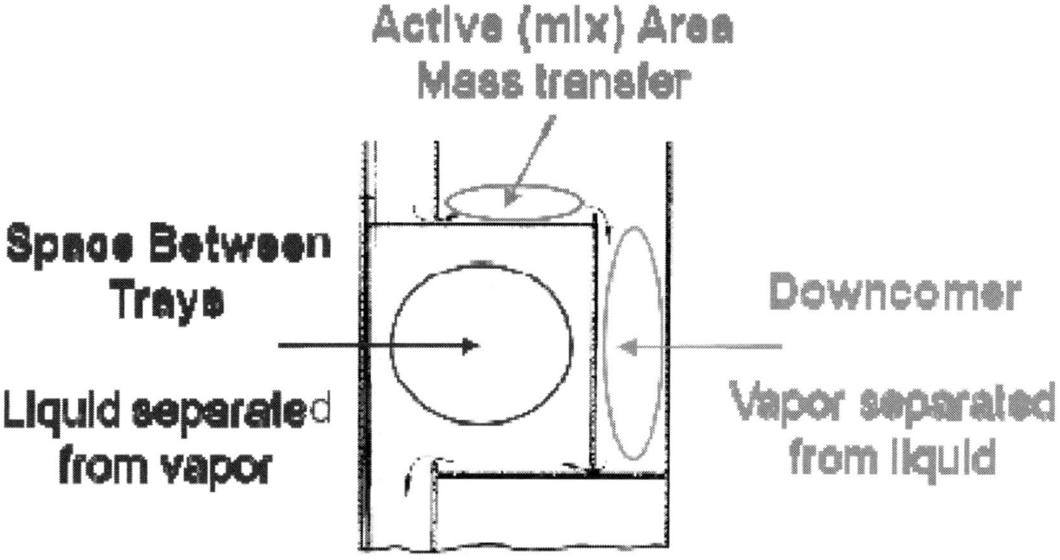

Figure 2 Functional Areas of a Conventional Tray

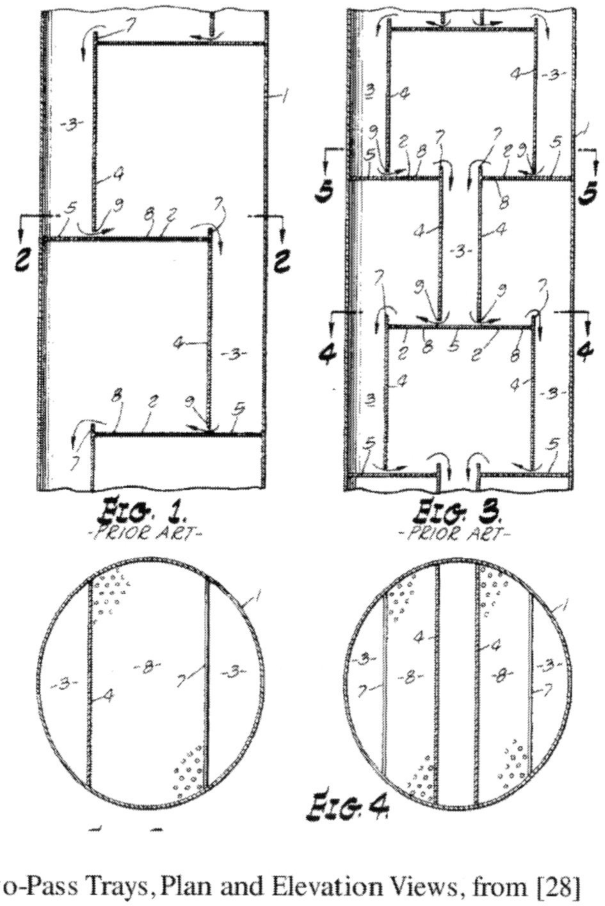

Figure 3 One-Pass and Two-Pass Trays, Plan and Elevation Views, from [28]

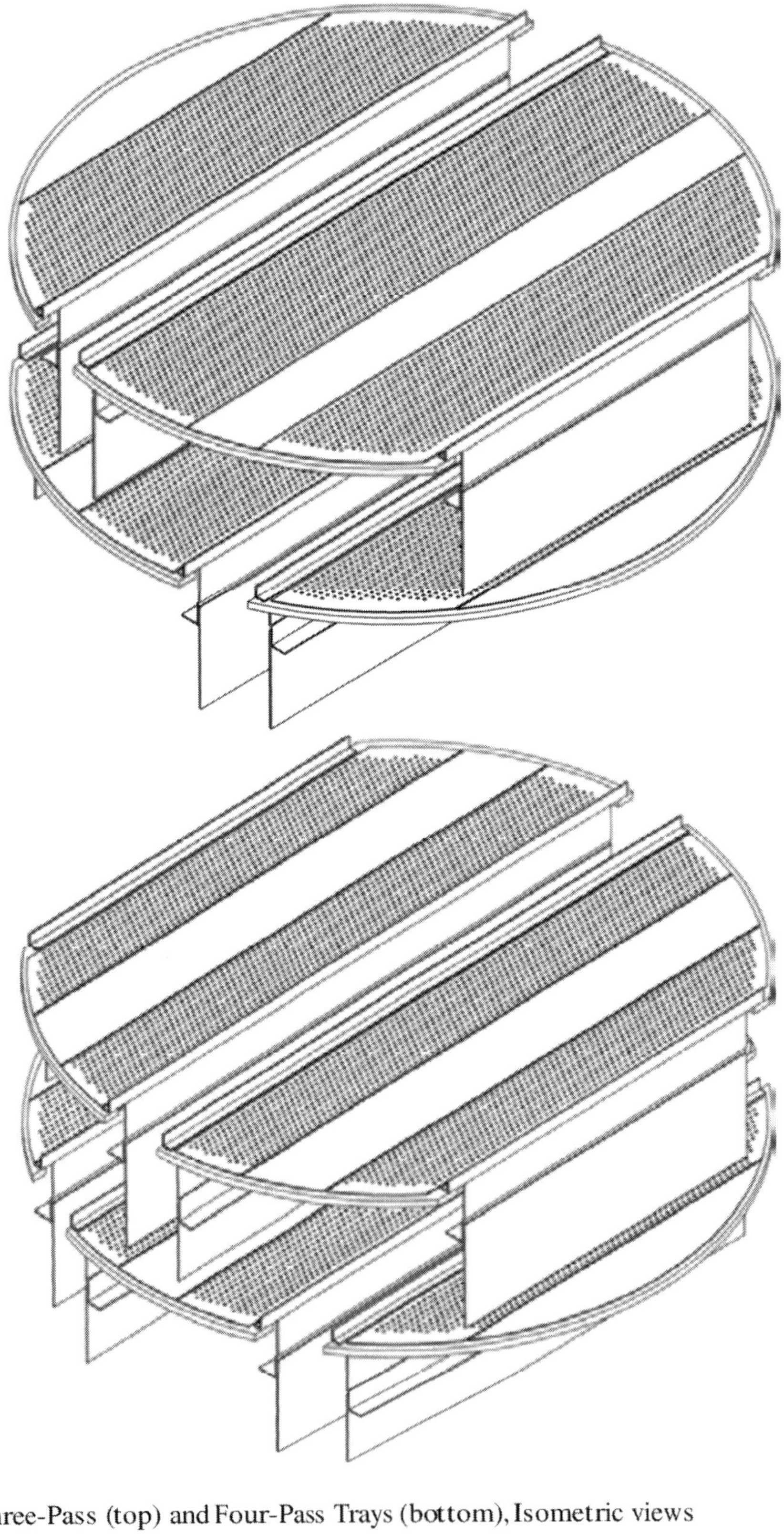

Figure 4 Three-Pass (top) and Four-Pass Trays (bottom), Isometric views

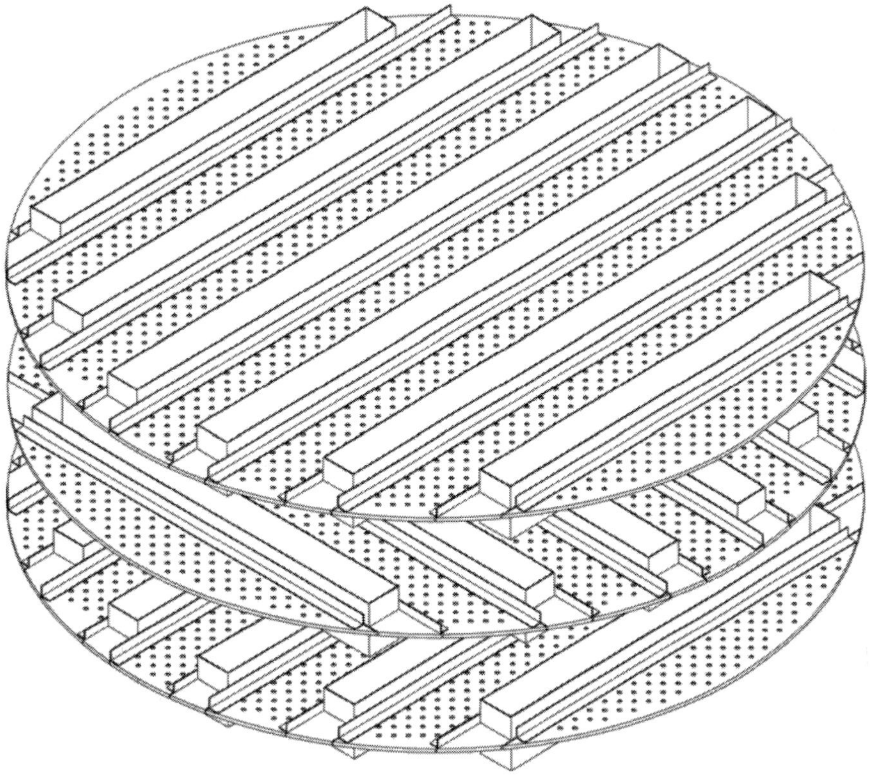

Figure 5 High-Capacity Tray With Extended Downcomer Length [29]

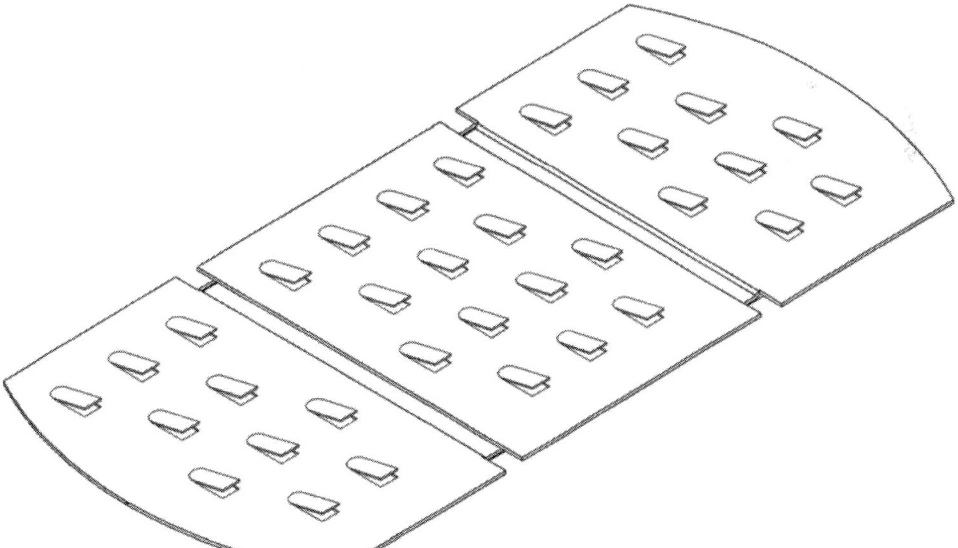

Figure 6 Jet Tray Deck [30]

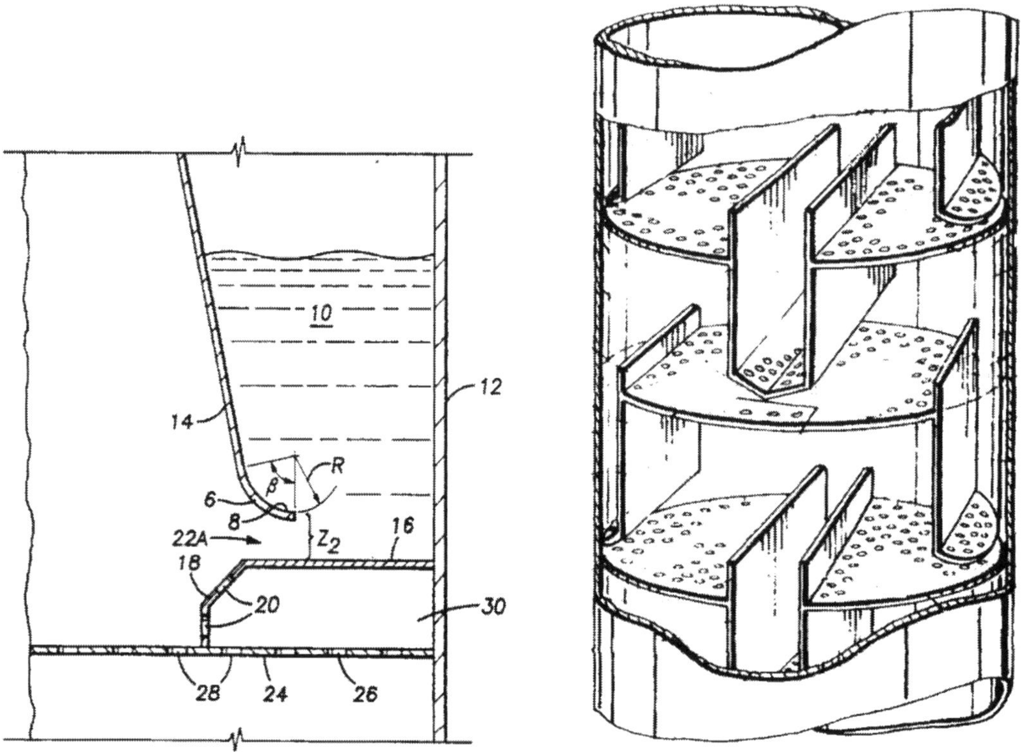

Figure 7 Hanging Downcomers, [31] left and [32] right. Left section shows a raised tray component (#16) with side vapor injection from holes (#20).

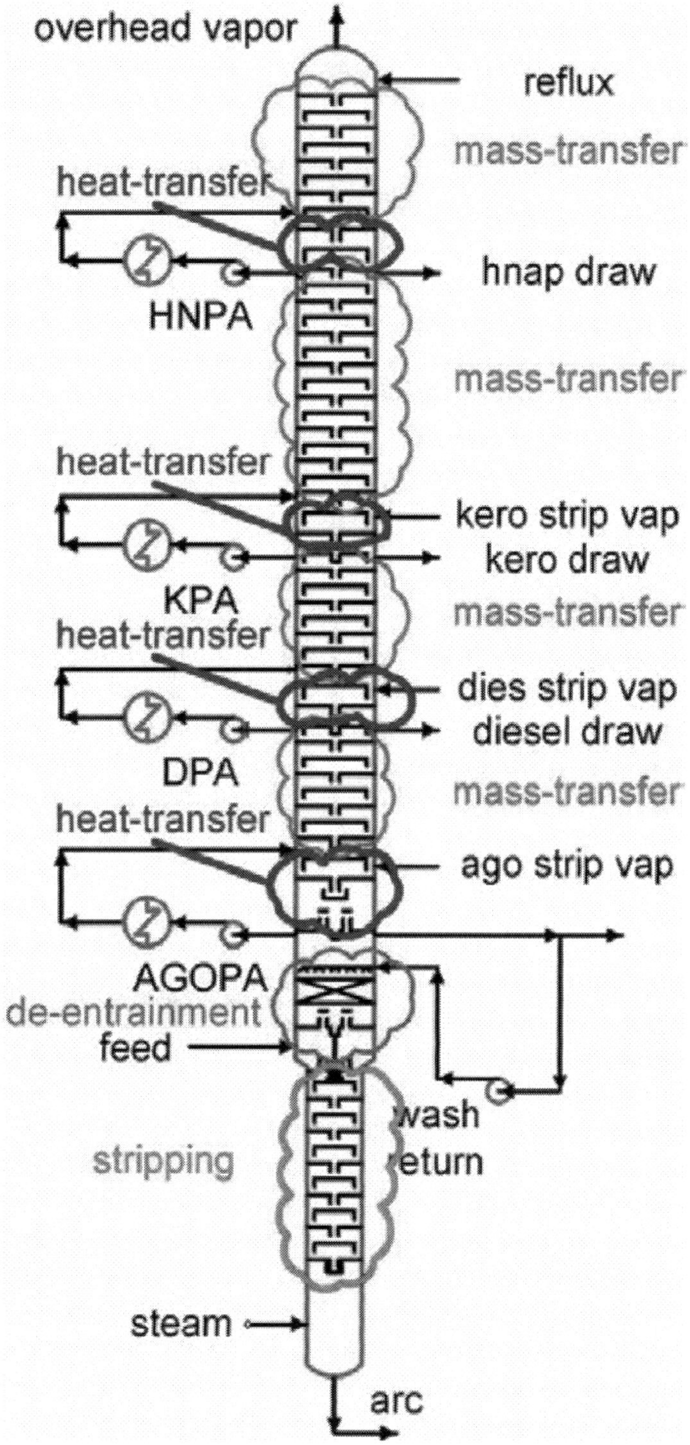

Figure 8 Atmospheric Tower Internals Services

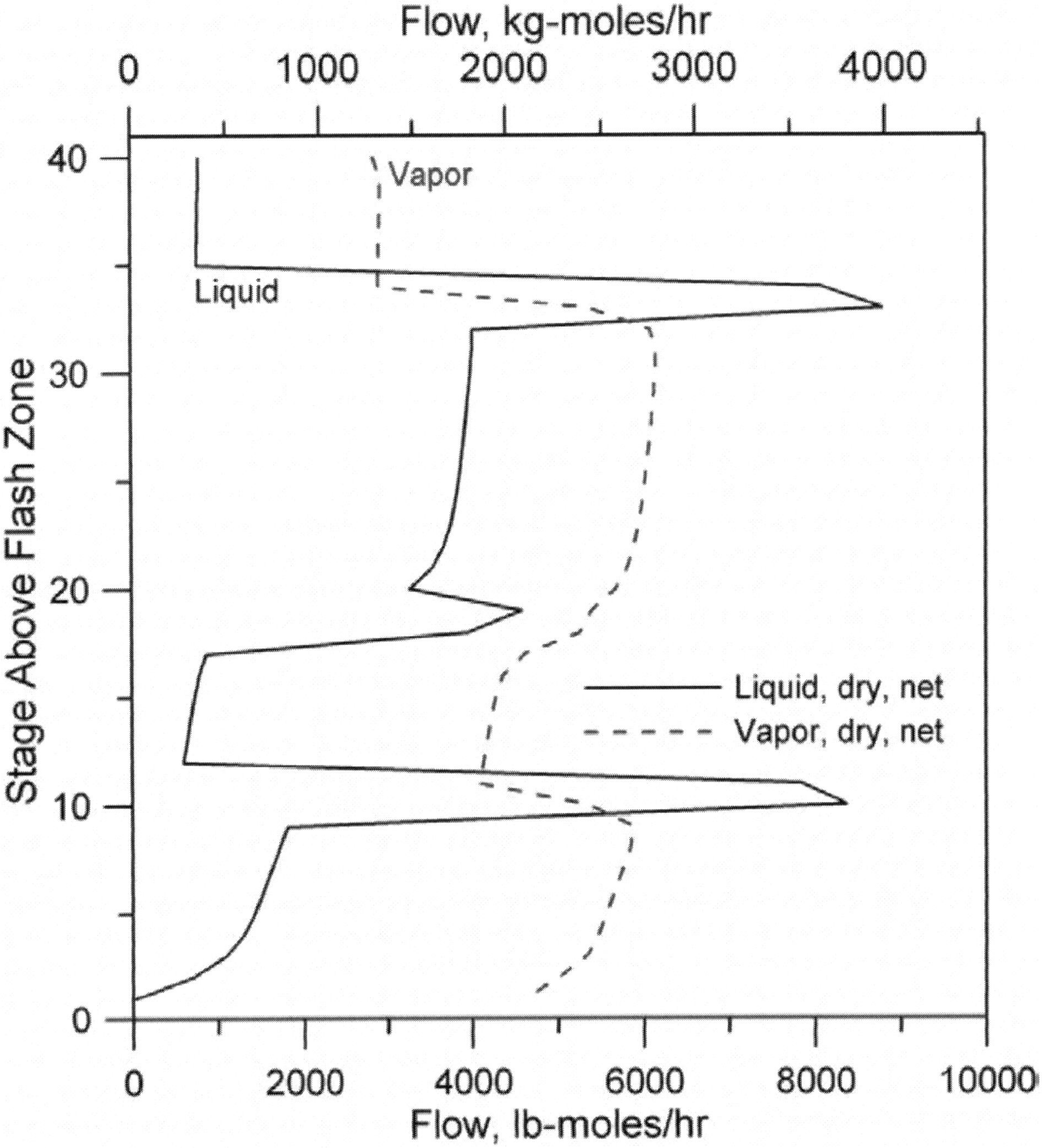

Figure 9 Tray Loads – For fractionation evaluation

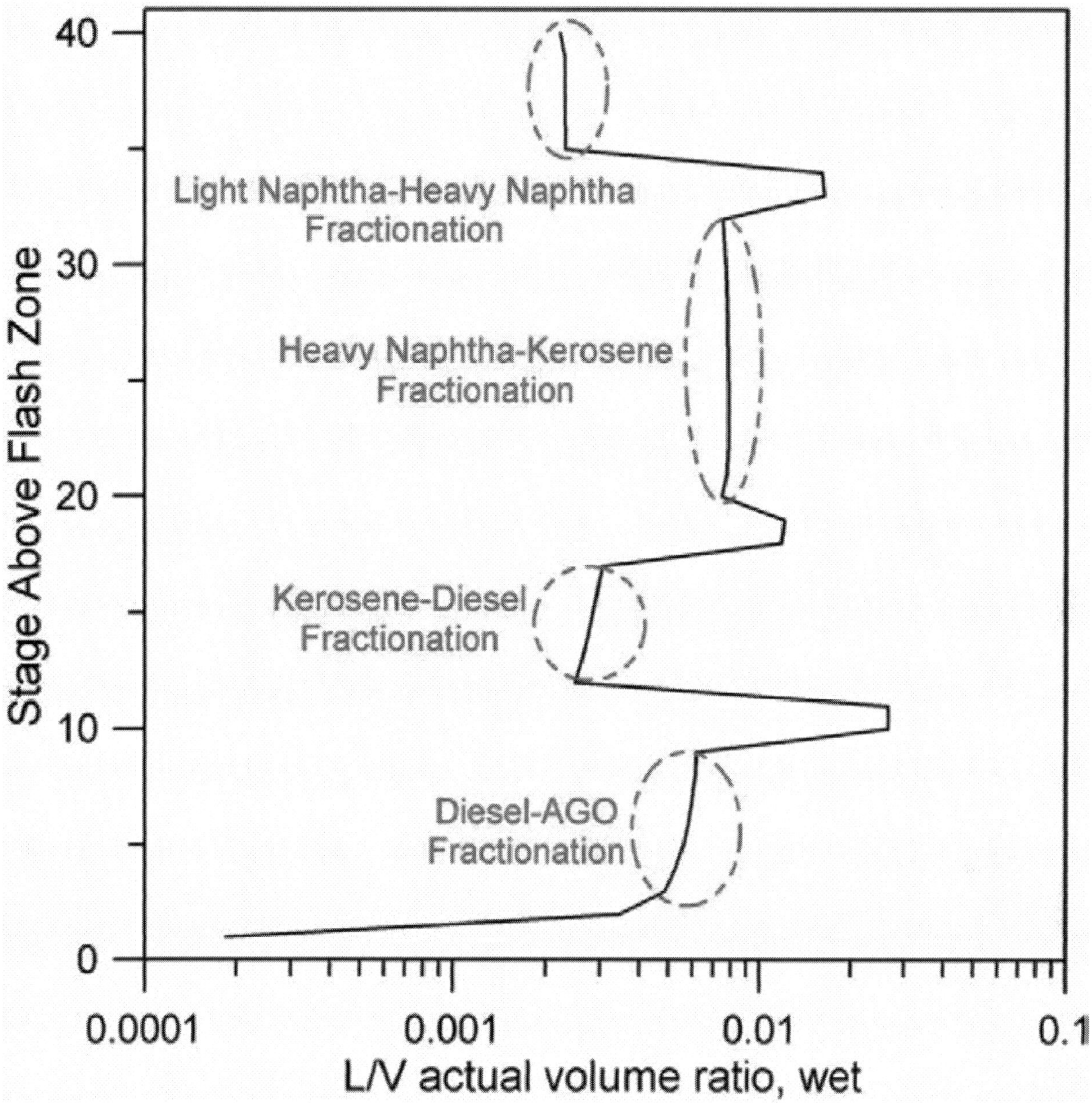

Figure 10 Tray Loadings, Volume Ratios

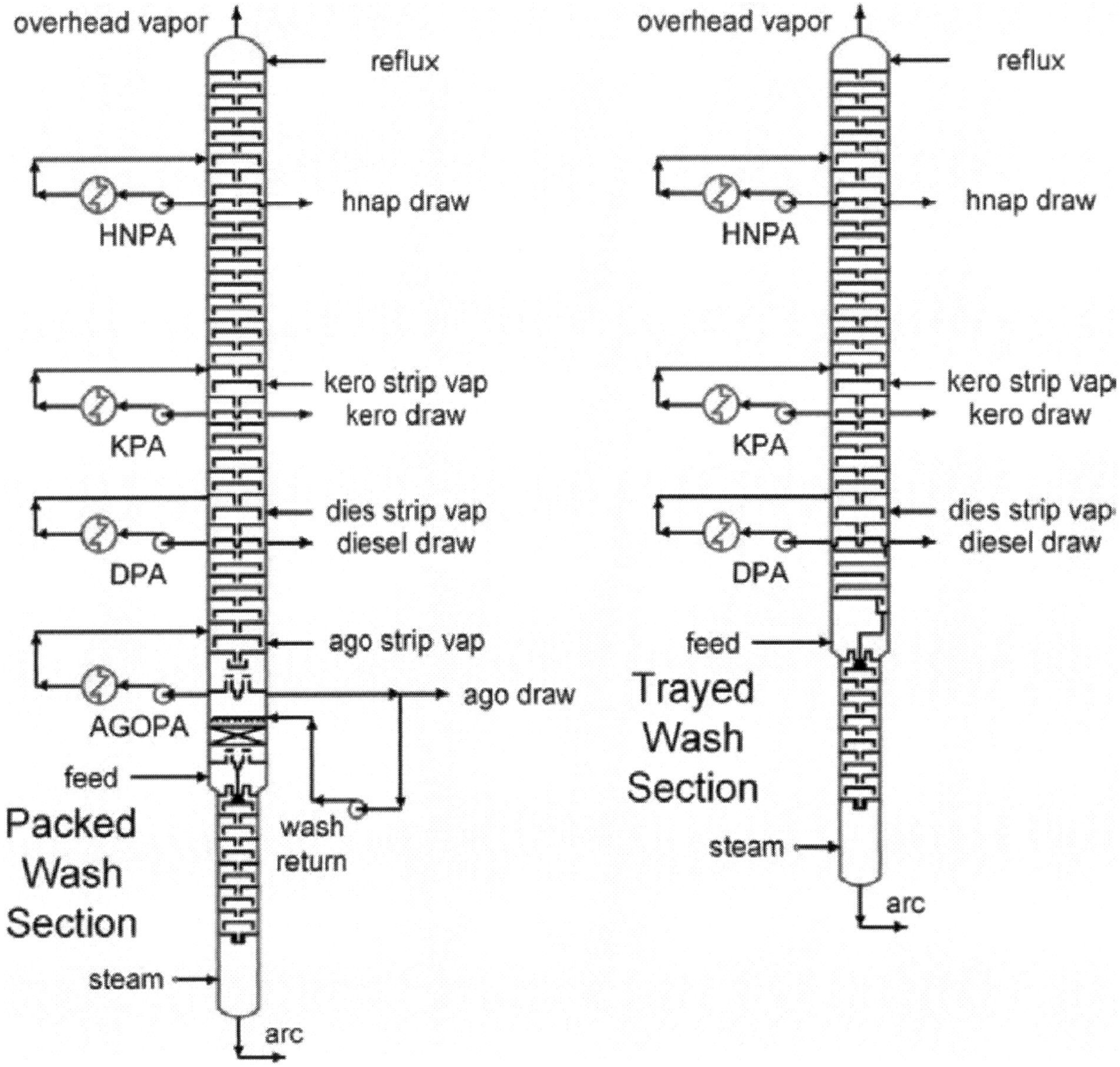

Figure 11 AGO versus Diesel as Lowest Side-Draw

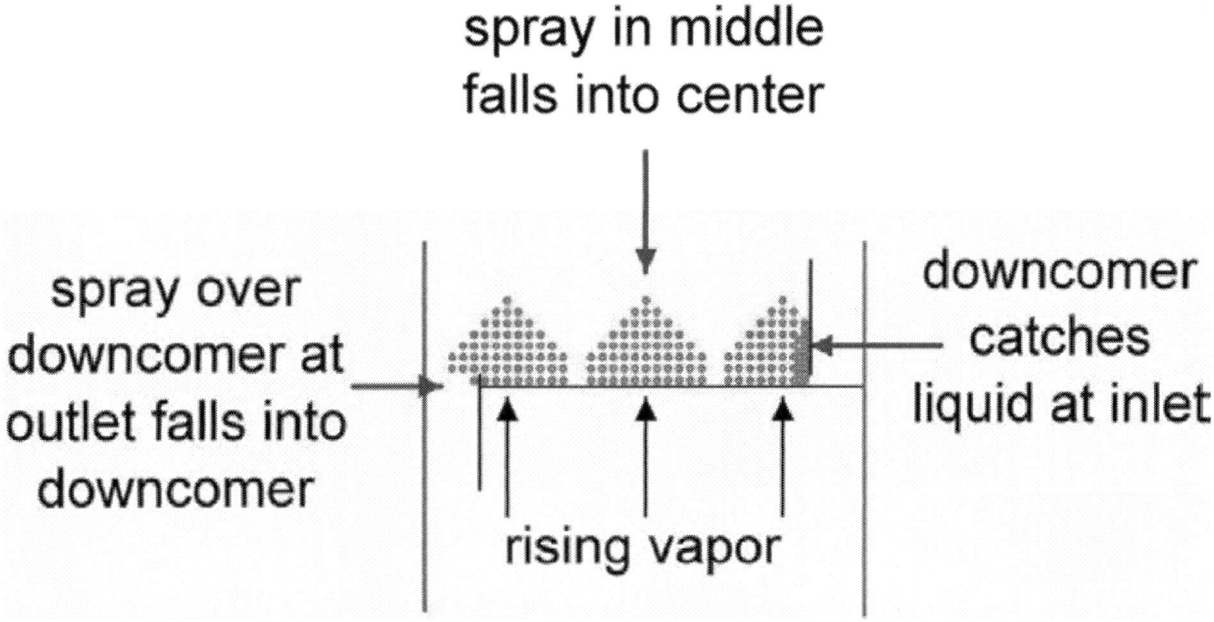

Figure 12 Blowing Operation in Spray Regime

Figure 13 Tray with Picket-Fence Weir (Courtesy Sulzer Chemtech Ltd.)

Figure 14 Tray with Channel Baffles (Courtesy Sulzer Chemtech Ltd.)

Figure 15 Tray with Channel Baffles and Trap Doors (Courtesy Sulzer Chemtech Ltd.)

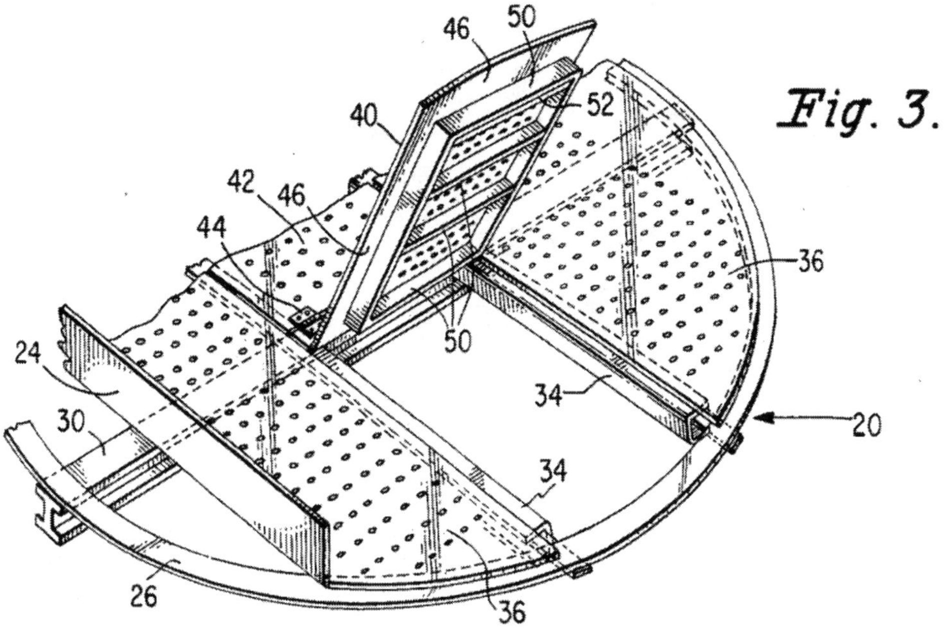

Figure 16 Trap Door Tray for Severe Service (from [23]).

Balanced Distillation Equipment Design for Crude Distillation Units

Sounho Lee – GTC Technology US LLC
2016 AIChE Spring Meeting / Distillation Symposium / "Distillation in Refining Session"
Houston TX April 11, 2016

GTC Technology
→ Engineered to Innovate

Crude Distillation Unit Distillation Equipment

GTC Technology
→ Engineered to Innovate

- Fouling and corrosion: major issues
- Distillation equipment fouling resistance & efficiency: inverse relationship
- Balancing / optimization between fouling resistance and efficiency requirement
- Poor process and/or equipment design know-how often downgrade column performance and reduce unit run length
- Two (2) actual retrofit case studies
  - Crude distillation units

AIChE 2016 Spring Meeting / Distillation Symposium

Case Study 1- Crude Distillation Unit
Unit Configuration

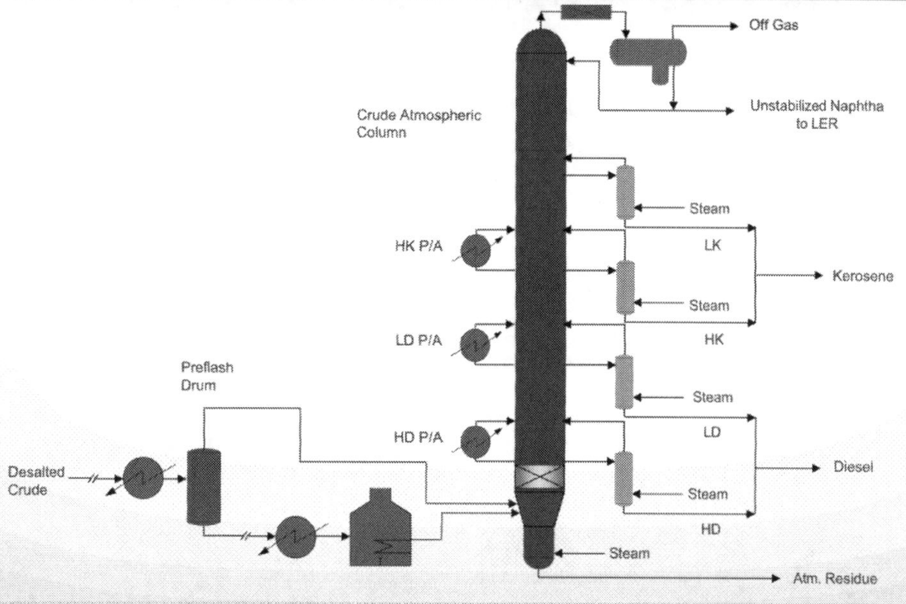

AIChE 2016 Spring Meeting / Distillation Symposium

Case Study 1- Crude Distillation Unit Limitation

- Valve / perforation hole wear & corrosion
  - Naphtha / light kerosene fractionation section
- Fouled trays
  - Light kerosene/heavy kerosene fractionation section
  - Heavy kerosene/light diesel fractionation section
- Side product draw limitation
  - Limited product yields : kerosene & diesel
  - Varied crude slate composition

AIChE 2016 Spring Meeting / Distillation Symposium

Case Study 1- Crude Distillation Unit Original Side Draw Configuration

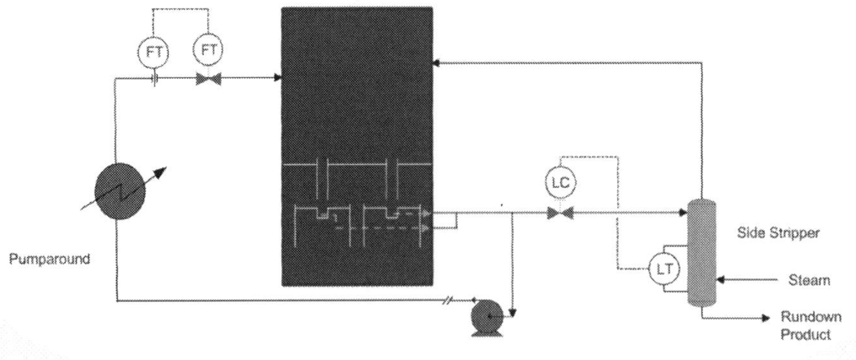

AIChE 2016 Spring Meeting / Distillation Symposium

Case Study 1- Crude Distillation Unit Modification

GTC Technology – Engineered to Innovate

- Fixed valve tray implementation
  - Improving fouling/wear resistance
- Intermediate product/pumparound draw:
  - Original: withdrawal from fractionating tray directly
  - Modification: conversion to chimney tray
- Losing one fractionating tray for each
  - Light kerosene/heavy kerosene fractionation section
  - Heavy kerosene/light diesel fractionation section
  - Light diesel / heavy diesel fractionation section
- Sensitivity Analysis
  - Heavy kerosene/light diesel fractionation section

Case Study 1- Crude Distillation Unit Sensitivity Analysis

GTC Technology – Engineered to Innovate

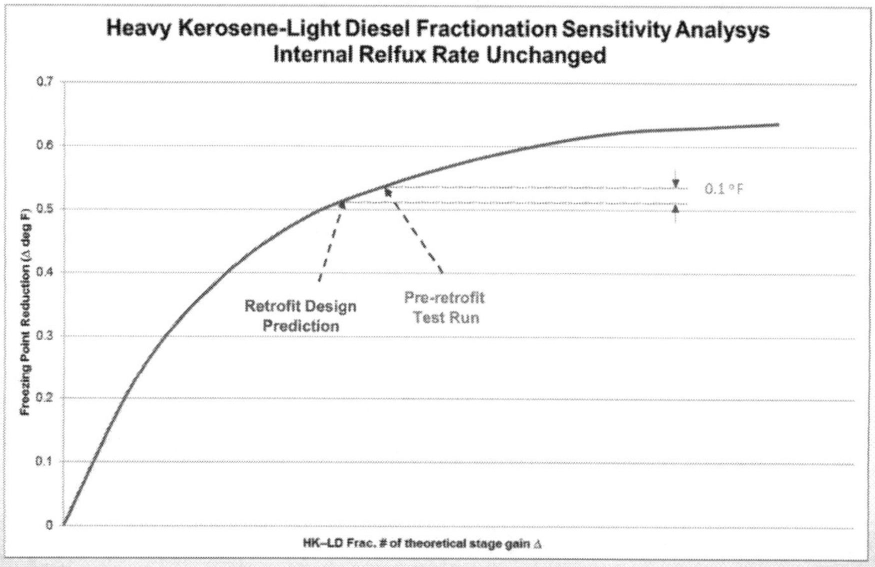

AIChE 2016 Spring Meeting / Distillation Symposium

Case Study 1- Crude Distillation Unit Sensitivity Analysis

GTC Technology - Engineered to Innovate

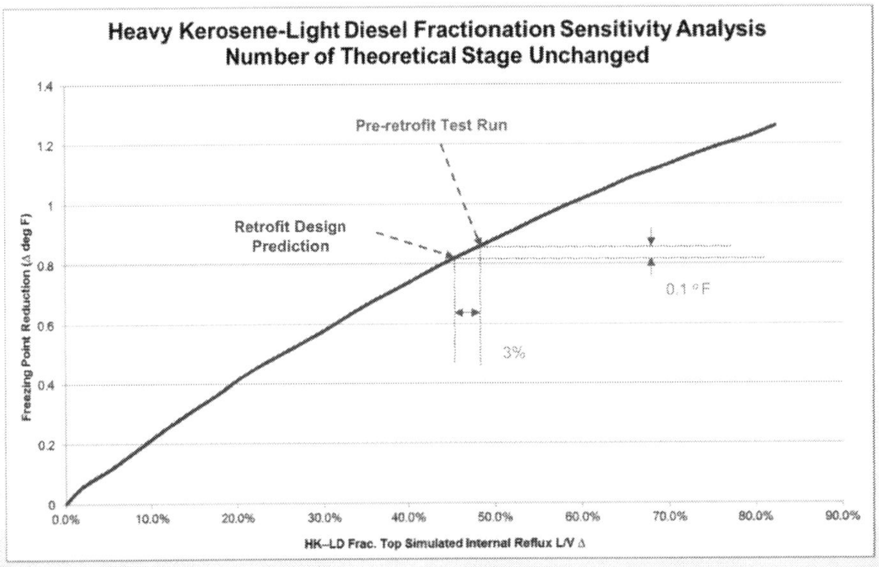

AIChE 2016 Spring Meeting / Distillation Symposium

Case Study 1- Crude Distillation Unit Performance Summary

GTC Technology - Engineered to Innovate

Parameter	Pre-retrofit Test Run	Post-retrofit Test Run
Crude Charge (BPD)	BASE	+ Δ 12 %
Unstabilized Naphtha Yield (LV%)	BASE	+ Δ 6.0 %
Kerosene Yield (LV%)	BASE	+ Δ 19.5 %
Diesel Yield (LV%)	BASE	+ Δ 12.1 %
Naphtha – Light Kerosene Gap, °F	BASE	+ Δ 1.0 °F
Heavy Kerosene – Light Diesel Gap, °F	BASE	- Δ 1.0 °F
Kerosene Flash Point, °F	BASE	Δ 0.0 °F
Kerosene Freezing Point, °F	BASE	+ Δ 0.2 °F
Heat removal – Overhead condenser (% of total BTU/hr)	72 %	62 %
Heat removal – Heavy Kerosene P/A (% of total BTU/hr)	4 %	6 %
Heat removal – Light Diesel P/A (% of total BTU/hr)	7 %	10 %
Heat removal – Heavy Diesel P/A (% of total BTU/hr)	16 %	21 %
HK–LD Frac. Top Simulated Internal Reflux L/V (wt)	BASE	- Δ 15 %

AIChE 2016 Spring Meeting / Distillation Symposium

Case Study 1- Crude Distillation Unit
Post Retrofit Sensitivity Analysis

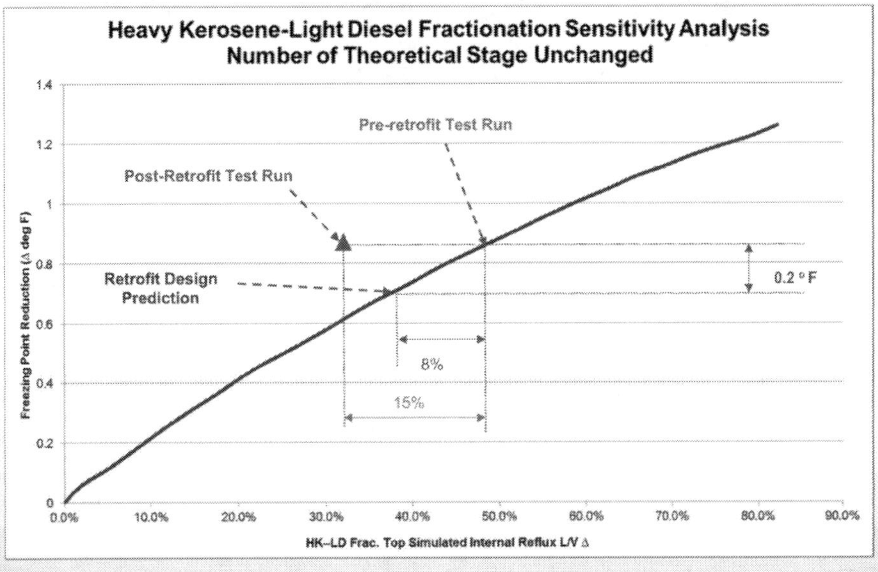

AIChE 2016 Spring Meeting / Distillation Symposium

Case Study 1- Crude Distillation Unit
Fractionation Efficiency Comparison

Heavy Kerosene / Light Diesel Fractionation	Pre-retrofit Test Run	Retrofit Prediction	Post-retrofit Test Run
Kerosene Freezing Point, °F	BASE	+ Δ 0.1 °F	+ Δ 0.2 °F
HK–LD Frac. Top Simulated Internal Reflux L/V (wt)	BASE	- Δ 3 %	- Δ 15 %
Simulated Section Efficiency [note 1]	~55%	~50%	~63%

Note
1. Overall efficiency (# of theoretical stage ÷ # of fractionating tray)

- New fractionating trays
  - Fixed valve
  - Equipped the performance enhancing features

AIChE 2016 Spring Meeting / Distillation Symposium

Case Study 2- Crude Distillation Unit Unit Configuration

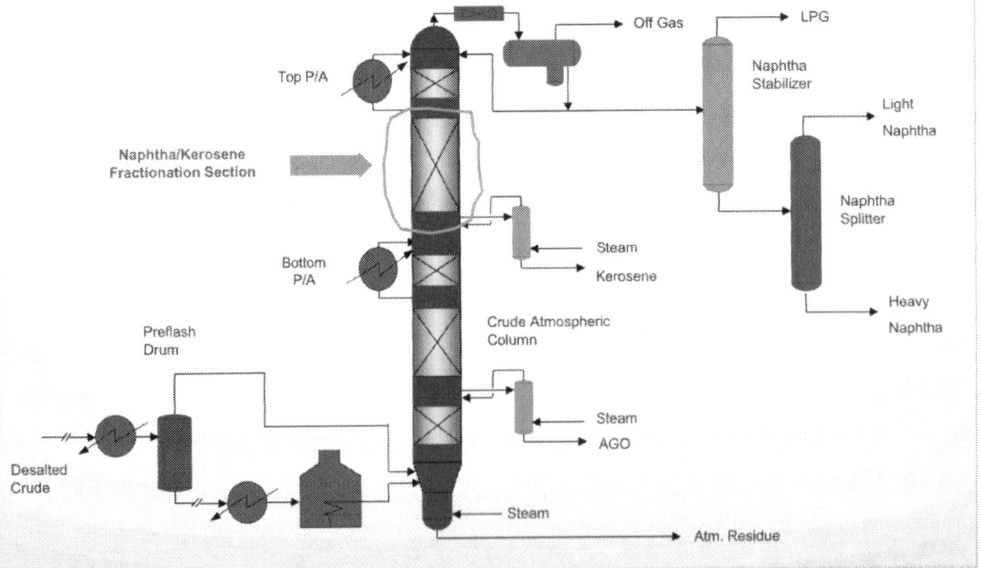

AIChE 2016 Spring Meeting / Distillation Symposium

Case Study 2- Crude Distillation Unit Limitation

- Downgraded degree of fractionation between naphtha & kerosene
  - After 4 month operation
- Unit trouble
  - Kerosene yield limitation
  - Downstream reformer performance
- Liquid distributor fouling
  - Naphtha/kerosene fractionation

Case Study 2- Crude Distillation Unit Root Cause Identification

- Ammonium salt formation
  - Transferred to naphtha/kerosene fractionation section
- Limited column inside wall cladding
  - Rust from the column shell (carbon steel section)
- Original liquid distributor design for naphtha/kerosene fractionation
  - High drip point density
  - Small drip hole size

Case Study 2- Crude Distillation Unit Gravity Distributor Optimization

GTC Technology – Engineered to Innovate

$$H = \left(\frac{L_v}{k \times N \times HA} \right)^2$$

H : Liquid Height ("Head") above Round Hole
L_v : Liquid Volumetric Flow
N : Number of Holes (Drip Point Density)
HA : Individual Hole Area (Drip Hole Size)
K : Orifice Coefficient

- Drip point density vs. Drip hole size
  - Higher drip point density reduces individual hole size @ total hole area
- Distributor operating range
  - Adequate liquid head should be maintained @ turndown rate

Case Study 2- Crude Distillation Unit Distributor Modification Summary

GTC Technology – Engineered to Innovate

Naphtha/Kerosene Fractionation Packing	Unit	Original	Modification
Nominal Crimp Size	inch	1"	Re-used
Nominal Specific Surface Area	ft^2/ft^3 (m^2/m^3)	38 (125)	Re-used
Inclination Angle	°	45°	Re-used

Naphtha/Kerosene Fractionation Distributor	Unit	Original	Modification
Distributor Drip point density	dp/ft^2	16	4.6
Distributor Operating Range	-	2.5 : 1	1.7 : 1
Drip Hole Elevation note 1	inch	1.5"	3"
Distributor Drip Hole size	inch	3 / 16"	7 / 16"

Note
1. From bottom of trough

Case Study 2- Crude Distillation Unit Performance Summary

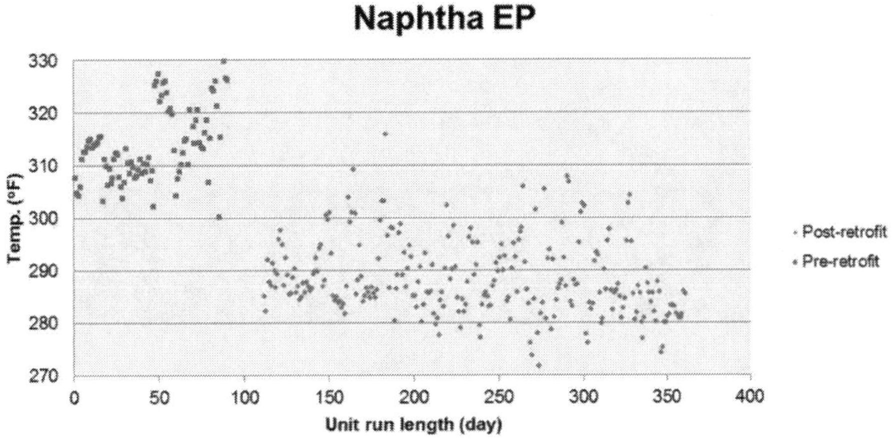

AIChE 2016 Spring Meeting / Distillation Symposium

Case Study 2- Crude Distillation Unit Performance Summary

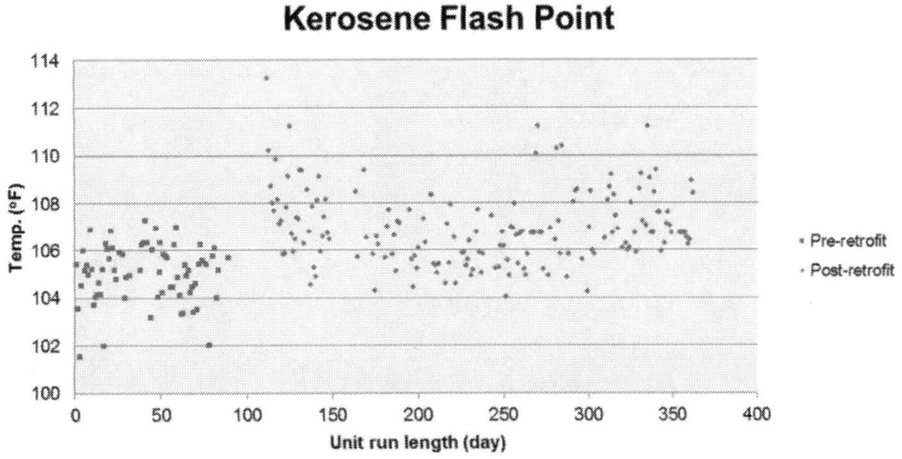

AIChE 2016 Spring Meeting / Distillation Symposium

Thank you

→ Engineered to Innovate

Thank you

445110 Energy and Hydrogen Systems REAL Time Optimization

Abstract:

Hydrogen management can have a significant effect on refinery utility supply through the integration with the energy system. Real-time optimization of hydrogen production in conjunction with the steam, power and fuels can yield significant savings opportunities for the refining industry.

Hydrogen is required in refining processes to produce low and ultra-low sulfur fuels. Hydrogen sources can be external or internal, of course. External sources are generally nearby third party industrial gas providers, who supply pure hydrogen streams, sometimes in exchange of fuel gas (FG), liquid petroleum gases (LPG) or natural gas (NG) for either or both feed and fuel. In some cases, impure hydrogen can also be provided by external sources, which can be purified in the refinery in the Pressure Swing Absorption (PSA) units, producing FG as a by-product.

Many refineries operate internal high purity hydrogen production units, generally based on gas reforming, using either NG or FG as feedstock. Gas reforming processes are endothermic, requiring FG for heating the reactors, and use steam as diluent of the feed and power to drive the feed charge and product compressors, which can be either electric or steam driven. The Steam Methane Reforming (SMRs) units usually produce steam and FG as byproducts.

When FG is used as a feedstock, critical decisions must be made about the best use of sources of the FG streams within the refinery units. When NG is used as a feedstock, many times it is considered a petrochemical feedstock rather than a fuel, having an incremental cost that could depend on its final use.

Any decision to import or internally produce hydrogen will affect not only the hydrogen system purity and availability, but also the FG system in addition; affecting the volumes of any externally purchased or internally supplied fuels (like NG or LPG).

Any attempt to reduce the costs from the utilities side (i.e., to optimize the energy system) should be based on accurate utility system models including steam, fuels, and power and calibrated with validated and consistent set of measurements. In order to consider the interaction of the hydrogen with the fuel system, a compositional model of the fuel must be included in the model. Solving and optimizing the energy and hydrogen balances at the same time is key to ensuring consistency with operations and constraints handling.

This paper describes how an integrated model optimizing the energy and hydrogen systems altogether has been implemented in real industrial environments.

444795 Optimization of Refrigeration and Storage Systems for Propane and Butane Export

Abstract:

The shale gas boom has changed the energy landscape, not only in the U.S. but also beyond our borders and is impacting World Markets. With the advent of horizontal drilling and "fracking" in North American Shale Plays, the U.S. has become a net exporter of natural gas liquids (NGL's), including propane and butane.

This presentation provides an overview of studies which compare options for refrigerating and storing propane and butane for loading on to ocean-going vessels for export from the U.S. The methods evaluated include using a single core-in-kettle chiller, phased chilling using multiple core-in-kettle chillers, an open loop refrigeration design and combining propane and butane chilling into a single refrigeration system. Storage arrangements such as off-site cavern storage and on-site refrigerated storage also have an impact on refrigeration system design, operation and economics. Very large amounts of power and energy are required for these chilling cycles and the options involved. This paper includes a summary of the optimal designs resulting from the studies, including an overview comparison of equipment costs, energy costs, and payback periods.

444569 Utilizing Vent Gases from Chemical Storage and Transfer to Produce Power and Heat

Abstract:

Rather than using a traditional thermal abatement methods for destruction of various chemicals that vent during storage or transfer, the Ener-Core PowerStation can utilize these gases to generate power and recover heat. The Ener-Core PowerStation's destruction efficiency is >99% and has virtually zero NOx emissions.

444315 Outlier Detection and Analysis in Batch and Continuous Processes

Abstract:

Title: Outlier Detection and Analysis in Batch and Continuous Processes

Author/Presenter: Peter J. Ryan, Ph.D., P.E.

Author/Presenter email: peter.ryan@responsepc.com

Company: Response Process Consulting LLC

Motivation

All manufacturing sectors, continuous or batch, gather process data for archiving and analysis purposes. Often the amount of data gathered, and the quality of data, makes it difficult to use this resource effectively. Major drawbacks in the quality of the available data include:

- large gaps in the process data
- noise, poor signal-to-noise ratios
- correlated data
- accuracy
- precision

New methods have been developed to handle these issues and to develop process models based on this resource. Specifically, a new approach to handling missing data and reconstructing quality data based on the observed (archived) data is presented. Once the models are developed, outliers can be identified in the continuous or batch data, and relationships between the Key Process Indicators (KPI's) and the upstream process variables can be established.

Approach

Often, steady state or dynamic models are available to describe a process. However, these first-principle models often do not have the granularity needed to describe product specifications such as color, turbidity or solvent loss due to entrainment in a separator. Machine learning can be used to examine large historical process data sets and determine the leading controlled variables of a process. The machines learning methods first fit the data to a specified model, and then use the model to explore the data space. Unsupervised learning gives the fitting method full ability to determine what is different in the data. Supervised learning requires the fitting method to consider both the archived process data and measured quality data (acquired off-line of the process). An example of unsupervised modeling is Principal Component Analysis (PCA). An example of supervised modeling is Partial Least Squares (PLS). Both of these methods can be used to model the historical process data and find outliers by plotting the two principal component elements (scores) that capture the most variability in the data. The score plots examine the data for clusters that classify the data as meeting product specifications and not meeting product specifications. The clusters that represent production runs not meeting product specifications can be further examined to discover the upstream process variables that are the cause of the product not meeting specification. While visual inspection has been described, statistical metrics such as the Squared Prediction Error (SPE) and Hotellings T^2 metrics can be calculated to find the same results in the data.

Results

Examples of modeling both continuous and batch processes are given. The continuous example is of a commodity chemical process where color is the product specification of interest. The batch example is of a nylon process where the relative viscosity is the product specification of interest. While the analysis methods are the same, one significant difference between handling continuous and batch data is that the batch data must first be "unfolded" before a model can be developed. Once the models are developed, clusters corresponding to successful and non-compliant products are found. The non-compliant product clusters are further examined, and the relationships between the product specification (KPI) and the upstream process variables causing the non-compliance are discovered. Figure 1 shows the results of the scores plot of the continuous commodity chemical example. Figure 2 is an example of the scores plot of the batch nylon example. In both cases, clusters of production activity where both in-spec and non-compliant quality production are observed. Focusing on the batch example, Figures 3 and 4 show the SPE and Hotelling T^2 charts of the initial process data. The outliers observed visually in Figure 2 are also detected numerically by calculating the Hotelling T^2 metric. The Hotelling T^2 metric finds points in the scores plot that are on the model plane but far away from the center-of-mass of the model. The SPE metric finds points in the clusters that are away from the model plane.

Contribution charts are used to discover the relationships between the KPI's and the upstream process variables, as shown in Figure 5. Note that in Figure 3 (SPE metric), batch 49 is far away from the model plane, even though its projection is in the cluster of points that represent batches with good product specification. The SPE contribution chart (Figure 5) reveals that deviations in five process variables caused the batch to product off-spec product. Specifically, the batch did not meet the turning points in the prescribed trajectories when the batch reached the 58th time interval of its run. An examination of the control system revealed that a control issue was accountable for missing the turning points.

These examples show how a very large set of process data – of varying quality (missing data, low signal-to-noise ratio, correlation, etc) – can be reduced in dimensionality and how the leading process variables can be identified and related to the downstream KPI's to improve product quality and consistency. The model has the granularity that is typically missing in first-principal models. The resource for this method of model building – historical process data – is readily available but seldom used. The resource is seldom used for process optimization and analysis because, without the methods needed to reduce the data-space dimensionality and cope with missing and correlated data, this resource presents too much raw, unconditioned information.

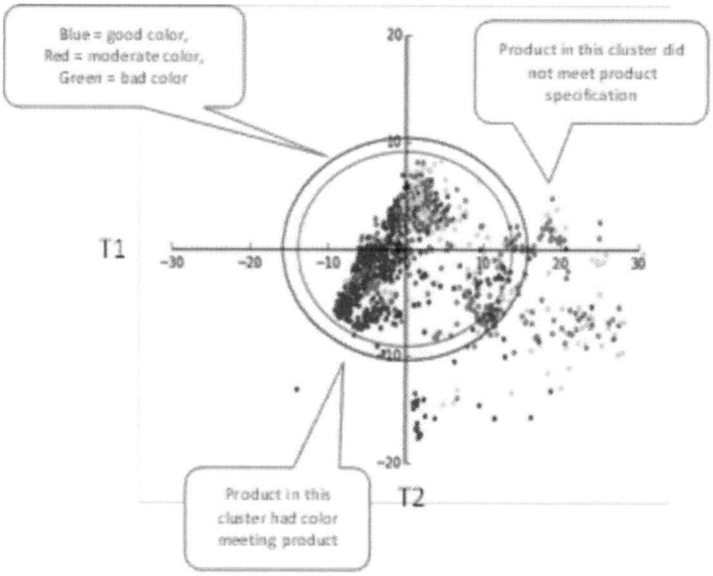

Figure 1: Model developed using **continuous** historical process data

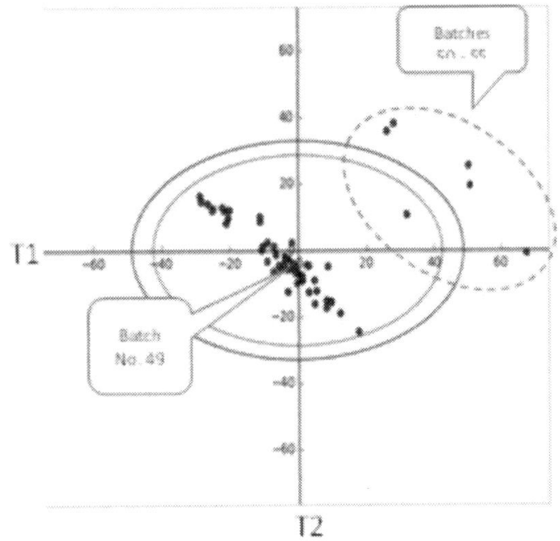

Figure 2: Model developed using **batch** historical process data

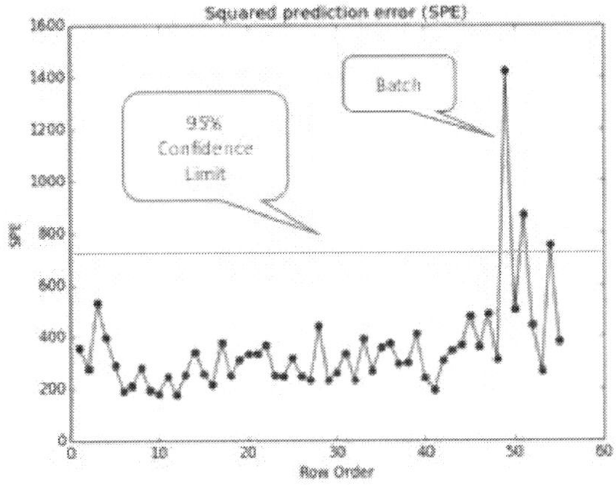

Figure 3: Squared Prediction Error (SPE) of the batch data (Nylon example)

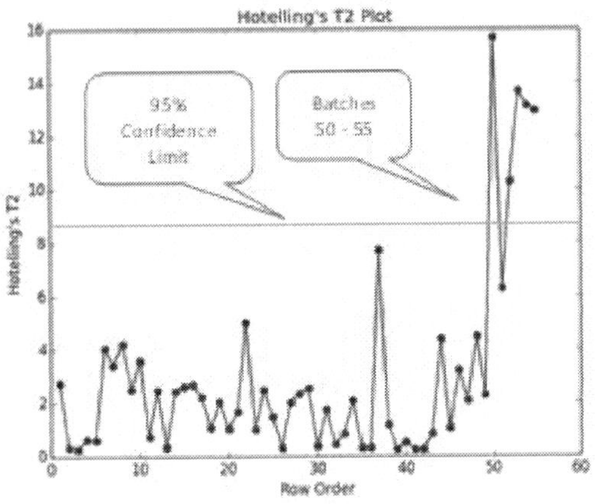

Figure 4: Hotellings T^2 metric of the batch data (Nylon example)

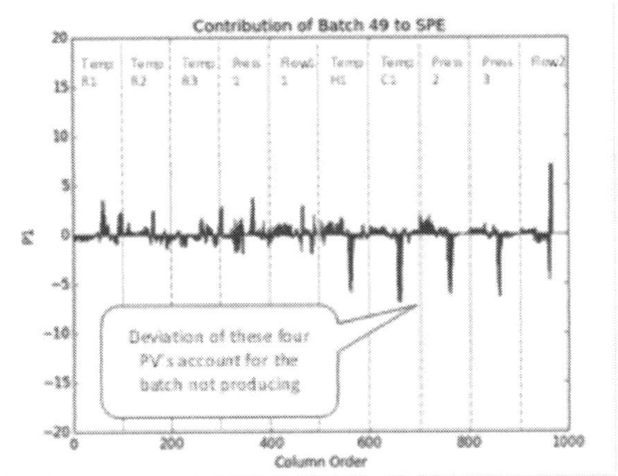

Figure 5: Contribution to the SPE of Batch 49 of the batch data (Nylon example)

(47b) Advances in Asset Health Monitoring and Predictive Analytics

By Tim Olsen from Emerson Process Management

Big data is a hot topic in many commercial and industrial applications, and it also applies to the operation of refineries and petrochemical plants through the use of sensor data to improve operations. Online data from sensors, often wireless, can be analyzed and used to provide plant operators with the actionable information to make timely decisions for improving asset reliability and efficient operations.

Big data from sensors can be analyzed by a variety of commercially available software programs, each of which can evaluate sensor data for abnormal operation and provide alerts. Properly trained operations are invaluable in this process, as they not only act on information from predictive analytics software, but often make decisions on their own based on sensor data.

Online data from sensors has been available for decades, but a transformation is now taking place due to the low cost and quick installation time for WirelessHART® (an international industry standard IEC 62591) sensors when compared to their wired counterparts. The term pervasive sensing describes the application of wireless sensors to areas which were formerly too expensive to monitor with wired instruments, leading to the creation of big data.

Examples of big data and predictive analytics applications include vibration sensing of rotating equipment, ultrasonic leak detection, steam trap monitoring, corrosion and erosion detectors, flows, pressures, and temperature measurements. These and other monitoring applications are providing major savings at many process plants in maintenance costs and energy while mitigating unscheduled downtime and potential ensuing safety and environmental incidents.

Here are some examples of monitoring applications utilizing asset health data and predictive analytics.

Vibration sensing

One of the earliest industrial applications of pervasive sensing was the use of small, battery-operated wireless vibration sensors to monitor the condition of motors, pumps, gearboxes and other non-critical rotating equipment. Readings from the sensors can be analyzed and alert the plant's asset management system when a piece of equipment shows abnormal operation or imminent failure.

A Midwest refinery needed to prevent hydrocarbon vapor cloud releases when pumps failed. Wireless vibration sensors now let operators know which pumps need service. They now monitor more than 150 pumps rather than manual checking non-critical pumps once a month. The system has provided early warning of three impending pump failures in its first year of operation.

Examining effects of corrosion

There are estimates that 36 percent of a typical refinery's maintenance budget is spent on corrosion remediation and repairs. Until recently, corrosion monitoring required a shutdown to install a monitoring device (e.g., detection coupons), and then subsequent shutdowns to inspect them. But the past few years have seen the introduction of wireless corrosion monitors that provide online data on corrosion rates that can be used to detect a spike in corrosion, control inhibitor injection, and correlate with other process data for root cause analysis. It is still recommended to continue using traditional coupons and probes along with inspections during plant shutdowns.

Monitoring heat exchangers

A Gulf Coast Refiner had been experiencing problems with a crude unit due to accelerated fouling within the preheat train. Installing wireless temperature and pressure instruments made it possible to use heat duty calculations to analyze heat exchanger rate of fouling and determining which bundle required cleaning; the return-on-investment was less than 90 days which didn't included the valuable knowledge learned about crude oil blend incompatibilities.

Monitoring of compressed air and gas leaks

The Department of Energy's Office of Energy Efficiency and Renewable Energy (EERE) states that leaks often waste as much as 20 to 30 percent of compressor output. EERE recommends the use of ultrasonic acoustic detectors to identify leaks; these can also detect leaks of process gases, including toxic and flammable gases, although in the latter two cases it is a good idea to back up the acoustic sensors with fixed gas detectors and optical flame detectors.

An excellent way to get a handle on waste in the plant is to monitor flow rates on all utility fluids: compressed air, steam, natural gas and water. Today this is easily done using wireless sensors, and the cost savings can be significant. In plants that already measure these flows, newer technology can reduce costs. Orifice plates in compressed air lines create permanent pressure losses; replacing them with averaging Pitot tube flowmeters saved a Brazilian chemical company $750,000 annually in electricity costs.

Conclusion

Advances in asset health monitoring and predictive analytics are being used today to mitigate unscheduled downtime and reduce maintenance costs by taking corrective action before an asset failure. Abnormal operation can quickly be identified and corrected with properly trained staff to act on this new timely analyzed information. Big data and predictive analytics will continue to have a role with shortages of skilled resources, thus having the staff be more efficient and effective at their role with additional information.

443916 Rigorous Simulation of a Crude Distillation Unit with Heat Exchanger Network

Abstract:

Crude distillation is one of the most important operations in oil refineries since it provides feed to downstream units for further processing. Because of the nature of the present complex and highly integrated refineries, it is critical to understand the operation of the crude distillation unit (CDU) in order to accurately predict the performance and thereby plan operations of the downstream units in response to market demand of the different products.

The present study focuses on simulation of a CDU using a commercially available process simulator Petro-SIM. A detailed model has been developed and validated using refinery operating data. A sensitivity analysis of the model has also been conducted for incorporating different property packages, and Peng-Robinson Equation of State has been found to provide excellent predictions for the CDU performance. Parametric studies have been performed for key operating scenarios in order to understand the relationship among process parameters, feed qualities and product profiles and properties. Different scenarios were studied to demonstrate effects of draw rates of different products on the respective cut properties. Such a robust predictive model opens up opportunities for CDU operational optimization, contributing to an overall refinery profit margin enhancement. The CDU consumes a significant amount of energy. In order to conduct energy optimization studies, e.g. using pinch analysis technique, a detailed accurate model such as the one being presented would be highly useful.

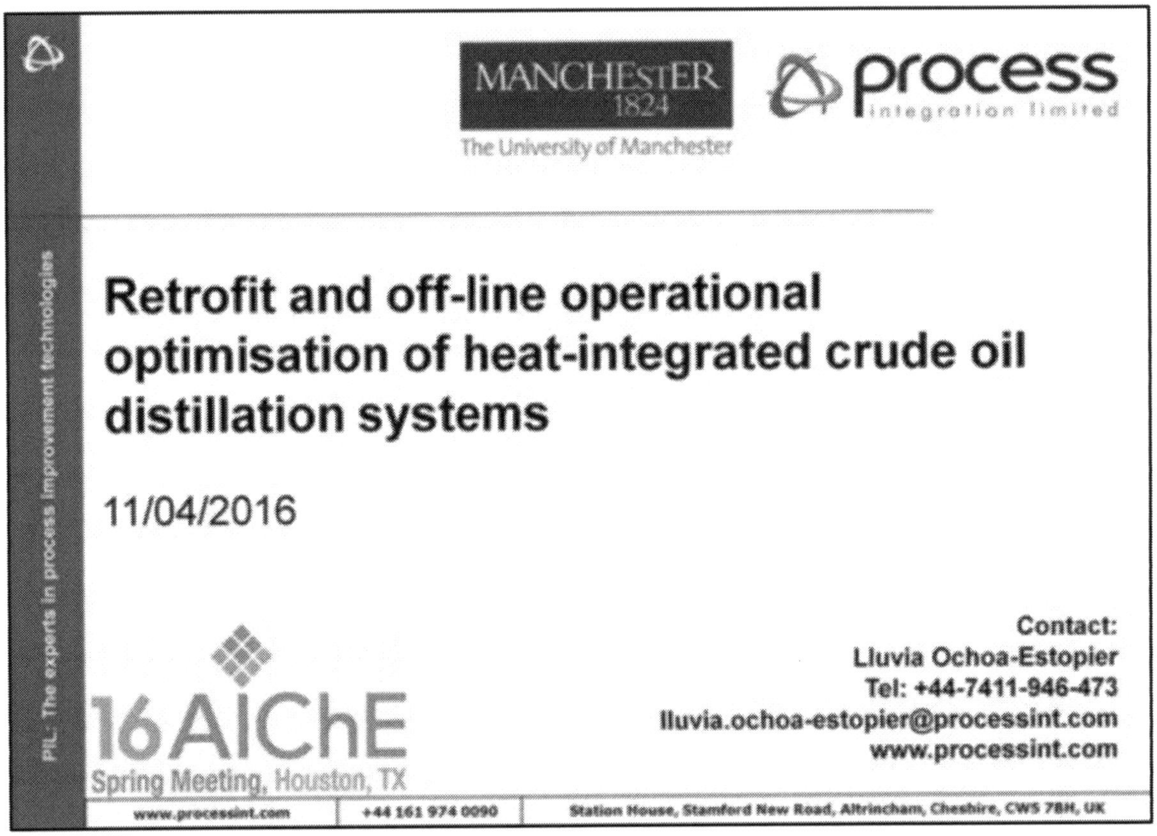
MANCHESTER 1824
The University of Manchester
process integration limited
PIL: The experts in process improvement technologies
Retrofit and off-line operational optimisation of heat-integrated crude oil distillation systems
11/04/2016
Contact:
Lluvia Ochoa-Estopier
Tel: +44-7411-946-473
lluvia.ochoa-estopier@processint.com
www.processint.com
16AIChE
Spring Meeting, Houston, TX
www.processint.com
+44 161 974 0090
Station House, Stamford New Road, Altrincham, Cheshire, CWS 7BH, UK

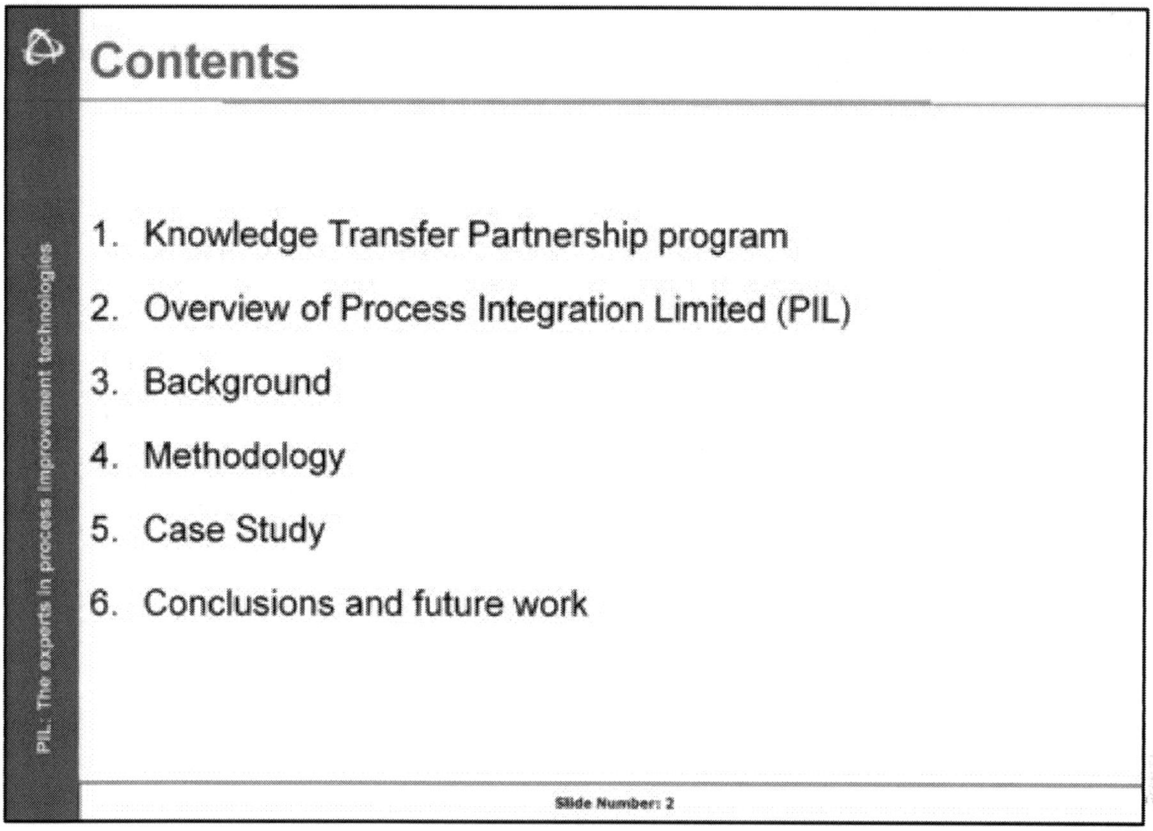
PIL: The experts in process improvement technologies
Contents
1. Knowledge Transfer Partnership program
2. Overview of Process Integration Limited (PIL)
3. Background
4. Methodology
5. Case Study
6. Conclusions and future work
Slide Number: 2

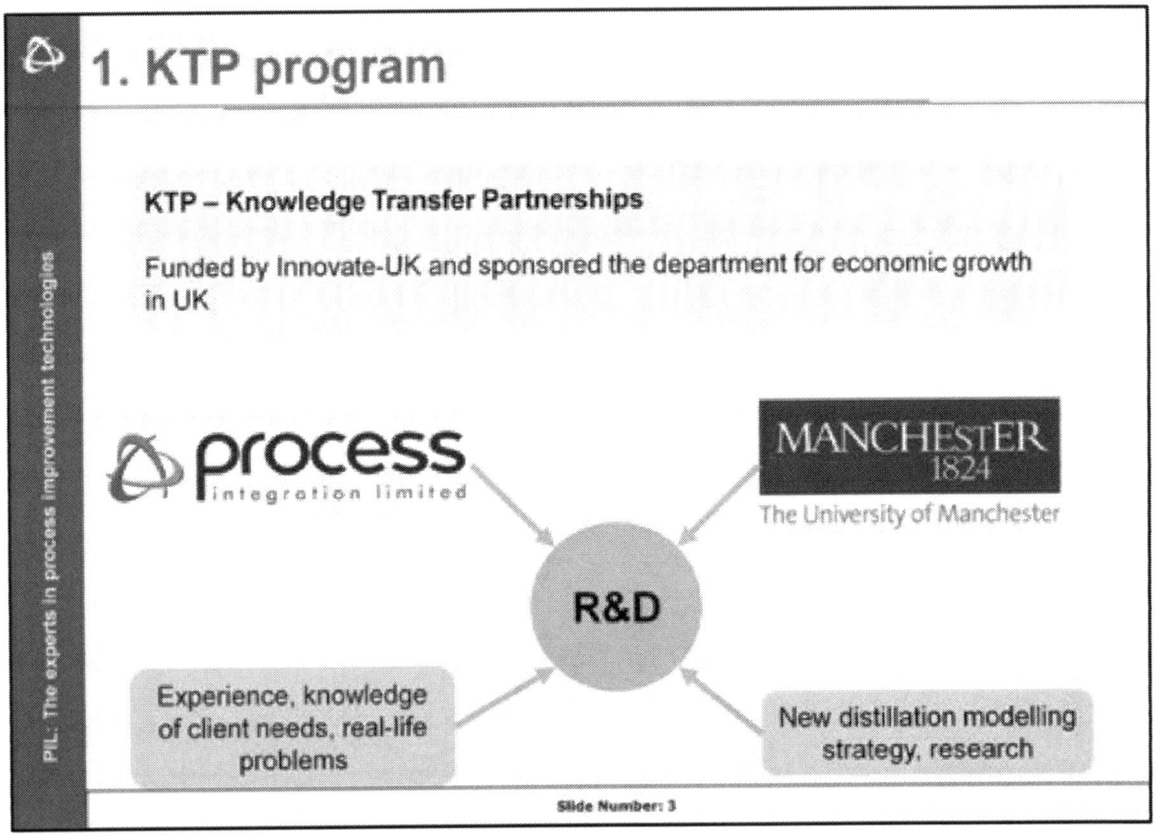

This research work is part of a 30-month Knowledge Transfer Partnership (KTP) program sponsored by the department of economic growth in UK.

"Knowledge Transfer Partnerships (KTP) is Europe's leading programme helping businesses to improve their competitiveness and productivity through the better use of knowledge, technology and skills that reside within the UK knowledge base." (https://connect.innovateuk.org/web/ktp, 17/03/2016)

The objective of this KTP is to develop and embed new mathematical approaches for the optimisation of refinery crude units into commercial tools and deliver a step change in the company's consultancy and software portfolio.

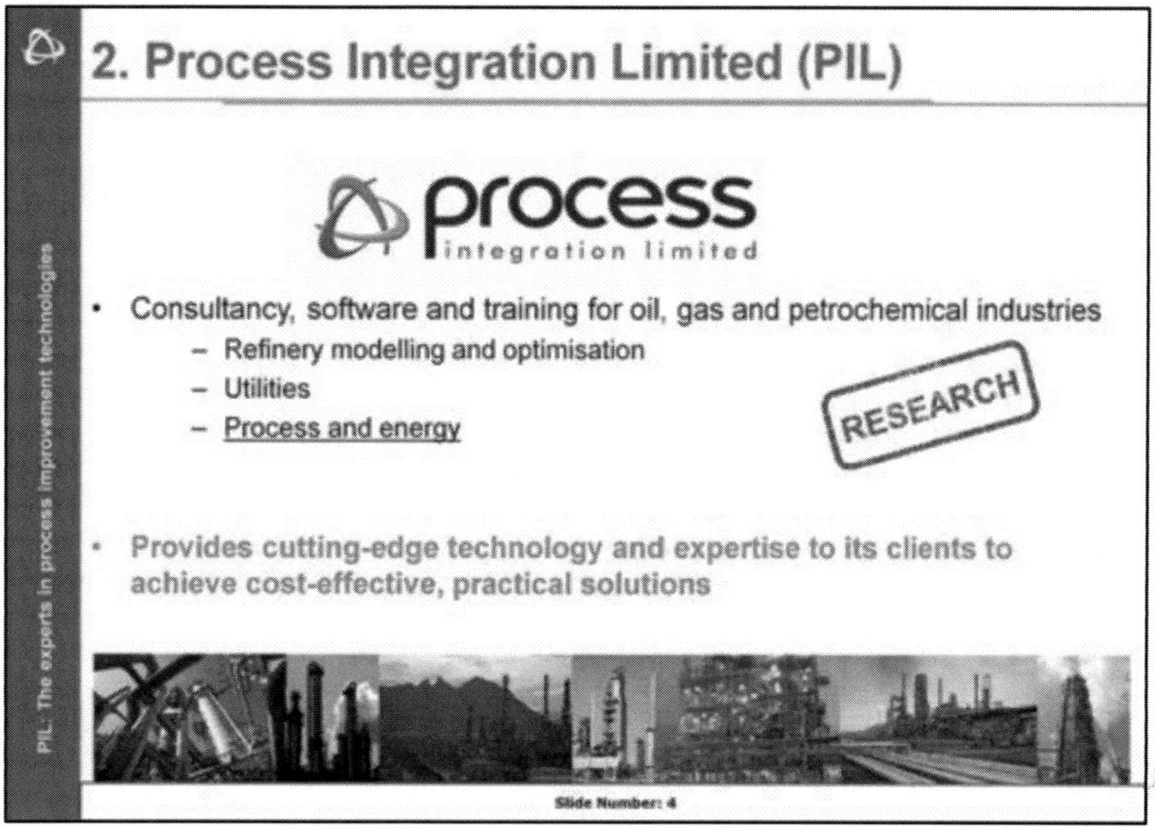

"Process Integration Limited (PIL) provides state-of-the-art process improvement technologies for the oil, gas, petrochemical and chemical industries by identifying innovative improvements in operation and design. The company's technology and know-how can effectively be leveraged to:

Increase energy efficiency
Decrease emissions
Improve yields
Increase capacity"

(http://www.processint.com/about-us, 17/03/2016)

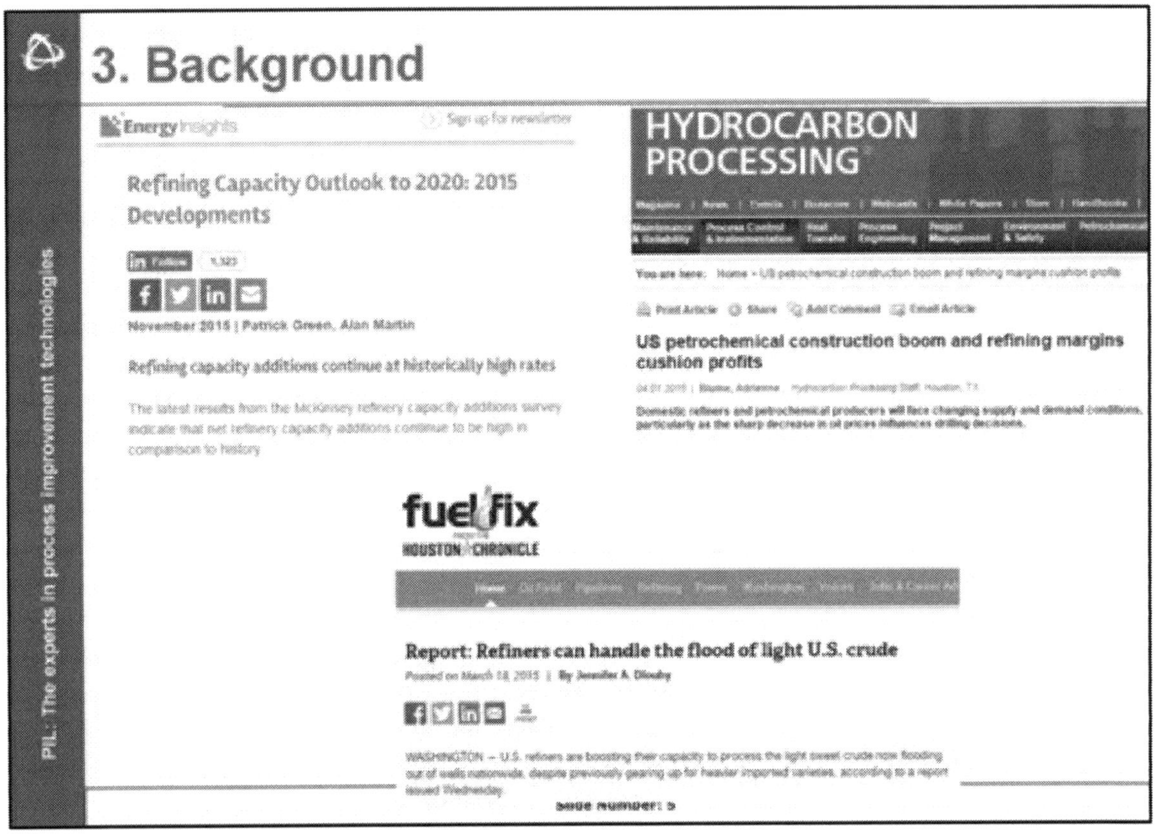

Refiners today are facing challenges related to costs, global and local demand, changes in crude oil feedstock, changes crude oil prices etc. The key to success is to maintain a cost-effective operation in these changing scenarios.

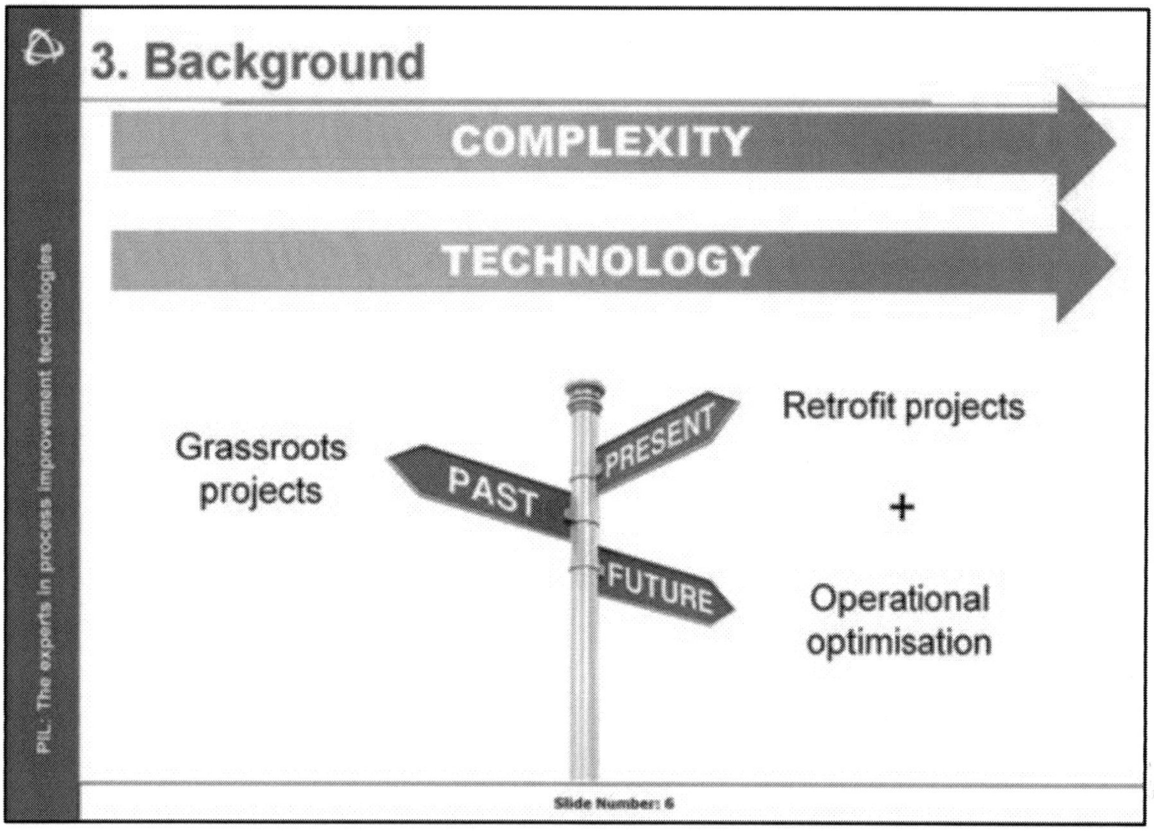

Refineries that develop research and implement projects to process crude oil in a cost-effective way, while meeting strict environmental regulations and product quality requirements, ensure a competitive position in the global market. In this context, the crude oil distillation process is key to improve the economic and environmental performance of refineries.

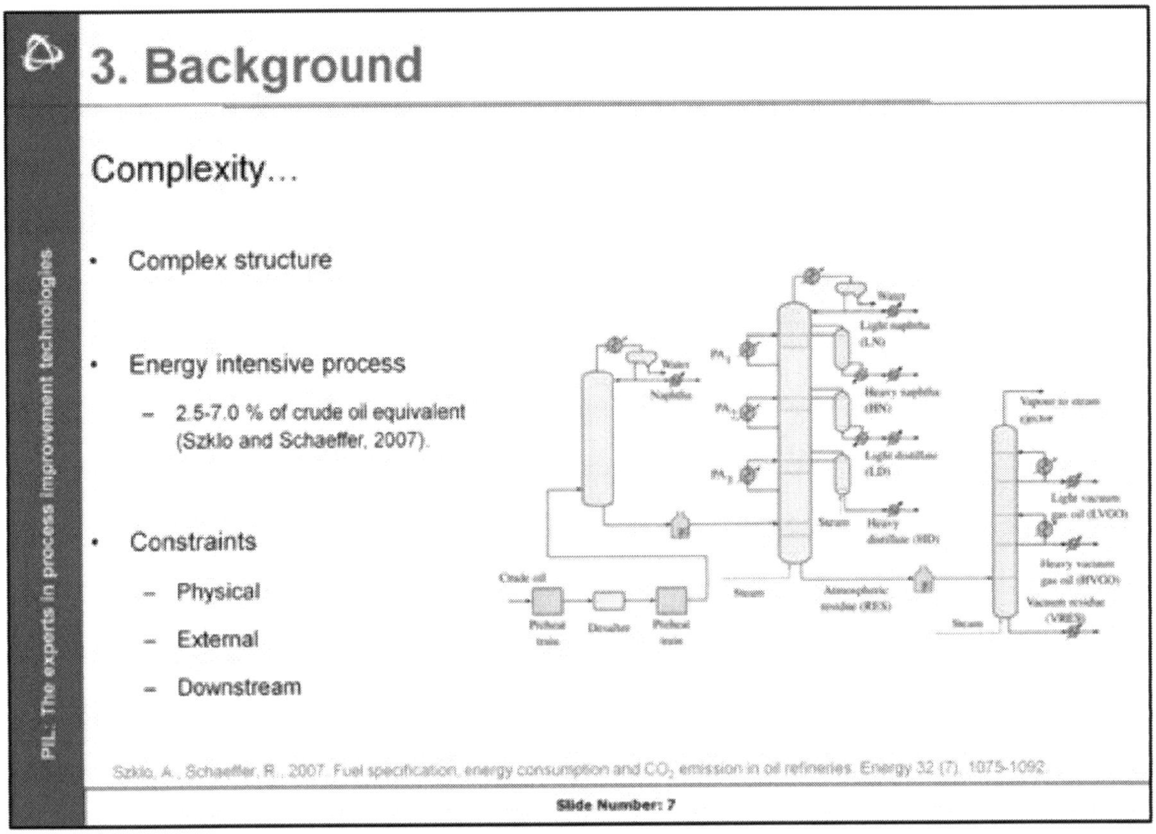

Energy intensive:

- Environmentally challenging
- Economically challenging

Complex structure

- Expensive to modify
- Strong interactions between the column and HEN
- Many degrees of freedom for design
- Practical constraints

Feedstock characteristics

- Variable conditions
- Large number of components

Several products

- Product specifications

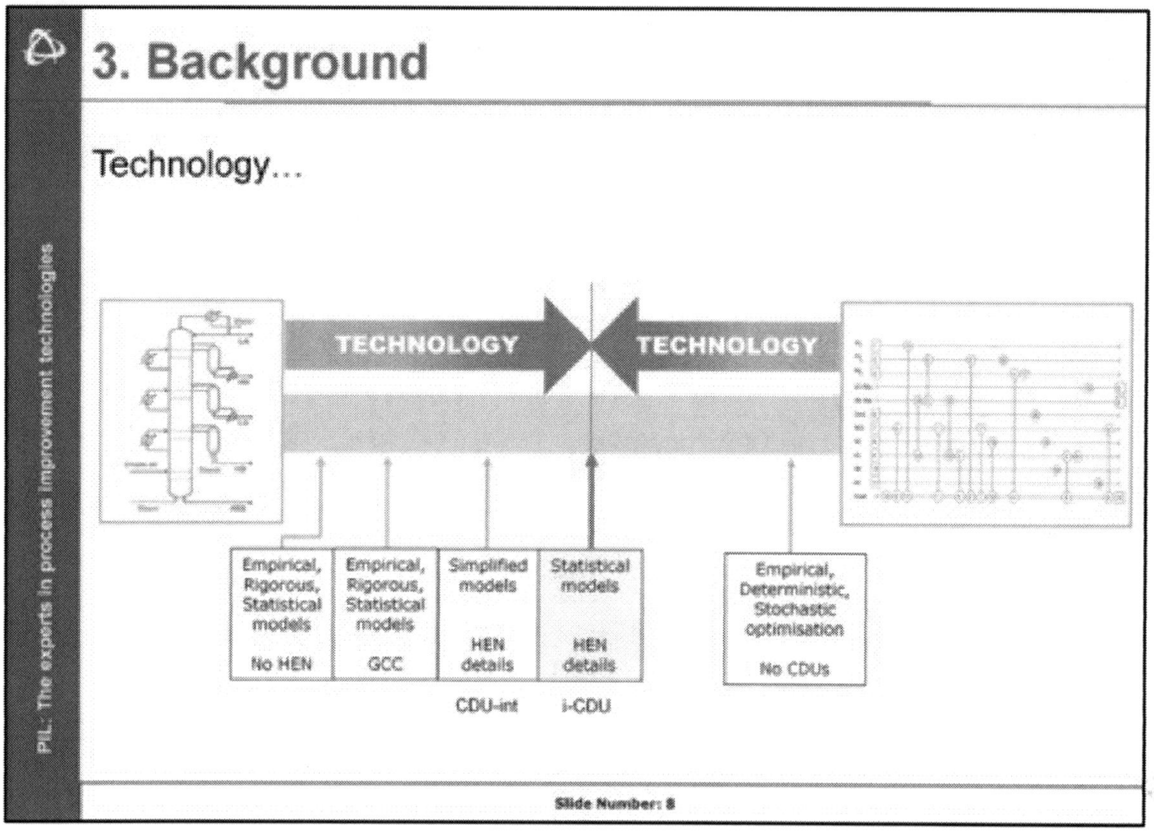

Over time, new computational tools have facilitated the implementation of optimisation-based design approaches, e.g. grass-roots design, retrofit and operational optimisation approaches. However, to develop such tools, several challenges need to be addressed. For heat-integrated crude oil distillation systems, the main challenge is finding models that are accurate enough to represent interactions between the crude oil distillation unit (CDU) and HEN, but at the same time are robust and computationally efficient enough to be implemented in optimisation frameworks.

Rigorous, simplified, hybrid and statistical models have been employed to represent crude oil distillation units. Each type of model has characteristics that define its suitability for a given design application. In particular, available degrees of freedom and model robustness are very important to select the best distillation model for each design purpose. For grass-roots design and retrofit, simplified models are most suitable, while for operational optimisation, statistical models are most suitable.

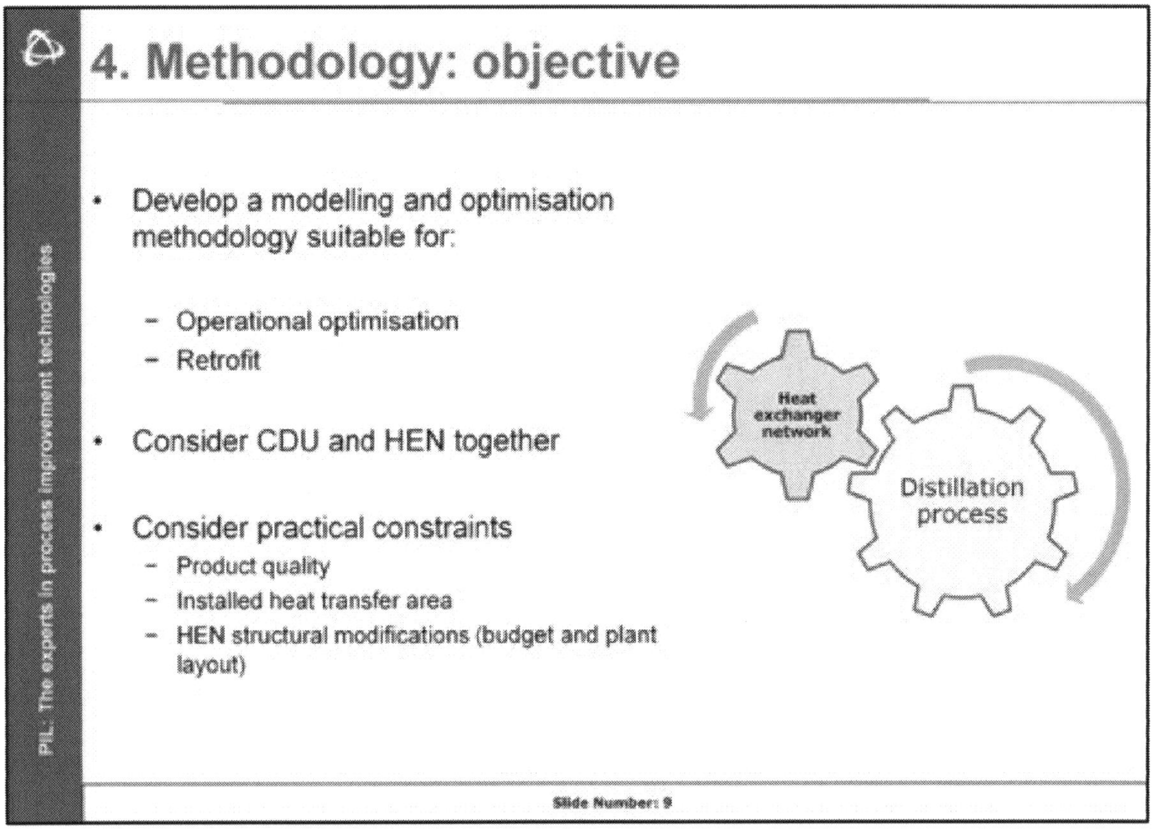

As discussed previously, many challenges arise in the design of heat-integrated crude oil distillation systems. In this project, a new optimisation approach is proposed to address some of these challenges. The new approach considers the distillation process and heat exchanger network together, which allows interactions within the overall system to be exploited. Operational optimisation is proposed for the distillation process, while retrofit of the heat exchanger network is allowed to accommodate the new operating scenarios. The new approach aims to produce improved designs in terms of product yields and operating costs, while taking into account process constraints (e.g. product quality, bounds on operational variables, etc.).

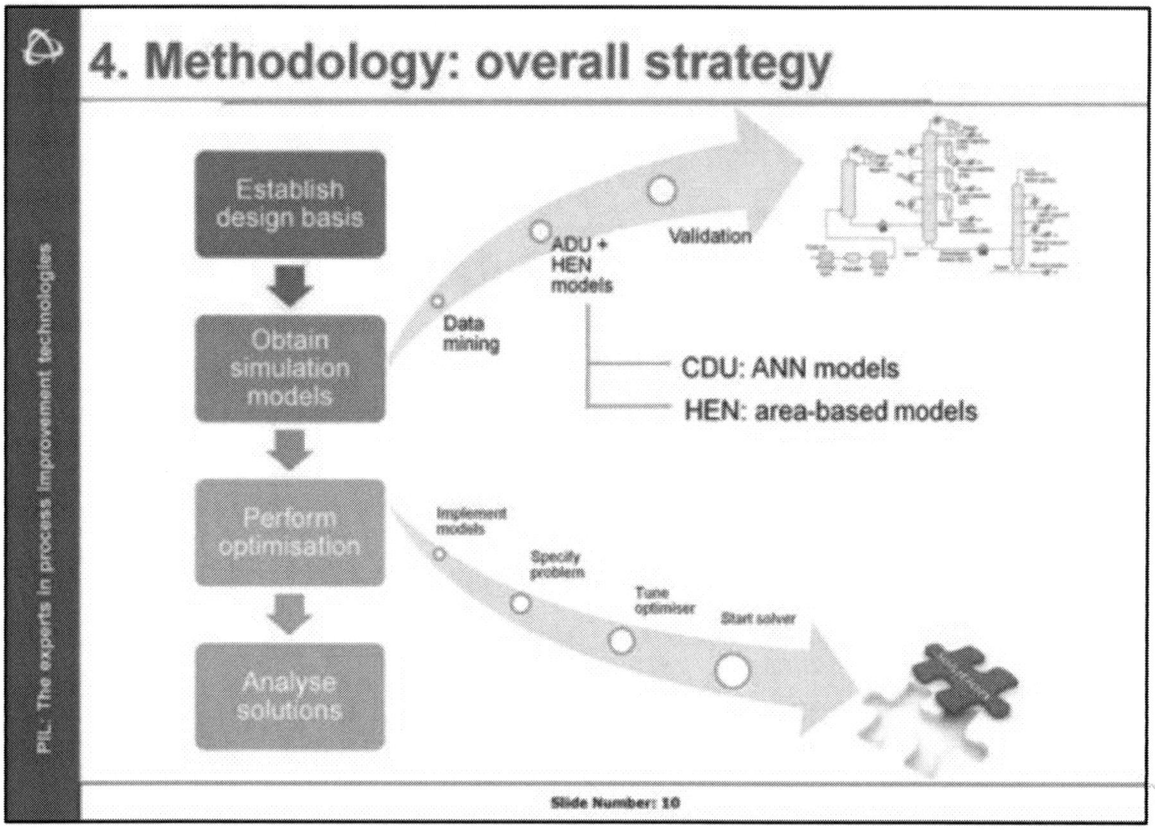

A new optimisation framework for heat-integrated crude oil distillation units using Artificial Neural Networks (ANNs) is proposed. In this approach, a statistical model for the atmospheric tower is constructed from data generated in rigorous models simulations. To represent the HEN, mass and energy balances are solved. The heat-integrated system is incorporated in an optimisation scheme that can optimise product profit and operating costs, while proposing retrofit modifications for the HEN. The optimisation problem is formulated to consider the dominant operational degrees of freedom (such as pump-around specifications or product yields) and constraints (e.g. product quality, heat exchanger area).

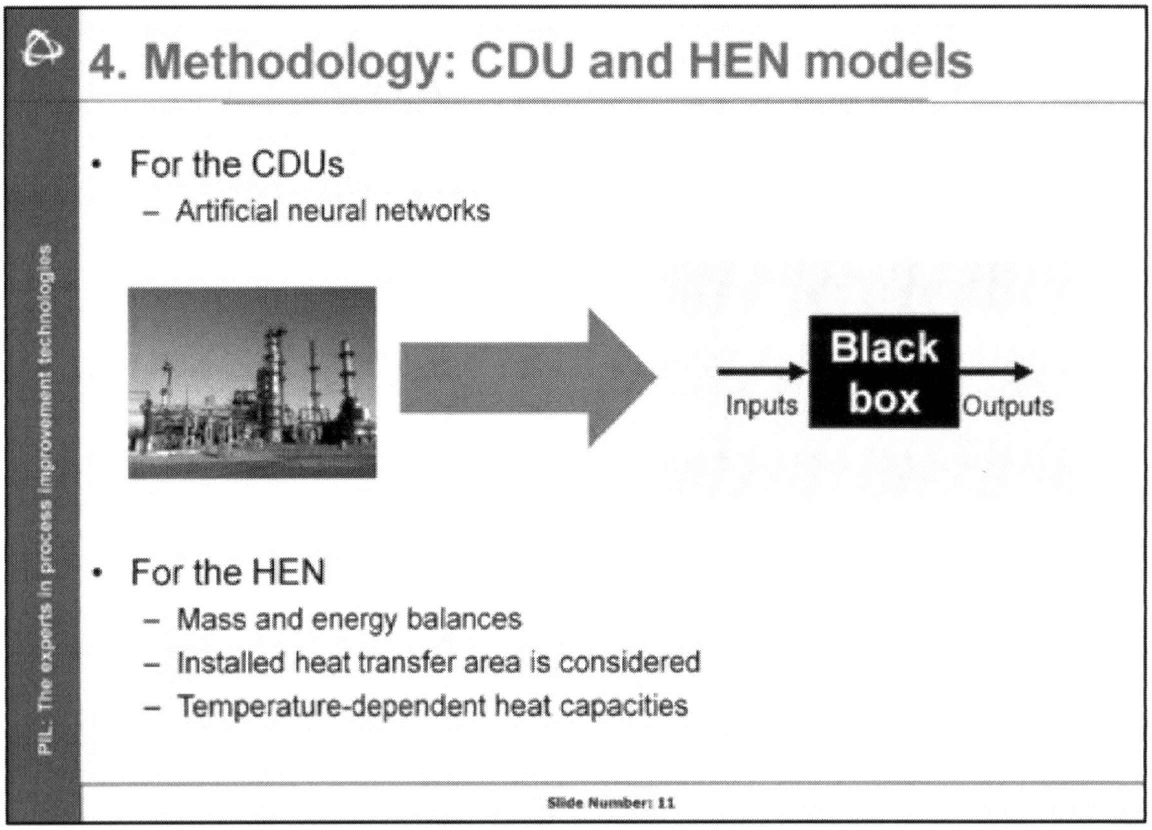

The main advantage of statistical models, such as ANNs, consists in their capability of capturing complex relations between great number of variables. They allow the designer to easily model systems that are often nonlinear, with strong interactions between their elements, such as CDUs. Another advantage is that they can build models from real plant data and are self-adapting to new information. Nevertheless, because ANNs are constructed using stochastic relations, their performance is entirely dependent on the quality of the information supplied to build the model, that is, the ANN is as good as the data used to represent the system.

iHeat is used to simulate the HEN. iHeat calculates the utility requirements and outlet temperatures of every heat exchange unit, splitter and mixer. iHeat also ensures that the calculated temperatures do not violate the stream energy balance, minimum temperature approach and installed heat transfer area constraints.

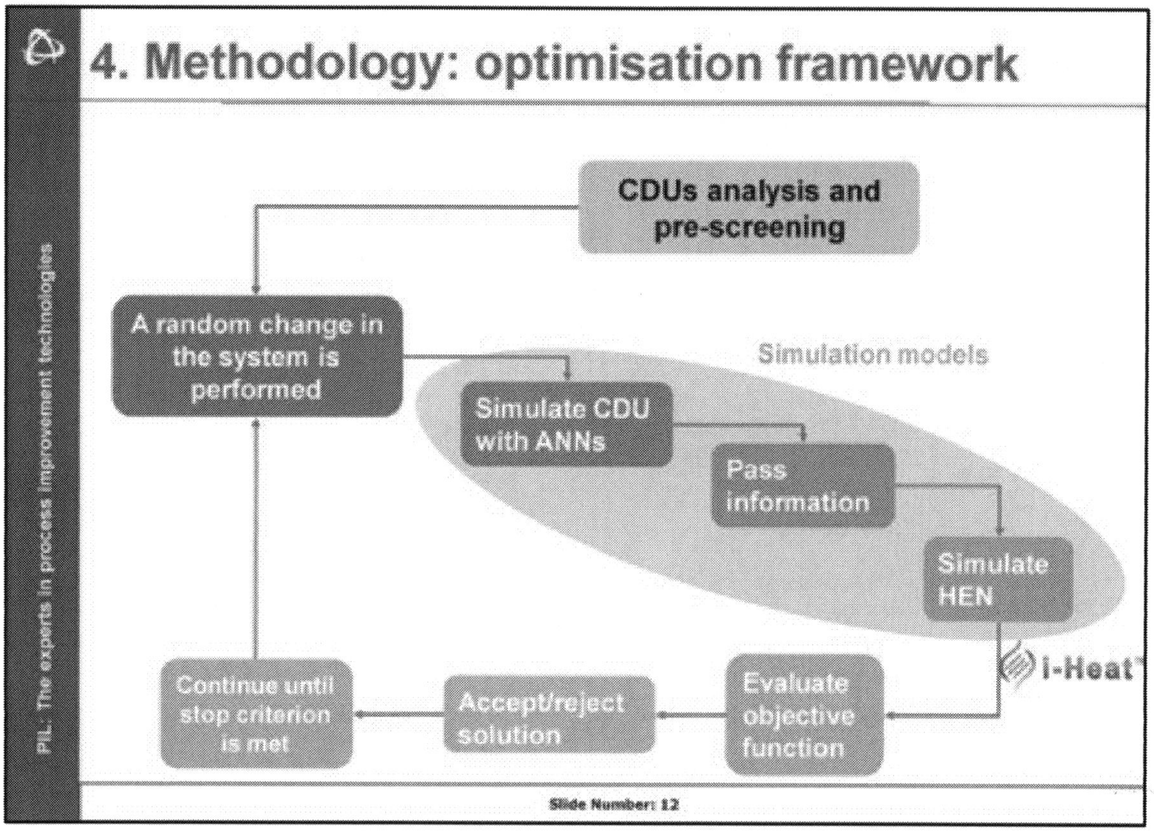

1. The optimiser starts with a feasible set of CDU operating conditions and HEN structure. The objective function for the initial distillation system is calculated.

2. For each iteration, the algorithm randomly performs a change in the distillation system. Changes can be implemented either on the operating conditions of the CDU or the HEN structure.

3. Once the CDU or HEN move is applied, the distillation system is simulated to compute the objective function. The CDU is simulated using ANNs while the HEN is simulated using i-Heat.

4. For a feasible set of operating conditions and HEN structure, the optimisation algorithm accepts or rejects each design of each iteration based on the value of the objective function. If the design is not feasible (i.e. violates constraints), the design is automatically rejected by the optimiser.

5. Steps 2 and 4 are repeated until the optimisation ends.

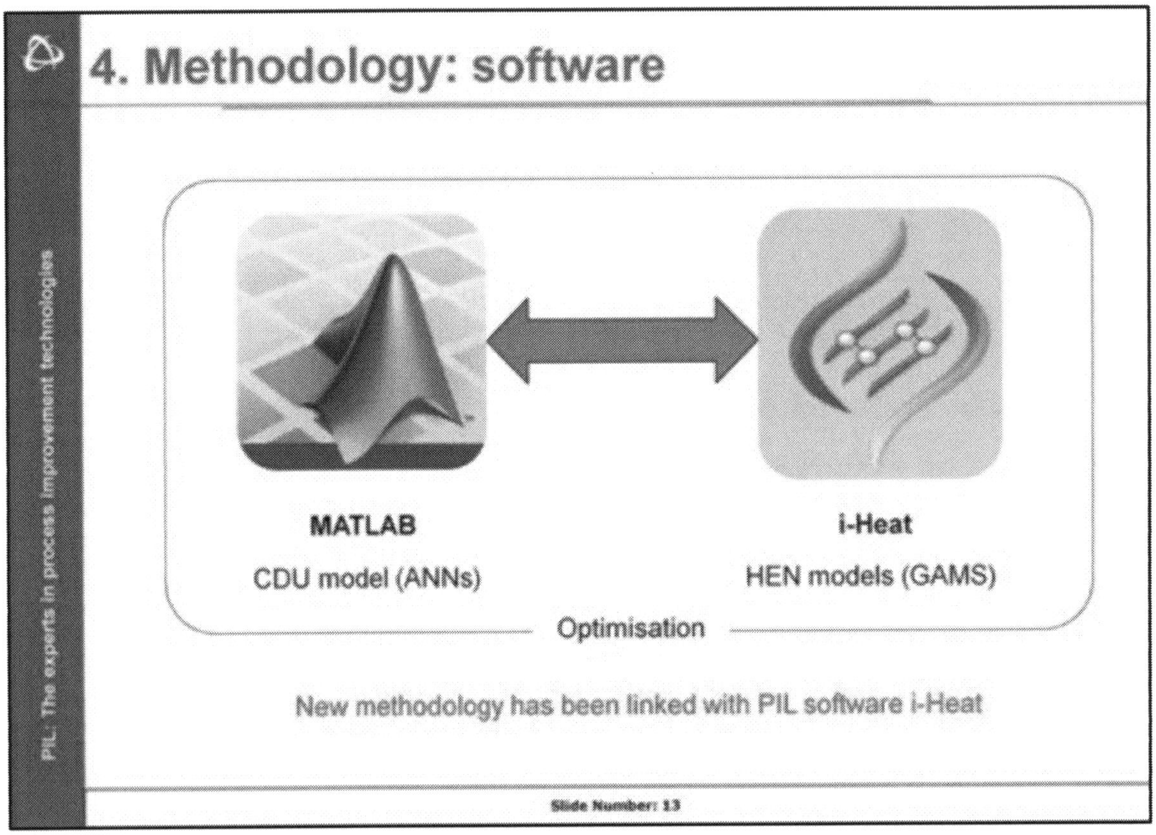

This slide show the software used to simulate the distillation system.

The CDU is simulated using ANNs developed in MATLAB while the HEN is simulated using i-Heat. MATLAB and i-Heat are linked during optimisation.

PIL: The experts in process improvement technologies

4. Methodology: challenges and solutions

Challenge	Solution
Consider CDU and HEN together	Methodology based on the work of Chen (2008) and Ochoa-Estopier (2014)
Detailed CDU models are highly non-linear and hard to converge	Build a simple CDU model using statistical models
HEN models are highly combinatorial and non-linear	Use simulated annealing optimisation to design the HEN and the CDU
CDU product quality needs to be kept	Screen scenarios that meet specifications using sensitivity analysis and neural networks
HEN installed areas need to be taken into account	HEN model considers heat transfer areas and are kept during optimisation
Pressure and temperature-dependent thermal properties	Simple first- and second-order models to represent this behaviour

Chen, L., 2008. Heat-Integrated Crude Oil Distillation Design. PhD Thesis, The University of Manchester, Manchester, UK.
Ochoa-Estopier, L. M., 2014. Optimisation of Existing Heat-Integrated Crude Oil Distillation Systems. PhD Thesis, The University of Manchester, Manchester, UK.

Slide Number: 14

PIL: The experts in process improvement technologies

5. Case Study

- **Modified from a Chinese refinery**
  - Reduction in energy consumption
  - Product yield improvement
- **Distillation system**
  - 2 prefractionators
  - 2 atmospheric units
  - 1 vacuum unit
  - 40 optimisation variables
  - 18 product quality constraints
- **HEN**
  - 50 process-to-process heat exchangers
  - 20 utility heat exchangers
  - Many practical constraints (e.g. installed area, EMAT, target temperatures)

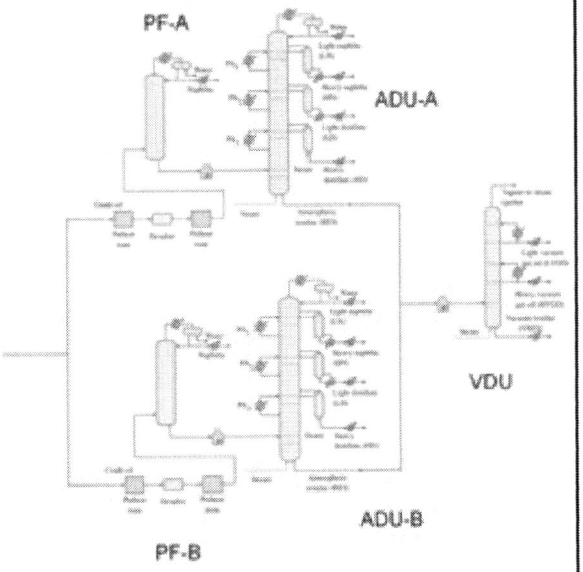

Slide Number: 15

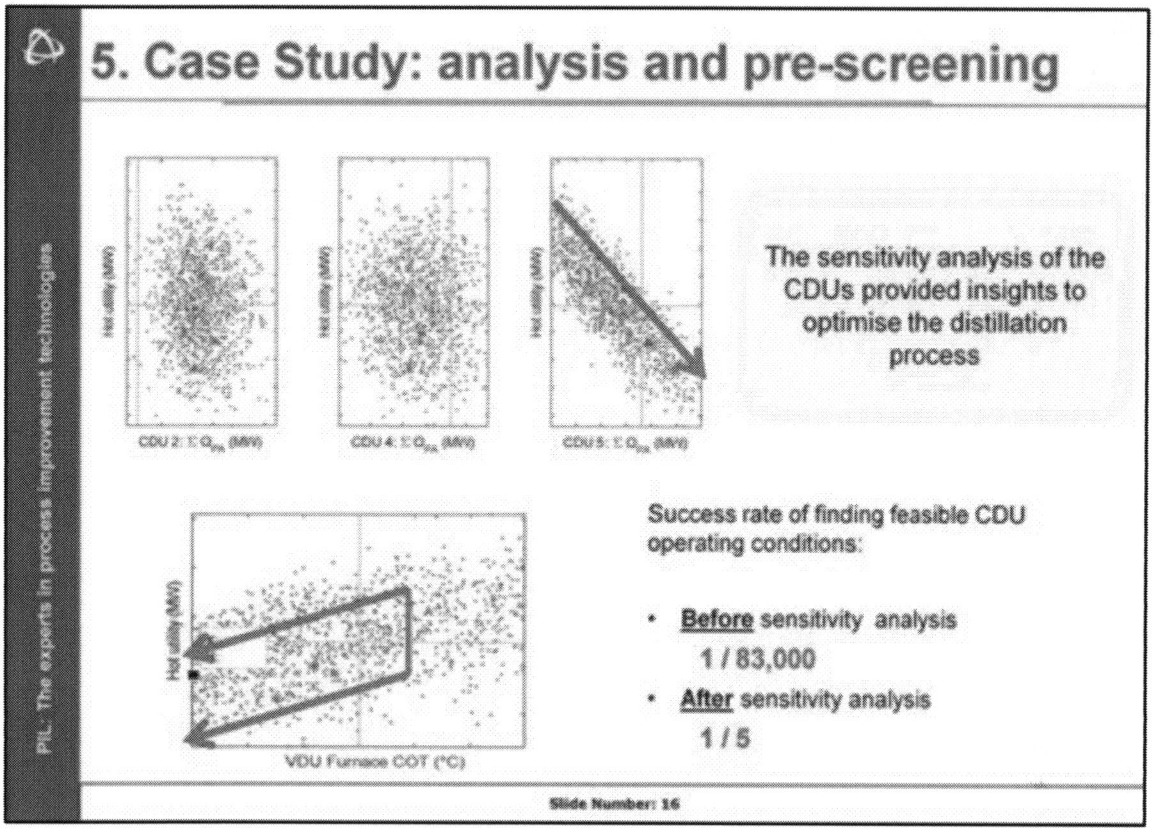

A sensitivity analysis of the CDUs is performed before developing the ANN models for the CDUs and before performing optimisation. This sensitivity analysis is very useful to gain insights of the distillation process, screen CDU operating conditions before optimisation, and identify the potential optimisation search space. This slide shows the analysis of some degrees of freedom of the CDUs and their effect on minimum hot utility requirements (calculated with the GCC). The horizontal and vertical lines (in green) indicate the base case values of these variables. The points are the samples obtained for the sensitivity analysis. The points form a "cloud" in each plot. The shape of this cloud is helpful to identify trends and operational limits. For example, in the first plot (top left), the cloud of points is far away from the lines that depict the base case values. This suggests that the feasible operating range in within the cloud rather than closer to the base case conditions. Thus, it is very likely that, during optimisation, the value for this variable lies inside the cloud.

The third (top right) and fourth (bottom) figures show trends (relationships) between hot utility requirements and the corresponding degree of freedom. The third figure indicates that increasing the total cooling duty of the VDU (i.e. CDU 5) reduces minimum hot utility requirements. On the other hand, the fourth figure suggests that decreasing the coil outlet temperature (COT) also reduces hot utility requirements. These trends can be applied during optimisation, for example, the upper bound of these degrees of freedom can be set to lower values to bias the optimisation to follow this trend.

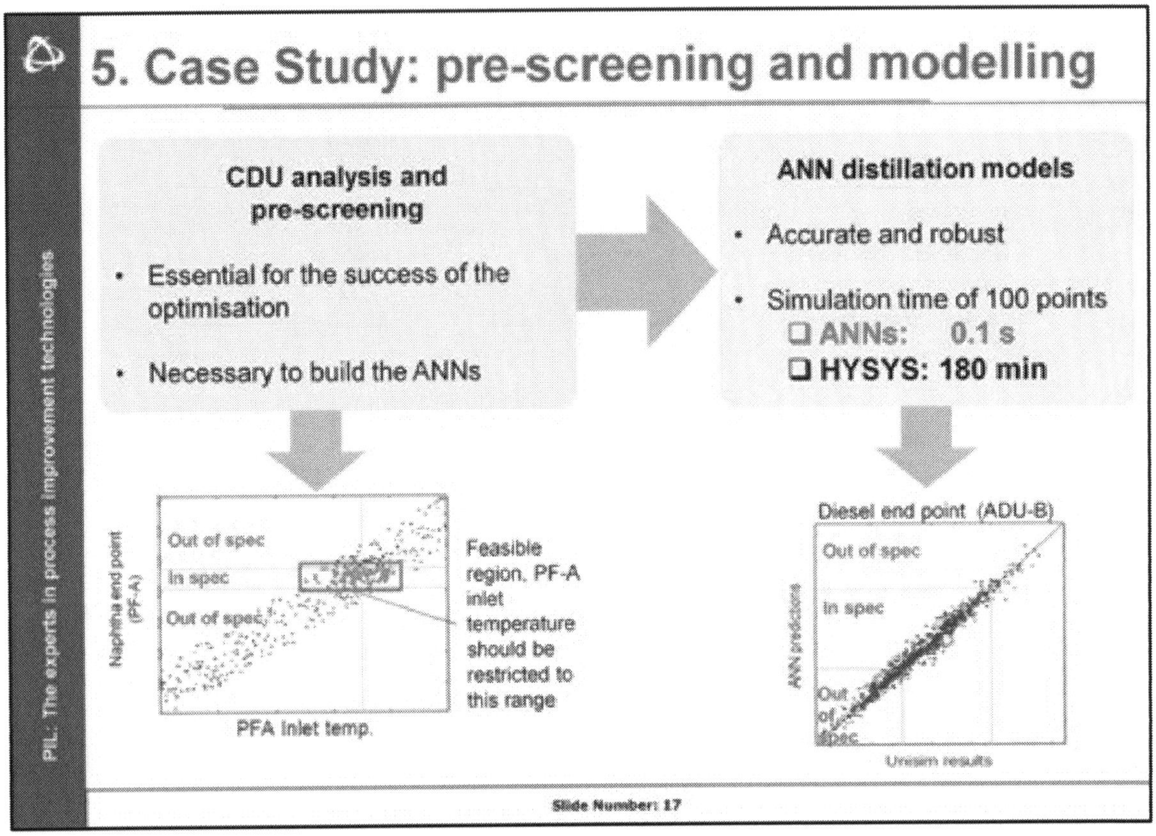

During the development of this project, it was found that the analysis and pre-screening of CDU operating conditions is essential for the success of the optimisation, especially for highly-constrained distillation systems. Analysis and pre-screening is also necessary to build accurate ANNs.

The figure in the left illustrates the sensitivity analysis of a CDU degree of freedom (PF-A inlet temperature) and one product quality specification (naphtha end point). Labels "in spec" and "out of spec" denote samples that meet and do not meet product quality specifications. It can be seen that the product meets constraints only in a limited range of PF-A inlet temperature values. This indicates that the PF-A inlet temperatures should be restricted to those values where product meets constraints.

The resulting ANN models are significantly faster in convergence and more robust than rigorous models built in Aspen HYSYS, with practically no detriment to ANN model accuracy. Once built, the ANN models can predict product quality and can also help to further screen operating conditions that do not meet product quality specifications, as shown in the figure in the right.

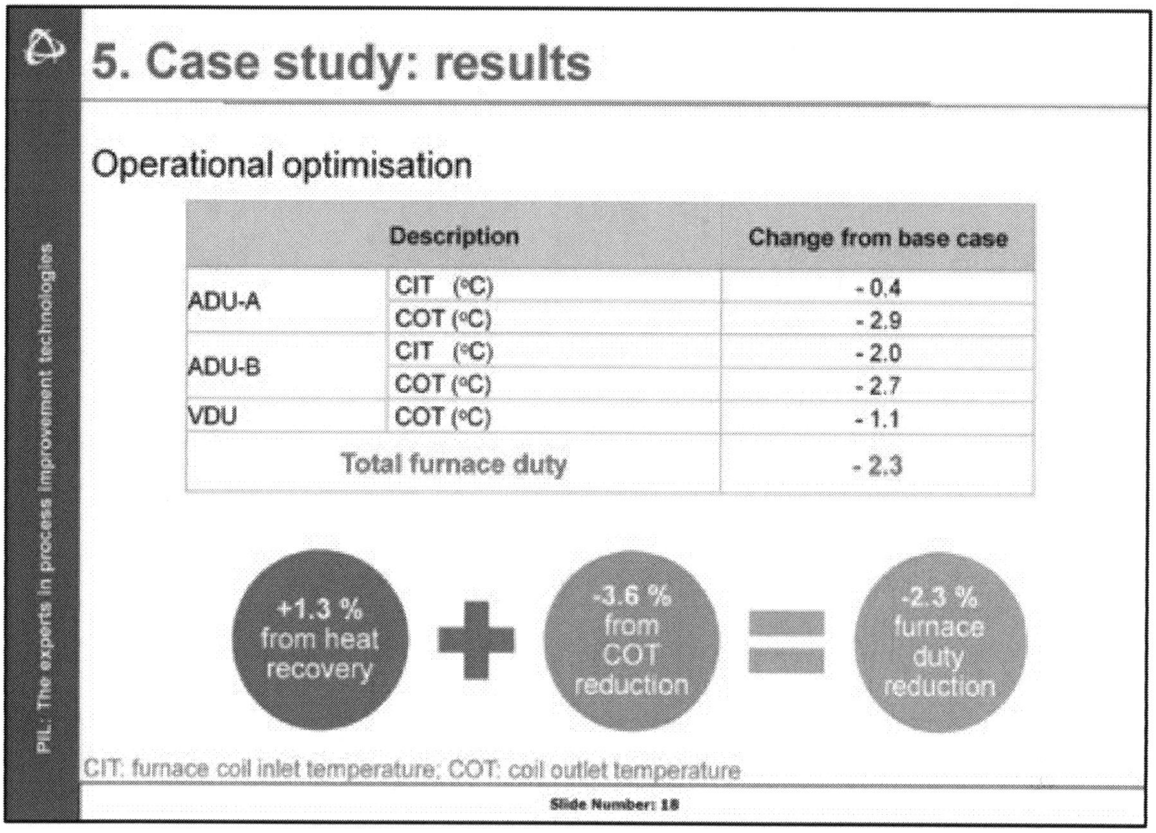

After the sensitivity analysis and pre-screening, the ANN models were implemented in the optimisation framework (using MATLAB and i-Heat at the same time). This slide shows some of the results for operational optimisation. For space saving and confidentiality, sensitive client information is removed from this presentation.

The furnace duty is reduced by 2.3 % for operational optimization. This reduction is indicated by a decrease in the coil outlet temperature (COT) that is considerably greater than the reduction in the coil inlet temperature (CIT).

The optimization framework is able to find a significant reduction in the furnace duty, which is directly related to fuel oil consumption and operating costs. For operational optimization, results can be readily implemented at no cost as no capital investment is needed.

6. Conclusions

- Recent developments in the optimisation of crude oil distillation systems include
  - New analysis tools to model CDUs and screen solutions
  - A robust, simple and accurate model to represent the complex CDUs
  - More realistic approach to represent the CDUs and HEN
  - An optimisation framework that considers practical constraints such as product quality, column flooding, installed heat transfer area, plant layout constraints, etc.

- New methodology is able to find reductions in energy consumption even for heavily constrained systems
  - Product quality kept during optimisation
  - Practicable solutions
  - Savings found for operational optimisation

PIL: The experts in process improvement technologies

6. Future work

Scope	Development status	Software package
Off-line operational optimisation (CDU+HEN)	Current work	i-CDU
CDU operational optimisation + HEN retrofit	Current work	i-CDU
On-line operational optimisation	Future work	e-CDU
CDU retrofit + HEN retrofit	Future work	i-CDU
Grassroots design (CDU + HEN)	Future work	i-CDU
Use of plant data for operational optimisation	Future work	e-CDU

Slide Number: 20

This table shows current and future work to be developed by PIL

443712 Pilot Experiments of a Novel Catalytic Distillation Process for Sulfur Removal of FCC Gasoline

Abstract:

The environmental restriction regarding sulfur content in transportation fuels is currently one of the most important issues. Since major sulfur compounds in gasoline come from the fluid catalytic cracking (FCC) naphtha cut, sulfur removal of FCC naphtha is, therefore, essential for attaining the clean gasoline.

This paper will present pilot experiments of a novel catalytic distillation process for sulfur removal of FCC naphtha in a column packed with structural catalytic distillation cells containing Amberlyst35 resins. In comparison to the conventional hydrodesulfurization, this process can be conducted under mild conditions (Pressure of the catalytic distillation column is 100kPa and temperature of the bottom is lower than 150) and will consume lesser hydrogen.

In this process, thiophene and methyl-thiophenic compounds react with low carbon alkenes to form higher boiling point thiophenic derivatives, as well as mercaptans are etherified to higher boiling point alkyl-sulfides. Then, the higher boiling point sulfides are removed from light fractions by distillation and concentrated in the heavy fractions. Consequently, the sulfur free light cut is obtained from the top of distillation column. the fraction cut point can be raised from 75 to 120 , which will prevent olefins (with boiling point below 120) from saturation at the condition of hydrodesulfurization and avoid octane number loss with deeper desulfurization. This study would be of useful references to further industrial application of desulfurization of FCC gasoline.

446770 Wireless RV Monitoring in Oil & Gas Facilities

Abstract:

This presentation will elaborate on the challenges and technical developments of wireless sensors to monitor relief valves which can improve environmental compliance, minimize product losses, and facilitate troubleshooting of processing units in oil and gas facilities.

445491 Boiler Emissions Reduction through Humid Combustion

Abstract:

This presentation will compare conventional emissions abatement techniques with alternative options utilizing direct contact heat exchangers, which will also boost efficiency beyond the constraints found in typical equipment. These solutions also feature low capex, high reliability and min disruption to running units.

448247 *Ballasted Technology to Overcome Clarification Bottle Necks*

Abstract:

Ballasted clarification offer a cost-effective method to reduce space requirements for new units or to increase production from existing ones. Technical description and case studies will be presented

Improvement of Energy Consumption in Crude Distillation Unit of Tabriz Oil Refinery Using Exergy Analysis

Mir Esmaeil Masoumi
Islamic Azad University,Tehran North Branch, Faculty of Engineering, Chem. Eng. Dept.
Shahid Babaie Highway (West to East), Hakimieh Outway, Shahid Sadoughi St., Vafadar Blvd.
m_masoumi@iau-tnb.ac.ir

Elham Hajighassem Sharbatdar
Islamic Azad University,Tehran North Branch, Faculty of Engineering, Chem. Eng. Dept.
Shahid Babaie Highway (West to East), Hakimieh Outway, Shahid Sadoughi St., Vafadar Blvd.
elham4169@yahoo.com

Prepared for Presentation at
American Institute of Chemical Engineers
2016 Spring Meeting
11th Global Congress on Process Safety
Houston, Texas
April 10-14, 2016

UNPUBLISHED

Improvement of Energy Consumption in Crude Distillation Unit of Tabriz Oil Refinery Using Exergy Analysis

Mir Esmaeil Masoumi
Islamic Azad University,Tehran North Branch, Faculty of Engineering, Chem. Eng. Dept.
Shahid Babaie Highway (West to East), Hakimieh Outway, Shahid Sadoughi St., Vafadar Blvd.
m_masoumi@iau-tnb.ac.ir

Elham Hajighassem Sharbatdar
Islamic Azad University,Tehran North Branch, Faculty of Engineering, Chem. Eng. Dept.

Keywords: energy saving, exergy analysis, distillstion unit, refinery

Abstract

Crude distillation unit is one of the most energy consumer units in the refinery. Optimization of this plant using exergy analysis has direct impact on the performance of refinery and product quality. In this study, crude distillation unit of Tabriz refinery simulated by Aspen Plus software. The analysis was indicated that the Atmospheric distillation column, LPG and Naphtha section has the lowest exergy efficiency. It is also known that the largest value of the exergy improvement potential has been in the atmospheric distillation column. Some solutions were suggested for increasing exergy efficiency which are including the effect of introducing the pre flash vapor directly to the atmospheric distillation column, decreasing pressure drop of furnace and flash drum, increasing the pump-arounds rate of the atmospheric distillation column and changing the pump-arounds return stages of the atmospheric distillation column. The aim of this work is the evaluating of the improvement of energy loss and also fuel gas rates reduction of crude distillation unit furnace with offering alternatives. The results show that the exergy efficiency increases from 0.1% up to 1.6% and furnaces' fuel gas rate consumption decreases from 0.65% up to 2.03% compare to the base state. Moreover, the best proposed modification is the combination of 15% rise in crude distillation column pump-around rates with an increase in the column's feed temperature and also a decrease in the condenser's duty that has caused the most reduction rate in energy loss about 3.22% as average for the unit.

1. Introduction

Crude distillation unit is one of the most energy consumer units in the refinery. Many researches and engineers conduct exergy analysis as a method for analyzing, designing and improving systems and processes [1].

Exergy is defined as the ability of energy to do work or quality of energy [2]. Exergy analysis is a powerful technique to analyse the exergy consumption of a refining process. This method is done by studying a system using the combination of 1st and 2nd law of thermodynamics. Through such an analysis, irreversible and dissipative parts of a system can be identified and overcome by mean of proper proposed solutions, whereas with energy analysis these results cannot be achieved [3].

Some of the previous works done on the application of exergy analysis in industrial processes has been reported on crude distillation unit, power plant, congeneration plant, cooling system and fuel cells [4-5]. Rivero et al studied exergy analysis on crude oil combined distillation unit and also did an exergy analysis of an MTBE production unit containing distillation, reaction and separation processes for improvement alternatives of this unit[6-7]. Nur Izyan studied on a typical refinery in Malaysia and determined from exergy composite curve, the highest exergy loss occurs in the furnace [1]. Doldersum applied exergy analysis on distillation process. The author found that the exergy loss occurred in the furnace and distillation column. Several process modifications to reduce exergy loss were proposed [8]. A research group in London University investigated the performance improvement of a Debutanizer column with exergy analysis in order to upgrade the energy efficiency of the system [9].

Most of the previous literature works in the refineries, have only studied the exergy efficiency and destruction rates. There are few works that have continued such a study in order to modify the system by proposing some practical solutions. Hence, in this study, it is attempted to improve the exergy efficiency of the crude oil distillation unit (CDU) of the Tabriz oil refinery by completely exergy analyzing and applying some process alternatives to the system.

2. Exergy analysis

Exergy is a measurement of how far a certain system deviates from a state of equilibrium with its environment [10]. All real process involves exergy destruction. Exergy partly consumed in irreversible processes [11]. One proposed method by literatures for analyzing a system is Gray-Box method. In this method, gray boundaries are considered for the analysis of the objects (system, subsystem or equipment). Hence, inside the system is obvious for analyzing and its subsystem can be considered as some boxes which contain exergy flows going inside and outside of them. The Gray box model is suitable for evaluation and analysis of the system or subsystem [7].

2.1 exergy calculation

Equation1 has been used for exergy analysis and calculated exergy losses.

$$\left(\sum_{Streams} Ex_{in} + \sum (Ex_Q)_{in} + \sum w_{in}\right) - \left(\sum_{Streams} Ex_{out} + \sum (Ex_Q)_{out} + \sum w_{out}\right) = Ex_{loss} \quad \text{[Eq. 1]}$$

Exergy of streams involve physical and chemical exergy. Physical exergy of a stream can be defined as the maximum work that can be obtained from it in taking it to physical equilibrium with environment. The dead state conditions in all calculations are a temperature T_0=25°C , a pressure P_0=1 atm and a composition X_0 equal to the composition used by the Szargut model for the standard chemical exergy of substances which can defined by eq. 2 [12].

$$Ex_{Ph} = (H - H_0)|_X - T_0(S - S_0)|_X \quad \text{[Eq. 2]}$$

Where the enthalpy and the entropy of the substance have to be evaluated at temperature and pressure conditions of process (T, P) and environment (T_0, P_0) [7].

Chemical exergy of a stream of matter can be defined as the maximum work that can be obtained from it in taking it to chemical equilibrium with the environment.

$$Ex_{Ch} = (H - H_0)|_{T_0,P_0} - T_0(S - S_0)|_{T_0,P_0} \quad \text{[Eq. 3]}$$

Where the enthalpy and the entropy have to be evaluated for the chemical composition of the substance (X) and environment (X_0). For many identified substances, the standard chemical exergy (i.e. at T_0 and P_0) can be found in the literature [12], however, for the pseudo-components (unidentified compounds of a crude oil and derived streams), the chemical exergy can be determined from heuristic empirical expressions, like eq. 4, as a function of the elementary composition and the heating value of each pseudo-component [7-13].

$$Ex_{CH,k} = NHV_k \beta_k + \sum Z_m Ex_{CH,m} \quad \text{[Eq. 4]}$$

Where z_m are the mass fractions of metals, Fe, Ni, V, and water in the pseudo-component k and $Ex_{CH,m}$ are their corresponding specific standard chemical exergies, obtained from the standard chemical exergy values of the Szargut reference model. NHV_k is the net heating value of the pseudo-component k and β_k is the chemical exergy correction factor as a function of its C, H_2, O_2, S, and N_2, mass fractions which define by eq. 5 [7].

$$\beta = 1.0401 + 0.1728\frac{z_{H_2}}{z_C} + 0.0432\frac{z_{O_2}}{z_C} + 0.2169\frac{z_S}{z_C}\left(1 - 2.0628\frac{z_{H_2}}{z_C}\right) + 0.0428\frac{z_{N_2}}{z_C} \quad \text{[Eq. 5]}$$

2.2 *exergy performance parameters*

2.2.1 exergy losses

The exergy losses of a system can be split in two parts. The first part resulting from the irreversibility of the processes taking place in the system (internal losses), and the second part resulting from an exergy discharge to the environment (external losses).

The irreversible exergy losses are the difference between the total input (Ex_{in}) and output exergy (Ex_{out}). These losses can also be calculated using the Gouy-Stodola theorem with eq. 6 [7].

$$Irr = \sum Ex_{input} - \sum Ex_{output} = T_0 \Delta S \quad \text{[Eq. 6]}$$

2.2.2 exergy efficiency

Usually the relation between total output exergy and total input exergy is considered as the exergy efficiency. In most industrial processes some parts of their outlet exergy cannot be used, like waste exergy to the environment [14-15].

$$E_{out} = E_{pr} + E_{waste} \quad \text{[Eq. 7]}$$

$$\eta_{ex} = \frac{E_{out} - E_{waste}}{E_{in}} \quad \text{[Eq. 8]}$$

In some cases a portion of a main exergy flow passing through a system may have no impact on the system. Transit Exergy (Ex_{tr}) is the name has been given for this part of the exergy flow. Therefore, another type of the exergy efficiency can be used by subtracting the transit exergy from both the input and output exergy flows.

$$\eta_{ex} = \frac{E_{out} - E_{waste} - E_{tr}}{E_{in} - E_{tr}} = \frac{E_{pr} - E_{tr}}{E_{in} - E_{tr}} \quad \text{[Eq. 9]}$$

In crude distillation processes, chemical exergy of crude oil being refined into petroleum products is considered as transit exergy [14-15]. The reason for this view is that the crude oil chemical exergy does not affect the system, and just its value has been changed into chemical exergy values of products by refining process. Although, it must be noticed that this consideration is just in distillation process especially in crude distillation units, and cannot be used for cracking and catalyst reaction units that chemical structure change is the base process of such a system [3].

2.3 exergy improvement potential

Total exergy losses and effectiveness are the quantitative and qualitative measures of energy degradation, respectively, and they can be combined to have a more complete parameter of the performance of the system, this parameter is called the exergy improvement potential [2].

$$Pot = Irr(1-\eta) + Efl \quad \text{[Eq. 10]}$$

As can be shown from eq. 10 the exergy improvement potential is formed by three contributions: the absolute potential (Irr), the relative potential ($1-\eta$), and the invironmental potential (Efl) [7].

3. Process description

Figure 1 shows the process flow diagram of the crude distillation unit of Tabriz oil refinery. The crude distillation unit is the first step in a refinery complex to separate crude oil into different fractions [1]. Crude is heated in the first section of the heat exchanger network that utilizes as heating stream the streams from the main column, in this way the crude oil reaches a temperature of about 120°C and is fed to the desalter. Then the desalted crude flows through the second section of the heat exchanger network. Due to the great attention on energy integration by maximizing heat exchanges between the crude oil and the product streams from the main column, the crude can reach a maximum temperature. This temperature is still too low to achieve the grade of crude vaporization necessary for the separation in the main column and thus a furnace is always necessary [16]. After desalter and second heat exchanger, crude oil enters to a preflash drum. The preflash drum bottom is further heated by heat exchanger before entering furnace. The preheat crude oil is mixed with outlet vapor of preflash drum then feed to the main column.

The unit is divided into 5 sections for more accurate exergy analysis: 1- Feed preheating section: contain the primary heat exchangers and crude oil desalter drum, 2- Preflash drum and furnace section, 3- Atmospheric distillation column section, 4- LPG and naphtha splitting section and 5-Vacuum distillation column section.

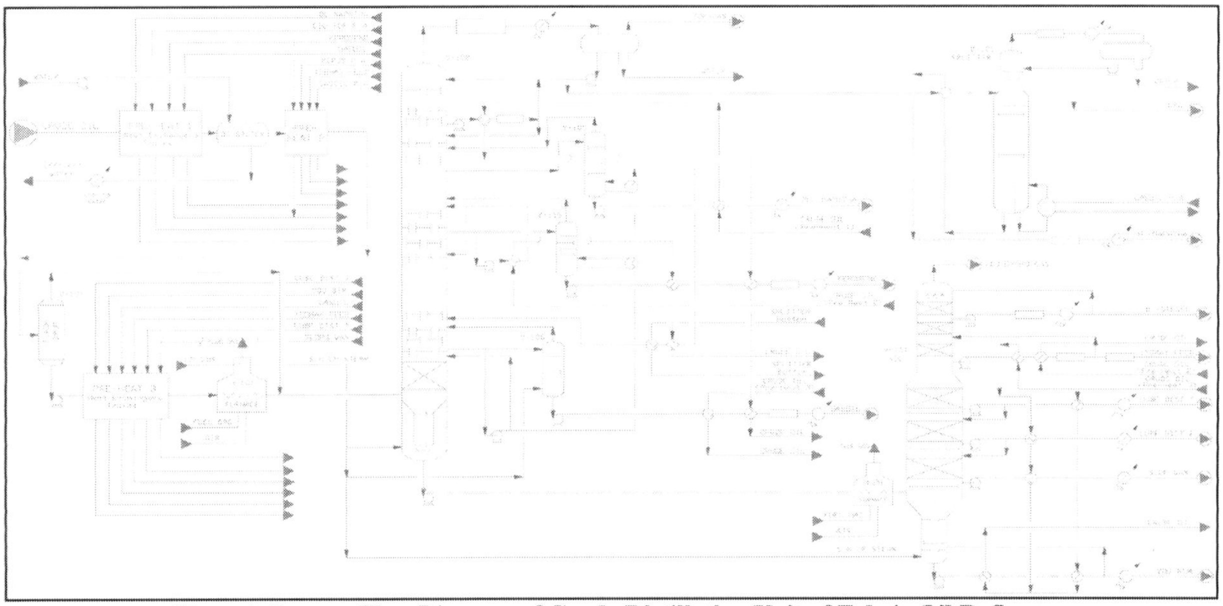

Figure1. Process Flow Diagram of Crude Distillation Unit of Tabriz Oil Refinery

4. Simulation of the plant

According to the provided data bank by AspenTech company in the Aspen Plus software, the BK10 thermodynamic package is suitable for simulating an oil refining unit. The BK10 property method uses the Braun K-10 K-value correlations. The BK10 property method is suit for vacuum and low pressure applications (up to several atm) [17]. After extraction of essential data from simulation results, exergy analysis of each section has been performed. Table 1 shows the results of exergy calculation for LPG and naphtha splitting section and all essential thermodynamics amounts for exergy calculations.

Table 1 Sample of Exergy Analysis Calculations for the LPG and Naphtha Section

Inlet	T (°C)	P (Barg)	Rate (kg/hr)	H (kj/kg)	S (kj/kg)	H_0 (kj/kg)	S_0 (kj/kg)	Energy (MW)	Energy (%)	Ex^{ch} (MJ/kg)	Ex^{Ph} (KJ/kg)	Ex (GW)	Ex^{Ph} (MW	Ex (%)
Naphtha Feed	57.2	6.4	101220	-2111.3	-6.8	-2183.0	-7.0	2.0	7.8	47.3	2.32	1.330	0.065	99.40
Gasoil P.A.	289.7	6.4	135309	-1295.9	-5.4	-1927.7	-6.8	23.7	91.7	0.00	209.28	0.008	7.866	0.59
E182 Aircooler.	25.0	0.0	240,000	-0.2	0.0	-0.2	0.0	0.0	0.0	0.00	0.00	0.000	0.000	0.00
E118 Aircooler.	25.0	0.0	140,000	-0.2	0.0	-0.2	0.0	0.0	0.0	0.00	0.00	0.000	0.000	0.00
Pumps Electricity								0.06	0.2			0.000	0.057	0.00
Aircooler Electricity								0.07	0.3			0.000	0.070	0.01
Total Rate			616529	Total Energy Inlet				25.9	100	Total Exergy Inlet		1.338	8.058	100
Outlet														
LPG	55.3	4.2	27,400.0	-2,501.5	-7.2	-2,358.5	-6.7	-1.1	-4.2	48.4	5.65	0.368	0.043	27.51
Water	55.3	4.2	42.8	-15,738.6	-8.7	-15,865.6	-9.1	0.0	0.0	0.00	6.49	0.000	0.000	0.00
H-Naphtha	38.0	7.0	73,777.2	-2,001.4	-6.8	-2,029.2	-6.9	0.6	2.2	46.9	1.68	0.962	0.034	71.85
Gasoil P.A.	206	5.4	135,309	-1,522.0	-5.9	-1,927.7	-6.8	15.2	58.9	0.00	111.9	0.004	4.207	0.31
E182 Aircooler.	90.7	0.0	240,000	65.8	0.2	-0.2	0.0	4.4	17.0	0.00	6.43	0.000	0.429	0.03
Heat Exch.187	45.0	4.7						1.2	4.7			0.000	0.047	0.00
Heat Exch.183	45.0	4.7						0.8	3.2			0.000	0.032	0.00
Total Rate			616529	Total Energy Outlet				24.1	93.2	Total Exergy Inlet		1.33	5.1	99.7
Loses								1.8	6.8			0.00	2.9	0.26
Unutilized outflow			380000					7.4	28.4			0.001	0.755	0.056
Energy Efficiency								93.2	Exergy Efficiency			99.69	54.14	
$Ex_{transit}$:								1.3	Exergy Eff. without $Ex_{transit}$:			50.96		

Table 2 shows the exergy efficiency, exergy losses and Pot% for 5 sections of crude distillation unit. The analysis was indicated that the preheat section has the highest exergy efficiency because of energy integration and the atmospheric distillation column. Also the LPG and naphtha section have the lowest exergy efficiency because of separation and aircoolers in these sections.

Table 2 Exergy Analysis Results for 5 Sections of the Crude Distillation Unit

Section	Exergy Efficiency%	Exergy Losses%	Pot%
Feed preheating	98.59	7.55	0.23
Preflash drum and furnace	67.49	27.34	27.12
Atmospheric distillation column	52.58	45.31	49.99
LPG and naphtha splitting	50.96	2.79	3.72
Vacuum distillation column	67.95	17.01	18.94

5. Proposed solutions for reducing exergy losses and their results

The proposed methods which are discussed in this work, are collected from previous literature works and some new ones. It has been attempted to offer them in the range of over-design percent of the unit.

5.1 Introducing the preflash vapor directly to the atmospheric distillation column

In this unit, the preflashed vapor is mixed again with the liquid stream after the furnace and before entering to the atmospheric distillation column. This mixing is against with aim of system and causes decreasing temperature of outlet furnace stream. One of the exergy loss reasons is mixing some streams with different physical and chemical characteristics. In this proposed method, vapor streams are introduced to main column at a point which has more physical and chemical conditions in common.

5.1.1 Introducing the preflash vapor directly to the atmospheric distillation column and decreasing furnace pressure drop about 15%

This solution is the combination of first suggestion with a decreasing of inlet or increasing of outlet furnace pressure drop about 15%. If furnace pressure drops increase, it will not attainable to reach the same temperature by burning a fixed amount of fuel. Reducing pressure on tubes furnace causes over firing problem that is a factor for formation of hot spots in the pipes [18].
When furnace outlet pressure increase, feed pressure of atmospheric column will increase too and it causes separation of light and heavy components will be easier than before condition by pressure drop in atmospheric column. Furthermore the products quality increase, performance of heat exchanger improve and efficiency of preheat section increase. Then furnace inlet temperature increase and it causes reduction in furnace fuel consumption.

5.2 Decreasing furnace pressure drop and increasing preflash inlet pressure

In this proposed method, pressure drop decreased in control valve before preflash drum, which it causes better separation in this drum and prevent heavy components of crude, went to preflash vapor. On the other hand, furnace pressure drop decreases so these solutions together increased theoretical flame temperature (TFT). This solution increase fuel consumption but TFT of this solution is higher than base case, so it should be decrease until base case. It causes furnace fuel consumption decrease about 0.24% in compare with base case. The result of fuel and TFT for this solution summarize in table 3.

5.3 Increasing the pump-arounds rate and decreasing condenser duty of the atmospheric distillation column

In this proposed method the flow rates of the pump-arounds of the main atmospheric distillation column are increased about 15%. Increasing their flow rates will transfer more heat from the column and thus more energy can be utilized in crude oil preheating section which can lead to some reduction in furnace duty [3].
In this proposed solution, condenser duty decreased about less than 10%. Furthermore in this method fuel gas decreased 2.03% by reducing TFT to base case. Their result for TFT and reducing fuel gas before and after changing TFT has been summarized in table 3.

Table 3 Comparison of Changing Fuel Consumption and TFT with Base Case (%) for Modification 2 & 3

solution	modification 5.2	solution	modification 5.3
increasing fuel rate	0.1	decreasing fuel rate	0.9
increasing TFT	0.31	increasing TFT	1.18
decreasing fuel rate after TFT reduction	0.24	decreasing fuel rate after TFT reduction	2.03

The advantage of this study is considering effect of proposed solution in whole of CDU which has shown the effect of one part to another part in this unit after proposed solutions. This method caused 4.51% decreasing exergy losses of atmospheric distillation column section.

5.3.1 Increasing the pump-arounds rate and decreasing condenser duty to optimum point

Decreasing of condenser duty has an optimum point which after that will causes rise in furnace duty and even bad fractionation in main column. So the optimum point of condenser duty has been determined and the duty has decreased until that point.

5.3.2 Increasing the pump-arounds rate, decreasing condenser duty to optimum point and increasing main column inlet temperature

In this solution, the temperature of the inlet stream to the atmospheric distillation column has been increased about 2°C. These solutions together concluded improvement in exergy efficiency.

5.4 Changing the pump-arounds return stages to one upper place in the atmospheric distillation column

In case of pump-arounds of the distillation column, the places of their return stage are important by the both process and exergy points of view. The return conditions of pump-arounds streams shall not be differed in temperature and composition a lot with the entering point to the column. By studying this issue in section 3 which contains 3 pump-arounds, changing their return stages to one upper place has been evaluated. It concluded improvements in exergy efficiency and increased the products quality.

The results of proposed solutions are summarized in table 4. This table shows the fact that the exergy efficiencies for all solutions have been improved. Increasing the pump-arounds rate of the atmospheric distillation column has the most effect on improvement of CDU.

Table 4. Comparison of Proposed Solutions Results

solution	increasing exergy eff %	decreasing energy losses%
Introducing the preflash vapor directly to the atmospheric distillation column	0.44	1.67
Introducing the preflash vapor directly to the atmospheric distillation column and decreasing furnace pressure drop about 15%	0.4	1.77
Decreasing furnace pressure drop and increasing preflash inlet pressure	0.06	0.01
Increasing the pump-arounds rate and decreasing condenser duty of the atmospheric distillation column	1.27	2.27
Increasing the pump-arounds rate and decreasing condenser duty to optimum point	1.39	2.9
Increasing the pump-arounds rate, decreasing condenser duty to optimum point and increasing main column inlet temperature	1.61	3.22
Changing the pump-arounds return stages to one upper place in the atmospheric distillation column	0.16	0.84

Table 5 shows percent of decreasing fuel consumption in proposed solutions. By comparing these results together, it has been observed that the increasing of the pump-arounds rate has the most effect on decreasing fuel.

Table 5. Comparison of Reducing Fuel Consumption with Base Case

solution	modif 1	modif 1-2	modif 2	modif 3	modif 3-1	modif 3-2	modif 4
fuel rate decreasing %	0.78	0.85	0.24	2.03	1.48	1.84	0.65

6. Conclusion

The exergy analysis of the crude oil distillation unit of Tabriz oil refinery, has been carried out in this research. The Gray box method has used for evaluation and analysis of this unit. It can be concluded that: The atmospheric distillation column, LPG and naphtha section have the lowest exergy efficiency. Therefore, some process alternatives have been proposed for these sections in order to improve the exergy efficiency. All proposed suggestions have led to exergy efficiencies improvement. Among these solutions, increasing the pump-arounds rate of the atmospheric distillation column about 15%, has the most effect on this unit and the best proposed modification is the combination of this solution with an increase in the column's feed temperature and also a decrease in the condenser's duty that has caused increasing exergy efficiency about 1.61%, the reduction rate in energy loss about 3.22% and the decreasing fuel consumption about 1.84%.

6. References

[1] Nur Izyan, Z; Shuhaimi, M; Exergy Analysis for Fuel Reduction Strategies in Crude Distillation Unit , Energy; Vol. 66, PP. 891-897; 2014.

[2] Rivero, R; Application of the Exergy Concept in the Petroleum Refining and Petrochemical Industry , Energy Conversion and Management; Vol. 43, PP. 1199-1220; 2002.

[3] Sajedi, S.N; Masoumi, M.E; Movagharnejad, K; Exergetic Improvement and Environmental Impact Assessment of Crude Oil Distillation Unit of Shazand-Arak Oil Refinery , International Journal of Exergy; Vol. 16, No. 4, PP. 464-480; 2015.

[4] Al-Muslim, H; Dincer, I; Zubair, S.M; Exergy Analysis of Single- and Two- Stage Crude Oil Distillation Units , Journal of Energy Resources Technology; Vol. 125, PP. 199-207; 2003.

[5] Hussain, M.M; Baschuk, J.J; Li, X; Dincer, I; Thermodynamic Modeling of a PEM Fuel Cell Power System , International Journal of Thermal Sciences; Vol. 44, PP. 903-911; 2005.

[6] Rivero, R; Exergy Analysis of a Reactive Distillation MTBE Unit , Int.J. Applied Thermodynamics; 4(2), pp. 85 – 92; 2001.

[7] Rivero, R; Rendon, C; Gallegos, S; Exergy and Exergoeconomic Analysis of a Crude Oil Combined Distillation Unit , Energy; Vol. 29, PP. 1909-1927; 2004.

[8] Doledersum, A; Exergy Analysis Proves Viability of Process Modifications , Energy Conversion and Management; Vol. 39, PP.1781-1789; 1998.

[9] Olakunle, M; Oluyemi, Z; Olawale, A; Adefila, S; Distillation Operation Modification with Exergy Analysis , Journal of Emerging Trends in Engineering and Applied Sciences; Vol. 2, PP. 56-63; 2011.

[10] Wall, G; Exrgy- a Useful Concept Within Resource Accounting , Institute of Theoretical Physics, Chalmers University of Technology and Universityöteborg ; Report No.77-42 Göteborg, Sweden ; 1977.

[11] Harati, F., Improvement Energy Efficiency of Natural Gas Refineries by Exergy Analysis, Master Thesis of Tehran University, Iran, 2010.

[12] Szargut, J; Morris, D.R; Steward, F.R; Exergy Analysis of Thermal, Chemical and Metallurgical Processes , New York: Hemisphere Publishing Corporation; 1988.

[13] Saidur, R; Ahamad, J.U; Masjuki, H.H; Energy , Exergy and Economic Analysis of Industrial boilers , Energy Policy; Vol. 38, PP. 2188-2197; 2010.

[14] Gong, M; Exergy analysis of a pulp and paper mill , International Journal of Energy Research; Volume 29, pp. 79 – 93; 2005.

[15] Gong, M; Wall, G; on Exergetics, Economics and Optimization of Technical Processes to Meet Enviromental Conditions , International Conference on TAIES 97, Beijing(china) ;PP.453-460; 1997.

[16] Errico, M; Tola, G; Mascia, M; Energy Saving in a Crude Distillation Unit by a Preflash Implementation , Applied Thermal Eng; Vol. 29, PP. 1642-1647; 2009.

[17] Sinnott, R.K; Coulson, J.M; Richardson, J.F; Chemical Engineering Design ,4th Ed, Oxford: Elsevier Butterworth- Heinemann; 2005.

[18] Golden, S.W; White, S; Increasing crude unit preheat, PTQ, issue Q1; 2006.

Using a Large Data Bank to Develop a Robust and Simple Model to Estimate the Saturation Pressure of Oil Reservoirs

By

Hassan Alkandari[1], Adel Elsharkawy[1], Adel Malallah, Osamah Alomair

Petroleum Engineering Department, College of Engineering & Petroleum
Kuwait University
Po Box 5969 Safat 13060-Kuwait

Abstract

Measurements of saturation pressure are crucial for all hydrocarbon reservoir fluids. Below the crude oil saturation pressure gas reaches a critical saturation; then, a two phase flow occurs and results in decreasing oil production and recovery. To maximize oil production and recovery, the reservoir pressure has to be maintained closer to the original saturation pressure. This pressure is normally measured using bottom-hole samples or surface recombined samples of oil and gas. Occasionally, the samples become unavailable and the pressure needs to be estimated using computational methods.

In this study, a large data bank of 231 crude oil compositions and saturation pressure measurements including literature data, unpublished data, and newly measured data were used to develop two empirical models to predict the saturation pressure of a variety of crude oils if the oil sample is unavailable or the experimental measurements are unreliable. The first proposed model utilizes the extended compositions of hydrocarbons up to the heptane plus fraction in addition to non-hydrocarbons. The second model uses the lumped compositions of light, intermediate, and heavy components in addition to non-hydrocarbon components as the input variables. The lumped model has the advantage of using fewer input parameters while maintaining the thermodynamics. The accuracy and validity of both models to calculate the saturation pressure for volatile oils, black oils, and heavy oils are presented using several compositional data. The models performance is also compared to the Peng-Robinson equation-of-state (PR-EOS), and the Soave-Redlich-Kwong equation-of-state (SRK-EOS) as well as all published methods that use compositions as input variables. The comparison indicates that the proposed models are much simpler and more accurate than the other computational methods. The proposed models treat the heptane-plus fraction as a single component, thus eliminating the splitting and characterizing the plus fractions and the binary interaction parameters needed for the EOS's calculations.

Key words: Saturation pressure (Ps), Bubble-point pressure (BPP), Equation of state (EOS), Correlations, and PVT Data

---

[1] Corresponding Authors: Hassan Alkandari , Hassan@hkandari.com, Adel Elsharkawy, adel.elsharkawy@ku.edu.kw

Introduction:

Reservoir fluid properties are an essential source of information for all calculations on hydrocarbon reserves. Accurate estimation of the pressure-volume-temperature (PVT) measurement is crucial. Of all the PVT properties, saturation pressure (Ps) is considered the most important parameter. It is well known that gas starts to flow once the reservoir pressure drops below the bubble point pressure and the gas saturation reaches a critical value. To minimize the gas flow and maximize the oil recovery, it is very important to maintain the reservoir pressure close to the original bubble point pressure.

Although the PVT properties are experimentally measured in laboratories and provide reliable results, this process is expensive, time consuming, and sometimes unavailable; therefore, measured field data (separator gas oil ratio, separator pressure, stock-tank oil specific gravity, and reservoir temperature) or reservoir fluid compositions are used to estimate the bubble point pressure.

There are two methods in the petroleum industry to calculate the saturation pressure of crude oil when experimental is unavailable; either using production data or compositional data.

Methods Based on Production Data (Composition is unavailable)

Numerous attempts have been made since the 1940s to develop correlations to estimate the saturation pressure from production data, such as producing gas oil ratio, oil gravity, gas gravity, reservoir pressure and temperature. Table 1 presents summary of all the methods that use production data to estimate the saturation pressure.

Methods based on Compositional data

Many authors have reported that the application of any correlation in other regions results in serious errors. These conclusions arise from the fact that regional crude oils represent a physical-chemical trend; therefore, all attempts to develop a general correlation using field measured parameters to estimate the saturation pressure from a large data bank have failed. For this reason, Elsharkawy (2003) made the first attempt to correlate the saturation pressure to the reservoir fluid composition instead of using the production data. In recent years, there have been several attempts to estimate the saturation pressure of oil reservoirs. These methods are reported in Table 1.

Elsharkawy (2003) presented a correlation of the bubble point pressure for the oil reservoir using the same input data needed for the equation-of-state (EOS) calculations, i.e., the compositional data and the reservoir temperature. Elsharkawy's model is based on the saturation pressure measurements of 60 crude oil samples from the Middle East and 75 collected from the literature. He demonstrates that the calculation of the saturation pressure depends on the type of EOS, the number of pseudo components, the method used to characterize the pseudo-components and the binary interaction number. Elsharkawy compared the performance of his model to the SRK-EOS and PR-EOS for black oils, volatile oils, heavy oils, and synthetic oils. He indicated that his model eliminates splitting the heavy plus fraction, characterizing the pseudo-components and inclusion of the binary interaction parameters needed for the EOS calculations.

Later, Tale et al. (2009) made the second attempt to model the saturation pressure of crude oil using 132 data samples published by Elsharkawy (2003). Tale's model, which is based on the alternating conditional expectation (ACE), has a set of 14 equations with 42 constants. The average absolute relative error (AARE) reported by Tale et al. indicates that the model did not perform well with the selected data. The ACE model has an AARE of 12.3% compared with AARE of 6.7% by Elsharkawy (2003) and 8.1% AARE for the SRK-EOS.

AlQuraishi (2009) attempted to model the saturation pressure using Elsharkawy's (2003) data. The model was developed using linear genetic programming. He correlated the saturation pressure to the formation temperature, methane content, and heptanes plus molecular weight. Thus, AlQuraishi's model neglected the effect of the intermediate and heptane plus content on the saturation pressure. The error analysis reported by AlQuraishi indicates that his model has a 5.8% error compared to 7.7% by Elsharkawy, 9.9% using the SRK-EOS, and 10.1% using the PR-EOS.

Bandyopadhyay and Sharma (2011) presented a model to predict the bubble point pressure, using temperature and fluid composition. They used 129 experimental and literature data originally published by Elsharkawy (2003) to develop their model. The model introduced a temperature interaction parameter that depends on compositions. The proposed model has a set of 24 coefficients and the estimated bubble point pressure is calculated as the sum of the series calculations. They reported an average absolute error (AAR) of 7.83% using their model compared to 8.3% using Elsharkawy's simple model (2003), 10.15% using SRK-EOS, and 10.58% using the PR-EOS.

Khamehchi et al. (2011) used an intelligent proxy simulator (IPS), which is based on an adaptive neural fuzzy inference system (ANFIS) and a structured artificial neural network (ANN) to predict the bubble point pressure and asphaltene onset pressure. The model uses a reservoir fluid composition and temperature to predict the Ps. It is not clear how the Ps is calculated from the proposed method, and they did not present the accuracy of the proposed method compared to the other published methods.

Farasat et al. (2013) presented a new mathematical approach to calculate the crude oil saturation pressure as a function of the temperature and the reservoir fluid composition. The model developed by Farasat et al used 130 experimental data, which were originally published by Elsharkawy (2003). Farasat used a support vector machine (SVM) to model the saturation pressure. They did not clearly demonstrate how to calculate the saturation pressure from the SVM model. Moreover, they did not indicate the overall average absolute deviation (AAD) for their model compared to other published models and both the PR-EOS and SRK-EOS for the 130 data.

Ahmadi et al. (2014) also presented an ANN technique called gene expression programming (GEP) to determine the bubble point pressure of the oil samples. Similar to Farasat et al. (2013), Ahmadi's model used most of the data previously published by Elsharkawy (2003) to construct a non-linear equation using reservoir fluid compositions and temperatures as input parameters. The proposed model is highly complicated and consists of 8 equations that have several constants compared to the one simple equation proposed by Elsharkawy (2003). They compared the performance of their proposed model to the SRK-EOS and PR-EOS; however, their paper does not mention how the heptanes plus fraction were treated and characterized for the EOS calculations. They reported an AAD of 4% using the proposed model compared to 8% using Elsharkawy's model (2003), 10.9% by PR-EOS, 10.5% by SRK-EOS, and 15% by Bandyopadhyay and Sharma (2011).

Gholami et al. (2014) estimated the bubble point pressure using a support vector regression SVR and a supervised learning algorithm SLT based on the statistical learning theory, and then the ACE was used. The model used reservoir fluid compositions and temperature to estimate the bubble point pressure. Gholami et al. used the same data originally published by Elsharkawy (2003). They presented three attempts based on genetic programming to correlate the bubble point pressure: SVR, ACE, and power-law committee machine (PLCM). The ACE-based method resulted in a set of three equations with a matrix of [8 by14] constants. Gholami et al. compared the accuracy of their model to the SRK-EOS and PR-EOS predictions reported by Elsharkawy (2003). Based on their AARD analysis, the PLCM model has the lowest errors followed by the SVR model, Elsharkawy's model, ACE model, PR-EOS, and SRK-EOS.

Lately, Jarrahian et al (2015) presented two empirical models to estimate the saturation pressure of black oils. The first model is based on production information. The second model is based on compositional information mostly from Elsharkawy (2003) and Heidaryan and Moghadasi (2011). They have not presented comparison of the accuracy of their model to Farasat (2013) model, Ahmadi (2014) model, and Gholami (2014) model.

This paper is concerned with the estimation of saturation pressure using compositional data. Therefore, the objective is to develop a robust yet simple model to estimate the saturation pressure for various types of reservoir oils using a larger data bank than that used by other published studies. The accuracy of the proposed models is compared to previously published models as well as SRK-EOS and PR-EOS.

Data bank

This study uses a large data set of 231 crude oil compositions and saturation pressure measurements. The data set consists of 130 data sets originally reported by Elsharkawy (2003); in addition to others collected from literature (Jacopy, 1958; Vogel, 1980; Coats, 1986; Droham, 1980; Dansh, 1992; Moharam, 1996; Rosenegger, 1999); 94 new data sets that have not been utilized before; as well as 7 newly measured samples at the Petroleum Fluid Research Center (PFRC) in Kuwait University. These data represents a wide range of compositions, such as volatile oils, black oils, and heavy oils. The data cover a pressure range of 537 to 5065 psia and a temperature range of 58 to 319 °F, crude oil compositions from C_1 to C_{7+}, non-hydrocarbons, such as N_2, H_2S, and CO_2, and the molecular weight and the specific gravity of the heptane plus fraction.

Data screening and classification

The compositions and the saturation pressure measurements in the data bank are subjected to a meticulous screening process to check the validity of the data. Our screening criteria uses the following steps: the sum of the molar compositions of all the components must add up to one, otherwise, the sample is eliminated. Samples with a missing molecular weight or specific gravity of the plus fraction are also eliminated.

After filtering, the data were reduced to 214 and divided into 179 measurements used to develop the new models, and 35 data sets are used for the blind test of the robustness of the proposed models. Statistical analysis of the input variables of the data set used in training and testing the extended and the lumped models are illustrated in Tables 2.

Several of the reported compositional data have extended analysis up to (C_{20+}). The following techniques are used for lumping or regrouping the extended analysis into single pseudo-fractions (C_{7+}). The mole fraction (X_i) of the individual components of the extended analysis is equal to the plus fraction (Ahmad, 2011).

$$X_{C_{7+}} = \sum_{i=7}^{n} X_i \qquad i=7, 8 \ldots n \qquad (1)$$

The average molecular weight (Mw) of the individual component of the extended analysis is equal to the molecular weight of the plus fraction.

$$Mw_{C_{7+}} = \sum_{i=7}^{n} \frac{X_i Mw_i}{X_{C_{7+}}} \qquad i=7, 8 \ldots n \qquad (2)$$

The density of the total plus-fraction can be calculated as follows:

$$\gamma_{C_{7+}} = \frac{X_{C_{7+}} Mw_{C_{7+}}}{\sum_{i=7}^{n} \frac{X_i Mw_i}{\gamma_i}} \qquad i=7, 8 \ldots n \qquad (3)$$

It is important to note the significance of the density of plus fraction because its density determines its paraffinic naphthenic aromatic (PNA) content.

Equation-of-state calculations

The EOS calculation of the saturation pressure is compared with the newly developed models as well as the other methods. The two EOS calculations considered in this paper are the ones most considered and widely used in the petroleum industry to evaluate the volumetric and phase behavior of reservoir fluid properties: the Soave-Redlich-Kwong EOS (SRK-EOS) and the Peng-Robinson EOS (PR-EOS). It should be noted that the accuracy of any EOS calculation of the saturation pressure largely depends on the number of pseudo-components and the splitting schemes used to divide the heptane plus fraction; as well as the characterization and calculation of the critical properties of the pseudo-components and the binary interaction parameters. Therefore, the two EOS calculations of the saturation pressure are not the same, even though the number of pseudo-components, characterization of the pseudo-components, and binary interaction parameters were the same.

Model development

As previously stated, the objective of this work is to provide the petroleum reservoir engineer with a simple yet accurate tool to estimate the saturation pressure for a variety of crude oils using compositional data when experimental is unavailable. Two models are presented in this study to calculate the saturation pressure of the crude oil samples. The first model uses all input parameters, such as the molar compositions of the hydrocarbon and non-hydrocarbon components similar to the EOS models. This model is referred to as the extended model. The second model uses fewer input parameters in which some of the hydrocarbon components have been lumped into a single group; this is designated as the lumped model. Both models are simpler than all published correlations, physically correct, and have an acceptable accuracy compared with the well-known EOS.

Figure 1 shows a correlation matrix of all the data available under consideration. This figure also indicates the relative importance of each of the input variables into the first proposed model. It agrees with most published correlations and the EOS that the methane content, heptane plus content and temperatures have the greatest impact on the saturation pressure followed by the intermediate content. The data understudy indicates that the non-hydrocarbon has the smallest impact. This result is simply a reflection of the molar contribution of each component comprising the crude oil composition.

Model I - Extended Model

The extended model uses 13 variables comprising the molar compositions of the pure components, heptane plus properties, and reservoir temperatures, which are similar to the input parameters for the EOS. Figure 2 shows the correlation coefficient graph of all the input parameters in the extended model. Thus, the elimination of the intermediate effect has a negative impact on the accuracy of the predictive model. Our proposed extended model has the following form:

$$Ps = 124.72N_2 + 17.57CO_2 + 22.55H_2S + 64.22C_1 - 9.80C_2 - 52.49C_3 + 6.16C_4 - 19.22C_5 - 23.63C_6$$

$$- 21.43C_{7+} + 435.31\gamma_{C7+} + 1.14Mw_{C7+} + 4.29T \qquad (4)$$

Where Ps is the saturation pressure in psia, T is the temperature in °F, and the hydrocarbon and non-hydrocarbon compositions are expressed in mole.

Model II - Lumped Model

The second model is much simpler and uses only 7 -variables (the lumped model). Several lumping schemes are adapted in the petroleum industry for phase behavior modeling and reservoir simulation of the petroleum reservoir fluids (Ahmed, 2011). Hydrocarbon components are grouped into a pseudo fraction based on their molecular weight (Whitson, 1980), the similarity of their physical-chemical properties (Lee, 1979), their carbon number, their true boiling point or their number of families based on the chemical structure (Becker et al., 2015). (Emera and Samara, 2005) proposed lumping the hydrocarbon components into volatile (C_1 & N_2), intermediate (CO_2, H_2S, C_2 to C_4) and heavy categories (C_{5+}) to calculate the minimum miscibility pressure. (Duan et al., 2013) proposed lumping several components into a single carbon number (SCN) with predetermined values for properties, such as critical pressure, temperature, and eccentric factor. Others have proposed optimal lumping schemes based on the objectives (Lin et al., 2008). In lumped model, $C_{1+}N_2$, C_2, CO_2, and H_2S, the intermediate (C_3 to C_6), heptanes plus (C_{7+}) and its properties, and temperature are the input variable. The correlation matrixes of the lumped components are provided in Figure 3. Even though the non-hydrocarbon has the least effect their contribution is important, especially the H_2S in the case of sour crudes. Both models presented by AlQuraishi (2009) and Bandyopadhyay and Sharma (2011) have eliminated the effect of H_2S.

The proposed model has the following form:

$$Ps= 82.115*(C_1+N2)-11.635*C_2+39.158*CO_2+ 38.244*H_2S -1.217*(C_3—C_6) - 890.701*(C_{7+}/MW_{7+}*\gamma_{7+}) + 4.217 *T-1042 \quad (5)$$

Where Ps is the saturation pressure in psia; and T is the temperature in °F.

Result and discussion

Model accuracy

The accuracy of the two proposed models used to calculate the saturation pressures of the previously described crude oil samples is shown in Table 3. This table shows reports the average relative error, average absolute relative error standard deviation and the correlation coefficients for all the methods considered in this study. Table 3 also reports the accuracy of the PR-EOS, SRK-EOS, Elsharkawy's model, AlQuraishi's model, and Bandoypandyay and Sharma's model and Jarrahian model. The methods proposed by Gholami et al., Ahmadi et al., and Jarrahian et al are included in Table 3, even though they produce a unreasonable estimate of the saturation pressures, as explained in the appendix.

This table indicates that the extended and the lumped models have an (AARE) in the order of 8% and 8.5%, respectively. It should be noted that the complications in the Bandoypandyay and Sharma model did not improve the accuracy of the saturation pressure, because it had a large error. It is also noted that Jarrahian et al model produces an average error of 35%. Compared with the EOS, the lumped model has a few constants and eliminates the splitting, characterizing the pseudo-components and binary interaction parameters necessary for the EOS calculations. The lumped model as well as the extended model are very simple and require one-step calculations compared with all the other methods reported in Table 3.

A cross-plot of the calculated saturation pressure using the extended model based on the training data versus the measured data is illustrated in Figure 4A. This figure indicates that most of the data falls on the unit slope line, and there is no over-fit. This trend is supported by the high correlation of regression of the model (97.6%) presented in Table 3. Figure 4B shows a cross-plot of the testing data (28 from literature and 7 new measured in PFRC at Kuwait University). The extended model indicates an excellent agreement with all of the 28 data sets chosen at random, with the exception of a few data points at higher pressures. Additionally, the model has a perfect match for the 7 newly measured samples.

Figure 5A shows the performance of the lumped training-model for capturing the measured saturation pressure for most of the 179 samples. A cross-plot of the calculated saturation pressure using the lumped model based on the training data versus measured data is depicted in Figure 5B. This figure indicates that most of the data scatters falls to the unit slope line. The lumped model also has a high coefficient of regression of 97.4%, which is indicated in Table 3.

Models Validity

The accuracy and validity of the models for estimating the saturation pressure of various types of crude oil are reported in Table 4. This table illustrates the capability of the newly proposed models to predict the saturation pressure for volatile oils, black oils, and heavy oils. Table 4; provides the compositions, properties of the heptane plus fraction, temperature, and measured saturation pressures of the samples. Additionally, the calculated saturation pressure using the PR-EOS, SRK-EOS, Elsharkawy's model (2003), AlQuraishi's model (2009), and Bandoypandhyay and Sharma's model (2011), Jarrahian et al (2015); as well as the extended and lumped models are provided in Table 4. It is clear from this table that the proposed models have a wide range of applications regardless of crude oil types. We have also included information regarding the 7 newly measured data; at the PFRC, compositional, and saturation pressure. At the bottom of Table 4, we have provided the unpublished data of the compositions; and the saturation pressures at low temperatures. It is well known that the saturation pressure of a crude sample is measured at low temperatures in addition to the reservoir temperature as a method to check the quality of the sample.

Figure 6A reports the error of all the methods considered in this paper to predict the saturation pressure for volatile oils (heptane plus content below 30%). Both the extended and the lumped models have an AARE of nearly 4% followed by Bandoypandhyay and Sharma's model (2011); and Elsharkawy's model (2003); the SRK-EOS has nearly twice the AARE of the newly proposed model. AlQuraishi's model (2009) has an AARE of 12% because the model is based on methane and neglects the effect of heavy and intermediate components, whose contribution is highly significant in the reservoir and production performance of volatile crudes. Jarrhaine et al (2015) model is not suitable for volatile oils. It shows a 35% error in estimated saturation pressure.

The performance of the extended and lumped models compared to the various methods for the black oil samples ($35< C_{7+} <55\%$) is provided in Figure 6B. The newly proposed models have an AARE close to 10%; followed by Bandoypandyay and Sharma's model 14%, the SRK-EOS and PR-EOS 16%, and Elsharkawy 17%. AlQuraishi's model has an AARE of 24%. Jarrhain et al (2015) model shows a 33% error in estimated saturation pressure even though the authors claim that the model is suitable for black oils.

Figure 6C shows the error level of all the methods as well as the proposed models for the newly measured crude oil samples at Kuwait University using the PFRC. Again, Jarrhain et al (2015) model show a 40% error in estimated saturation pressure.

The performance of all published models at low temperature is illustrated in Figure 6D. The low-temperature calculation indicates that Bandoypandhyay and Sharma's model (2011) has accuracy problems and is not valid at low temperatures because the authors proposed a temperature-composition related coefficient. Also, Jarrhain et al (2015) model is not suitable at low temperature. It shows a 40% error in estimated saturation pressure.

The ability of the proposed models to capture the thermodynamic is thoroughly checked against its independent variables. The validity of the model is compared to the experimentally measured data as well as the predictions by other methods used to capture the physics of increasing the saturation pressure as a function of temperature is illustrated in Figure 7. Two main points are observed here; first, the model matches the measured saturation pressures much better than the other methods. Second, it follows the physical trend predicted by the EOS's. Figure 8 shows the model performance for capturing the effect of changing the saturation pressure as a function of the molecular weight of the heptane plus fraction compared with the SRK-EOS and PR-EOS. The extended model follows the trend of the PR-EOS in increasing the saturation pressure. Elsharkawy's model (2003) was unable to predict the correct trend as predicted by the EOS's. The model sensitivity in capturing the change in the saturation pressure as a function of increasing the methane content is shown in Figure 9. The proposed model as well predicts the correct trend of increasing the saturation pressure when increasing the methane content.

Conclusion

In this paper, two models are presented to estimate the saturation pressure of a variety of crude oils when measures are unreliable or samples are unavailable. The two models are developed using compositional data and saturation pressure measurements from a large data bank comprising literature data, unpublished data, and newly measured and reported data of 231 crude oil samples. The input parameters in the first model are the extended hydrocarbon and non-hydrocarbon molar percentages and reservoir temperatures. The second model uses lumped hydrocarbon fractions of intermediate, and heptane plus fractions in addition to its properties; and the reservoir temperature as input parameters. The accuracy of the two models has been tested and compared to that of the SRK-EOS and the PR-EOS; as well as published correlations that use compositional data as the input parameters. The comparison indicates that the two models are much more accurate than all the published methods. The validity of both models has also been tested using various types of crude oils samples and compared to experimentally measured saturation pressures and estimates using the EOS as well as the published correlations. The proposed models are physically correct and much simpler than all the published methods. The models do not require tedious computations and eliminate splitting of the plus fraction, characterization of the pseudo-components; as well as the binary interaction parameters that are required for the EOS calculations.

Acknowledgement

The authors thank the General Facility Lab at Kuwait University for their cooperation and conducting the compositional and saturation pressure measurements of the 7 new samples described in this paper using Petroleum Fluid Research Center (PFRC) project no. GE 01107.

References

Agarwal, R.K., Li, Y.K., Nghiem, L., 1990. A regression technique with dynamic parameter selection for phase – behavior matching. SPE Res. Eng. February, 115 – 120.

Ahmadi, M. A., Zendehboudi, S, James, L., Elkamel, A., Dusseault, M., Chatsiz, I., Lohi, A., 2014, New Tools to determine Bubble Point Pressure of Crude Oils: Experimental and Modeling Study, *Journal of Petroleum Science and Engineering, v.123, 204-216.*

Ahmed, T., 2011, Splitting and lumping schemes of the plus fraction, working guide to vapour-liquid phase equilibria calculations, Chapter 9, Hydrocarbon Phase Behaviour, Gulf Publishing Co. 111-126.

Al-Gathe, A., Abd-el Fattah, Kh., El-Mtwally, K., 2014, New artificial intelligent approach for bubble point pressure, Int. J innovation and Sci. Research, Vol.4, No.2, July, 121-135.

Al-Marhoun, M.A. 1985. Pressure-Volume-Temperature Correlations for Saudi Crude Oils. Presented at the SPE Middle East Oil Technical Conference and Exhibition, Bahrain, 11–14 March. SPE-13718.

Almehaideb, R.A. 1992. Improved PVT Correlations for UAE Crude Oils. Presented at the Middle East Oil Show and Conference, Bahrain, 15-18 March. SPE-37691.

Al-Najjar, H.S., Al-Soof, N.B.A., and Al-Khalisy, K.M. 1988. Correlations for Bubble-Point Pressures, Gas Oil Ratios and Formation Volume Factors for Iraqi Crude Oils. *Journal of Petroleum Research* (June 1988): 13.

AlQuraishi, A. A., 2009, Determination of crude oil saturation pressure using linear genetic programing, Energy & Fuel, 23, 884-887.

Al-Shammasi, A.A. 2001. A Review of Bubblepoint Pressure and Oil Formation Volume Factor Correlations. *SPE Res. Eval. & Eng.* 4 (2): 146-160. SPE71302.

Asgarpour, S., McLauchlin, L., Wong, D. et al. 1989. Pressure-Volume-Temperature Correlations for Western Canadian Gases and Oils. *J Can Pet Tech.* 28 (4): 103.

Asoodeh, M., Bagheripour, P.,2012, Estimation of bubble point pressure from *PVT* data using a power-law committee with intelligent systems, *J. Pet. Sci. and Eng., Vol. 90–91, July 2012, Pages 1-11.*

Bandyopadhyay, P., Sharma, A., 2011, Development of a new semi analytical model for prediction of bubble point pressure of crude oils, *Journal of Petroleum Science and Engineering, Volume 78, Issues 3–4, September 2011, Pages 719-731.*

Becker, P. J., Celse, B., Guillaume, D., Dulot, H., Costa, V., 2015, Hydrotreatment modelling for variety of VGD feedstocks: a continuous lumping approach, Fuel, 139, 133-143.

Coats, K.H., and Smart, G.T., 1986, Application of regression base EOS PVT Program to laboratory data, SPERE, May 277-299.

Dansh, A. Xu, D. H., Todd, A. C., 1992 A grouping method to optimize reservoir description for compositional simulation of gas injection process, SPERE, Aug. 343-348.

De Ghetto, G. and Villa, M. 1994. Reliability Analysis on PVT Correlations. SPE-28904, Presented at the European Petroleum Conference, London, United Kingdom, 25-27 October.

Dindoruk, B. and Christman, P.G. 2001. PVT Properties and Viscosity Correlations for Gulf of Mexico Oils. Presented at the SPE Annual Technical Conference and Exhibition, New Orleans, 30 September-3 October. SPE-71633

Dokla, M.E. and Osman, M.E. 1992. Correlation of PVT Properties for UAE (United Arab Emirates) Crudes. *SPE Form Eval.* (March 1992): 41.

Droham, J.K., 1988, Enhancing the evaluation of PVT data, OSEA paper presented the 7th Offshore South East Asia, Feb. 2-5, Singapore.

Duan, J., Wang, W., Liu, H., Cong, J., 2013, modelling the characterization of the plus fraction by using continuous distribution function, Fluid Phase Equilibria, 345, 1-10.

Elam, F.M. 1957. *Prediction of Bubble Point Pressures and Formation Volume Factors from Field Data.* MS thesis, University of Texas at Austin, Austin, Texas.

Elmabrouk, S., Zekri, A., Shirif E., 2014, The prediction of bubble point pressure and bubble point oil formation volume factor in absence of PVT analysis, PET. Sci. & Tech., 32, 1168-1174.

Elsharkawy, A. M., 2003, An empirical model for estimating the saturation pressures of crude oils, Journal of Petroleum Science and Engineering 38 (2003) 55-77.

Elsharkawy, A.M. and Alikhan, A.A. 1997. Correlations for predicting solution gas/oil ratio, oil formation volume factor, and undersaturated oil compressibility. *J. Pet. Sci. Eng.* 17 (3–4): 291-302.

Emera, M., Samara, H., 2005, Use of genetic algorithm to estimate CO2-oil minimum miscibility pressure- a key parameter in design of CO2 miscible flood, J. Pet. Sci. Eng., Vol. 46, issue 1-2, 37-52.

Farasat, A., Shokrollahi, A., Arabloo, M., Gharagheizi, F., Mohammadi, A., 2013, toward an intelligent approach for determination of saturation pressure of crude oil, Fuel Processing Technology 115 (2013) 201-214.

Frashad, F., LeBlanc, J.L., Garber, J.D. et al. 1996. Empirical PVT Correlations for Colombian Crude Oils. SPE-36105 Presented at the SPE Latin American and Caribbean Petroleum Engineering Conference, Port of Spain, Trinidad and Tobago, 23–26 April. -MS.

Gharbi, R.B.C., Elsharkawy, A.M., 1999, neural network model for estimating the PVT properties of Meddle East crude oils, SPE Reservoir Eval. Eng. 2 (3), 255-265.

Gholami, A., Asoodeh, M., Bagheripour, P., 2014 How committee machine with SVR and ACE estimates bubble point pressure of crudes, *Fluid Phase Equilibria, Volume 382, 25 November 2014, Pages 139-149.*

Glasø, Ø. 1980, Generalized Pressure-Volume-Temperature Correlations. *J Pet Tech.* 32 (5): 785-795.
Heidaryan, E., Moghadasi, J., 2011, Crude saturation pressure through imperial correlation, Thermodynamics 2011, Athens, 31 August- 3 September 2011.
Hasan, M., 1993, PVT Correlations for Indonesian Crude Oils. *Proc.* 22nd Annual Convention of the Indonesian Petroleum Association, Indonesia, Vol.1, IPA93-2.1-070.
Jarrhian, A., Moghadsi, J., Heidaryan, E., 2015, Empirical estimating of black oil bubblepoint (saturation) pressure, J. Petroleum Science and engineering, (126), 69-77.
Jacopy R.H., Berry V.J.,1958, A method for predicting the pressure maintenance for reservoir producing volatile oil, AIME 213, 59-65.
Kartoatmodjo, T., Schmidt, Z. 1994, Large Data Bank Improves Crude Physical Property Correlations. *Oil Gas J.* 92 (27): 51–55.
Kartoatmodjo, T.R.S. and Schmidt, Z. 1991, *New Correlations for Crude Oil Physical Properties*, SPE 23556.
Khairy, M., El-Tayeb, S., Hamdallah, M., 1998. PVT Correlations Developed for Egyptian Crudes. *Oil Gas J.* 96 (18): 114.
Khamekhcki, E., Behvandi, R., Rashid, F., 2011, Prediction of bubble point pressure & asphaltene onset pressure during CO2 injection using ANN & ANFIS models, Pet. Sci. & Tech. Vol. 1 No.2, 35-45.
Labedi, R.M. 1982, *PVT Correlations of the African Crudes*, PhD thesis, CSM.
Labedi, R.M. 1990, Use of Production Data to Estimate the Saturation Pressure, Solution GOR, and Chemical Composition of Reservoir Fluids, SPE-21164-MS. Presented at the SPE Latin America Petroleum Engineering Conference, Rio de Janeiro, Brazil, 14-19 October,
Lasater, J.A. 1958, Bubble Point Pressure Correlations. *J Pet. Tech.*10 (5): 65–67. SPE-957-G.
Levitan, L.L. and Murtha, M. 1999. New Correlations Estimate Pb, FVF. *Oil Gas J.* 97 (10): 70.
Lee, S., 1979, Experimental and theoretical studies on the fluid properties required for simulation of thermal process, Paper SPE 8393.
Li, Y.K., Nghiem, L.X., Siu, A., 1985. Phase behavior computations for reservoir fluids: effect of pseudo-components on the phase diagrams and simulation results. J. Can. Pet. Technol. Nov. – Dec., 29 – 36
Lin, B., Leibovici, C. F., Jorgensen, S. B., 2008, optimal component lumping: problem formulation and solution techniques, Computer & Chem. Eng., 32, 1167-2272.
Macary, S.M. and El-Batanoney, M.H. 1992, Derivation of PVT Correlations for the Gulf of Suez Crude Oils. *Proc. of the* 11th EGPC Pet. Exp. & Prod. Conference, Vol. 1, 374, Cairo, Egypt,
Moharam, H.M., 1995. Prediction of viscosity of heavy petroleum fractions and crude oil using a corresponding states method. Ind. Eng. Chem. Res. 34, 4140 – 4144.
Moradi, B., Malekzadeh, E., Amani, M., Boukadi, F.H., Kharat, R., 2101, Bubblepoint pressure imperial correlation, SPE 132756.
Obomanu, D.A. and Okpobiri, G.A. 1987, Correlating the PVT Properties of Nigerian Crudes. *J. Energy Res. Tech.* 109 (4): 214-217.
Okoduwa, I. G., Ikiensikimama, S.S., 2010, Bubble point pressure correlation for Niger Delta Crude Oils, SPE 136968.
Omar, M.I. and Todd, A.C. 1993. Development of New Modified Black Oil Correlations for Malaysian Crudes. SPE-25338 Presented at the SPE Asia Pacific Oil and Gas Conference, Singapore, 8-10 February.
Owolabi, O.O. 1984. *Reservoir Fluid Properties of Alaskan Crudes*. MS thesis, University of Alaska, Fairbanks, Alaska (May 1984).

Pedersen, K.S., Thomassen, P., Fredenslund, Aa., 1988a. Characterization of gas condensate mixtures. AIChE Spring National Meeting, March 6 – 10, New Orleans, LA.

Peng, D. y., Robinson. D. B., 1976. A new two-constant equation of state , Ind. Eng. Chem. Fund. 15, 59-63.

Petrosky, G.E. Jr. 1990. *PVT Correlations for Gulf of Mexico Crude Oils*. MS thesis, University of Southwestern Louisiana, Lafayette, Louisiana.

Petrosky, G.E. Jr. and Farshad, F. 1998. Pressure-Volume-Temperature Correlations for Gulf of Mexico Crude Oils. *SPE Res Eval & Eng.*1 (5): 416-420. SPE-51395.

Riemens, W.G., Schutle, A.M., Jong, L.N.J., 1988. Birba field PVT variations along the hydrocarbon column and confirmatory field tests. J. Pet. Technol. Jan., 83 – 88

Rosengger, L., Wu., R.1990, Integrated PVT data characterization- lessons from four case histories, J. Can. Per. Tech., 38, (13), (paper 97-05)

Shojaei, M., Bahrami, F., Barati, P., Riahi, S., 2014, Adaptive neuro-fuzzy approach for reservoir oil bubble point pressure estimation, *J. of Natural Gas Sci. and Eng., Volume 20, September 2014, Pages 214-220.*

Soave, G. 1972. Equilibrium constants from a modified Redlich-Kwong equation of state, Chem. Eng. Sci. 27, 1197-1203.

Standing, M.B. 1947. A Pressure-Volume-Temperature Correlation for Mixtures of California Oils and Gases. *API Drilling and Production Practice* (1947): 275-287.

Tale, M., Alquraishi, A., Al Mushaigeh, E. 2009, New model for crude oil saturation pressure prediction using alternation conditional expectation, Offshore Mediterranean Conf. & Exhibit. Italy 25-27, 2009.

Talebi, R., Ghiasi, M. Talebi, H., Mohammadyian, M., Zendehboudi,S., Arabloo, M., Bahadori, A., 2014, Application of soft computing approaches for modelling saturation pressure of reservoir oil, *J. Nat. Gas Sci. and Eng., Vol. 20, Sep. 2014, 8-15.*

Vazquez, M., Beggs, H. D. 1980, Correlations for Fluid Physical Property Prediction. *J Pet. Tech.* 32 (6): 968-970. Vazquez, M.E. 1976. *Correlations for Fluid Physical Property Prediction,* MS thesis, University of Tulsa.

Velarde, J., Blasingame, T.A., and McCain Jr., W.D. 1997. Correlation of Black Oil Properties at Pressures below Bubble Point Pressure - A New Approach. Presented at the Annual Technical Meeting of CIM, Calgary, Alberta, 8–11 June. PETSOC 97-93.

Vogel, J.R., Yarborough, L. 1980, the effect of nitrogen on phase behavior and phase properties of reservoir fluids, SPE 8815.

Whitson, C. H., 1980, Characterizing hydrocarbon plus fraction, SPEJ, 24 (4), 683-694.

Williams, C.A., Zana, E.N., 1980. Use of the Peng-Robinson equation of state to predict hydrocarbon phase behavior and miscibility for fluid displacement. SPE 8817, SPE/DOE on Enhanced Oil Recovery, April 20–23, Tulsa, OK.

Appendix

In this appendix, sample calculations are presented to show the deficiency of some of the published methods.

Input parameters:

N_2	CO_2	H_2S	C_1	C_2	C_3	C_4	C_5	C_6	C_{7+}	GR.C7+	MWC7+	T, °F	Ps, psia
0.65	0.02	0	45.02	12.45	8.93	6.03	3.02	1.44	22.44	0.81	184	140	3002

Ahmadi et al (2014)

Ahmadi et al. proposed an 8 steps calculation (copied from their paper) to estimate the BPP as follows:

$$BPP = A \times Mw_{C7+} - \frac{B}{C_{7+}} + C \times Vol./Inter. + D \times SG_{C7+} \frac{E}{SG_{C7+}} + \frac{F}{Mw_{C7+}} + T \times 505.967 + \frac{2.45789 \times 10^6}{T} + \frac{G}{Vol./Inter.} \quad (5)$$

$$A = 1488.46 - SG_{C7+} \times 1087.64 - \frac{601.512}{SG_{C7+}} + 1670.58 + T \times 0.40771 + \frac{6345.73}{T} + \frac{8.72863}{Vol./Inter.} \quad (6)$$

$$B = \frac{4132.53}{Vol./Inter.} + \frac{6.1977 \times 10^7}{Mw_{C7+}} + T \times 3181.39 + \frac{1.06393 \times 10^8}{T} \quad (7)$$

$$C = \frac{20964.7}{C_{7+}} - \frac{1.76162 \times 10^6}{Mw_{C7+}} - T \times 19.1963 - Mw_{C7+} \times 21.5427 + \frac{12851.5}{SG_{C7+}} - \frac{712613}{T} - 3477.49 + SG_{C7+} \times 9532.73 \quad (8)$$

$$D = 1.05345 \times 10^{+06} - \frac{1.50748 \times 10^8}{Mw_{C7+}} - T \times 676.048 - \frac{1.20252 \times 10^7}{T} - \frac{12178.2}{Vol./Inter.} \quad (9)$$

$$E = 641357 - \frac{1.02891 \times 10^8}{Mw_{C7+}} - T \times 301.034 - \frac{1.91446 \times 10^6}{T} - \frac{3873.62}{Vol./Inter.} \quad (10)$$

$$F = 2.51471 \times 10^{+08} + T \times 25332.5 + \frac{3.47893 \times 10^8}{T} - \frac{244573}{Vol./Inter.} \quad (11)$$

$$G = 20125.6 - T \times 23.9271 - \frac{703193}{T} \quad (12)$$

Ahmadi et al. define the Vol. /Inter. as the mole percentage ratio of the volatile (C_1& N_2) to the intermediate components (C_2 to C_6, CO_2 and H_2S). Application of the above equations from their paper using the input parameters results in the following:

Vol./Inter.	A	B	C	D	E	F	G	BPP
1.43211	1643.9	1545062	-271.127	45122.71	23642.47	2.57E+08	11753	1,794,123

The calculated saturation pressure (BBP) from Eq.5 is 1,794,123 psia which is unreasonable. This indicates a deficiency in the model.

Gholami et al (2014)

Gholami et al. presented 3 steps calculation (copied from their paper) to estimate the BPP as follows:

$$Tr(M) = \sum_{i=0}^{5} B_i M^i \quad (15)$$

Application of Eq. 15 resulted in the following

N_2	0.00584
CO_2	0.060917
H_2S	0.050636
C_1	-0.47658
C_2	-0.41547
C_3	-0.45083
C_4	-0.02003
C_5	0.068769
C_6	0.273921
C_{7+}	1.814523
SG.C7+	-0.0197
MWC7+	37.03525
T., F	-0.11139

Application of Eq. 16

$$Tr(P_b) = Tr(N_2) + Tr(CO_2) + Tr(H_2S) + Tr(C_1) + Tr(C_2) + Tr(C_3) + Tr(C_4) + Tr(C_5) + Tr(C_6) + Tr(C_{7+}) + Tr(SG_{C7+}) + Tr(MW_{C7+}) + Tr(T) \quad (16)$$

Resulted that Tr(Pb) = 37.8159

$$P_b = (57.0840387741 \times (Tr(P_b))^2) + (1166.9102304199 \times (Tr(P_b))) + 2207.3573204358 \quad (17)$$

Application of Eq. 17 resulted in saturation pressure (P_b) = 127,967 psia, which is unreasonable, compared to the measured one is 3002 psia. This indicates that there is a deficiency in the model.

Jarrahian et al (2015)

Jarrahian et al. proposed the following Eq. (copied from their paper) to estimate the saturation pressure:

$$P_{sat} = A_9 \exp\left(A_1 z_{N_2} + A_2 z_{CO_2} + A_3 z_{H_2S} + A_4 z_{C_1}^{A_5} + A_6 \sum_{2}^{6} z_{C_i} + + A_7 z_{C_{7+}}^{A_8} \gamma_{C_{7+}}^{A_9} Mw_{C_{7+}}^{A_{10}} + A_{11}(T - 459.67)\right) \quad (2)$$

Application of Eq. 2 resulted in saturation pressure (Ps) = 1812 psia, compared to the measured one is 3002 psia. This indicates that model has inaccuracy problem. This has been the pattern in all the crude oil samples shown in Table 4 and figures 6A through 6D.

Table 1- Summary of attempts to calculate the saturation pressure of crude oils

Method based on production information	Method based on compositional data
Standing, 1947; Elam, 1957; Lasater, 1958; Vasquez, 1976; Vasquez and Beggs, 1980; GlasØ, 1980; Labedi, 1982&1992; Owolabi, 1984; Al-Marhoun, 1985; Obomanu and Okpobiri, 1987; Asgarpour et al., 1989; Al-Najar et al., 1988; Dokla and Osman, 1990; Petrosky, 1990; Kartoatmodjo and Schmidt, 1991&1994; Macary and El-Batanoney, 1992; Omar and Todd, 1993; Hasan, 1993; De Ghetto and Villa, 1994; Farashad et al., 1996; Almehaidaeb, 1997; Elsharkawy, 1997; Velarde et al., 1997; Kairy et al., 1998; Petrosky and Farshad, 1988; Al-shammasi, 2001; Gharbi & Elsharkawy 1999 Levitan and Murtha, 1999; Dindrouk and Christman, 2001 Moradi et al., 2010 Okoduwa and Ikiensikimama 2010 Asoodeh and Bagheripour, 2012 Shojaei et al., 2014 Talebi et al., 2014 Elmabrouk et al., 2014 Kanu and Ikiensikimama, 2014 Jarrahian et al 2015	Elsharkawy , 2003 Tale et al., 2009 AlQuraishi, 2009 Bandyopadhyay and Sharma, 2011 Khamehchi et al., 2011 Farasat et al., 2013 Ahmadi et al., 2014 Gholami et al. 2014 Jarrahian et al 2015

Table 2- Description of the data used to develop and test the extended model

Input	Training model (179 Data)			Testing (28 Data)			Testing (7 new experimental)		
	Min	Ave	Max	Min	Ave	Max	Min	Ave	Max
N_2	0.00	0.35	1.67	0.00	0.27	1.67	0.10	0.26	0.35
CO_2	0.00	1.27	11.37	0.00	1.17	11.37	0.14	0.41	0.83
H_2S	0.00	0.15	3.68	0.00	0.15	3.22	0.00	0.00	0.00
C_1	13.16	33.86	74.18	19.50	34.79	73.36	20.64	28.25	42.79
C_2	3.36	8.07	13.71	4.64	8.26	11.63	6.69	7.88	10.82
C_3	0.89	6.59	11.87	2.46	6.76	12.03	6.05	6.59	7.10
C_4	0.95	4.46	8.4	1.66	4.41	6.58	4.03	5.05	5.93
C_5	0.40	3.21	5.95	1.00	3.11	5.23	3.36	4.33	5.52
C_6	0.00	3.14	6.37	0.99	3.15	5.46	2.88	3.74	4.33
C_{7+}	10.72	38.89	57.73	11.18	37.88	51.22	27.61	43.49	51.31
$SG.C_{7+}$	0.74	0.87	0.959	0.70	0.86	0.93	0.85	0.89	0.92
MWC_{7+}	134	240	368.9	105	223	292	216	262	298
T, °F	58	156	319	58	165	261	121	157	230
Ps, psia	537	2198	5065	1025	2307	5000	1227	1779	3344

Table 3: Error analysis of the proposed models in comparison with all the investigated methods

Method	ARE	AARE	SD	R^2%
PR-EOS	6.24%	10.19%	1028	97.9
SRK-EOS	-3.12%	10.36%	1024	96.7
Elsharkawy (2003)	0.85%	8.57%	1015	96.4
Alquraishi (2009)	-6.06%	10.16%	1088	96.7
Bandyopadhyay (2011)	8.05%	12.28%	1054	96.8
Farasat (2013)	NA	NA	NA	NA
Gholami et al. (2014)	?	?	?	?
Ahmadi et al. (2014)	?	?	?	?
Jarrahian et al (2015)	35.52	35.56	983	95.2
New Extended Model	-0.37%	8.20%	1030	97.6
New Lumped Model	-0.50%	8.46%	1036	97.4

ARE = Average Relative Error

AARE = Average Absolute Relative Error

SD = Standard Deviation

R^2 = Correlation Coefficient

NA = Calculation was not possible, because the weight and bias were not provided.

? = The method produces unreasonable results as indicated in the appendix.

Table 4- Comparing model accuracy for calculations of saturation pressures for various crude oils

Volatile-Oil-Samples (C7+<30)																						
Fluid Source	index	N2	CO2	H2S	C1	C2	C3	C4	C5	C6	C7+	GR.C7+	MWC7+	T, °F	Ps, psi	SRK	PR	Elsh	Alq	Bandy	Jarr	Extd	Lump
Rosenegger	7	0.03	8.39	0	47.4	10.3	6.12	4.2	2.89	2.05	18.6	0.83	180	295	4000	3933	3879	4460	4420	4165	3317	4139	4178
Vogel-80	13	0.65	0.02	0	45	12.5	8.93	6.03	3.02	1.44	22.4	0.81	184	140	3002	2624	2548	2742	2895	2643	1812	3019	2996
Dansh-92	18	0	0	0	46.8	8.77	7.44	4.01	2.56	4.02	26.4	0.77	158	212	2941	2550	2594	3248	3617	3269	1944	3244	3374
	31	0.44	0.38	0	49.1	7.6	6.13	3.84	2.52	2	28	0.84	231	199	3739	3522	3507	3418	3951	3508	2309	3625	3644
Middle East	45	0.03	0.97	0	41.6	10.8	7.52	4.5	3	2.43	29.1	0.85	208	241	3069	3109	2979	3080	3513	3138	1962	3127	3140
Coats-86	51	0.55	1.03	0	36.5	9.93	8.85	6	3.78	3.56	30.4	0.84	200	234	2746	2537	2472	2714	2945	2643	1682	2777	2720

Black-Oil -Samples (35<C7+<55)																						
Fluid Source	index	N2	CO2	H2S	C1	C2	C3	C4	C5	C6	C7+	GR.C7+	MWC7+	T, °F	Ps, psi	SRK	PR	Elsh	Alq	Bandy	Jarr	Extd	Lump
Coats-86	73	1.64	0.08	0	28.4	7.16	10.5	8.4	3.82	4.05	36	0.84	252	131	1694	1674	1618	1615	1738	1644	1084	1713	1713
Middle East	88	0.08	1.37	0.69	36	8.67	5.94	2.97	1.5	3.1	39.7	0.9	274	168	2505	2710	2318	2247	2608	2321	1476	2441	2445
Middle East	114	0.29	0.48	0	28.4	8.29	7.38	5.06	3.42	4.41	42.3	0.88	252	133	1632	1672	1477	1544	1748	1517	984	1587	1598
Middle East	122	0	1.25	0	33.4	9.24	6.07	2.4	1.62	2.78	43.3	0.85	252	230	2110	2313	2278	2424	2737	2438	1410	2388	2412
Middle East	134	0	0.17	0	28.6	8.63	7.43	3.81	2.61	4.11	44.7	0.86	249	134	1548	1551	1429	1477	1764	1433	937	1503	1567
Middle East	146	0.44	0.83	0	27.8	7.54	6.9	4.61	3.1	3.11	45.7	0.9	272	133	1565	1799	1505	1558	1730	1528	1002	1602	1590
moharam-96	149	0.21	0.34	0	20	7.93	8	6.6	5.87	5.08	45.9	0.86	230	235	900	1353	1290	1531	1514	1225	773	1303	1294
Middle East	162	0.35	0.47	0	26.5	7.71	6.05	4.03	4.08	4.22	46.6	0.89	253	135	1500	1613	1380	1427	1631	1391	923	1478	1456
Middle East	176	0.12	0.51	0	28.8	7.91	6.46	3.17	2.07	3.25	47.7	0.88	250	134	1615	1665	1476	1574	1783	1502	981	1571	1614
Middle East	194	0.21	0.15	0	27.8	7.68	6.19	3.41	1.81	2.84	49.9	0.86	228	134	1590	1532	1377	1545	1680	1426	923	1466	1493
Rosenegger	207	0.37	0.02	0	17.3	6.11	8.23	5.47	3.76	3.96	54.8	0.85	242	156	537	938	888	914	1064	836	558	661	731

Newly-Measured Kuwait University Laboratory

Fluid Source	index	N2	CO2	H2S	C1	C2	C3	C4	C5	C6	C7+	GR.C7+	MWC7+	T, °F	Ps, psi	SRK	PR	Elsh	Alq	Bandy	Jarr	Extd	Lump
KU-1	209	0.33	0.22	0	25.6	6.87	6.39	5.61	4.68	4.33	46	0.879	222	133	1595	1360	1530	1442	1514	1394	860	1357	1337
KU-2	210	0.35	0.47	0	26.5	7.71	6.05	4.03	4.08	4.22	46.6	0.893	253	135	1500	1400	1620	1427	1631	1391	924	1478	1456
KU-3	211	0.27	0.74	0	23.1	7.2	6.45	5.93	4.66	4.16	47.5	0.906	290	138	1283	1140	1360	1284	1448	1296	800	1287	1219
KU-4	212	0.22	0.26	0	20.6	6.69	6.65	5.11	5.52	3.6	51.3	0.919	298	171	1227	1120	1300	1188	1415	1170	742	1176	1131
KU-5	213	0.19	0.14	0	27.8	8.37	6.88	5.28	4.33	3.85	43.2	0.89	278	121	1524	1300	1560	1402	1659	1409	928	1534	1489
KU-6	214	0.33	0.24	0	31.3	7.48	6.58	4.86	3.7	3.17	42.3	0.9	274	171	1978	1800	2200	1975	2242	1985	1224	2060	2026
KU-7	215	0.1	0.83	0	42.8	10.8	7.1	4.51	3.36	2.88	27.6	0.85	216	230	3344	3070	3400	3083	3568	3139	2003	3213	3200

Low-Temperature Oil Samples (59°F<T<71°F)

Fluid Source	index	N2	CO2	H2S	C1	C2	C3	C4	C5	C6	C7+	GR.C7+	MWC7+	T, °F	Ps, psi	SRK	PR	Elsh	Alq	Bandy	Jarr	Extd	Lump
Middle East	101	0	1.12	0.21	27	10.8	9.23	4.97	2.62	3.33	40.8	0.88	249	71	1247	1090	1033	1040	1235	564	808	1148	1205
Middle East	117	0.36	0.17	0	27.2	8.93	8.6	6.07	3.71	2.55	42.4	0.88	271	68	1130	1028	1245	1094	1248	766	820	1239	1228
Middle East	135	0	0.17	0	28.6	8.63	7.43	3.81	2.61	4.11	44.7	0.86	249	70	1270	1055	1255	1130	1311	783	826	1229	1297
Middle East	138	0.14	0.54	0	29.4	7.96	6.19	3.96	3.25	3.84	44.7	0.9	290	59	1265	1065	1350	1191	1301	880	894	1408	1390
Middle East	147	0.44	0.83	0	27.8	7.54	6.9	4.61	3.1	3.11	45.7	0.9	272	67	1320	1125	1400	1201	1268	921	879	1319	1312
Middle East	173	0.06	0.81	0	31.3	7.42	5.68	2.82	1.63	2.64	47.6	0.89	270	68	1460	1215	1535	1441	1463	1026	975	1548	1576
Middle East	193	0.16	0.3	0	24.7	7.51	6.92	4.76	2.65	3.43	49.6	0.88	239	71	1050	975	1156	1008	1107	701	714	958	986

Elsh: Elsharkawy (2003) Alq: AlQuraishi (2009) Band: Bandyopadhyay and Sharma (2011) Jarr: Jarrahian et. al (2015)
Extd: Extended model Lump: Lumped model

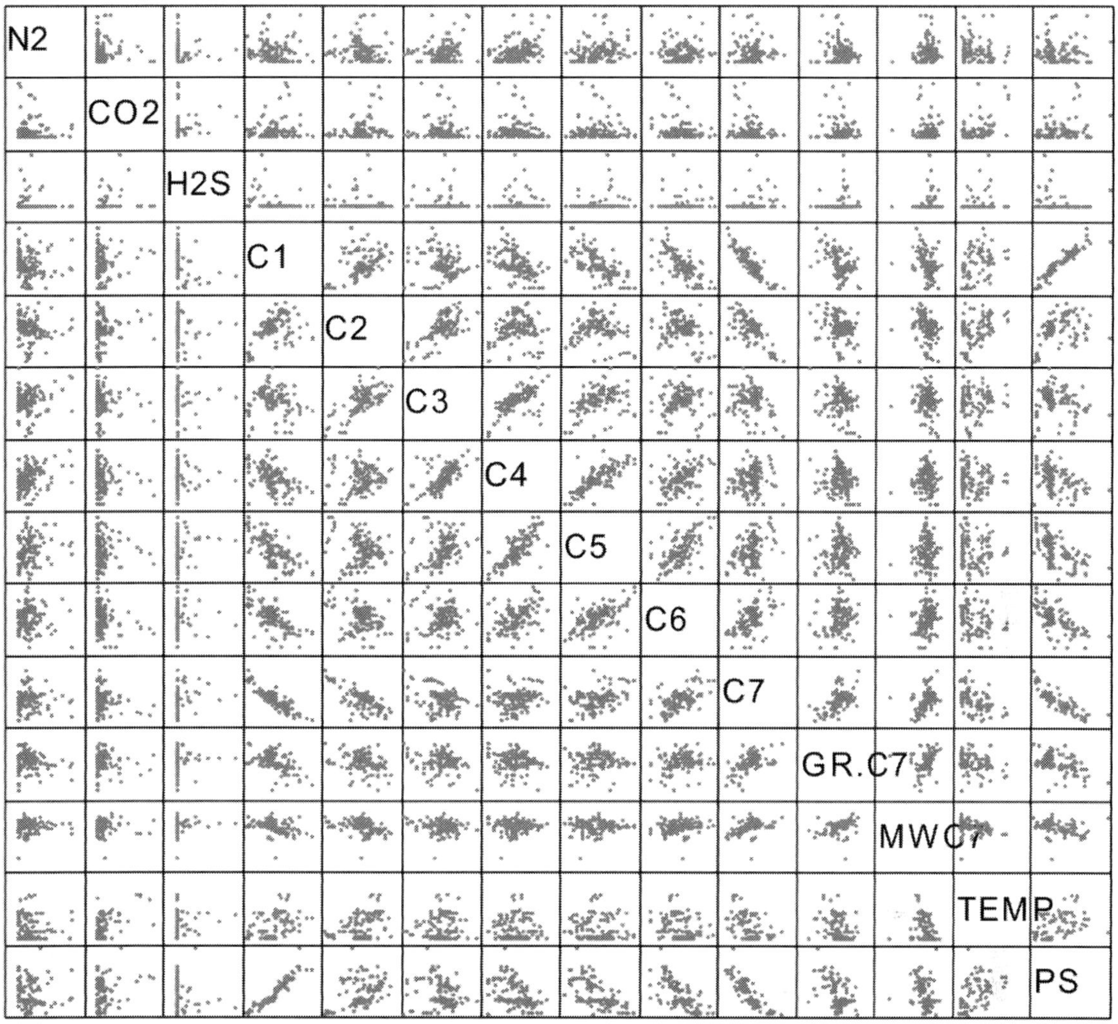

Figure 1: Correlation matrix for all the variables in the oil data bank

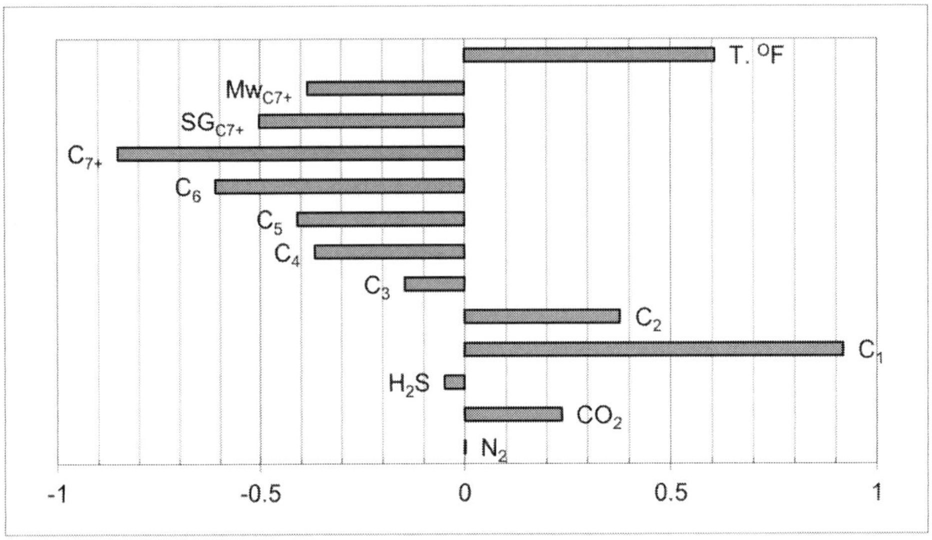

Figure 2: Correlation coefficient for all the variables in the training model (179 data)

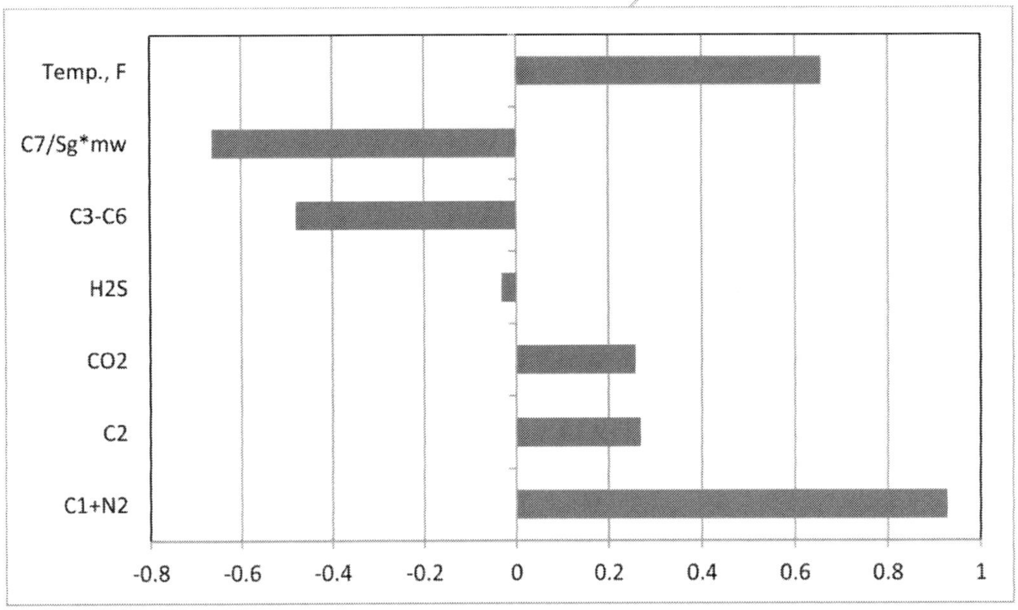

Figure 3: Correlation matrix for the lumped model (179 data)

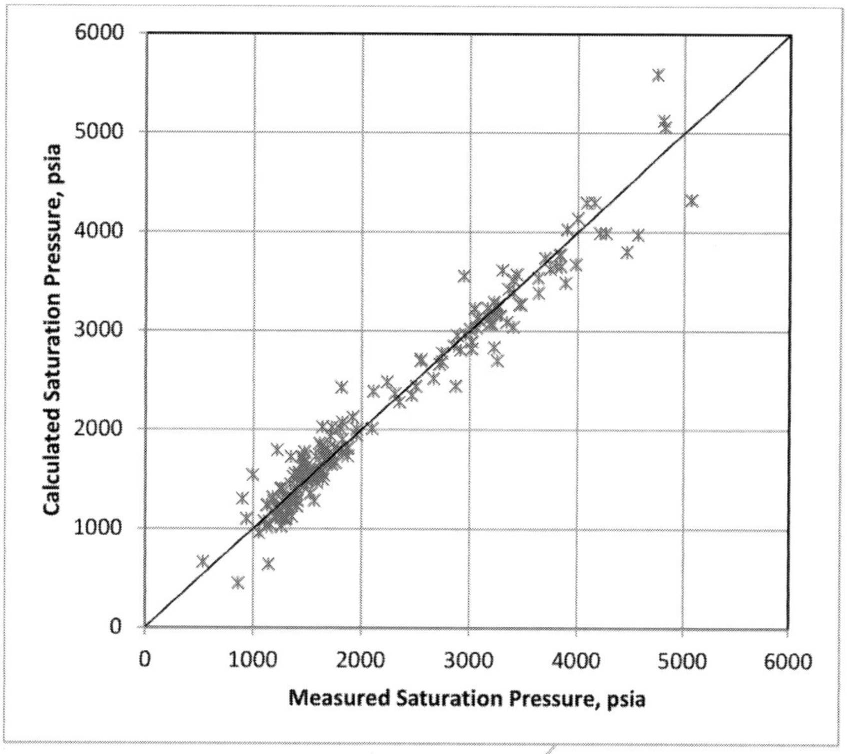

Figure 4A: Cross plot of the extended model using 179 training data

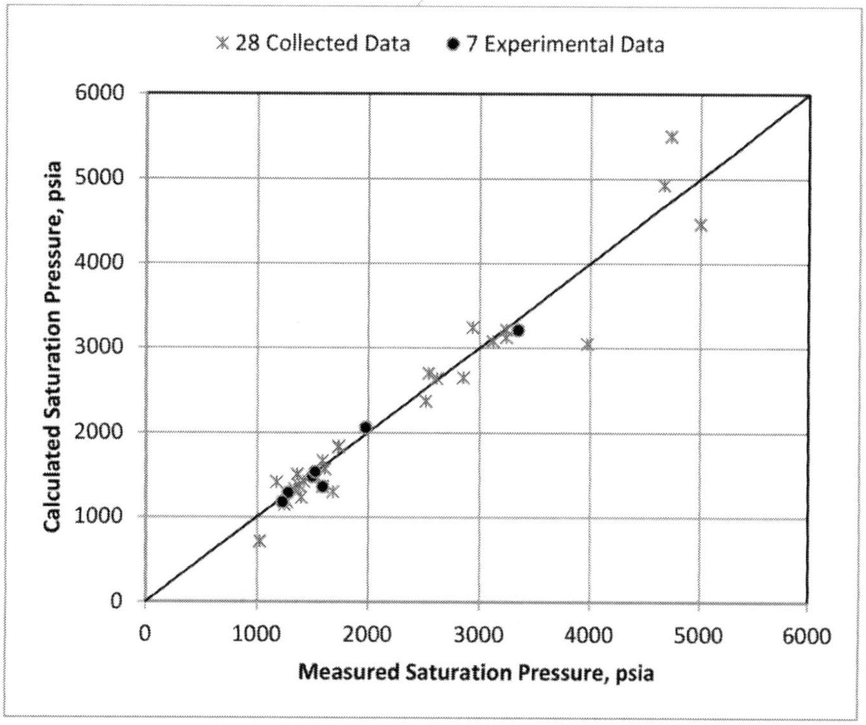

Figure 4B: Cross-plot of the extended model using 35-testing data

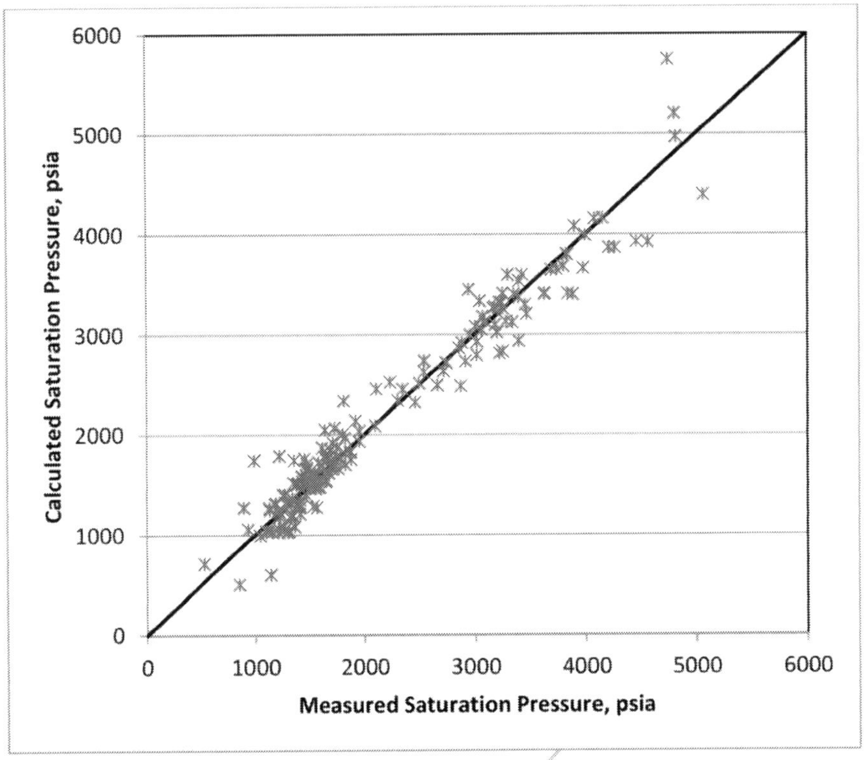

Figure 5A: Cross-plot of the lumped model using 179 training data

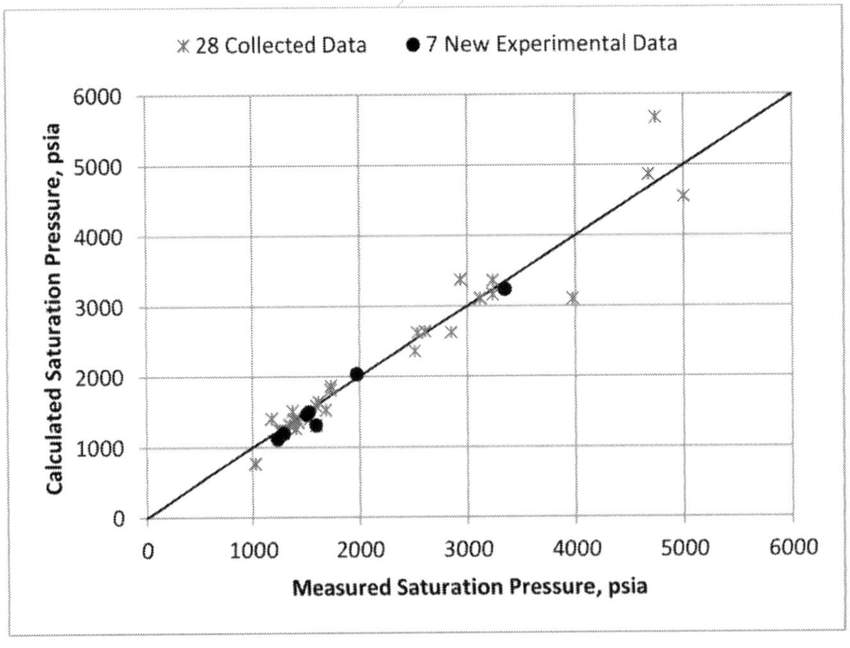

Figure 5B: Cross-plot of testing data for the lumped model

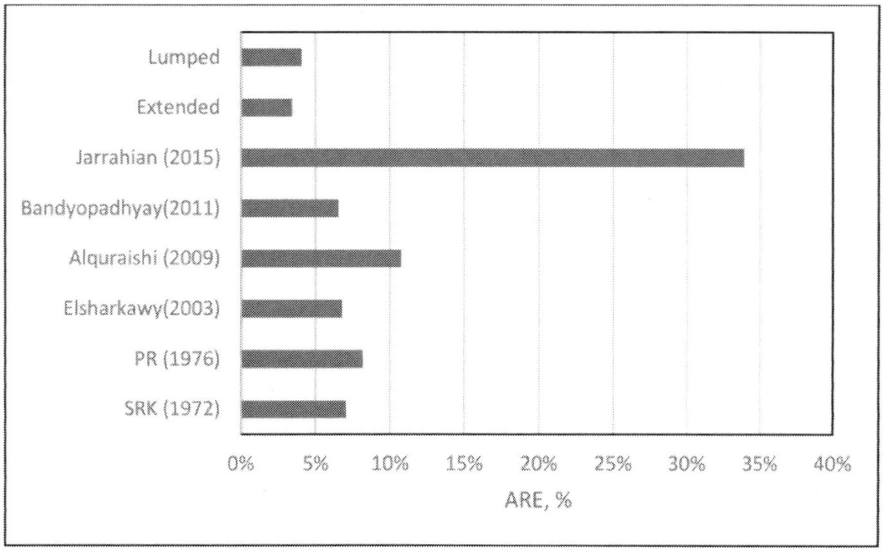

Figure 6A: Comparison of all the models' accuracy for the volatile oil samples

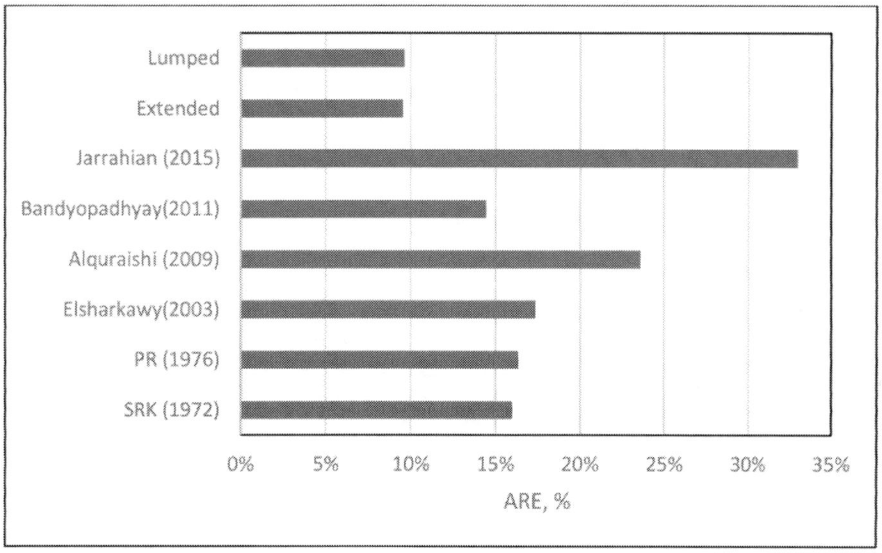

Figure 6B: Comparison of all the models' accuracy for the black oil samples

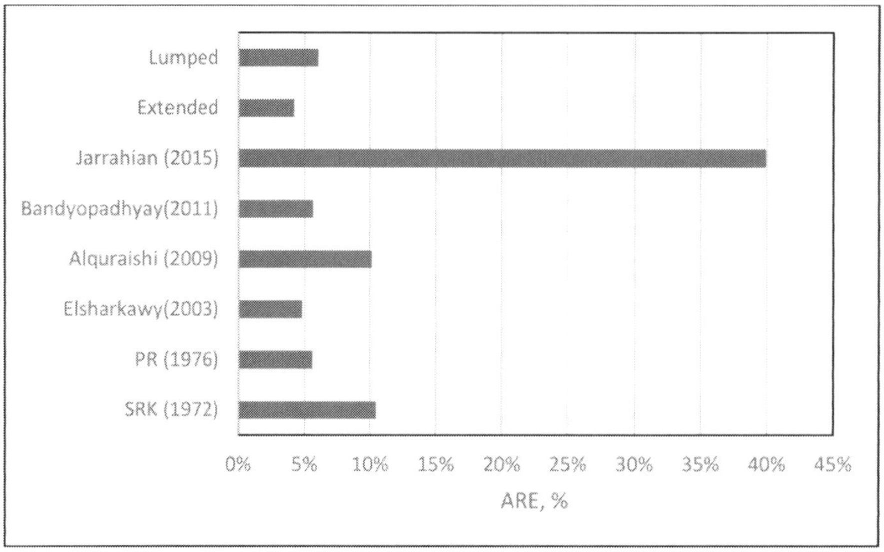

Figure 6C: Comparison of all the models' accuracy for the newly measured samples at KU

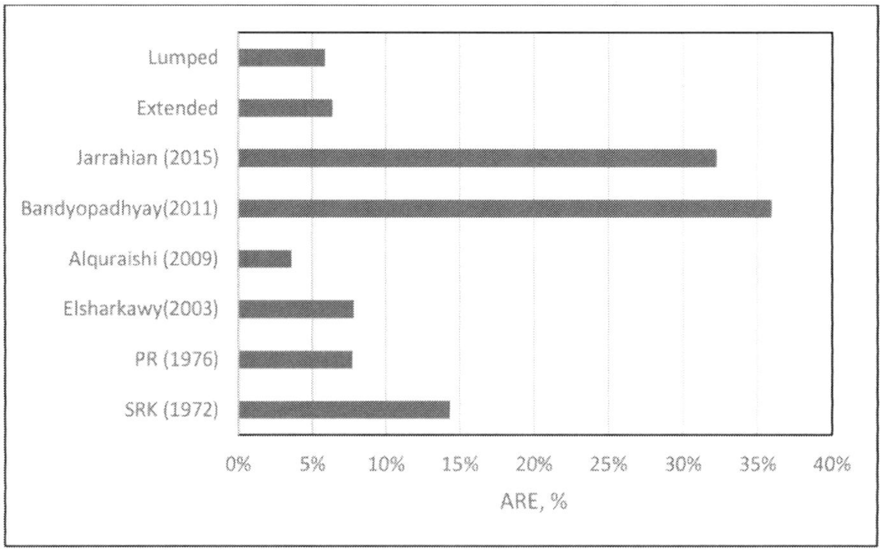

Figure 6D: Comparison of all the models' accuracy for low-temperature oil samples

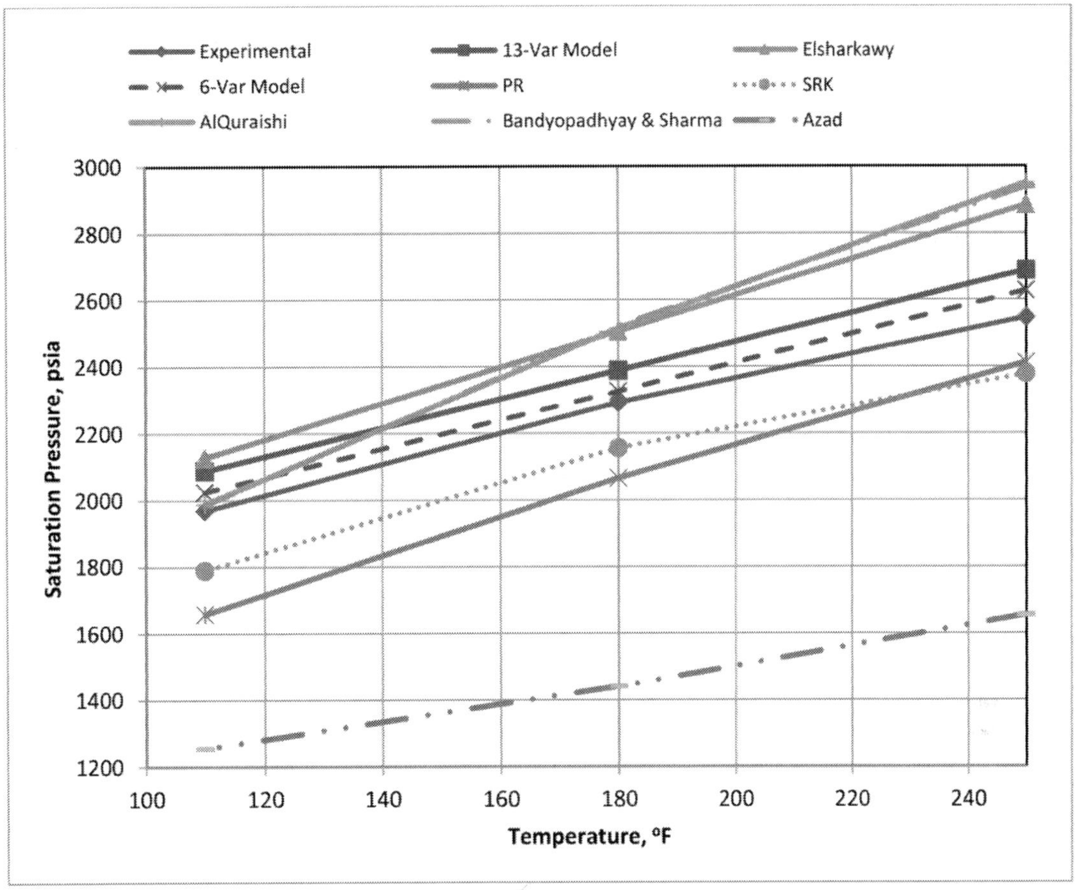

Figure 7: Changes in saturation pressure as a function of temperature for the different models (Sample #70)

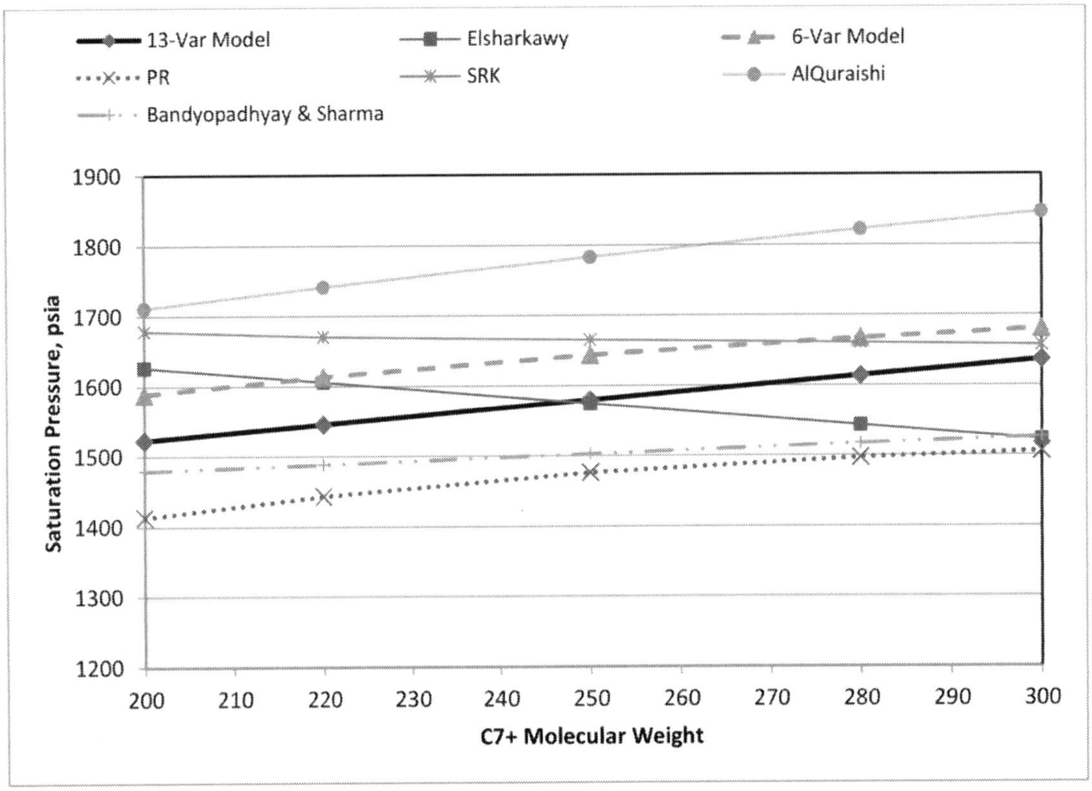

Figure 8: Comparing model performances: Changes in BPP as a function of C_{7+} molecular weight at 134°F (Sample #176)

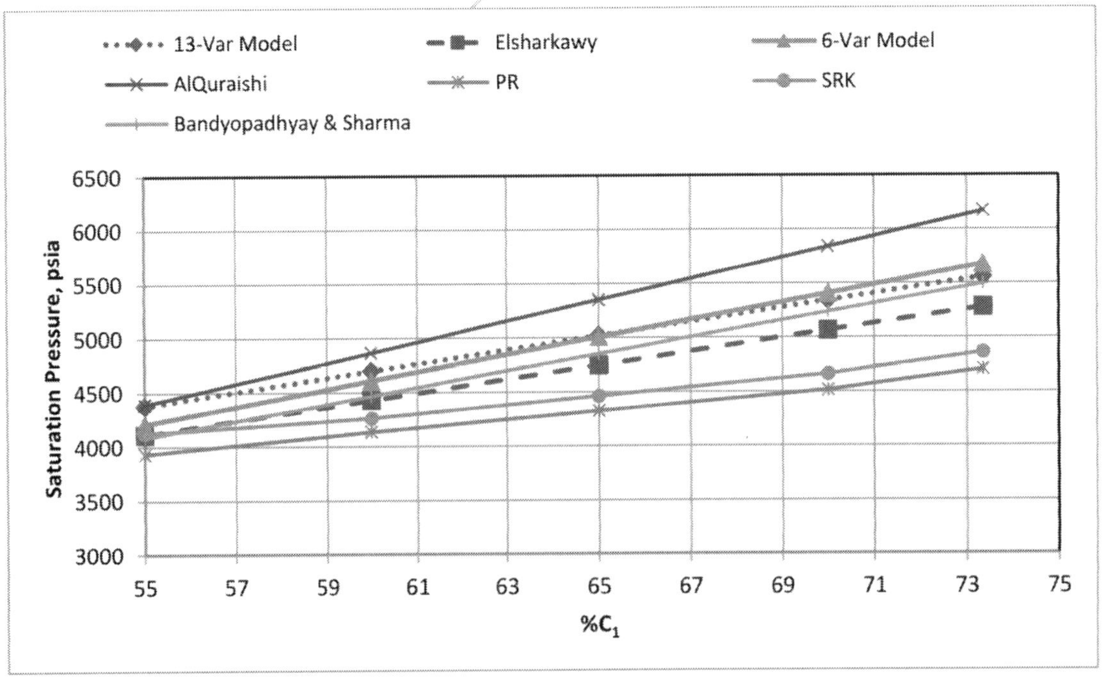

Figure 9: Sensitivity of saturation pressure to methane content at 212°F (sample #2)

439811 Purification of Ibuprofen Enantiomers By a New Technique Combining Distillation and Crystallization

Abstract:

Stripping crystallization (SC) is a new separation technique which combines distillation and crystallization. Basically, SC is operated at a triple-point condition, in which the liquid is simultaneously vaporized and crystallized due to the three-phase equilibrium. In essence, the process is continued until the liquid phase is completely eliminated and only the pure solid crystals remain in the feed. Thus, crystal washing is not required since only pure solid crystals remain in the feed and no impurities are adhered on the crystal surfaces at the end of SC. In the previous research, SC has been successfully applied to separate p-xylene from the mixed xylenes by Shiau et al. (2005, 2008, 2013).

Ibuprofen, also called 2-(isobutylphenyl)-propionic acid, is a non-steroidal anti-inflammatory drug widely used to treat headaches and minor pains. Although it is known that S-ibuprofen is responsible for the anti-inflammatory effects while R-ibuprofen might cause some long-term side effects, ibuprofen is still sold as a racemic mixture in the market due to the complexity involved in the separation of enantiomers. A thermodynamic model is developed in this work to simulate the three-phase equilibrium conditions during the SC operation. The experiments show that seeding with ultrasound mixing or magnetic stirring can be efficiently employed to purify S-ibuprofen from ibuprofen enantiomers. The experimental results, including the final enantiomeric purity and recovery ratio of S-ibuprofen, are consistent with the simulation results predicted by the model.

References:

1. Shiau, L. D.*, Wen, C. C. and Lin, B. S. "Separation and purification of p-xylene from the mixture of m-xylene and p-xylene by distillative freezing", *Ind. Eng. Chem. Res*, **44(7)**, p2258-2265, 2005.

2. Shiau, L. D.*, Wen, C. C. and Lin, B. S. "Separation of p-xylene from the multicomponent xylene system by stripping crystallization", *AIChE Journal*, **54(1)**, p337-342, 2008.

3. Shiau, L. D.*; Liu, K. F. "Investigations into the effects of the cooling rate on stripping crystallization", *Ind. Eng. Chem. Res*, **52**, p1716-1722, 2013.

441535 Hydrogen Disposal Flare System Hazard Identification and Mitigation Analysis

Abstract:

A hazard assessment on a hydrogen disposal flare system was conducted in preparation for a hydrogen turbopump test at the Air Force Research Laboratory (AFRL), Aerospace Systems Directorate at Edwards AFB, CA. The test operations required the disposal of large quantities of liquid and gaseous hydrogen through a commercial flare system to prevent the formation of an explosive environment. As part of the assessment, the primary steps in the test operation process were evaluated for the formation of a potential hazard by looking at each associated failure mode and the resulting affect on personnel and facilities. Evaluation of each operational step was conducted in a Process Failure Modes and Effects Analysis (PFMEA) that identified the probability and severity of the identified effects and associated hazards. If the probability and severity were unacceptable, mitigating steps were applied until the overall hazard risks were reduced to acceptable levels. The recommendations of this assessment were incorporated into both the flare system design and the operational procedures. This paper outlines the results of the PFMEA which include identification of the effects/hazards, available risk-mitigating techniques, and implementation of these techniques used for the mitigation of the identified hazards. Based upon the findings significant design modification to the vender's commercial equipment was needed to comply with recent changes in the code. Most of the identified hydrogen flare system hazards/effects were categorized into four groups: internal explosions, obstructions, external explosions, and maintenance degradation. Many corrective actions were implemented to mitigate both the probability and severity of the identified hazards. In addition, this paper discusses the installation of oxygen monitoring systems on the flare system, determination of appropriate system purge requirements, installation of purge system status alarms, incorporation of a triple redundant flare ignition system with status indicators, and a number of other hazard risk mitigation techniques. This comprehensive description of the hazards and mitigation of a hydrogen disposal system is a plant is vital part to safe operation of both hydrogen commercial and aerospace facilities.
<align="center">

442068 Enhanced Furfural Production By an Innovative Bottom Dividing Wall Column with a Decanter Configuration

Abstract:

Enhanced Furfural Production by an Innovative Bottom Dividing Wall Column with a Decanter Configuration

Nguyen Van Duc Long, Le Cao Nhien, Le Quang Minh and Moonyong Lee†

School of Chemical Engineering, Yeungnam University, Gyeongsan 712-749, South Korea

Furfural, which are used to derive other industrial chemicals, has been identified as one of the major bio-based platform chemicals that can compete with petroleum-based chemicals. However, the current commercial furfural process has low yield and requires intensive energy. This paper reports the simulation, detailed design and optimization of an industrial process recovering furfural from lignocellulosic biomass, a renewable and inexpensive feedstock. The aqueous feed mixture from the acidic hydrolysis of biomass was fed to the separation process to recover furfural as the main product, along with methanol as valuable byproducts. The purification process was intensified by applying an innovative bottom dividing wall column with a decanter configuration (BDWC-D). The response surface methodology (RSM), which allows the interactions between variables to be identified and quantified, was used to optimize the BDWC-D structure. The predictions by the RSM showed satisfactory agreement with the rigorous simulation results. The heat pump technology and heat integration were then employed to improve the purification process. The results showed that the proposed sequence could save the overall energy requirements and total annual cost significantly as compared to the conventional sequence. This sequence can be employed for both grass-root and retrofit designs.

Keywords: Lignocellulosic Biomass; Furfural; Energy Efficiency Improvement; Dividing Wall Column; Response Surface Methodology

Acknowledgments

The authors wish to acknowledge the financial support from the R&D Convergence Program of NST (National Research Council of Science & Technology) of Republic of Korea and KITECH (Korea Institute of Industrial Technology) (ES150001).

442557 Flow Field Simulation and Distribution Performance Analysis of a Gas Pipe Distributor in Fccu

Abstract:
A fluidized catalytic cracking (FCC) is an important petroleum refinery process, which, aided by highly selective zeolite catalysts, cracks large-molecule low-value heavy oil feedstock into small-molecule highvalue
products. When spend catalyst is fed into the regenerator for reactivating, it is fluidized by the gas distributor at the bottom of regenerator. The gas distributor is designed to uniformly distribute gas to maintain the fluidized state of the fluidized bed and to provide the reaction gas to guarantee the burning coke deposited on the spend catalyst. Therefore, the gas distribution performance of the gas distributor influences the quality of initial gas distribution directly. So the pipe distributor used in a industrial FCC unit, which is composed by main pipes and a number of branch pipes with many nozzle tubes, was investigated by means of flow field calculation. The gas distribution performance of the pipe distributor was analyzed based on the simulated flow field. The simulation result shows that there are a obvious nununiform
gas distribution from the different nozzle gas flow rate. The static pressure increases gradually along the flow direction in the nozzle tube while the axial flow rate decreases, which leads to the nozzle gas flow rate increasing gradually. Furthermore, an obvious bias flow occurs at the inlet of the nozzle tube, which influences the gas flowing into the nozzle. The pressure drops performance of nozzle and pressure varying in the branch pipe are analyzed comprehensively. Due to the nonuniform
gas distribution by the gas pipe distributor in the practical usage, the problems such as overlarge temperature difference among the fluidized bed layers in the regenerator have often happened, which will not only reduce burning coke efficiency and but also decrease activity of the regenerated catalyst. Meanwhile, the erosion of the distributor also occurs, and in consequence the function of the distributor fails, which even causes the nonscheduled shutdown of the FCC unit and leads to tremendous economic loss. The support is provided for optimizations of both structural parameters and operation parameters of the gas pipe distributor and improvement of the gas distribution performance. Furthermore, improvement suggestions were provided based on the above reasoning. The results are expected to be useful for the distributor design.

442925 Identifying Gas Liquid Contactor Performances for Optimized Natural Gas Desacidification Amines Units - from Process Calculations to Experimental Strategy

Abstract:

Amine units are widely used for gas sweetening. The absorber performances of these reactive absorption processes are typically mass transfer limited. The proper selection of gas liquid contactor is thus critical to optimize the size of towers, the solvent flowrate and the duty of the reboiler. This work proposes an approach to identify the most suitable packings for a given application.

The case-study selected here to illustrate the approach deals specifically with a CO_2removal application where the sour gas is at 7 % vol, and the sweet gas specification is 2% vol. The gas flowrate is fixed at 0,18 MMSCFD with a solvent flowrate fixed at the minimum of the thermodynamic constraints (via equilibrium curves). An in-house rate-based simulator with accessible forms of mass transfer correlations, based on a double film model, is used to calculate the absorber height and diameter which are needed to reach the sweet gas specifications. A reference structured packing is compared with standard valve trays.

The calculations show that the bed height required with the first contactor is considerably higher than with the second one while the column diameter is reduced. In this case the use of standard valve trays is more attractive for cost optimization. Interestingly, because mass transfer is accelerated, the absorption rate with the structured packing is found to be thermodynamically limited due to a more severe temperature increase which reduces the driving force in the bottom section of the absorber.

The study determined that the optimum design for the investigated case is a compromise between a full trayed tower, that generally imposes a larger diameter, and a packed tower where the higher mass transfer forces result in a temperature rise in the bottom of the tower which in turn imposes a height increase and a much higher solvent flowrate. . To determine the mass transfer characteristics of a gas-liquid contactor providing this optimum, a sensitivity analysis has been conducted. Mass transfer parameters, i.e., liquid and gas side mass transfer coefficients and interfacial area, were varied by a factor 2 relative to the ones obtained with the reference structured packing. Calculations show that one should use a contactor that reduces liquid phase resistance, specifically, the mass transfer conductance $k_L a_e$ should be increased by at least 50%. This target could be reduced to +30% if one could simultaneously reduce the gas phase conductance $k_G a_e$ by 20%. This result can be explained by the link between mass and heat transfer (Chilton-Colburn analogy) whereby reducing $k_G a_e$decreases the temperature level which in turn thermodynamically increases the mass transfer driving force.

The next step will be to review the literature more extensively to make a short list of pre-selected random and structured packings that best meet the targets. Experimental tests will then be run on IFPEN facilities to measure flooding limits and mass transfer parameters in a small 150 mm diameter column. This step is required to confirm packing's behavior, considering the disparity of the values reported by different authors in the literature. The last step is to consolidate the existing correlations for the most promising packing, and mitigate differences between various authors by performing characterization tests on larger diameter test columns. The final correlations will be implemented in our in-house rate based simulator for industrial use.

The present approach can be applied to determine the required mass transfer characteristics for optimizing other gas treatment applications. However, the conclusions reached here as regards to mass transfer targets and the problematic temperature levels, are valid only to the present case -study, and are not directly applicable to other gas sweetening applications (selectivity, LNG, ...). For example, in a case where H_2S and CO_2 are treated simultaneously the mass transfer limitations could be very different.

*** Formatting Issue - Best Available Paper/Graphic ***

443365 Technical Reconstruction of Double Effect Distillation for Methanol Process

Abstract:

Abstract: Methanol is one of important basis raw materials of chemical industry and also an excellent environment-friendly fuel, which make it as a significant commodity in the global market. Distillation is an essential unit operation for purified production of methanol. Product quality, production cost, environmental impact and energy consumption are largely related to distillation process.

Our research effort, under this circumstance, has been devoted since 2002 to energetically optimize 4-column methanol distillation system that is widely adopted in methanol units from coal, with reduced cycling water requirement. This patented process technology saw its industrial application in Chinese largest single unit of 1800Kt/a of Baofeng Group located at North China, with the lowest steam consumption of 0.9/1 t(steam)/t(methanol) compared with normally 25% higher for both simulative case study and on-site operation. Methanol product is in accordance with O-M-232G and the recycling water is reduced significantly from 67 t(CW)/t(methanol) to 52 t(CW)/t(methanol).

The flow diagram of this advanced double effect distillation process is showed in Fig.1. Its advantages arise for limitations found in current units: (1) heat exchange network failed to fully recover energy causing large CW flow to remove extra heat; (2) for pressured column, bubble point reflux need larger reflux ratio than the sub-cooled; (3) the heat of distillate from pressured column was normally ignored; (4) the low concentration of methanol of feed inlet to recovery column leads to higher reflux radio, responsible for unsatisfied unit energy efficiency.

Fig.1 The double effect distillation process.

(Column1.Light ends column; Column 2.Pressured column;

Column 3.Atmospheric column; Column4.Recovery column)

As a solution, a new flow diagram is built up and showed in Fig.2. (1) HEN totally reorganized (not completely show); (2) sub-cooled reflux to pressured column caused 10% reduction to the reflux ratio, economizing steam consumption; (3) top product of pressured column give heat to crude methanol before feeding light ends column; (4) change product and by-product position of recovery column to give efficient unit energy consumption.

Fig.2 The reconstruction of distillation process.

(Column1.Light ends column; Column 2.Pressured column;

Column 3.Atmospheric column; Column4.Recovery column)

It is estimated that the production of methanol will be over 60 million tons in 2014. If this reconstruction could be applied in the whole methanol industry, 1.3 million tons of standard coal would have been saved. This is equivalent to reduce 5 million tons emission of carbon dioxide, so the environmental impact can be largely decreased. Field operation shows reconstruction perfect substitution worth to spread to the whole methanol industry, energetically efficient and low environmentally impacting.

443095 Experimental and Numerical Study of Transport Phenomenon of Falling Film Flow with Different Heating Methods

Abstract:

Falling film flow is a common phenomenon in chemical industry. It is widely used in chemical instruments, such as structured packed columns and falling film evaporators. An experiment was implemented to study the mass transfer phenomenon of binary-mixture (ethanol-water) falling film flows with different heating methods (uniform or concentrated heating). In order to study the detailed flow and mass transfer behaviors, a 2-D multiphase computational fluid dynamics (CFD) model based on volume of fluid (VOF) was developed. This model divides mass transfer into two parts: one is caused by the vapor-liquid equilibrium (VLE) and the other is caused by the heat flux. Results from simulation are inconsistent with the experimental data. Both experiments and simulations show that the ethanol concentration of the vapor with concentrated heating is 1.5% higher than the one with uniform heating.

443101 Thermodynamic Modeling of HNO_3-H_2so_4-H_2o Ternary System with Symmetric Electrolyte NRTL Model

Abstract:

Nitric acid concentration/sulfuric acid concentration (NAC/SAC) process has been widely used for concentrating dilute aqueous nitric acid and recovering spent sulfuric acid. Nitric acid is concentrated by employing sulfuric acid to bind water and break nitric acid-water azeotrope at approximately 68 wt. % nitric acid. To support heat and mass balance calculations and process simulation for NAC/SAC processes, we develop a comprehensive thermodynamic model for nitric acid-sulfuric acid-water ternary system based on previously published thermodynamic models of nitric acid-water [Wang et al., Fluid Phase Equilibria, (2015) (DOI: 10.1016/j.fluid.2015.04.015)] binary and sulfuric acid-water [Que et al., Journal of Chemical & Engineering Data, 56 (2011) 963-9772] binary systems developed with symmetric electrolyte NRTL (eNRTL) activity coefficient model. The present work takes into account one-step hydration of nitric acid molecules, two-step hydration of sulfuric acid molecules, and further hydration of hydronium ions. With up to three temperature coefficients for each eNRTL binary interaction parameter and up to two temperature coefficients for each chemical equilibrium constant, the ternary model is validated with boiling point data from 383.15 to 471.45 K and activities at 273 K. Furthermore, contour plots of boiling points, vapor phase composition, and specific heat capacity of the ternary system are generated for engineering use.

Keywords:

Nitric acid, Sulfuric acid, Electrolyte NRTL model, Vapor-liquid equilibrium, Boiling point, Heat capacity, Nitric acid concentration/sulfuric acid concentration (NAC/SAC)

443503 A Computational Analysis of Fire Scene for Safety in a Compartment

Abstract:

This article details numerical prediction of fire spread carried out to study the distribution of carbon monoxide concentration, smoke dispersion, soot concentration and temperature in laboratory compartment, using a commercial fire modeling software tool. The heat release rate curve predicted from the simulation was compared with literature values carried out in a road tunnel. The time variations of carbon monoxide transportation, smoke concentration and temperature spread in compartment room are reported to evaluate the effect on humans. The soot concentration was described and used throughout this study as an indicator of safety in the compartment room. The effects of vent dimension and ventilation capacity were analyzed by smoke detector response times. This work provides improvements derived from reinterpretation of actual fire scene considering the ability of mechanical exhaust system and ventilation in compartment room to mitigate the hazard of fire.

443640 Preparation and Characterization of (*FURCRAEA spp*) as Green Support of Iron Nanostructured Catalyst

Abstract:

PREPARATION AND CHARACTERIZATION OF (*FURCRAEA spp*) AS "GREEN" SUPPORT OF IRON NANOSTRUCTURED CATALYST

Karen G. Bastidas G[1], Anamaria Barrera B[1], Cesar A. Sierra A[2], Hugo R. Zea R[3]

[1] Department of Chemical and Environmental Engineering. National University of Colombia (Bogota).

[2] Associate professor. Director of the Research Group of Macromolecules. National University of Colombia (Bogota).

[3] Associate professor. Director of the Research Group of Materials, Catalysis and Environment. National University of Colombia (Bogotá).

Abstract.

The availability of natural water resources has declined over time, becoming one of the biggest problems today. Environmental risk in Colombia's rivers has increase due to the widespread use of harmful chemicals in a variety of processes, ranging from illegal mineral extraction to industry wastewater, e.g. Medellin's river, due to its strategic location along a textile industrial corridor might be subject to dye contamination; the gold and platinum mining zones in Choco and Antioquia regions are being heavily affected by the use of mercury in the extraction processes. In 2009, 520 tons of mercury were imported by miners working in this area, amount that exceeds several times the mercury used by regulated industry in previous years (Ministry of Mines and Energy of Colombia, 2007). Not only the environmental risk is a concern but also the economic conditions of many low-income families whose livelihood in many cases depends exclusively of the of the mining.

Due to the issues raised above, the main goal of this study is to research, develop and implement innovative technologies in the synthesis of highly efficient, low cost and environmentally friendly iron based catalysts capable of degradate pollutants, ensuring operational stability. A native fiber (Fique, FURCRAEA spp) has been selected as a solid support for these catalysts due to its abundance, low cost and special physical and chemical characteristics.

The combination of cellulosic fibers with iron oxide nanoparticles could provide exceptional biodegradable composite materials for the treatment of organic dyes and heavy metals present in wastewater. Likewise, the use of iron particles immobilized on fique fibers allows an easy separation catalyst-treated water, turning the biocatalyst in highly recyclable and reusable material.

Fique fiber was initially characterized by determining the percentage of ashes, total lignin, soluble and insoluble lignin, cellulose, hemicellulose, total solids and humidity, obtaining 0.94, 22.58, 0.5, 22.8, 36.3, 27.5 0.701 and 2.1%, respectively. Additionally, the isoelectric point (pzc), acidic and basic sites were measured.

Then, the fiber was subjected to a pretreatment in order to expose more active sites which on time serve as nucleation points for the deposition of nanoparticles. This was done by means of an ultrasound-assisted procedure, varying the sonication time (30, 60 and 90 minutes). The effect of the sonication was followed by measuring the contents of lignin, cellulose and hemicellulose within the fiber after each time. 60 minutes of sonication was found to be most beneficial time since the amount of lignin and hemicellulose present in the fiber was reduced, while conversely the amount of crystalline cellulose exposed increases.

In order to improve the chemical ability of the nucleation sites to interact with iron and then to form stable and size controllable nanoparticles, a cationization procedure was performed on the fiber; this was achieved by the exposure of the fiber to acidic and basic conditions promoting cationization. Three cationization times (1, 2 and 3 hours) were studied and their effect on the fiber chemical surface was evaluated using FTIR-ATR and XRD analysis. The results showed 3 hours as the optimal cationizing time.

Subsequent to the cationization of the fiber surface, a procedure of ionic/cationic exchange was implemented by exposing the fiber to a solution of iron salts in order to replace the cations in the fiber surface and anchoring the iron nanoparticles. Different conditions were studied in order to synthetize zero valent iron nanoparticles (nZVI) and others with different oxidation states. Finally, to track the impregnation process on the fiber, samples from the precursor solution were taken at different times and its iron concentration was determined by atomic absorption spectrophotometry (AAS). Complementary characterization performed by X-ray diffraction, BET surface area and electronic microscopy, among others.

Key words: Natural support (fique), iron nanoparticles, bionanocatalyst.

443764 Optimization of Heat Exchanger Network in Steam Assisted Gravity Drainage Oil Sands Surface Facilities

Abstract:

Steam assisted gravity drainage (SAGD) is the most viable and environmentally safe recovery technology for extracting heavy oil and bitumen. It is a thermal in-situ production process where parallel wells are drilled horizontally into an underground bitumen reservoir. Steam is produced at the central processing facility (CPF) and injected into the reservoir through the wells. The steam heats the bitumen to a point where gravity allows it to flow down to the lower well where the mixture of bitumen and water is then pumped to the surface. The water and bitumen are separated at the CPF. Typically, more than 90% of the energy consumed in producing a barrel of bitumen using SAGD is used to generate the steam that is injected into the bitumen reservoir. Fossil fuels is employed in the process, which causes GHG (greenhouse gas) emissions. The usage of hot and cold utility is reduced by increased heat exchange and optimization of the heat exchanger network, which results in reduction of GHG emissions. In this work, pinch analysis is performed to minimize energy consumption.

443774 Thermodynamic Anaysis of Dehydration of n-Propanol Using Heterogeneous Azeotropic Distillation

Abstract:

Heterogeneous azeotropic distillation is a widely popular process in chemical process industries for the separation of binary azeotropic mixtures into their constituent pure components. It involves the addition of an entrainer which is partially miscible and forms a heterogeneous azeotrope preferably with one of the original components. This is a powerful technique which combines several physical phenomena such as presence of azeotropes and liquid-liquid immiscibilities, and is used widely in industries to enhance difficult or intractable distillation separations. For the design of heterogeneous azeotropic distillation columns, a detailed study of the heterogeneous region is crucial to plan the distillation sequence, which requires experimental vapor-liquid-liquid equilibrium (VLLE) data.

The present study focuses on the investigation of the heterogeneous region of the well known industrially available azeotropic mixture n-Propanol-Water system using Propyl acetate as entrainer. The VLE data of the system was generated by means of modified Othmer still and the still is modified using a probe type ultrasonic homogenizer to account for the heterogeneity. Isobaric vapor- liquid- equilibria and vapor - liquid - liquid - equilibria data at atmospheric pressure have been obtained for the ternary system n-Propanol-water-propyl acetate and the constituent binary systems. The equilibrium data has been correlated using NRTL, UNIQUAC and UNIFAC models and compared with the experimental data.

443776 Study of Ionic Liquids As Entrainers for the Extractive Distillation of *N*-Propanol and *N*-Propyl Acetate Binary Azeotropic Mixture and the Isobaric VLE Studies of the System

Abstract:

Distillation is a widely accepted technique for the separation of binary liquid mixtures. Separation by simple distillation is not possible for binary mixtures that form azeotropes. One of the commonly used methods to separate azeotropes is extractive distillation. It is the process of adding an entrainer that can alter the relative volatility of the mixture and thereby breaking the azeotropic point. Ionic liquids on account of their physiochemical properties as well as the recyclability have been used as a preferred replacement for the conventional entrainers. Extractive distillation with ionic liquids has the advantages of both liquid solvent (easy operation) and solid salt(high separation ability). The corrosion and dissolution problems by using inorganic salts can be rectified by these non-volatile, chemically stable, high electrical conductive molten salts.

The azeotropic system under consideration in the present study is *n*-propanol - *n*-propyl acetate mixture. These compounds which constitute the products/reactants of esterification reaction of n-propanol form a minimum boiling azeotrope at 94.7°C. *n*-propyl acetate has been included in a list of volatile acetates as an ecologically benign CFC-free solvent for aerosol sprays.

Vapour liquid equilibrium(VLE) measurements of *n*-propanol and *n*-propyl acetate are performed at atmospheric pressure in a modified Othmer still which is a continuous distillation still in which vapour phase is recirculated. The most suitable ionic liquids for shifting of azeotropic composition of the system are selected using COSMO-RS model with suitable combinations of anions and cations. The isobaric VLE of the system with [EMIM][HSO_4] is obtained and compared with the existing data. The influence of alkyl chain length on the performance of ionic liquids is also analysed with regard to the previous studies. The effect of polarity is the most important factor on the behaviour of ionic liquid solvation in alcohols. Experimental VLE data were correlated using modified UNIFAC model.

443784 Measurement and Correlation of the Mutual Solubility of Diisopropylamine + Water and Triethylamine + Water Systems at High Pressure

Abstract:

The mutual solubility data of diisopropylamine + water and triethylamine + water systems were measured at the pressure up to 100 MPa and the temperature range was 280.2~316.2 K. Experiments were performed in a high-pressure autoclave with a moveable piston. The experimental data were correlated satisfactorily with NRTL activity coefficient model and the binary parameters were obtained.

443819 Improving High Shear Rate Friction Reduction Behavior of Polyacrylamide-Based Fracturing Fluids Using Xanthan Gum

Abstract:

The main component of slickwater in fracturing low permeability formations is a high molar mass polymer such as polyacrylamide (PAM), which acts as a drag reduction agent in turbulent flow. Polyacrylamide does not degrade in the environment and also suffers from mechanical degradation at high turbulent flow rates. Five different concentrations of PAM as well as mixtures of polyacrylamide/xanthan were prepared to investigate the possibility of improving shear stability of PAM. Drag reduction measurements were performed in a flow loop and friction factors at different flow rates were calculated using rheological parameters. Based on the observations, Xanthan (XG) is not as good a friction reducer as PAM, even at three-fold higher concentration. XG slightly improved the shear stability in mixed polymer solutions at low concentrations. But at high concentrations, XG significantly improves the friction reduction behavior of PAM at nonionic and ionic environments.

443836 Oxidative Dehydrogenation of n-Butane on Carbon Catalysts

Abstract:

Butenes and butadiene, which are one of the important industrial chemical products to produce synthetic rubber , are sythesized traditionally by oxidative dehydrogenation (ODH) of n-butane over various mixed metal oxides. Metal-free carbon nanomaterials has recently received much attention by researchers as a catalyst for the ODH reaction due to its environmental acceptability, corrosion resistance, and unique surface properties. For the reason, the reaction system was modelled in this study for the sythesis of butene and butadiene. For the modelling of the reaction system, the catalytic performance of several carbon catalysts, such as graphite and activated carbon, were initially tested. Then, the reaction mechanisms and kinetics were developed based on experimental results. Lastly, the mathematical modeling of the reaction system was optimized to maximize an overall yield of the main products. This approach allowed us to develop the optimized reaction system more efficiently.

This research was respectfully supported by Engineering Development Research Center (EDRC) funded by the Ministry of Trade, Industry & Energy (MOTIE). (No.N0000990)

443855 Modeling the Thermal Behaviors of an Ultracapacitor Module

Abstract:

Ultracapacitors, also known as supercapacitors, have the potential to meet the increasing power requirements of the sustainable energy systems for the grid-scale energy storage systems and electrified vehicles. As compared to batteries, ultracapacitors offer a higher power density, higher efficiency, and longer shelf and cycle life. Because the performance of an ultracapacitor depends on temperature, it is important to calculate accurately the thermal behavior of a single ultracapacitor cell and the module composed of multiple ultracapacitor cells for the efficient and reliable systems integration of an ultracapacitor in grid and transport applications.

In this work, modeling is performed to study the effect of the layout configugation of the ultracapacitor cells on the thermal behaviors of an ultracapacitor module. The ultracapacitor module is subject to the charge/discharge cycling with constant-current between the nominal voltage of the ultracapacitor module and the half of this voltage. The charge/discharge current values examined are 50, 100, 150, and 200 A. The validation of the modeling approach is provided through the comparison of the modeling results with the experimental measurements.

443856 Modeling the Electrolyte Flow Behaviors of a Zinc/Bromine Flow Battery Module

Abstract:

The zinc/bromine (Zn/Br_2) flow battery is a promising rechargeable power source for energy storage systems because of its inherent chemical simplicity, high degree of reversibility at the electrodes, good energy density, and abundant low-cost materials. The Zn/Br_2 flow battery system, intended for the grid-scale energy storage system, is to be made in modular form. A Zn/Br_2 flow battery module is composed of the multiple stacks consisting of bipolar carbon electrodes. Since the stacks require electrolyte circulation on both the negative and positive sides, two plumbing loops are provided. The flow loops contain an electrolyte storage tank and a circulation pump. A heat exchanger may be included for temperature control in the flow loop. The design of the electrolyte circulation plumbing determines the distributions of electrolyte flow and pressure within the stacks and the efficiency of the stack is dependent on the flow distribution throughout the stack. The modeling of the distributions of electrolyte flow and pressure within the module is needed to optimize the plumbing for electrolyte circulation and thereby to improve the performance of the module. In this work, modeling is performed to study the effect of the layout configuration of the plumbing for electrolyte circulation on the distributions of electrolyte flow and pressure within a 50 kWh Zn/Br_2 flow battery module composed of four stacks. It is seen that the plumbing layout for electrolyte circulation can have a significant effect on the flow rate distribution of electrolyte through the stacks.

443837 Blowdown Dynamic Modeling and Analysis of Pressure and Temperature

Abstract:

Vessel blowdown is one of the safety operation procedures that discharges flammable gas or liquid to flare system during emergency situation such as fire. When blowdown occurs, the liquefaction and freezing effect cause a pressurized vessel to be brittle, exposing it to possible hazard situation. Also the blowdown systems are recommended to be accordance with API(American Petroleum Institute) 521 code. API 521 requires pressure in vessel should be half of the initial pressure or be lower than 100 psig in 15 minutes after commencing blowdown in fire case. Therefore, accurate estimation over the change of temperature and pressure of the vessel during blowdown period is crucial for the design of blowdown system. For the reason, researchers have developed a few dynamic blowdown models so far. However, the model was found to be rather inaccurate under certain conditions. Therefore a new dynamic blowdown model was developed in this work. For the accurate analysis, the model algorithm considers both vapour and liquid phase in non-equilibrium condition that the temperatures of vapour and liquid phases are different. Futhermore, heat transfer from the vessel wall to discharge gas at either laminal or turbulent flow in the vessel was considered in the model to increase the accuracy of blowdown estimation. For thermodynamics, both Peng-Robinson and Soave-Redlich-Kwong equations of state were applied to the model, and the results were compared each other. Finally, the simulation results were compared with Haque et al's experimental and steady-state simulation results.

443974 Observer-Based Model Predictive Control for Time-Delay Systems

Abstract:

Model predictive control is one of the most widely used advanced process control technologies in industry.[1] It has many capabilities such as dealing with input/state constrains and uncertainties. Observer-based model predictive control was developed to estimate states information when the states are not measurable directly. Many chemical processes have time delays in either states or measurements, such as polymer production and some bioprocesses. [2] In this presentation, we are proposing an unknown input observer based model predictive control algorithm for time-delay systems. Unknown input observer is an observer design to eliminate certain types of unknown inputs in the system. Stability of unknown input observer-based model predictive control without time delay was provided by our previous study. Terminal cost and terminal region are designed for a class of non-linear time delay systems. The close-loop stability of the controller is provided. [3][4]

An example of bioreactor is provided which has uncertain time delays in measurements but with a known upper bound. Two measurements are provided which are online biomass and offline ethanol measurements. Model-plant mismatch is considered as unknown inputs in the system. State estimation and control are calculated using the observer and an optimization algorithm.

Reference:

[1] Marcus Reble et al., Model predictive control of constrained non-linear time-delay systems. IMA Journal of Mathematical Control and Information, 2010, 1-19

[2] Seung Cheol Jeong and PooGyeon Park, Constrained MPC algorithm for uncertainty time-varying systems with state delay, IEEE TRANSACTIONS ON AUTOMATIC CONTROL, 50, 2005, 257-262

[3] Mayuresh Kothare et al., Robust constrained model predictive control using linear matrix inequalities. Automatica, 32, 1996, 1361-1379

[4] Wook Hyun Kwon et al., A simple receding horizon control for state delayed systems and its stability criterion, Journal of Process Control, 13, 2003, 539-551

444020 Thermodynamic Modeling of Aqueous Lithium Sulfate and Its Solubility in the Presence of Na+, K+ and Mg2+ Sulfates

Abstract:

This work is an extension of the previous works done by our group towards development of a rigorous and comprehensive thermodynamic model for high salinity produced water that is essential for mass and heat balance calculations and simulation and design of novel treatment processes. Specifically this work presents a thermodynamic model that accurately represents all thermodynamic properties of aqueous lithium sulfate solution including the solubility of Li2SO4 in water and also in the presence of Na+–K+–Mg2+ sulfates. By explicit consideration of solution chemistry of each subsystem, symmetric electrolyte NRTL model is used to correlate available experimental data of all pertinent thermodynamic properties including activity coefficient, osmotic coefficient, excess enthalpy, and liquid heat capacity over the entire concentration range. This facilitates quantification of eNRTL binary molecule–molecule, molecule–electrolyte, and electrolyte–electrolyte interaction parameters. Also, solubility data are used to regress the thermodynamic constants required for modeling the liquid–solid equilibrium of lithium sulfate in water and in the presence of Na+–K+–Mg2+ sulfates. Displaying a significant prediction ability, the model accurately represents the thermodynamic behavior of the system from infinite dilution up to saturation and temperatures from 298.15 K to 573.15 K.

444044 Organic Solar Cells

Abstract:

This paper describes highlights the need of next-generation "organic solar cells". There is a continuing need for organic solar cells and arrays with very high specific power densities (1000-5000 kW/kg) for generating power in a new generation of ultra light space payloads example in NASA. An emerging technology with promise to meet these ambitious goals are organic solar cells based on very thin films of discrete layers. Theoretical efficiencies for multi junction solar cells are more ~70% as compared to ~30% for single junction devices. There is continuous development going on high mobility, band gap engineered organic donors and acceptors, as well as development of three junction devices and scaled up devices on lightweight flexible polymer substrates. The main interest in this paper is organic semi-conducting crystal: Pentacene, which has photo-conducting states. Organic solar cells manufactured today have low efficiency other than their stability issues. The efficiency of organic solar cells can be improved by either designing polymer materials that absorbs more of solar spectrum or by minimizing the charge collection efficiency by controlling the morphology of polymer films. Once fully developed, organic solar cells would revolutionary power the industry due to their flexibility and minimal weight

444055 A Refined Non-Random Two-Liquid Segment Activity Coefficient Model for Solubility Modeling

Abstract:

Accurate correlation and robust prediction for complex molecule solubility or solid-liquid equilibrium (SLE) in solvents and solvent mixtures have drawn much interest recently [Tanveer et al., *Chem. Eng. Prog.* **2014**, 110 (9), 37-47]. Solubility models such as the Non-Random Two Liquid Segment Activity Coefficient (NRTL-SAC) model [Chen and Song, *Ind. Eng. Chem. Res.* **2004**, 43, 8354-8362] are now routinely used in the industries. However, an important unfilled need is to model the solubility in oligomers/polymers/surfactants such as asphaltene nanoaggregates in petroleum fluids and drug molecules in lipid-based or solid dispersion formulation [Tung, *Org. Proc. Res. Dev.* **2013**, 17, 445-454]. This paper presents a refinement to the very successful NRTL-SAC model to better account for the contribution due to size differences of molecules. Specifically, the Flory-Huggins combinatorial entropy expression for conceptual segments in the model is substituted with the Staverman-Guggenheim expression for real molecules. We present the rationale for the refined model, the updated conceptual segment numbers for common solvents, and the newly identified conceptual segment numbers for common drug excipients. The model performance is illustrated by prediction of drug molecule solubility in various lipid solvents and solvent mixtures, compared to experimental data and previous work by other models. The refined NRTL-SAC model shows enhanced capability to correlate and predict drug molecule solubility in common solvents and excipients used in drug formulation.

444057 Explaining the Paradox of Asphaltene Precipitation with n-Alkanes

Abstract:

Explaining the paradox of asphaltene precipitation with n-alkanes
Md Rashedul Islam and Chau-Chyun Chen
Department of Chemical Engineering, Texas Tech University
Abstract
Paraffins, specifically alkanes, promotes asphaltene precipitation when they are blended with asphaltene-containing petroleum fluids. The alkane/oil volume ratio required for onset of asphaltene precipitation increases with the carbon number and yields a maxima. Experimental study indicates that the maxima lays at C7[1] or between C8 -- C10 [2]. However, precipitation yield decreases asymptotically with increase of the carbon number even if excessively high alkane/oil volume ratio is used [3]. For example, n-pentane be a better solvent for asphaltene than n-hexane in regard of both onset of precipitation and precipitation yield. However, high carbon number alkane beyond the maxima, n-hexadecane for instance, can be a better solvent at onset of precipitation point and a poorer solvent in regard of precipitation yield than n-pentadecane. A thermodynamic explanation for this paradox is necessarily important to find the compatibility of bitumen or crude oil with paraffinic hydrocarbons. In this work, this paradox is explained by the aggregation thermodynamics of asphaltene precipitation [4]. Both the enthalpy and entropy of mixing are investigated to explain the effect of paraffin carbon number on asphaltene precipitation. Athabasca and cold lake bitumen samples are studied.
References
[1] G. Hotier and M. Robin, "Action de divers diluants sur les produits pÃ©troliers lourds: mesure, interprÃ©tation et prÃ©vision de la floculation des asphaltènes," Oil & Gas Science and Technology, vol. 38, pp. 101-120, 1983.
[2] I. A. Wiehe, H. W. Yarranton, K. Akbarzadeh, P. M. Rahimi, and A. Teclemariam, "The paradox of asphaltene precipitation with normal paraffins," Energy & fuels, vol. 19, pp. 1261-1267, 2005.
[3] D. L. Mitchell and J. G. Speight, "The solubility of asphaltenes in hydrocarbon solvents," Fuel, vol. 52, pp. 149-152, 1973.
[4] M. Wang, Y. Hao, M. R. Islam, and C. C. Chen, "Aggregation Thermodynamics for Asphaltene Precipitation," AIChE Journal, Submitted.

444086 Optimization of Layered Graphene Oxide Nanostructures through Controlling the Membrane Deposition Rate Towards Size-Sieving Separation

Abstract:

Single-layer graphene oxide (GO) flakes were coated on anodized aluminum oxide (AAO) substrate through a simple vacuum filtration process, the effect of the deposition rate on the nanostructures of obtained membrane was investigated though pure water flux and organic vapor diffusion, which indicated GO membrane assembled with slow deposition rate might have more ordered lamellar structure and, correspondingly, have smaller effective pore size. In contrast, under fast deposition rate, GO flakes assembled in a more random way. Therefore, through controlling the membrane deposition rate, the nanostructure of layered GO membrane could be optimized for size dependent separation. We found after thermal reduction, 100 nm GO membrane (prepared @ 0.02 nm/s) has effective pore size smaller than 6.1 Å, which exhibit great potential in n-hexane/2,2-dimethybutane (DMB) separation.

444142 Robust Design of a Radiant Wall Burner for a High Temperature Environment Utilizing Computational Fluid Dynamics Simulations

Abstract:

Radiant Wall Burners are often utilized in Ethylene Cracking Furnaces. The conditions and fuel gases for a specific unit can vary considerably within the unit and across similar units. With burner air preheated to high temperatures, premixing the fuel can result in dangerous and destructive flashback, while complete nozzle-mixing of the fuel can result in burner instability at startup or turn-down conditions. These challenges can be solved through the generation of a robust radiant wall burner design, which is facilitated by executing Computational Fluid Dynamics simulations of many different geometric configurations.

444161 Novel Microwave Synthesis Carbon Dots from PINE Bark and Its Application for Heavy Metal Sensing

Monday, April 11, 2016
Exhibit Hall E (George R. Brown)
Saheed Sanni, Chemistry, Vaal University of Technology, Vanderbijlpark, South Africa, Puthalapattu ReddyPrasad, Chemistry, Vaal University of Technology, Vanderbiljpark, South Africa and Augustine E. Ofomaja, Chemistry, Vaal university of Technology, Vanderbiljpark, WV, South Africa

NOVEL MICROWAVE SYNTHESIS CARBON DOTS FROM PINE BARK AND ITS APPLICATION FOR HEAVY METAL SENSING

Sanni S.O., ReddyPrasad P., Ofomaja A.E.

Biosorption and Waste water Treatment Research Laboratory, *Department of Chemistry, Faculty of Applied and Computer Sciences, Vaal University of Technology, P. Bag X021, Vanderbijlpark-1900, South Africa.*

INTRODUCTION

Microwave pyrolysis for the production of carbon has the advantages of being cheap, produces high yield of carbon and is absolutely a green synthetic method [1]. In recent times, the use of microwave pyrolysis have been extended to the preparation of carbon dots. Carbon nanodots (C-dots) falls under the class of carbon nanoparticles with size below 10 nm[2]. C-dots are carbon materials which processes high water solubility and strong fluorescence depending on their size, edge, shape, surface ligands, and defects. Since C-dots preparation can be achieved from difference carbonaceous sources and by varying a range of parameters during their synthesis, the properties of the final product may vary depending on these factors. C-dots have sp^2 characteristics and possess several functional groups on their surfaces that makes them highly hydrophilic and easily functionalized with various organic, polymeric, inorganic, or biological spices[3]. Several synthesis methods that have been explored includes electrochemical synthesis, combustion/thermal/hydrothermal/acidic oxidation, supported synthesis, microwave/ultrasonic, arc discharge, laser ablation/passivation, and plasma treatment[4]. C-dots processes excellent such as water dispersibility, chemical stability and photostability, ease of modification, excitation-dependent multicolor emission, low toxicity, and good cell permeability, great efforts have been paid to their potential applications in biosensing, bioimaging, drug delivery, and other biological related aspects.

METHODS

The carbon dots (C-dots) with green luminescence was synthesized from microwave assisted pyrolysis of pine bark waste our carbon source at 800, 1000 1200 W for 1 Hr. The dots produced were characterized using FTIR, TGA, XRD, UV-Vis and Raman spectroscopy. The dots were applied for heavy metal sensing.

RESULTS AND DISCUSSION

The functional groups present on the C-dots (**Fig.1**) includes -OH stretching band mode at 3413 cm^{-1}, C-H stretching at 2934 and 2845 cm^{-1}, stretching peaks of C=O, C=C at 1731, 1604 and 1437 cm^{-1}. The XRD pattern (**Fig 2a &b**) of produced C-dots depicts sharp peak at 20.9° and 26.8°, which is highly crystalline cores and possess amorphous nature of C-dots. This peculiar peak is ascribed to 002 of the graphitic carbon, which in line with inter-planar spacing of 3.77 Å with higher spacing than bulk graphite at 3.34 Å. The thermogravimetric and differential thermal analysis display (**Fig 3**) that the C-dots possess thermal stability more than 100 °C and mass losses occurs afterwards. The thermal analysis depicts 3 endothermic peaks at 331, 542 and 735 °C, which are ascribed to oxygenated groups such as carboxyl, hydroxyl, carbonyl and phenols. The optical properties of the synthesize C-dots were explored using UV-vis absorption and photoluminescence (PL) spectroscopy. The UV-vis spectra display (**Fig 4**)intense absorption peak in the range of 260 to 290 nm, which is ascribed to π-π^* transition of the aromatic sp2 domain and also presence of multiple polyaromatic chromophores. A shoulder peak at 330 nm which is ascribed to n-π^* transitions of C=O and structural defects of the C-dots. PL spectrum of the C-dots (**Fig 5a**)exhibits an intense maximum emission peak at 432 nm upon excitation at 330 nm, which also shows characteristic yellowish green color under sunlight and intense blue coloration UV lamp (365 nm). The PL properties of as prepared C-dots was varied at (**Fig5b**) different excitation wavelength from 300 to 405 nm. There was red-shifting in emission peak as the excitation peak was increased from 300 to 405 nm and also simultaneously decrease in fluorescent intensity, which is ascribed to quantum effect and surface defects. This behavior of the prepared C-dots is dependent on excitation-fluorescence, which is accordance with other report [13]. The quantum yield of synthesized C-dots was measured to be 17% at maximum excitation at 330 nm

Fig.6 shows the Raman spectrum of the PB-GQDs, which has been usually used to confirm the quality of the prepared PB-GQDs. There is two major Raman peaks at 1370 cm^{-1} (D band), 1580 cm-1 (G band). The peak of G band corresponds to the Sp^2 bonded carbon atoms in a two-dimensional (2D) hexagonal lattice and the peak of G band is associated with the vibrations of carbon atoms with dangling bonds in the termination plane of tangled graphite or glassy carbon. The D/G ratio band ratio is ID/IG: 0.68 is confirm the formation of graphene like structure from synthesized pine bark.

Fluorescence detection of Cu^{2+} ion

The detection of Cu^{2+} was performed at room temperature in phosphate buffer (PBS) solution pH 4.3. 1.0 mL of PB-CQDs solution was added into 0.5 mL of PBS buffer, followed by the addition of 1.0 mL different concentrations of Cu^{2+}. The fluorescence emission spectra were recorded after reaction for 5 min at room temperature (~28 °C).

Selectivity and sensitivity studies

Moreover sensitivity and selectivity is additional significant parameter to assess the performance of the sensing system. Consequently, we examined the fluorescence intensity changes in the presence of representative metal ions such as including Cu^{2+}, Hg^{2+}, Ca^{2+}, Al^{3+}, Ba^{2+}, Mn^{2+}, Ni^{2+}, Pb^{2+}, Co^{2+}, Zn^{2+}, Fe^{3+}, $Cs^{2+}$and Mg^{2+}, of each at a concentration of 15.0 µg mL^{-1} were reacted with a PB-CQDs solution. Fig. 7 A displays the relative change in PL intensity of PB-CQDs in the presence of various metal ions. Fe^{2+}, Mn^{2+} and Hg^{2+} ions cause the slight PL changes (defined as the relative change of PL intensity in 60–80% as compared to blank) which can be attributed to the nonspecific interactions between the carboxylic and/or amine groups and the metal ions. Among these metal ions, Cu^{2+} displays the strongest PL quenching effect on PB-CQDs, assigning to the special coordination between the Cu^{2+} ions and the phenolic hydroxyl and/or amine groups of PB-CQDs which has been widely used for the detection of Cu^{2+} ions or colored reactions in traditional organic chemistry. Such a specific fluorescence quenching effect may originate from the strong interactions between Cu^{2+} ions and the surface groups of PB-CQDs which transfer the photoelectrons from PB-CQDs to Cu^{2+} ions. These results clearly demonstrate that the PB-CQDs-based Cu^{2+} sensor is highly selective to Cu^{2+} over the other metal ions. Fig.8. Fluorescence responses of CCDs to the addition of different metal ions under acetate buffer (pH 4.3). The concentration of each metal ion is 15 µg $mL^{-1.}$ F_0 and F correspond to the fluorescence intensities of PB-CQDs at emission wavelength of 440 nm excited at 320 nm in the absence and presence of metal ions, respectively.

As shown in Fig. 3, the fluorescence intensity of PB-CQDs gradually decreased with the Cu^{2+} concentration. The fluorescence intensity displayed liner responses with the Cu^{2+} concentrations, with two linear ranges of 0.0□14.0 µg mL^{-1} and 15.0□25.0 µg mL^{-1} (Fig.3B). The linear equations are $F/F0$ =0.9827□0.8485C_{Cu2+} (R = 0.993) and $F/F0$ = 0.7565–0.0033 C_{u2+} (R = 0.996), respectively.

REFERENCES

[1] W. B. Lu, X. Y. Qin, S. Liu, G. H. Chang, Y. W. Zhang, Y. L. Luo, A. M. Asiri, A. O. Al-Youbi, X. P. Sun, Economical, green synthesis of fluorescent carbon nanoparticles and their use as probes for sensitive and selective detection of mercury(II) ions, Anal. Chem. **2012**, 84, 5351-5357.

[2] J. Jiang, Y. He, S. Li, H. Cui, Amino acids as the source for producing carbon nanodots: microwave assisted one-step synthesis, intrinsic photoluminescence property and intense chemiluminescence enhancement. Chem Commun **2012**, 48, 9634–6.

[3] Y. Wang, A. Hu, Carbon quantum dots: synthesis, properties and applications, J. Mater. Chem. C **2014,** 2, 6921–6939.

[4] C. Liu, P. Zhang, X. Zhai, F. Tian, W. Li, J. Yang, et al., Nano-carrier for gene delivery and bioimaging based on carbon dots with PEI-passivation enhanced fluorescence, Biomater. **2012,** 33, 3604-13.

Figures:

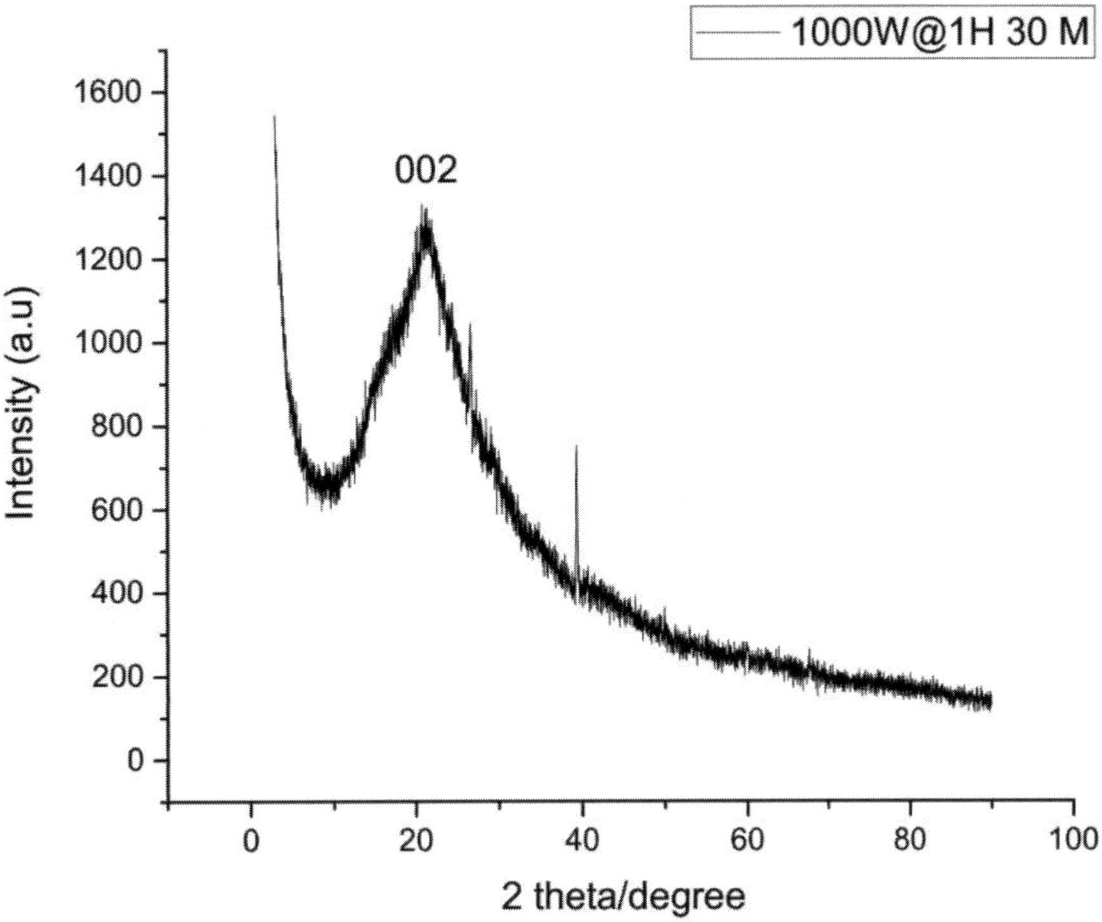

1000W@1H 30 M
002
Intensity (a.u)
2 theta/degree
0
200
400
600
800
1000
1200
1400
1600
20
40
60
80
100

Fig 1. FTIR spectra of as-prepared C-dots from Pine Bark

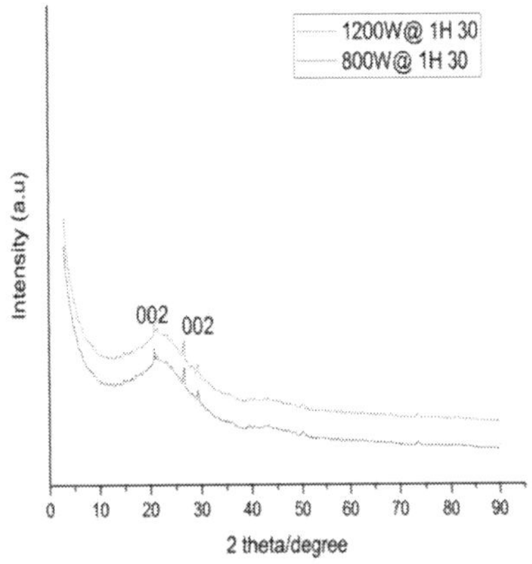

(a)

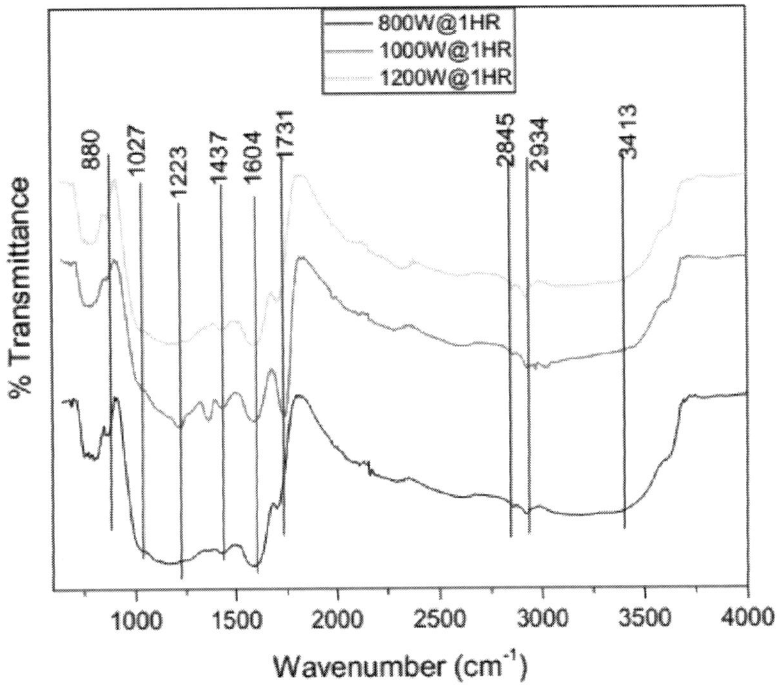

(b)

Fig. 2a & b. XRD pattern of C-dots.

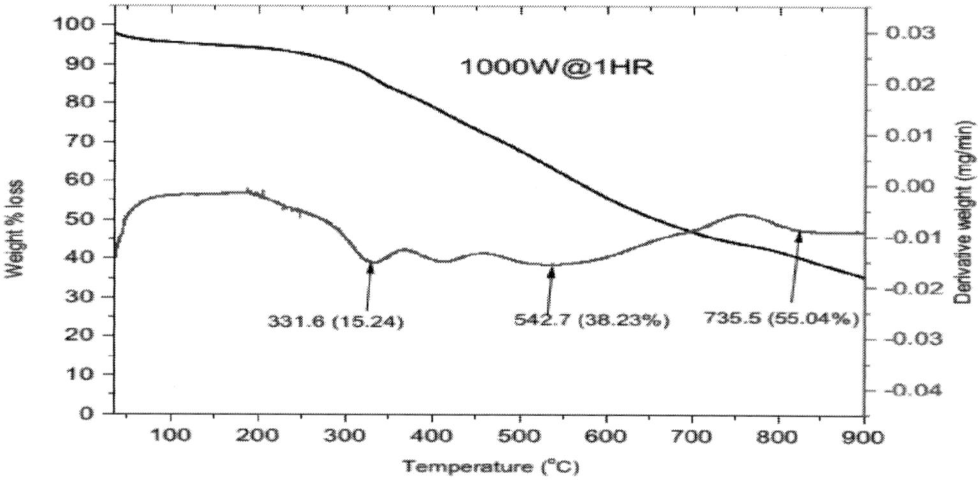

Fig 3. Thermal curves for C-dots under N_2 atmosphere.

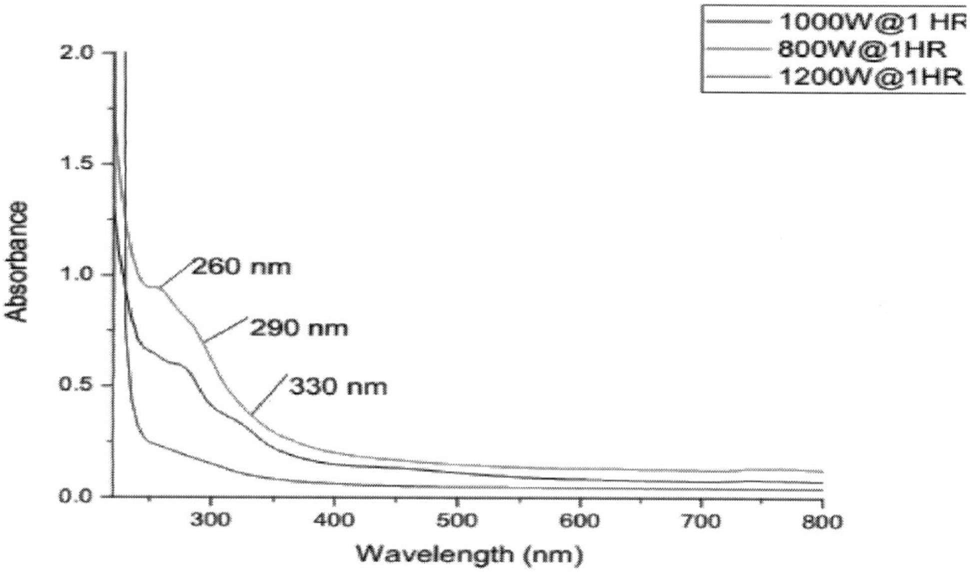

Fig 4. UV-vis absorption of C-dots.

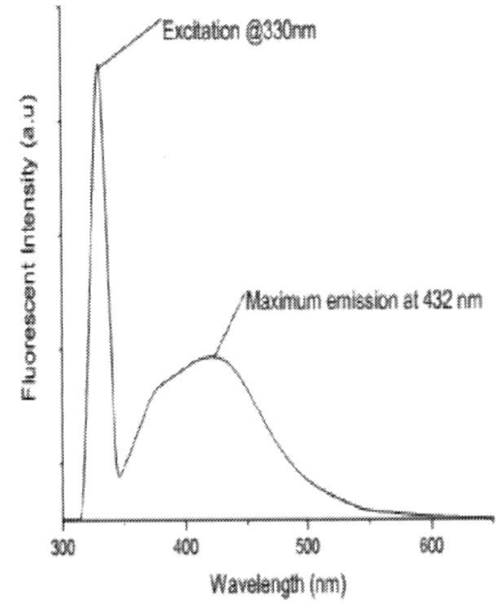

(a)

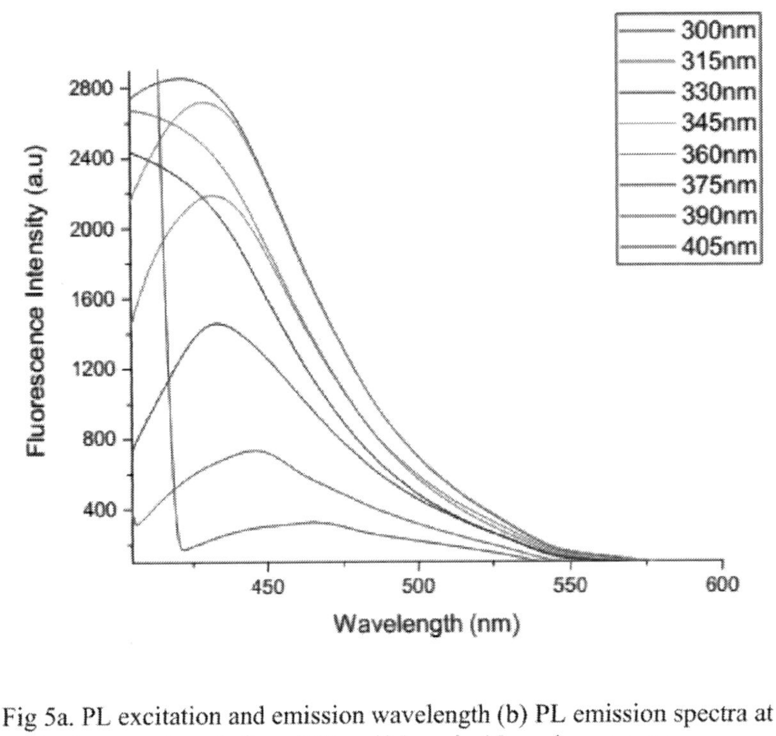

(b)

Fig 5a. PL excitation and emission wavelength (b) PL emission spectra at different excitation wavelength from 300 to 405 nm in 15 nm increments.

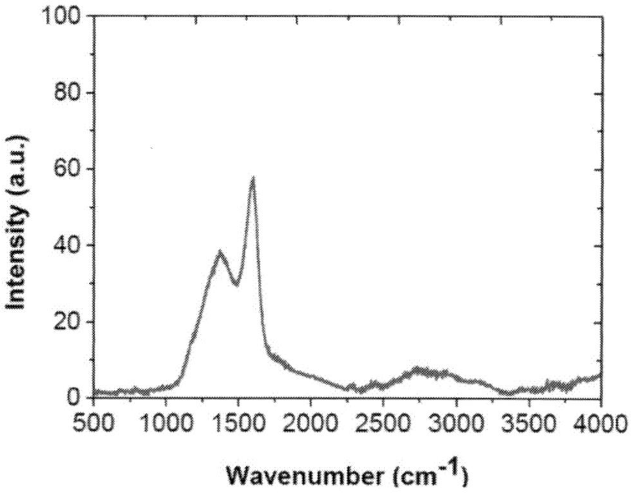

Fig. 6. Raman Spectra of PB-CQDs

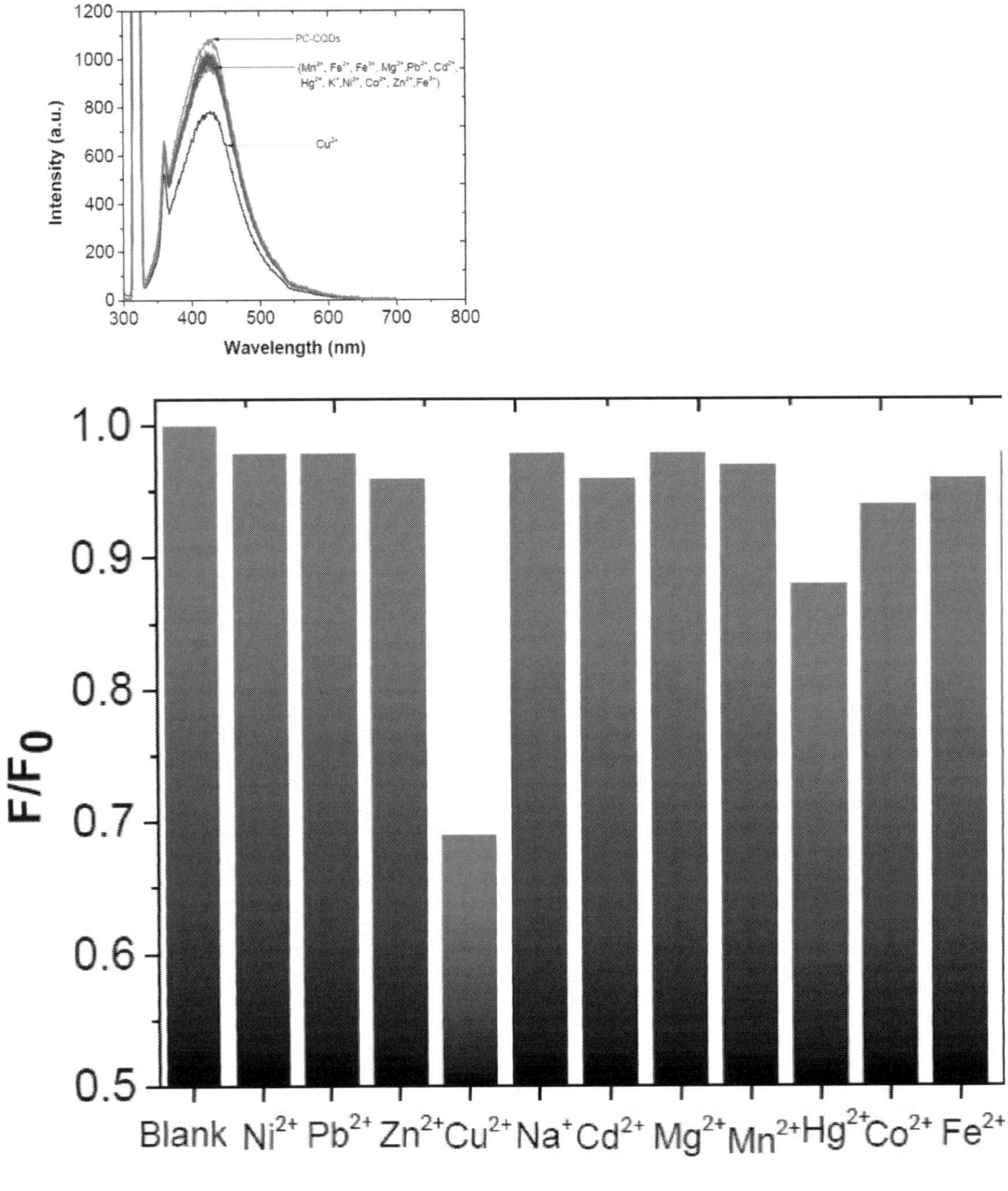

Fig.7. Fluorescence responses of CCDs to the addition of different metal ions under acetate buffer (pH 4.3).

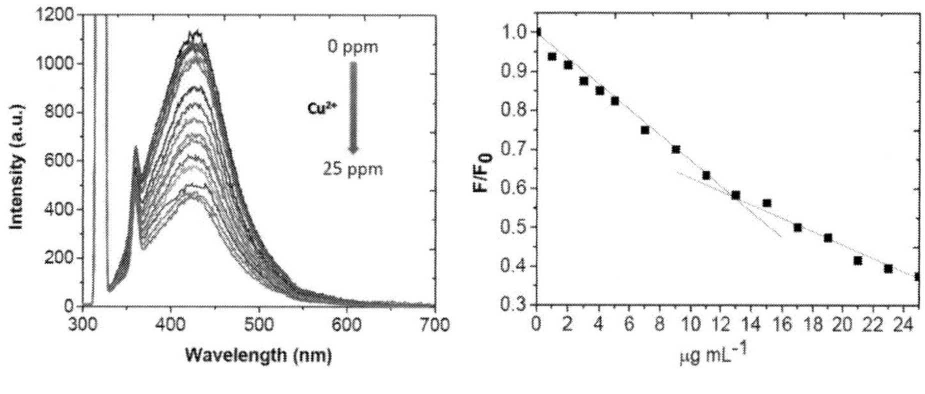

Figure 8. Fluorescence titrations of the y-CDs with Cu^{2+} from 0 to 25 μg mL^{-1}. (Influence of Cu^{2+} concentration on the fluorescence intensity of PB-CQDs (top to bottom: 0, 1, 2, 3, 4, 5, 7, 9, 11, 13, 15, 17, 19, 21, 23 and 25.0 μg mL^{-1})

444269 Functionalization of OH-Terminated Nanoparticles By Cyclic Azasilanes

Abstract:

Colloidal suspensions of functionalized nanoparticles have been suggested for enhanced oil recovery, formation damage remediation, flow assurance, and water treatment operations. Surface modified nanoparticles can also be used as tracers in reservoir studies, active media in thermal management fluids, and fillers in high performance elastomers. Conventional functionalization of nanoparticles with alkoxysilanes and chlorosilanes is limited by needs for catalytic amount of water for the reaction to proceed, requirement of multiple hours to fully saturate the surface, and release of by-products upon reaction completion. Monolayer deposition with a high density of functional groups represents a particularly difficult challenge.

Our ongoing studies indicate that application of cyclic azasilanes that have a strained ring structure with adjacent Si and N centers provides highly efficient functionalization of hydroxylated silica, alumina, magnetite, and diamond nanoparticles which is catalyst-free and by-product-free. The data on synthesis and characterization of functionalized nanoparticles as well as possible reaction mechanism of functionalization will be discussed.

444093 New Development on Hoist Scheduling with the Relaxation of Cyclic Precondition

Abstract:

Multi-Stage Material Handling (MSMH) processes are designed to manufacture materials requiring a series of treatments and in industry, they are broadly used due to their capability to handle large work orders and compatibility with materials with different recipes. A typical example is electroplating process. A hoist handles the transportation of each job among units (stages) according to its processing recipe. The hoist movement plan is predetermined through optimization and this plan is repeated to perform continuity. Obviously, the ultimate goal is to maximize the productivity within each movement plan, which is equivalent to minimize the cycle time to finish one work order. The optimization process is generally called Cyclic Hoist Scheduling (CHS), since every movement plan is a cycle due to its repeatability. This precondition in term of hoist operation and unit usage is general for most cases. However, it could block the really optimal results.

In this study, this precondition has been relaxed for multi-capacity units, thus the unit usage can be more flexible. Thus, the real optimality can be found through CHS. To deal with the cycle connection issue from the relaxation of the precondition, a new module Cycle Connection Identification (CCI) is proposed. Obtained hoist schedule will be checked. Once a schedule with cycle connection problems is identified, a new cycle will be scheduled forcing the smooth connection with the first cycle. The objective of scheduling a new cycle is to match the beginning of the first cycle and the ending of the second cycle. If the two-cycle group is still with cycle connection issue, more cycles will be scheduled till the cycle group is free from cycle connection problems. Each cycle is called a sub-cycle and the constructed cycle group without cycle connection problems is called a full cycle. Note that in each sub-cycle, the work order and all transportation operations are completed.

In this paper, hoist scheduling models with/without such a precondition have been developed and proposed as mixed integer linear programming. The performance of the two models is tested through case studies and the difference is compared and demonstrated in detail.

444359 Development of Novel Emulsified Acid Using Waste Oil

Abstract:

In carbonate reservoirs, hydrochloric acid (HCl) is commonly used for stimulation purposes. The main disadvantage for using HCl is its high and rapid reactivity allowing for less control on acid reaction with formation. In addition, HCl may react with the tubing causing severe corrosion and damages. Several slow-reacting acid systems, though HCl-based, have been developed to retard the HCl reaction and control its reactivity to allow deeper penetration in carbonate reservoirs before it breaks out and reacts with formation rock. However, the selection of the hydrocarbon phase of emulsified acid is critical to the creation and stability of the emulsified acid system. Diesel oil, crude oil and solvents such as xylene have been reported to be successful in the field.

In this work, alternative and less expensive source of oil is developed to replace the current practices. Emulsified acid is a water-in-oil emulsion where oil is the continuous phase and acid is the dispersed one. Industrial and filtered waste oil was used to prepare the emulsified acid of this work. Extensive work has been carried out to study the chemical composition of the waste oil, namely SARA and to select the proper emulsifier. Emulsions were prepared using 15% HCl with emulsifier concentration ranging from 0.5 to 2.0 vol% and 70:30 of acid-to-oil ratio. Droplet size distributions were measured and used to benchmark with other emulsions. Emphasis was given to the stability of the emulsion especially at high temperatures. In order to apprehend the effect of corrosion inhibitor on the emulsion stability, its concentration was varied (up to 0.3 vol%). Later, rheological analyses of the waste oil emulsion were compared to the commonly used diesel-based emulsified acid.

Results showed the potential use of waste oil to prepare high temperature emulsified acid (up to 100°C). The addition of corrosion inhibitor to the emulsified acid caused a negative effect on stability resulting in a less stable emulsion. On the other hand, addition of cationic emulsifier helped to formulate a more stable emulsion. Apart from that, the optimum concentration of the emulsifier was at 2.0 vol%. The rheological data confirms the stability of the new emulsified formulation at higher temperatures. This paper summarizes the findings of using waste oil emulsion and recommends it for field applications. The newly proposed emulsified acid introduces waste oil as a replacement of diesel.

The novelty of using waste oil will help to significantly reduce the expenses of formulating emulsified acid. Since waste oil is basically waste it has a good impact on the environment, that being said, instead of throwing it, the industry will be able to recycle it.

444364 Molecular Simulation of Controlling Ions and Their Role in Low Salinity IOR for Carbonate Reservoirs

Abstract:

Carbonate geology represents around 50% of hydrocarbon-bearing reservoirs in the world and more than 70% in some areas such as the Middle East. Water flooding has not been very efficient in carbonate reservoirs due mainly to the low water wetness. Modifying the ionic composition and salinity of the injected seawater demonstrated that the oil recovery from these carbonate reservoirs could be improved. Although the underlying mechanisms are still a subject of debate, the improved oil recovery has been mostly attributed to the increased water-wetness of the rock through the interactions of specific ions like Ca^{2+}, Mg^{2+}, SO_4^{2-}, and Cl^- , which are present in seawater with the rock surface under favorable conditions

The study of hydrated gas-phase metal ions provides a link between intrinsic chemistry of the isolated ion and its chemistry in solution. The structures and binding enthalpies of a variety of gas-phase and solution mono- and divalent ion hydrates containing up to 12 water molecules have been studied computationally. A DFT and molecular dynamics simulations were employed to study hydrated structures of the form $[\mathbf{M}(H_2O)_n]^{2+}.mH_2O$ for calcium and magnesium and $[\mathbf{M}(H_2O)_n]^{+}.mH_2O$ for sodium, in which there are n water molecules in the first coordination shell and m water molecules in the second coordination shell. The hydration, dehydration, and interactions of metal ions with Cl^{-1} and SO^{-2}_4 have been studied to better understand the mechanism of low salinity IOR. Density functional (DFT) theory using the B3LYP functional, and 6-311++g** basis set has been used. The charge of the metal ion decreases monotonically as the number of water molecules increases in the complex. Extensive calculations of the energetic contribution of individual structural characteristics as well as the energetic trends in binding energy are developed, which will lead to better understanding of how low salinity water IOR works.

Keywords: Hydration, carbonate reservoir, DFT, binding energy, Low Salinity

*** Formatting Issue - Best Available Paper/Graphic ***

444278 Insight into the Mechanism of Toxic Chromium (VI) from Wastewater Using Magnetite Coated PINE Powder

Abstract:

INSIGHT INTO THE MECHANISM OF TOXIC CHROMIUM (VI) FROM WASTEWATER USING MAGNETITE COATED PINE POWDER

Pholosi A., Ofomaja A., Naidoo E.B.

Biosorption and Wastewater Treatment Research Laboratory, Department of Chemistry,

Faculty of Applied and Computer Sciences, Vaal University of Technology, P. Bag X021, Vanderbiljpark, 1900 South Africa.

INTRODUCTION

A knowledge of the interaction between pollutant molecules or ions and active sites on an adsorbent surface is an important phenomenon in many industrial processes such as catalysis, adsorption and other separation techniques. The use of lignocellulosic materials for Cr(VI) adsorption has gained interest over the years **(Miretzky and Cirelli, 2010)**. The adsorption properties of lignocellulosic materials has been attributed to acidic functional groups on their surfaces which can take part in ion-exchange, complexation, chelation and hydrogen bonding with pollutant species in solution **(Vithanage et al., 2015)**. Two mechanisms for lignocellulosic adsorption of Cr(VI) have been proposed; (i) an electrostatic mechanism in which at low pH, the protonated acidic functional groups on the biomaterial carrying a positive charge attracts the negatively charged Cr(VI) species **(Demirbas, 2005)** and (ii) the adsorption-coupled reduction mechanism in which Cr(VI) is reduced to Cr(III) either in solution (Direct reduction) or after being adsorbed (Indirect reduction) by electron-donor groups of the biomaterial that have lower reduction potential values than that of Cr(VI) **(Gao et al., 2008)**. The Cr(III) ions formed in solution may then form complexes with Cr(III)-binding groups on the biomaterial surface or released into bulk solution by repulsion between Cr(III) and other positively charged groups on the biomaterial **(Miretzky and Cirelli, 2010)**.

METHODS

Sodium hydroxide treated pine coated with magnetite (NTP-NC) was prepared by co-precipitation method. The effect of solution pH was performed in a batch system in which 0.5 g of the adsorbent was contacted in six 250 cm^3 beakers containing 75 cm^3 of 75 mg/dm^3 Cr(VI) solutions set at pH 1, 2, 4, 6, 8 and 10 at 26 °C and agitated at 200 rpm for 2 h and the clear solution analysed for Cr(V/I) left using UV-Vis spectrophotometry at 460 nm

RESULTS

The results of the effect of initial solution pH on the adsorption of Cr(VI) onto NTP-NC is shown in Fig. 1. To confirm the operating mechanism, the change in $[H^+]$ concentration at each initial pH, the oxidation/reduction potential before and after adsorption and the amounts of Total Cr, Cr(VI)and Cr(III) present in solution at each pH was monitored for NTP-NC and the results displayed in Fig1 . The change in $[H^+]$ concentration ($\Delta[H^+] = [H^+]_{final} - [H^+]_{initial}$) at the end of the adsorption process for each initial solution pH was recorded and plotted against the initial solution pH. The results showed that the change in $[H^+]$ concentrations at the end of the adsorption process where higher for lower initial solution pH's (where lower amounts of Cr was found in solution) than for higher initial solution pH's (where higher amounts of Cr was found in solution) and the change in $[H^+]$ concentration became small and almost constant after initial solution pH 4. This is consistent with the assumption that the removal of Cr from solution is associated with $[H^+]$ ion consumption and that as the concentration of $[H^+]$ ions reduces in solution, the Cr removal reduces. Cr(VI) ion in acidic solution has a positive redox potential value (above 1.3 V at standard conditions) and in the presence of an electron donor becomes unstable and can be reduced to Cr(III) **(Park et al., 2008)**. The results show that the initial ORP values (before adsorption) for the Cr solutions at pH's of 1, 2, 4, 6, 8, and 10 were 339, 280, 165, 55, -53 and -172 mV for NTP-NC respectively. The positive values of ORP at lower solution pH's indicates higher possibility for reducing Cr(VI), while the lower and negative ORP values indicates lower possibility for reducing Cr(VI). On addition of the adsorbents into the solutions, a change in ORP were observed . The values of ORP at the end of the adsorption process became 330, 268, 30, 9 -29 and -61 for NTP-NC respectively. A reduction in ORP at the end of the adsorption stage signifies that reduction has occurred in the system. **Yu et al. (2014)** showed the amount of Cr(VI) removed from aqueous solution in contact with zero-valent iron (nZVI) could be related to the change in ORP and pH of the solution. The authors observed a decrease in ORP from 400 to 100 mV and a corresponding 10 unit pH increase as Cr(VI) was removed from solution.

Finally, analysis of the prepared Cr solutions at the various solution pH's showed that the concentration of Total Cr obtained from ICP and that of Cr(VI) using UV-vis measurements were very similar, suggesting that Total Cr was almost equal to Cr(VI) in solution. After contacting the various solutions with the adsorbents, the concentration of Total Cr in the pH 1 solution was observed to have reduced to 21.56 mg/dm^3 for NTP-NC, while the Cr(VI) concentrations was 10.66 mg/dm^3 for NTP-NC. With solution pH 2, the concentration Cr(III) left in solution was lower than that of solution pH 1 for both samples. The reason is that at pH 1 there is an abundance of H^+ ions in solution which accumulates on the functional groups on the adsorbent making the surfaces of the adsorbent positively charged and repelling away the positively charged Cr(III) ions from the adsorbent surfaces **(Suksabye et al., 2009)**. At solution pH 2, it is believed that there is less accumulation of H^+ ions at the adsorbent surfaces which led to reduced electrostatic repulsion and increased adsorption of Cr(III) from solution. This increase in Cr(III) adsorption is also confirmed by the reduction in Total Cr in solution between solution pH 1 and 2. The concentrations of Cr(VI) left in solution at pH 2 was higher than at pH 1 for NTP while for NTP-NC the Cr(VI) concentrations were lower at pH 2. For solution pH's higher than 2 and up to 10, the residual Cr(III) in solution becomes very low and almost constant. This is consistent with the small change in $[H^+]$ in solution which is almost constant at this pH range.

FTIR Evidence for the adsorption of Cr(III)

When Cr(III) is formed in solution via adsorption-coupled reduction mechanism it may undergo hydrolysis and/or complexation depending on the solution pH and the total Cr(III) concentration. Hydrolysis may lead to the formation of various Cr(III) species with an increase in solution pH which gives rise to a reduction in its solubility and increasing the chances of precipitation of the Cr(III) species. At solution pH 1 – 4, Cr^{3+} and $CrOH^{2+}$ are the major species which can be adsorbed by negatively charged ligands on the biomaterial surface. On the other hand, at solution pH range of 4 – 10, Cr(III) species (hydroxo complex), such as $CrOH^{2+}$, and $Cr(OH)_3$, are the dominant forms in the aqueous system and at high pH is readily transformed into soluble complex, . When NTP-NC was contacted with Cr(VI) solution for 2 hr, the final solution pH was increased to pH 2.53. The FTIR spectra for NTP-NC (Fig. 2) shows the presence of characteristic Fe-O peaks of bulk Fe_3O_4 at 553.26 cm^{-1}, 564.36 cm^{-1}, 570.21 cm^{-1} and 576 21 cm^{-1}. Functional groups associated with the pine includes the broad band at 3250.99 cm^{-1} representing –OH of cellulose, 2905.17 cm^{-1} representing C-H stretching, while the peaks at 1627.69 and 1602.87 cm^{-1} shows the COO^- and C=O bonds respectively. After 2 hr contact with Cr(VI) solution, the peak at 564.34 cm^{-1} was observed to have split into several new peaks (Fig. 3b (ii)). These emerging peaks are characteristic of Fe-O bonds of maghemite (γ-Fe_3O_4), for example the characteristics peaks of maghemite was observed at 595.21, 685.65 cm^{-1} and 582.70, 634.91 cm^{-1} suggesting a change in the iron oxide phase (Fe_3O_4 to γ-Fe_2O_3) due to the conversion of Cr(VI) to Cr(III) by magnetite nanoparticles. According to **Peterson et al. (1997)**, Cr(III)(hydr)oxide produced is precipitated on the magnetite surface. In this study, α-Cr_2O_3 peak were observed at 604.63 cm^{-1} indicating that Cr(III) was formed during the adsorption process. Finally it was also observed that the carboxylate group at 1635.55 cm^{-1} was shifted to 1637.73 cm^{-1} while a new peak was observed at 1737.72 cm^{-1} suggesting that complexation between Cr(III) and Carboxylate ions also occurred.

REFERENCE

Demirbas, A. Energy Source 27 (2005) 1449–1455.

Miretzky, P., Cerelli, A.F. A review. J. Hard. Mater. 180 (2010) 1-9.

Vithanage, M., Rajapaksha, A.U., Ahmad, M., Uchimiya, M., Dou, X., Alessi, D.S., Ok, Y.S. J. Environ. Manage. 151 (2015) 443-449.

D. Park, C.K. Ahn, Y.M. Kim, Y.-S. Yun, J.M. Park, J. Hazard. Mater. 160 (2008) 422–427.

P. Suksabye, A. Nakajima, P. Thiravetyan, Y. Baba, W. Nakbanpote.. J. Hazar. Mater. 161 (2009) 1103–1108.

R.F., Yu, F.H., Chi, W.P., Cheng, J.C. Chem. Eng. J. 255 (2014) 568–576

M.L. Peterson, A.F. White, G.E. Brown, G.A. Parks Environ. Sci. Technol. 31 (1997) 1573–1576.

$Cr(OH)_4^-$

Fig. 1 Effect of solution pH on the adsorption of Cr(VI)

*** Formatting Issue - Best Available Paper/Graphic ***

$Cr(OH)_2^+$

Fig, 2: FTIR of NTP-NC

444280 Copper-Phyllosilicate Core-Shell Nanocatalysts with Balanced Active Sites for Carbon-Oxygen Hydrogenolysis Reactions

Abstract:

Copper-Phyllosilicate Core-shell Nanocatalysts with Balanced Active Sites for Carbon-Oxygen Hydrogenolysis Reactions

Hairong Yue, Weimao Fang, Changmin Liu, Jiankang Ying

College of Chemical Engineering, Sichuan University, Chengdu, China

Email: hryue@scu.edu.cn

Abstract

Hydrogenolysis of carbon-oxygen (C¨CO) bonds (e.g., esters, ethers, furfural, and CO_2) has emerged as a versatile synthetic tool in organic methodology, as it could produce a variety of products (e.g., chemicals, fuels, and polymers). Copper-based catalysts have been intensively explored for hydrogenation reactions as the copper sites account for the selective hydrogenation of carbon-oxygen bonds and relatively inactive for the hydrogenolysis of carbon-carbon bonds. Work on understanding the active sites of copper catalysts for hydrogenation reactions indicted that both Cu^0 and Cu^+ species were crucial to the activity of Cu-based catalysts. However, researchers were not able to establish a relationship between activity and Cu^0/Cu^+ active species since the conventional Cu-based catalysts is the deactivation by metal particle growth and unstable surface Cu^0 and Cu^+ active species in the strong reductive H_2 and oxidizing carbon-oxygen atmosphere.

We renctly present on efficient approaches for fabrication of a series of copper-phyllosilicate core-shell nanocatalysts and nanoreactors with balanced and stable Cu^0 and Cu^+ active species. We chose the hydrogenation of dimethyl oxalate, ethylene cyclicarbonate and as probe reactions for a better understanding of properties of active sites and the structure-activity relationship. The results indicated a synergy between copper and the oxide components and the balanced surface Cu^0 and Cu^+ species can greatly improve the catalytic C-O hydrogenolysis performance of the catalysts. These nanocatalysts with balanced and stable Cu^0 and Cu^+ active species, confinement effect, intrinsic high surface of Cu^0 and Cu^+ and unique tunable tubular morphology, has potential applications in the high-temperature hydrogenation reactions.

Keywords: core-shell nanocatalysts, Hydrogenation reaction; copper catalyst

References

1. Yue, H.; Zhao, Y.; Ma, X.; Gong, J.: Ethylene glycol: properties, synthesis, and applications. *Chem Soc Rev* 2012, 41, 4218-4244.

2. Gong, J.; Yue, H.; Zhao, Y.; Zhao, S.; Zhao, L.; Lv, J.; Wang, S.; Ma, X.: Synthesis of Ethanol via Syngas on Cu/SiO2 Catalysts with Balanced Cu^0-Cu^+ Sites. *J Am Chem Soc* 2012, 134, 13922-13925.

3. Zhao, S.; Yue, H.; Zhao, Y.; Wang, B.; Geng, Y.; Lv, J.; Wang, S.; Gong, J.; Ma, X.: Chemoselective synthesis of ethanol via hydrogenation of dimethyl oxalate on Cu/SiO_2: Enhanced stability with boron dopant. *J Catal* 2013, 297, 142-150.

4. Yue, H.; Ma, X.; Gong, J.: An Alternative Synthetic Approach for Efficient Catalytic Conversion of Syngas to Ethanol. *Acc Chem Res* 2014, 47(5):1483-1492

5. Yue, H.; Zhao, Y.; Zhao, L.; Lv, J.; Wang, S.; Gong, J.; Ma, X.: Hydrogenation of dimethyl oxalate to ethylene glycol on a Cu/SiO2/cordierite monolithic catalyst: Enhanced internal mass transfer and stability. *AIChE J* 2012, 58, 2798-2809.

6. Yue, H.; Zhao, Y.; Zhao, S.; Wang, B.; Ma, X.; Gong, J.: A copper-phyllosilicate core-sheath nanoreactor for carbon¨Coxygen hydrogenolysis reactions. *Nat Commun* 2013, 4, 2339.

Stimuli-responsive Rubber in Oil and Gas Applications

Rostyslav Dolog, Sankaran Murugesan, and Valery Khabashesku

Baker Hughes, 14990 Yorktown Plaza Dr, Houston, Texas, 77040, USA

ABSTRACT

A problem that exists today in the oil & gas industry is the loss of mechanical properties by reactive rubber seals due to swelling, and the limited space in downhole well applications to accommodate larger swellable packing elements.

This paper presents cost-effective, reinforcing, stimuli-responsive fillers, such as certain classes of fly ash, for use in rubber composite materials in various oil and gas applications. Apart from cost effectiveness, these fillers can also enable stimuli-responsive effects, as well as improved mechanical properties (e.g. modulus) for rubber composites when swelled in water. Fly ash (Class C) is a common component of Portland cement which was used in the literature to achieve a stimuli-responsive effect when added to Hydrogenated Nitrile Butadiene Rubber (HNBR) as filler. Exposure to water brines can cause a filler to promote an increase in pH which enables the hydrolysis of HNBR nitrile groups to carboxyl groups, thus creating an ionomer. This process is followed by neutralization of the new ionomer with metal ions, and formation of ionic crosslinks, further improving mechanical properties of the rubber.

This paper demonstrates that incorporation of fly ash as a reinforcing filler allows development of swellable seals that can retain, or even increase, their modulus and stiffness upon swelling. These characteristics allow avoiding the loss of mechanical properties by seals due to swelling and also eliminating the need for large swellable packing elements that occupy additional space downhole.

KEYWORDS: Elastomers, stimuli-responsive, fly ash, composites

INTRODUCTION

There is a growing demand for high performance elastomers in the oil & gas industry. Rubber compounds are used for various sealing applications, packer elements, stators for pumps, etc. In general, rubber formulations are comprised of elastomer matrix, curing system, antioxidants, plasticizers, along with reinforcing filler, which provides the compounds with their strength. The reinforcing filler is usually represented by carbon black and, to a lesser extent, silica. Along with other factors, filler mechanical properties and concentration generally define hardness, modulus and tensile strength of the elastomer compound. Throughout the years, many different materials have been tried as alternative fillers for rubber composites, including nanomaterials (CNTs, graphene, nanofibers), recycled rubber particles, and ash/fly ash. In the case of ordinary carbon black, the filler serves not only for reinforcement, but also as a cost reducer, because raw elastomer is often a more expensive component. Mechanical properties of the composites are affected by the aspect ratio of the filler, quality of the dispersion, and interface between blend components. The latter determines the effectiveness of load transfer from polymer matrix to the reinforcing filler. High performance compounds can be designed by incorporating well dispersed fine particles with a high aspect ratio and good compatibility with the polymer matrix.

In this context, the fly ash is a cheaper, low reinforcing filler alternative to carbon black, which will help reduce cost of the compounds for "non-demanding" sealing applications requiring large amounts of raw materials. Fly ash is produced in large quantities as a byproduct of coal burned at power plants, and can be considered as a recycled material. Fly ash costs several cents per pound, while carbon black, depending on the grade, is usually one order of magnitude more expensive. Hemmings et al[1] demonstrated that fly ash can be used as filler for polymeric composites with enhanced mechanical properties. Adoption of this technology in the oil and gas industry could change the design approach for reactive rubber compounds. Further functionalization of fly ash through a cost-efficient process could ensure compatibility with the rubber matrix, while promoting filler dispersion and ensuring an efficient load transfer. Such functionalization's include, but are not limited to, surface modification with fatty acids, dicarboxylic acids or phosphoric acid. The reasoning for surface modification is to attach functional groups with aliphatic chains to fly ash particles to make them compatible with aliphatic rubber chains, or to attach acidic groups on their surface that can interact with nitrile groups of HNBR. Initial attempts towards chemical functionalization of fly ash have been made by Seal et al for treatment of offshore oil spills.[2]

Apart from cost effectiveness, composites with fly ash can show stimuli-responsive properties. It has been shown that adding Portland cement to the HNBR matrix improves its mechanical properties in the swollen state by water uptake.[3] Exposure to water causes Portland cement to grow in volume and mass, and promotes an increase in pH which enables the hydrolysis of HNBR nitrile groups to carboxyl groups, thus creating an ionomer. This process is followed by neutralization of the new ionomer with metal ions and formation of ionic crosslinks, further improving mechanical properties of the rubber. Fly ash (Class C) is a common component of Portland cement, which contains a large variety of metal oxides, and is capable of increasing the

[1] Filler Comprising Fly Ash For Use In Polymer Composites. Hemmings et.al., US 6,916,863 B2, July 2005.
[2] Hydrophobic High Surface Area Zeolites Derived from Fly Ash for Oil Spill Remediation, Seal et.al., Environ. Sci. Technol., 2013, 47 (11), pp 5843–5850
[3] Stimuli-Responsive Cement-Reinforced Rubber, Musso et. al., ACS Appl Mater Interfaces. 2014 May 14;6(9):6962-8.

pH of the water solution by hydrolysis, which has a potential to achieve a stimuli-responsive effect when added to HNBR as filler.

Degradation of tensile properties is a common problem for swellable packers and other sealing elements. By selectively using certain classes of fly ash as effective reinforcing filler, one can develop swellable seals that can retain or even increase their modulus and stiffness upon swelling. Therefore, there is a potential to minimize or even avoid the loss of mechanical properties by water swellable reactive packing elements due to swelling. This provides an opportunity to decrease the size of these elements, which are often represented by rather long sections, and hence occupy a significant amount of space downhole. In this paper we have studied the reinforcing effects of fly ash fillers with HNBR compounds as well as stimuli responsive properties of these compounds.

EXPERIMENTAL SECTION

Rubber compounds T1, T2, T3, and T4 are elastomers based on a custom-built peroxide curing system. Formulation F2 was prepared by adding 15 wt. % of fly ash class C to compound F1, which is a commercial swellable NBR sulfur-cured compound.

T series compounds were prepared using a C.W. Brabender® internal mixer operated at 60 RPM for 20 minutes at ~160°F and then molded into sheets and cured at 340°F for 30 minutes in a hot press. Compound F2 was prepared by mixing fly ash with compound F1 in the Brabender mixer for 6 minutes at 60 RPM at ~100°F. Compound F1 is a commercial water swellable formulation that was used as received. F series compounds were cured at 300°F for 60 minutes in a hot press.

Tensile properties of rubber samples were tested using dog bone specimens, according to the ISO37 standard, at room temperature and at 350°F using an MTS® testing system.

Swelling tests in 3% KCl brine and LVT 200 oil were performed to assess the effect of fillers on the chemical resistance of elastomers. Standard one-inch rubber buttons were tested at 200°F at ambient pressure for 6 days. Thermal properties were studied using TGA Q50 and DSC Q2000 instruments.

To analyze the stimuli-responsive effect introduced by fly ash fillers, rubber buttons were aged for 10 days in 3% KCl in the oven at 200°F and then removed from the solution, dried for 30 minutes in the oven and then for 3 hours at room temperature. Samples were later tested in the MTS testing system in compression mode.

Viscoelastic properties of stimuli-responsive samples, based on water swellable compounds, were accessed using the RSA G2 dynamic mechanical analysis (DMA) instrument. This experiment was conducted in a temperature ramp mode of 2°C/minute at 1Hz frequency and 0.2% amplitude strain. A Nicolet 6700 Fourier Transform Infrared (FTIR) spectrometer was used to confirm neutralization of carboxylic groups by metal cations.

RESULTS AND DISCUSSION

The reinforcing properties of fly ash were evaluated at room temperature (Table 1) and at 350°F (Table 2) and compared with N550 carbon black. Sample T3, reinforced with fly ash, demonstrated inferior tensile properties and hardness compared to a compound filled with carbon black. However, at room temperature, fly ash was capable of delivering a reinforcing effect to the rubber compound without sacrificing its elasticity, e.g. the ultimate elongation of elastomer compound remained virtually unchanged after incorporation of 50 parts per hundred rubber (PHR) of fly ash, while its modulus and tensile strength have increased. Therefore, potentially, one can substitute carbon black with high loadings of fly ash in certain applications, thus further decreasing the cost of the rubber product. The summary of mechanical properties is provided in Table 1, E_x denotes modulus of a compound at certain strain (x), i.e. stress at strain.

Sample	N550 CB (PHR)	Fly ash Class F (PHR)	Tensile strength (psi)	E_{25} (psi)	E_{50} (psi)	E_{100} (psi)	Elongation (%)	Hardness Shore A
T1	0	0	829	71	97	121	472	48
T2	50	0	3402	165	273	744	290	68
T3	0	50	1176	94	127	166	467	54

Table 1. Room temperature tensile properties and hardness of elastomer samples

The relatively low increase in modulus and hardness of T3 compared to T1 at room temperature can be explained by a low aspect ratio of fly ash and weak interactions between blend components. The latter problem can be solved by functionalization of fly ash particles.

The general data trends from the high temperature tensile tests closely resemble the results of the tensile test conducted at room temperature. However, at high temperature, rubber specimens break at much lower elongation, which leads to a great decrease in the measured tensile strength. This behavior is typical for elastomers and limits their application to a certain maximum service temperature and is also due to the fact that high temperature tests are conducted at conditions that push the limits of the material. The modulus values at 15% (E_{15}) and 20% (E_{20}) strain are not provided herein, because half of the tested specimens were broken before they reached such strains.

The superiority of carbon black as an efficient filler became even more pronounced at high temperature. However, incorporation of fly ash doubled the tensile strength and increased the elongation at break by one-third when compared to the unfilled elastomer.

Sample	E_5 (psi)	E_{10} (psi)	E_{15} (psi)	E_{20} (psi)	Tensile strength (psi)	Elongation (%)
T1	31	46			91	23
T2	61	130	194	261	832	39
T3	42	73	98	119	173	30

Table 2. Summary of high temperature (350°F) tensile data for elastomer samples

Thermal properties of samples T1-T3 were assessed with Differential Scanning Calorimetry (DSC) and Thermogravimetric analysis (TGA). TGA allows evaluating thermal stability of samples and to verify the actual composition of a sample from the ash content. The results of the DSC and TGA analysis are summarized in Table 3. During TGA, samples were heated to 600°C at 10°C/minute rate in a nitrogen (N_2) environment, and subsequently heated at 10°C/min to 600 °C in air to burn off carbon-based fillers. This allows decomposing of organic matter first by applying heat, and then burn the remaining carbon-based material in air atmosphere.

The initial degradation of the polymer started at 200°C and progressed to complete decomposition of aliphatic chains in the temperature range of 420-500°C with a highest rate of decomposition at ~460°C. Sample T2 incurred the second rapid mass loss after 600°C due to burning of the carbon black, which comprised most of the ash content at 600°C in a N_2 environment. The DSC experiment demonstrated the glass transition temperature of the elastomer remained largely independent of the type of filler - the respective DSC heating scan was conducted with a 10°C/minute temperature ramp after precooling the sample to -50°C.

Sample	W_{600} (%)	W_{800} (%)	$T_{deg}1$, °C	$T_{deg}2$, °C	T_g, °C
T1	2.4	2.4	452.8	-	-29
T2	29.8	2.7	458.7	616.5	-29
T3	32.8	32.7	447.3	-	-28

Table 3. TGA/ DSC results for elastomer samples

Table 3 contains ash contents (W) at 600°C and 800°C, degradation temperatures (T_{deg}) which represent the temperature of the highest rate of decomposition during TGA experiment, and glass transition temperature (T_g) of the samples, as determined by DSC scan.

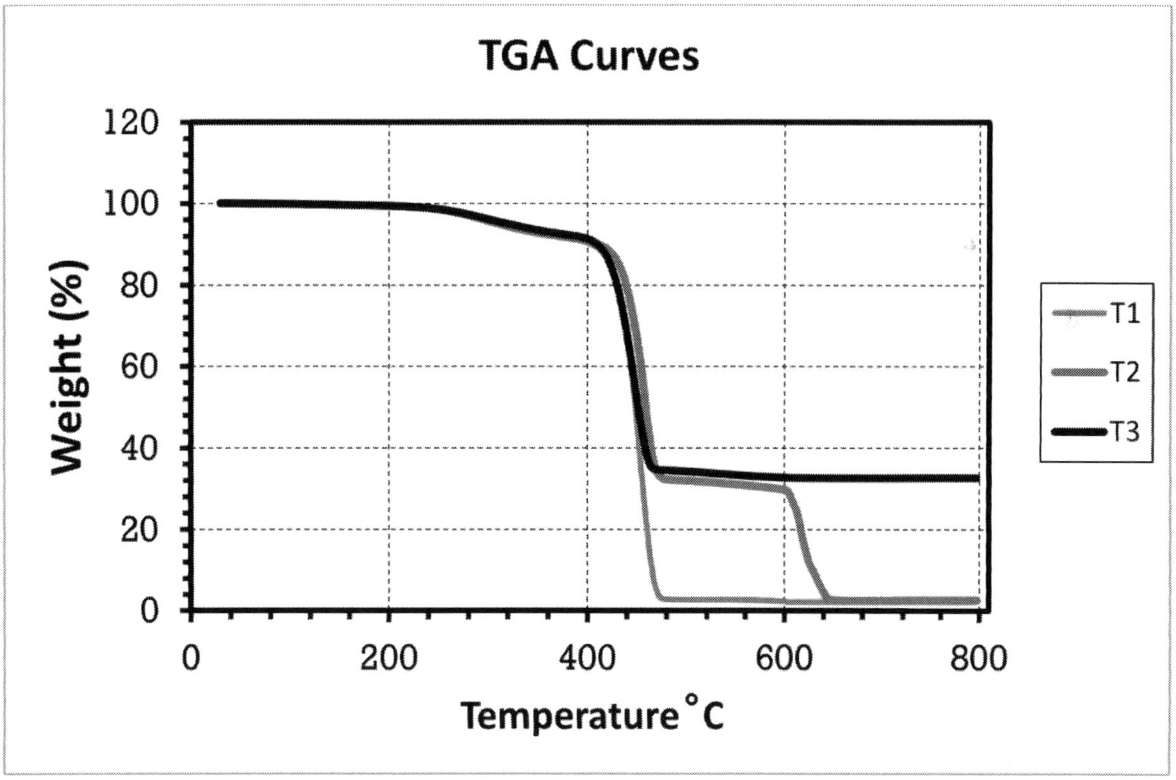

Figure 1. TGA Curves for samples T1, T2, and T3

Sample \ Testing fluid	3% KCl brine	LVT 200	Specific gravity
T1	0.6	13.0	0.96
T2	0.3	11.6	1.13
T3	0.8	9.1	1.18

Table 4. Volume increase (%) of elastomer samples due to swelling for 6 days at 200°F

Musso et. al. developed a stimuli-responsive HNBR composite reinforced with Portland cement that was increasing its stiffness and modulus when swelled in water[3]. Fly ash (Class C) is a common component of Portland cement, which contains the components necessary to achieve a similar effect. Therefore, one can avoid the loss of mechanical properties by water swellable packing element seals upon activation. To test this hypothesis, we studied the effect of aging in water brines on the properties of elastomers with fly ash fillers.

Rubber samples were molded and cured into standard size rubber buttons (cylinders of ~0.25" height and ~1" diameter). These buttons were aged for 10 days in 3% KCl brine in the oven at 200°F, removed from the solution, dried for 30 minutes in the oven and for 3 hours at room temperature. The swelling of both compounds was close to ~1% by volume (Table 4). Samples were later tested in an MTS machine in compression mode. Each button was compressed to 70% strain, buttons remained intact during the experiment, and visually all appeared to have a certain level of compression set at the end of the experiment. The results of the experiments are summarized in Table 5 and Figure 2. Only one button was tested for each sample. The provided compression modulus values (C_n) represent values of stress at a percentage strain (n), e.g. C_5 is stress at 5% compression strain.

Compound T4 was reinforced by fly ash Class C, which contains larger weight fraction of CaO than fly ash Class F (~25% vs ~10-15% respectively) and is expected to produce a higher pH and closely resemble cement that was used to prepare stimuli-responsive rubber in the original paper by Musso et.al.[3] The results demonstrate that exposure to water (3% KCl) at high temperatures can improve the modulus for the compounds filled with fly ash Class C (T4), while the control sample (T1) showed the opposite behavior. After aging, the compressive modulus at 50% strain (e.g. stress at 50% strain) of control sample T1 decreased by 19%, as compared to the unaged sample, while compressive strength of sample T4, reinforced by fly ash, increased by 15%. The difference in modulus values is even more drastic at 70% compression strain (-35% and 31%, respectively). This is in agreement with the hypothesis stated earlier.

Sample	Filler	Compressive modulus (psi)				
		C_5	C_{10}	C_{25}	C_{50}	C_{70}
T1	None	19	42	127	468	1810
T1 swollen	None	16	36	109	380	1185
T4	Fly Ash class C (50 PHR)	21	46	134	434	1628
T4 swollen	Fly Ash class C (50 PHR)	29	57	153	501	2125

Table 5. Summary of compressive modulus properties after 3% KCl soaking and 200°F drying

To further investigate the stimuli-responsive property of fly ash filler, the latter was incorporated into a water swellable compound, which provided for significantly more interaction of filler with water to occur. Sample F2 comprises unmodified commercial water swellable NBR formulation F1 mixed with 15 wt.% of fly ash Cass C. Sample F2 was processed into a 0.04" slab, then a rectangular sample was cut from the slab and swollen in 3% KCl solution until 100% volume increase. The F2 sample was then dried overnight in the oven, which lead to de-swelling of the sample due to solvent evaporation. Mechanical properties of the F2 sample were assessed with DMA prior to the swelling test, in a swollen state, and after drying - the data summary is shown on Figure 3.

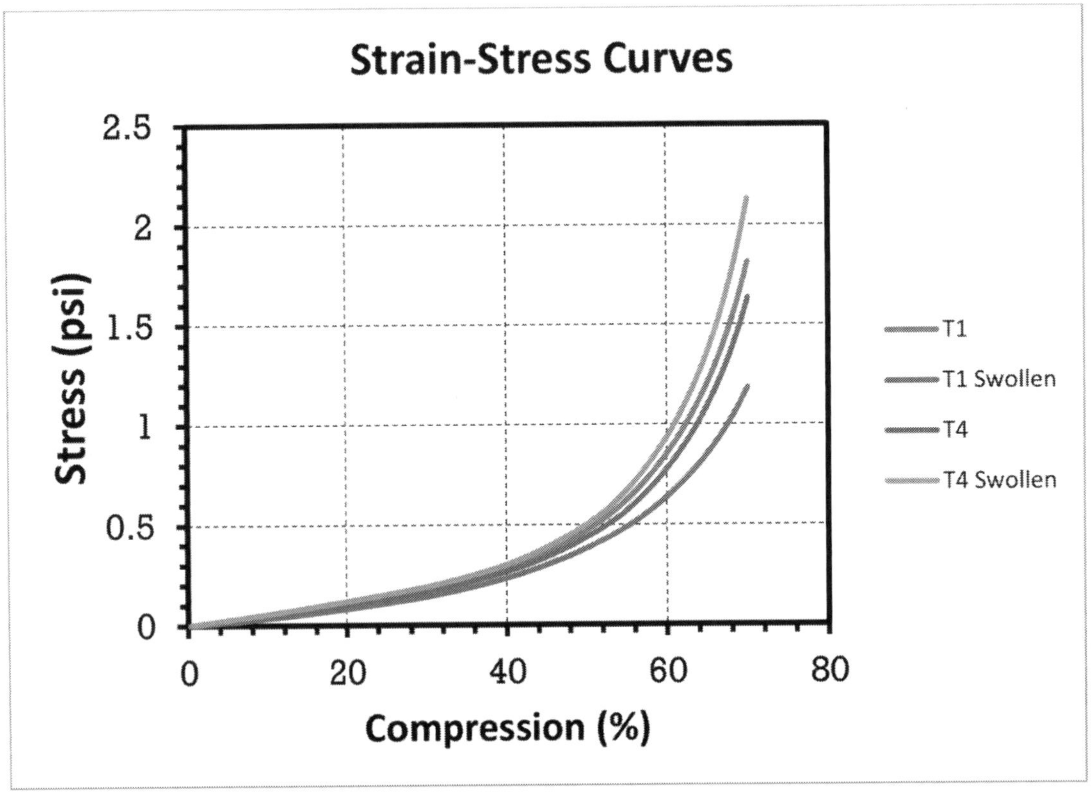

Figure 2. Stress-strain curves

The storage modulus of sample F2 after swelling and subsequent drying was about an order of magnitude higher than that of original sample F2 measured prior to swelling experiment. The storage modulus of the dried F2 sample was also less sensitive to the increasing temperature, contrary to the weakening original sample F2. The increase in the storage modulus of swollen F2 with temperature is simply due to the drying of the sample. The obtained result demonstrated that fly ash can essentially act as a stimuli-responsive material.

The significant improvement in the modulus is believed to be due to the neutralization of the swellable component of the commercial F1, and potential hydrolysis with subsequent neutralization of NBR elastomer. This is confirmed by the shift in T_g to a lower temperature range after drying as illustrated in Figure 3.

Carboxylic groups were detected in the original sample F2 prior to swelling, hence, they cannot be attributed solely to hydrolyzed HNBR, but rather to the super-absorbent polymer present in water swellable rubber compound (see Figure 4). FTIR confirms neutralization of carboxylic groups because a significant increase of the peak near ~1540 cm^{-1} and a lowering of peak intensity at ~1720 cm^{-1} have been observed. This neutralization is believed to promote the formation of ionic clusters in the compound with fly ash considered to be the source of cations. The neutralization of carboxylic groups could as well be attributed to K^+ cations present in the KCl brine solution, however, in that case, the respective peak in FTIR spectra would appear at higher wave numbers, approaching ~1600 cm^{-1}.

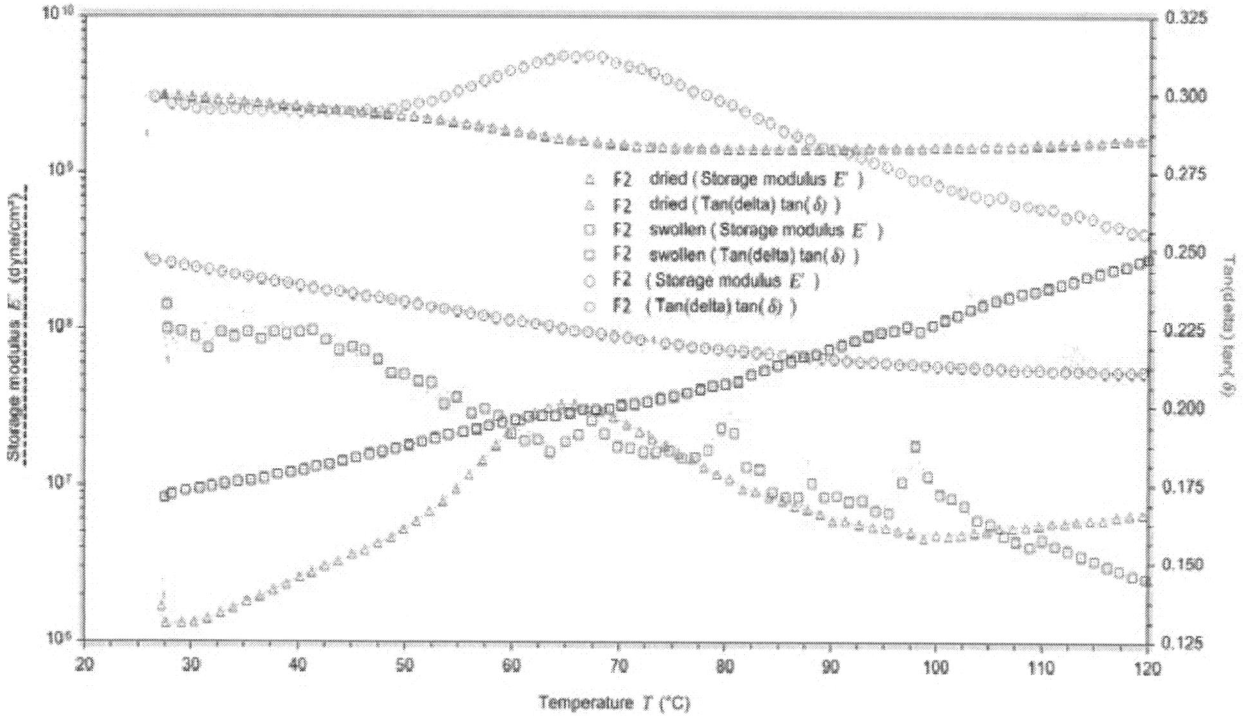

Figure 3. DMA data on F2 sample

It is important to mention that sample F2 exhibited a certain degree of brittleness after drying, which is undesired in rubbers, but this can be managed by incorporation of common ionomer plasticizers.

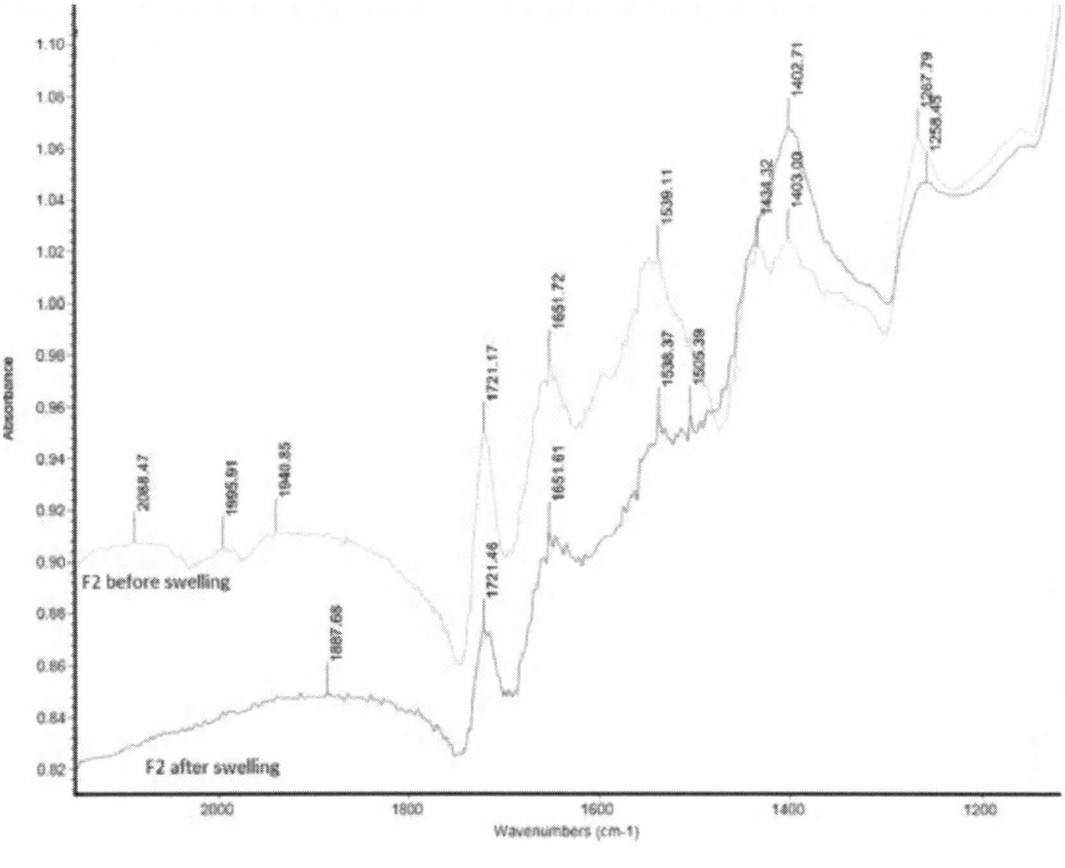

Figure 4. FTIR spectra of F2 sample taken before and after swelling

Conclusions

Fly ash demonstrated a great potential for consideration as cost effective, low reinforcing filler for downhole oil and gas applications. Although the reinforcing properties of fly ash are inferior to carbon black, incorporation of fly ash can improve the mechanical properties of HNBR elastomer compared to the unfilled raw crosslinked elastomer at both room temperature and at 325°F.

Furthermore, composites of HNBR with fly ash Class C demonstrated a stimuli-responsive effect by significantly increasing their modulus upon swelling in water brine. This effect was even more pronounced in composites of fly ash with swellable water compounds, which provided an easy access of filler to water upon swelling. A ten-fold increase in modulus was achieved in such fly ash reinforced sample, after swelling in 3% KCl water brine and drying in the oven overnight. The FTIR confirmed neutralization of carboxylic groups present in the investigated system, which contributed to overall improvement of mechanical properties of the compound upon swelling.

444310 Formal Modeling of Accidents and Automatic Generation of Tutoring Scenarios for Intelligent Safety Training of Process Operators

Abstract:

With the emphasis of accident-free operations and prevention of catastrophic events, safety-training demands for the chemical plants have been on the increase globally. However, the research results on modeling of accidents and involving human operators and the advanced theory and technology for (safety) operator training have been shared limitedly among domain experts and commercial developers. Even with the recent trend of simultaneously delivering the operator training system with the construction of a new process plant, in commercial operator training systems (OTS), safety training used to implement quite limited and obvious scenarios (and the burden was almost up to the trainer), partially due to the difficulty in modeling propagations of probable accidents. By the regulatory requirements or proactive investment by companies, effective scenarios for training of safety and abnormal situation management are necessary to be improved in the aspects of the size of volume as well as the training depth. Thus, considering the emergence of enhanced virtual reality technologies and inexpensive gadgets like HMD, supporting the immersive training environment, the future efficiency of the training heavily relies on the quantity and quality of safety scenarios, which would be essential in guiding the entire training process.

In this research we first overview a government-sponsored research project developing an enhanced safety training system for chemical process operators, using haptic devices in addition to enhanced audio and visual equipment. The initial version of the system runs with a fixed set of training scenarios developed for predetermined accident scenarios. However, with a limited set of training scenarios, the trainees simply memorize and rote learn the steps, required to be mastered to proactively solve the problems with an advent of possible accidents. In addition to that, without formal modeling, the system cannot give timely feedbacks and the performance of the trainee cannot be measured in real-time. Then to enhance the contents and usefulness of the training, the knowledge of process operations is collected and systematically stored as a form of knowledge base, in addition to the accident scenarios obtained by executing Fault Tree Analysis and HAZOP study.

By using a formal modeling language like UML and Petri nets, the accident scenarios and operators' problem-solving tasks are modeled hierarchically and classified into distinguishable subtasks. Then a new training scenario is generated on-line through automatic synthesis based on planning using the knowledge base. Safety training scenarios are generated intelligently for the tutoring of a single operator's job function as well as effective collaboration among a group of operators for real-scale malfunction diagnosis and intervention at the initial propagation stage of an accident. As case study, safety trainings for city gas governing stations are executed based on systematic scenario generation and analyzed for improved effectiveness and field applicability of the system. Reduction of time and cost in the training scenario production and real-time customized training based on new operator's level of knowledge and skill are also discussed in the framework of intelligent tutoring systems.

444376 Ozone Source Apportionments of Local Anthropogenic Emissions

Abstract:

Ozone is one of six principal air pollutants designed by US EPA. NOX and VOC emissions from anthropogenic sources are the most important factors to control ozone pollutant.

In this report, photochemical dispersion models are used to estimate the effect of ozone from local anthropogenic emission sources. EKMA-type diagrams are generated for NOX- and VOC- effects. Furthermore, APCA (Anthropogenic Precursor Culpability Assessment) technology is used to estimate ozone source apportionments.

EKMA-type diagrams show that 10-30% emission reductions could reduce ozone concentrations to meet standard. Further APCA results show the average ozone contributions from regions and emission categories. APCA results explain different effects for different monitoring stations.

444397 Flare Minimization and Emission Characterization during Start-Ups of Chilling Train in Ethylene Plants

Abstract:

Flaring is a crucial safety operation to protect chemical plant personnel and equipment during plant start-up. However, it generates huge amounts of off-spec products, and causes negative environmental and social impacts, as well as results in tremendous raw material and energy losses which could be saved to generate much more needed products. Thus, cost-effective start-up flare minimization strategies which is a double-win practice, simultaneously benefiting both the industrial and environmental sustainability, are becoming more and more important and attractive to the industries.

The chilling train is a very important section for the ethylene plants and accounts for more than 50% of the total start-up time and flare emissions of the whole ethylene plants. Within this section, the charge gas from the upper stream process is refrigerated to a very low temperature (lower than -160 °C) to separate hydrogen and methane. Meanwhile, the left charge gas is liquidized for cryogenic separation. The chilling train system consumes over 40% of the refrigerating capacity of an ethylene plant. During the chilling train start-up, different operational conditions with different compositions and flowrate of precooling process will affect both product separation efficiencies, start-up stability time and flare emission.

Traditional start-up scenarios are developed based on experience and improved by trial and error, which is inefficient and dangerous. In this study, a general methodology on flare minimization for chilling train start-up operations via rigorous dynamic simulation is presented. The methodology starts with setup and validation of chilling train steady-state and dynamic simulation models. The initial state of start-up in the real plant is achieved through systematically transforming from the validated dynamic model, and several operation conditions with different compositions and flowrate of precooling stream are examined to identify an optimal initial state of start-up based on the obtained initial state. Thereafter different flare minimization strategies for the chilling train start-up are conducted on the optimal initial state model. Also, emission source characterization under different flare minimization strategies for chilling train start-up is investigated. Deep insights of the start-up strategies, emission source distribution and dynamic emission profiles are provided. This study provides detailed technical support for both the industry and environmental agencies on evaluating and developing cost-effective flare minimization strategies in the future.

Experimental Study on Najas Minor of Producing Bio-Oil

Abstract: Due to the diminishing of petroleum reserves and the environmental degradation caused by consumption of fossil energy , many attentions have been paid to the application of biomass energy which is acceptable as an abundant renewable energy resource. Najas minor grows in dense clusters with highly branched stems of which a small fragment could propagate into a thrive one.In this study, we researched the properties of Najas minor to make biofuels .Firstly,the najas minor after heavy metal ions degradation were fed into pyrolyser to prepare biofuels. Significant differences of the product was observed compared with only Najas minor pyrolysed.
KEY WORDS: Najas Minor; catalytic pyrolysis; bio-oil

1.Introduction

Energy demands and environmental problems caused by fossil energy have been a matter of growing concern.Bio-oil has been considered to be a promising way for producing renewable bio-fuels or chemicals[1-3].In recent years,many countries pay attention to the importance of biomass that is an abundant renewable energy resource.Biomass can be defined as any organic material mainly consisting of carbon, hydrogen,oxygen, nitrogen, and some other components in small proportions[4].Although the biomass has been studied for nearly a century,most studies mainly concentrated in Lemna minor, Lemna gibba, Spirodela polyrhiza and Landoltia punctata.We still know little about the biological characteristics and potential application value of other varieties of biomass[5].Najas minor,a submersed herb,grows in dense clusters with highly branched stems of which a small fragment could propagate into a thrive one.Najas Minor is native to Europe, Asia and North Africa from the Netherlands to Morocco east to Japan and the Philippines,including China,Siberia, Central Asia, Iran, Turkey, Ukraine, Germany, France Italy and a host of other Countries.

Catalysts used for catalytic cracking are basically divided into the following categories: molecular sieve, alkali metal salt, metal oxide, etc.Different catalysts have different catalytic effects.The use of metal nanoparticles has shown to be very important in recent industrial applications[6].The application of precious metal nanometer materials in catalysis,which has caused the extensive concern of researchers in recent 30 years,has become a hot research topic in the field of catalysis.Gold nanoparticles have received a great deal of attentions due to their special optical and electrical performances, catalytic properties and remarkable quantum size effect, surface effect etc[7]. The synthesis of metal nanoparticles can be classified broadly into wet and dry methods[8]. The cost of production is relatively high both materially and environmentally. There is an unequivocal need to develop more cost-effective and environmental benign methods to alternate these existing methods[9].Researchers in this field turn to use organisms to produce inorganic nanostructures extracellularly[10].

Although there has been a large volume of work on the synthesis of gold nanoparticles[11-13], and obtainning bi-oil by abundant biomass, there are relatively fewer attempts to combine these two methods.In this work, gold nanoparticles were obtained by aqueous chloroauric acid solution with the of Najas Minor at room temperature. The direct pyrolysis and catalytic pyrolysis experiments of Najas Minor were carried out to observe the effects of catalysts on pyrolysis process of Najas Minor.

2. Materials and Methods

2.1. Experimental samples

The Najas Minor collected from Tianjin city and the solutions of Au(III) were used in this study.The stock solutions of Au(III) were prepared by dissolving analytical grade chloroauric acid

($AuCl_3 \cdot HCl \cdot 3H_2O$, Kewei Reagent, China) in deionized water.

2.2.Experimental apparatus and procedure

Pyrolyzer with GCMS,shown in Fig. 1,is mainly used for analyzing the composition of polymer.Pyrolyzer with GCMS includes three parts :quartz pyrolysis tube,pyrolyzer and GCMS.

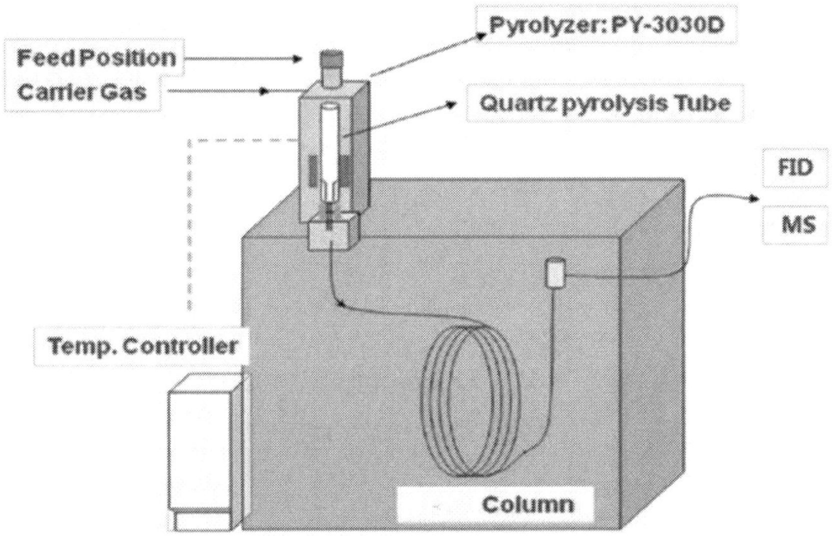

Fig. 1 Diagram of Pyrolyzer with GCMS

The quartz pyrolysis tube (shown in Fig.2) length is 12cm.The sample end diameter is 5 mm,and the pyrolysis products end diameter is 3mm.This special structure is beneficial to the fixation of the reactant in the tube.The top and bottom of the reactants are both supported by quartz wool to prevent reactants blowed by the carrier gas ,or fine particles into the chromatographic column.All pyrolysis experiments were carried out on the EGA/PY-3030D pyrolyzer(Frontier Lab,Japan) .The samples were filled in the special quartz pyrolysis tube to reduce the amount of valuable catalyst and reduce the cost of research work.

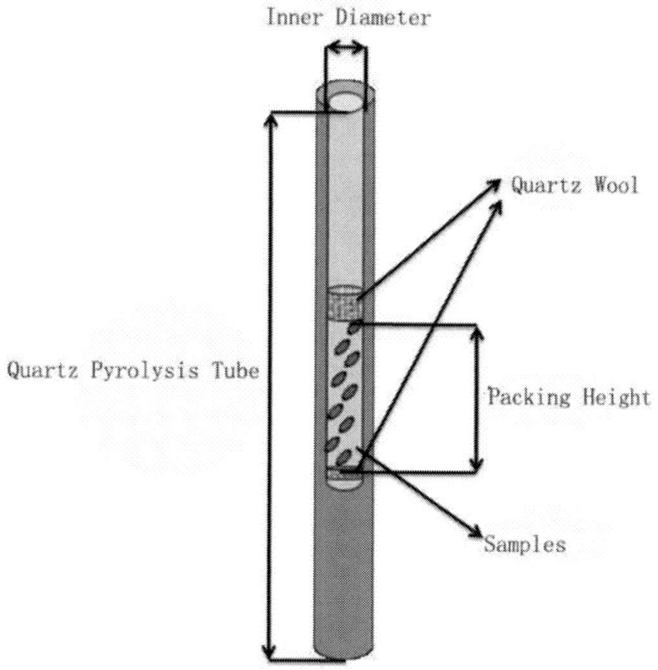

Fig. 2 Diagram of Quartz Pyrolysis Tube

Approximately 0.5g Najas Minors were used in each experiment.The samples were impregnated in 50ml deionized water and 50ml solution of chloroauric acid(0.0793mmol/L) for a night,followed by drying at 80℃ until the weight didn't change.Each experiment used 2.5mg Najas Minors as reactants.Each sample was heated, under the control of heating controller, from ambient temperature to 600 ℃ repeatly and kept at 600 ℃ for 2 h.

3. Results and discussion

3.1.MS Analysis of the Products during the Pyrolysis

Under the same pyrolysis conditions, bio-oil obtained by direct pyrolysis and catalytic pyrolysis were analyzed by GCMS,and the main chemical compounds were shown in Table1 and Table2.

Table 1 Indentification and quantification of main chemical compounds by direct pyrolysis

NO.	peak area%	Formula	Compound
1	12.67	$C_3H_6O_2$	2-Propanone, 1-hydroxy-
2	3.67	C_7H_8	Toluene
3	3.28	C_7H_8O	p-Cresol
4	3.81	$C_5H_{10}O$	Pentanal
5	3.53	C_8H_8O	Benzofuran, 2,3-dihydro-
6	2.17	C_8H_8N	Indole
7	5.93	$C_{20}H_{40}O$	3,7,11,15-Tetramethyl-2-hexadecen-1-ol
8	2.52	$C_{22}H_{42}O_2$	Phytol, acetate
9	5.25	$C_{16}H_{32}O_2$	n-Hexadecanoic acid
10	8.99	$C_{18}H_{30}O_2$	9,12,15-Octadecatrienoic acid, (Z,Z,Z)-

Table 2Indentification and quantification of main chemical compounds by catalytic pyrolysis

NO.	peak area%	Formula	Compound
1	6.01	C_7H_8O	p-Cresol
2	1.75	$C_7H_7NO_2$	3-Hydroxypyridine monoacetate
3	4.58	C_8H_7N	Indole
4	3.05	$C_{10}H9N$	1-Phenyl-1-cyclopropanecarbonitrile
5	3.90	$C_{12}H_{22}O_{11}$	.alpha.-D-Glucopyranose, 4-O-.beta.-D-galactopyranosyl-
6	8.04	$C_{20}H_{40}O$	3,7,11,15-Tetramethyl-2-hexadecen-1-ol
7	2.82	$C_{20}H_{40}$	2-Hexadecene, 3,7,11,15-tetramethyl-, [R-[R*,R*-(E)]]-
8	3.99	$C_{15}H_{28}$	1-Pentadecyne
9	2.12	$C_{15}H_{28}$	3,7,11-Trimethyl-2,4-dodecadiene
10	8.75	$C_{16}H_{32}O_2$	n-Hexadecanoic acid

3.2 discussing

According to the chemical category and common characteristics, the products were divided into hydrocarbons, alcohols, acids, ketones, nitrogen compounds and other compounds.Nitrogen compounds detected by GC/MS analysis ,mainly come from the pyrolysis of amino acids in Najas Minor.Table 3 showed the distribution of pyrolysis products.As shown in the table,the hydrocarbons ,alcohols and nitrogen compounds in biological oil of catalytic pyrolysis were significantly increased compared with direct pyrolysis.The content of hydrocarbons increased from 12.4% to 34.6%.The content of alcohols increased from 9.1% to 16.6%.The gold nanoparticles loaded in surface of Najas Minor, was conducive to the generation of hydrocarbon and alcohol in pyrolysis.By analyzing the GC/MS results, the main hydrocarbons in bio-oil by catalytic pyrolysis were long chain alkanes,and olefins.The main components in gasoline and diesel is hydrocarbons. As a result, whether the biomass can be effective as future power fuel oil, depends largely on the hydrocarbon content of bio-oil after further treatment.

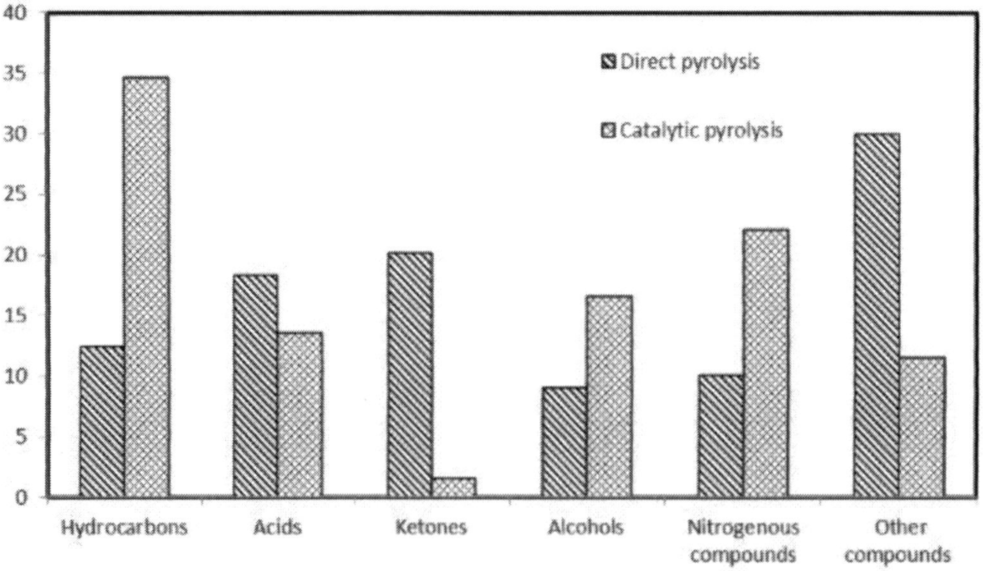

Fig.3 Distribution of pyrolysis products

4.Conclusion

Biomass was explored as a potential biomass energy resource. The experimental study on the direct and catalytic pyrolysis of Najas Minor had been presented, using Pyrolyzer with GCMS,to observe the catalytic effects of gold nanoparticles.In this study,remarkable catalytic effects for pyrolysis were observed.The pyrolysis of Najas Minor with gold nanoparticles,provided more hydrocarbons, alcohols and nitrogen compounds.

参考文献：

[1]Gong F, Yang Z, Hong C, et al. Selective conversion of bio-oil to light olefins: Controlling catalytic cracking for maximum olefins[J]. Bioresource Technology, 2011, 102(19):9247-54.

[2]Zhang Z, Bi P, Jiang P, et al. Production of gasoline fraction from bio-oil under atmospheric conditions by an integrated catalytic transformation process[J]. Energy, 2015, 90:1922-1930.

[3]Hui S, Yuqi D, Jinzhao D, et al. Transesterification of sunflower oil to biodiesel on ZrO2 supported La2O3 catalyst.[J]. Bioresource Technology, 2010, 101(3):953–958.

[4]Yaman S. Pyrolysis of biomass to produce fuels and chemical feedstocks[J]. Cheminform, 2004, 35(31):651-671.

[5]Hejný S. Ellias Landolt The family of Lemnaceae—A monographic study—Vol.1 biosystematic investigations in the family of Duckweeds (Lemnaceae) (vol. 2)[J]. Folia Geobotanica Et Phytotaxonomica, 1993, 28(1):50-50.

[6]Gardea-Torresdey J L, Tiemann K J, Gamez G, et al. Gold Nanoparticles Obtained by Bio-precipitation from Gold(III) Solutions[J]. Journal of Nanoparticle Research, 1999, 1(3):397-404.

[7]Marie-Christine D, Didier A. Gold nanoparticles: assembly, supramolecular chemistry, quantum-size-related properties, and applications toward biology, catalysis, and nanotechnology.[J]. Chemical Reviews, 2004, 35(16):293-346.

[8]Liu B ,, Xie J ,, Lee J Y, et al. Optimization of high-yield biological synthesis of single-crystalline gold nanoplates.[J]. Journal of Physical Chemistry B, 2005, 109(45):15256-63.

[9]Raveendran P, Fu J, Wallen S L. Completely "Green" Synthesis and Stabilization of Metal Nanoparticles[J]. Journal of the American Chemical Society Jacs, 2003, 125(46):págs. 13940-13941.

[10]Absar Ahmad , Satyajyoti Senapati , MIK, et al. Extracellular Biosynthesis of Monodisperse Gold Nanoparticles by a Novel Extremophilic Actinomycete, Thermomonospora sp.[J]. Langmuir, 2003, 19(8):3550-3553.

[11]Selvakannan P, Mandal S, Phadtare S, et al. Water-dispersible tryptophan-protected gold nanoparticles prepared by the spontaneous reduction of aqueous chloroaurate ions by the amino acid[J]. Journal of Colloid & Interface Science, 2004, 269(1):97-102.

[12]Gittins D I, Caruso F. Biological and Physical Applications of Water-Based Metal Nanoparticles Synthesised in Organic Solution[J]. Chemphyschem, 2002, 3(1):110 - 113.

[13]And A G, Murphy C J. Seed-Mediated Synthesis of Gold Nanorods: Role of the Size and Nature of the Seed[J]. Chemistry of Materials, 2004, 16(19):3633-3640.

444353 A Combined Rheology and Stability Study on Emulsified Acid for High Temperature Applications

Abstract:

HCl emulsion has been a popular formulation to stimulate carbonate reservoirs due to the various advantages that emulsified HCl has over ordinary HCl; including but not limited to deeper penetration and less corrosion damages. Emulsified HCl has even more potential advantages if utilized properly. The reaction kinetics, stability and rheology of emulsified HCl plays a significant role on achieving successful acidizing treatment.

The objective of this paper is to experimentally study the factors affecting the stability of emulsified HCl. The emulsion was prepared using 15wt% HCl and diesel with cationic emulsifier. The acid solution-to-diesel ratio was 70:30. The effect of stirring intensity, concentration of the emulsifier, addition of corrosion inhibitor, temperature, addition rate of acid phase to oil phase and mixing time of the emulsion have been studied. Stirring intensity was varied from 600 to 1500 rpm. Emulsifier dosage ranged between 5 to 20 gpt. Temperature used for heating, to mimic reservoir condition, was as high as 120°C.

The results showed that there is an optimum condition for emulsifier dosage and for mixing time of the emulsion. The optimum emulsifier dosage was found to be 10 gpt with 5 minutes of mixing time. As for the addition rate of acid phase to oil phase, 1.4 ml/min of drop-wise addition was found to give a very stable emulsion. The effect of stirring intensity on the emulsion stability was proportional, the higher the rpm speed the better the stability. Corrosion inhibitor, on the other hand, played a role as a destabilizer. With regards to heating temperature, it was found out that the higher the temperature, the faster the emulsion breaks.

The novelty of this study is the comprehensive approach in combining stablity and rheology of emulsified acid for high temperature applications. The particle size analysis and solution morphology help in better understanding HCl emulsion and how it works.

COMPARATIVE ANALYSIS OF MOLECULAR STRUCTURE IDENTIFIABILTY BASED ON SIGNATURES AND DESCRIPTORS

Zelimir F. Kurtanjek
University of Zagreb, Croatia

Abstract

This work is focused on molecule computer representation aimed for molecule predictive model (QSAR) and inverse modelling. Two different approaches are compared, i.e. the molecule structure coding based on graph theory (Faulon et al.[1] extended valence) and the chemical based descriptors[2]. Applied are 1-3rd level molecule graph fragments and 630 2D molecular descriptors. The molecule test sets are alkanes, alkenes, acetones, aromatics, organic acids and halogenated hydrocarbons in the range of C_1-C_{12}, selection of biomolecules (including aminoacids) and a set of binary ionic liquids (cations: imidazole, pyridinium, quinolinium, ammonium, phosphonium). The comparative comparison of molecule description based prediction potential are statistically evaluated by octanol-water (PWA) partition coefficients (for hydrocarbons), viscosity of ionic liquids and biological impact on acetycholinesterase inhibition. The evaluation is based on two distinct levels. On the first level the comparative analysis based on internal and external correlation between "scores" in multidimensional spaces of the separate first, second and third order valance spaces, and the molecular signature space. Firstly, the potential of information content applicable for inverse modeling is tested on ability for molecule classification by use of linear discrimination (LDA) based on the singular value decomposition (SVD) of corresponding molecular fragments and descriptors. For chemical-biological property inverse modelling proposed is application of constrained genetic algorithm (GA). Tested are linear structure multivariate (chemometric) models such as principal component regression and partial least squares and nonlinear decision trees (random forest). Applied are singular value decompositions for reduction of the molecule signature and the descriptor spaces up to the significance level of p=0.05. The selected level of significance is presumed to be in coherence with experimental data of molecule physical and chemical properties. The main focus is on extraction of the main molecular fragments (partial graphs extracted from molecular signatures) and specific molecule descriptors. Importance of application of open source software modules provided by Bioclipse[3] and R software packeges[4] for computational-statistical numerical algorithms is emphasized.

Key words: Molecule signatures, descriptors, inverse modelling, decision tree

Introduction

Advancement of high-throughput experiments have lead to „big data" databases applied in pharmaceuticals, medicine, new material science, catalysis, and new industrial processes. The role of chemoinformatics as a science for computer encoding of molecule properties and algorithms for data mining and inference are the essential steps in the development. Computer encoding of molecular properties has been a part of chemical engineering from the beginning of process design. It is mostly based on molecule group contributions, but has been extended by application of molecular descriptors (1-2-3 D), molecule signatures, and application of molecular quantum mechanics. Modeling is faced with the problem of „high dimension curse". For example, about 5 000 various molecular descriptors can be determined for each molecule, although most of them are highly correlated. It results in an ill-conditioned problem of modelling with large uncertainties in parameter and poor model prediction accuracy outside modelling training sets. Great improvement has been achieved by application by use of model ensembles, like random forest of many individual decision trees. These models are highly flexible, can be trained on discontinuous and nonlinear functional relations, can encompass various data formats like numerical and categorical variables, and due to large number of randomly selected individual predictors in principle do not have tendency to overfitting. However, when the object of modelling is extraction of important molecular features, due to autocorrelation between descriptors random forest, produced lists of importance are large sets of poorly distinguishable variables. The focus of this work is to apply feature regularization for optimization and reduction of list of variable importance. Compared are two methods for structure encoding, molecular descriptors and molecular signatures of first and second order.

Modeling and Factor Regularization

In this work are selected two distinct data sets of organic molecules from open source data bases (J. Sangster, 1989) and ionic liquids ILs (The UFT/ Merck and NIST, 2015), and lists of the selected molecules are given in Table 1. As target properties for organic molecules are selected the corresponding molecular classes and partition octanol-water (POW) coefficients, and dynamic viscosity in the range of 30-80 ^{0}C of ILs is selected The molecules are coded into SMILES and used as input lists into software for evaluation of molecular signatures. Most popular open source software tools are DRAGON (R. Todeschini, et al., 2009) and PADEL (Chun Wei Yap, 2010). Here are used 1-2 D descriptors generated by PADEL. For the same list of molecules are determined first and second order molecular signatures or extended valances (J.P. Faulon, 2003). The signatures are two dimensional graph based structural molecular fragments. The vertices are atoms and the edges are corresponding bonds (lengths or heights), and are here only used of the first and second level. A molecule is characterized by a complete set of signatures as input data for modelling. Since the models are based on large number of variables (molecular features) possibly lager than number of samples (molecules), usually denoted as p>>n models. They are intrinsically „ill conditioned" leading to good predictions on train data but very poor predictions on test data. Also, produced list of variable importance are very dense "clusters are not distinguished" with large uncertainties (broad confidence intervals). This makes the problem of inverse modelling aimed to predict a set of candidate molecules for given set of target properties, such as in a case of high dimension combinatorial task for selection of ILs cations and anions, very unreliable. In order to reduce model dimensionality and improve robustness and parsimonious of important molecular features here is applied modelling with feature regularization. The model is based on ensemble of random trees (CART models, L. Breiman et al. 1998) with included feature regularization.

Data sets of organic molecules		Range of number of atoms			
Molecular classes	C	H	O	Cl	F – I - Br
Alkanes	5-12	12-30			
Alkenes	4-9	6-18			
Acetones	6-24	6-26			
Halogenated hydrocarbons	3-10	6-22	1		
Organic acids	1-16	2-32	2		
Esters	4-11	6-22	2		
Aromatics	1-12	1-21		3	2 – 2 - 1
Data sets of ionic liquids (ILs)					
Cations		**Anions**			
1,3-Dimethylimidazolium (+)		PF_6 (-) Cl (-) CH_3COO (-)			
1-Methylpyridinium (+)		BF_4 (-) $MeSO_4$ (-) $EtSO_4$ (-)			
1,1-Dimethylpyrrolidinium (+)		Tf_2N (-) CF_3SO_3 (-)			

Table 1. The molecular groups of the organic and ionic liquid molecules selected for random forest models and extraction of the most important molecular features. The properties for the organics are obtained from J. Sangster (1989) and for the ionic liquids from Merck and NIST data bases (2015). Molecular descriptors and signatures of ILs ions are recorded separately and adjoined as a single record.

The modelling is formulated as an optimization (functional and multivariable minimization) problem with two parts of the objective function (T. Hastie et al., 2009).

$$\min F_{object.}(X,T,\omega) = SSE(y,\hat{y}) + \phi(T,\omega) \tag{1}$$

The first part is the sum of square errors (SSE) based on difference between observed y and prediction $\hat{y}$ based on input data by an ensemble T of interrelated CART models (L. Breiman et al. 1998), The second part is a selected metrics ϕ for the model complexity depending the sum of leaf probabilities for population of individual CART models and their total number N_T.

$$\phi(T,\omega) = \gamma \times N_T + \frac{1}{2} \times \lambda \times \sum_i^{N_T} \omega_i^2 \tag{2}$$

The ensemble output prediction is a sum of predictions of individual T_i tree CART models:

$$\hat{y} = \sum_i^{N_T} T_i(X) \tag{3}$$

A very efficient open source code Extreme Gradient Boosting XGB (T. Chen, 2016) is available in R language (R Developm-ent Core Team, v. 3.2.3, 2015). The optimization algorithm is consecutive, starting with first CART model which is appended with a next tree boosted by errors in predictions of previous tree. At each step optimization is solved by linearization of the objective function to quadratic terms. Number and structure of trees as well as the parameters for regularization need to be found by trial and error and minimal number of trees by folding and cross validation error on the training set. The critical step is validation on untrained set to have acceptable model error and avoid model overfitting with minimal model complexity.

Results and Discussion

Tested are models for prediction of the seven molecular classes (Table 1.) based on molecular descriptors and molecular signatures (first and second level). Results for molecular descriptor feature importance are shown on Fig. 1. The effect of regularization is very pronounced. Out of hundreds of descriptors for the molecular classification needed are only few, the most important being BCUTw.1h. On contrary, when an ensemble of independent CART trees is applied, i.e. random forest model, level of accuracy of prediction is the same, but the list of variable importance is densely populated with numerous descriptors.

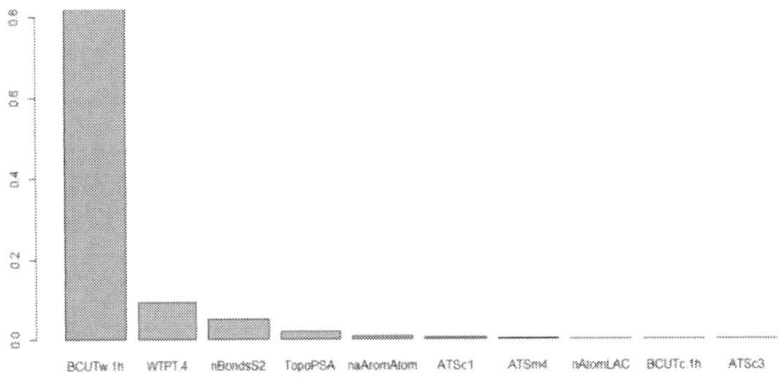

Fig 1. Variable importance molecular descriptor factors for the inverse modeling of the selected molecular classes. The first five descriptors are: BCUTw.1h (based on a weighted version of the Burden matrix which takes into account both the connectivity as well as atomic properties of a molecule), WTPT.4 (Randić's weighted path), nBondS2 (bond count), and naAromAtom (number of aromatic atoms).

The same classification problem was tested by the first and second order molecular signatures. The important signatures are depicted on Fig. 2-3. Accuracy of prediction on trained sets of molecules compared to descriptors is about the same, though the importance factors are not as distinctive as in case of the descriptors.

significantly better when molecular descriptors are applied. The

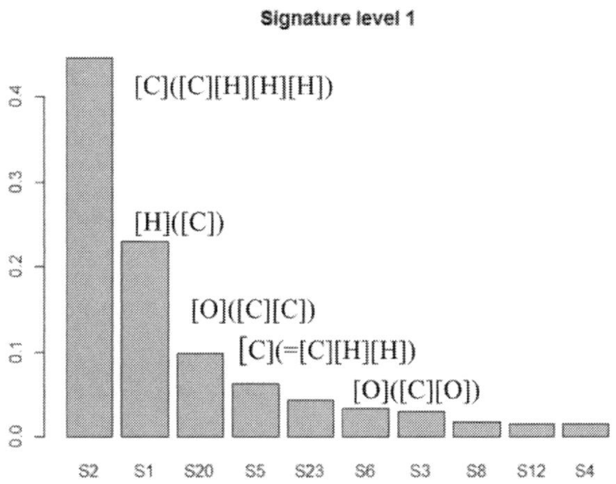

Fig 2. Relative importance of the individual first level signatures for inverse prediction of the selected molecular classes. The first five most important signatures are coded according to SMILES (Simplified Molecular Input Line Entry System).

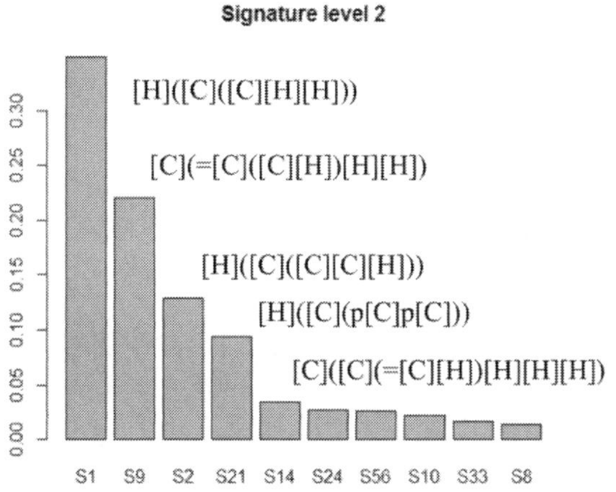

Fig 3. Relative importance of the individual second level signatures for inverse prediction for the selected molecular classes. The first five most important signatures are coded according to SMILES (Simplified Molecular Input Line Entry System).

However, model accuracy on untrained set of molecules, depicted in Fig. 4, is significantly lower for the model with molecular descriptors compared to signatures. The average error with descriptors is about 5% while for signatures about 11%. The "box and whiskers" plots are obtained by error evaluation of ten randomly generated untrained sets.

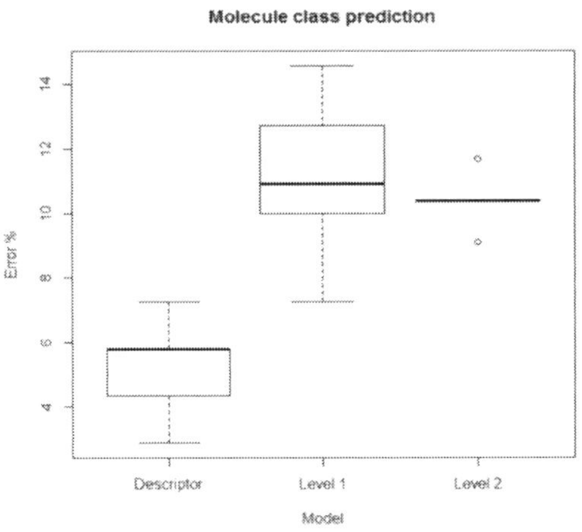

Fig 4. "Box and whiskers" plot for the model validation (relative errors) for inverse prediction of the molecular classes based on the set of untrained molecular descriptors and the first and second level of molecular signatures.

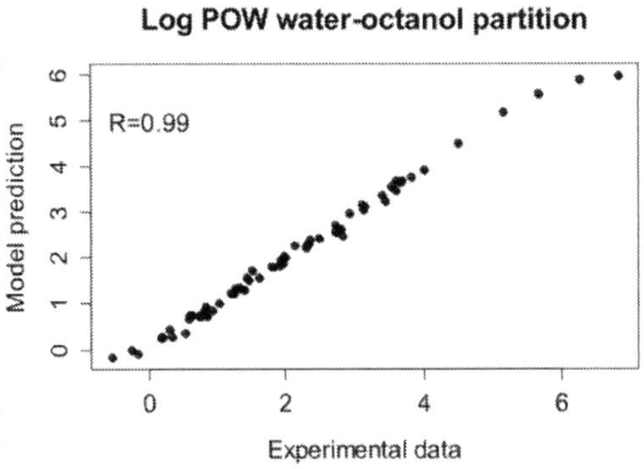

Fig 5. Predictions of the partition octanol-water coefficients POW, for the specified molecular classes of organic molecules, are based on the random forest model with the molecular descriptors.

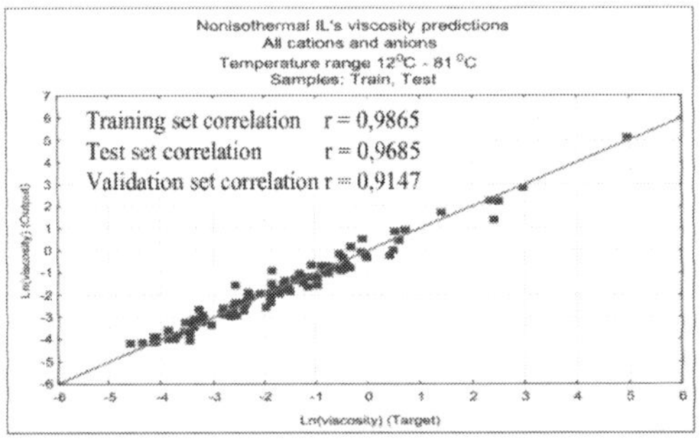

Fig 6. Predictions of dynamic viscosities for the specified groups of ionic liquids (ILs) based on the random forest model with the molecular descriptors. The viscosities cover the temperature range from 12 – 81 ^{0}C and five orders of magnitude.

In Fig. 4 can be observed that dispersion of errors (quantiles) for the second level molecular signatures is negligible, though two outliers are present.

For prediction of molecular properties applied are random forest models based on molecular descriptors. For all of the organic molecules, included in the seven classes, evaluated are octanol-water partition coefficients POW, shown in Fig. 5. The range of POW values is very high, on log scale fro 0 to 6, but the obtained accuracy of the model predictions is high (Pearson correlation R=0.99). The same approach is applied for prediction of ionic liquid viscosity data in temperature range from 12 to 81 ^{0}C. Since temperature effects on viscosity are intrinsically nonlinear, they represent a modelling challenge. However, the decision trees are very efficient for accounting nonlinearities, though nonlinearity is reflected in non-orthogonal decision planes, rendering trees with significantly increased number of branches. For the range of 6 order of magnitude the accuracy of prediction on trained set is R = 0.98 and on untrained set R = 0.91.

Conclusions

Compared are ensemble decision trees models for molecular feature class and property prediction based on extensive list of molecular descriptors and first and second level of extended valances or signatures.

The aim was to test effects of regularization on ensemble of CART decision tree models. Proven is effectiveness of the extreme gradient boosting algorithm in cases with extensive list of (molecular) features.

Obtained is increased accuracy on untrained sets of data, and more importantly, obtained are parsimonious lists for ranking of variable importance lists. This enables further significant reduction in development of robust and simple decision tree models.

The reduction in number of molecular features applied for decision tree models may greatly simplify the challenging problem of inverse modeling based on molecular descriptors and/or signatures.

References

1. J. Sangster, J. Phys. Chem. Ref. Data, **18**(3)(1989) 1111-1227
2. The UFT/ Merck Ionic Liquids Biological Effects Database, www.il-eco.uft.uni-bremen.de
3. NIST, Ionic Liquids Database Standard Reference Database #147, <ilthermo.bouder.-nist.gov /ILThermo/mainmenu.uix>
4. R. Todeschini and V.Consonni: "Molecular Descriptors for Chemoinformatics", (2 volumes), WILEY-VCH,Weinheim (Germany) 2009.
5. Chun Wei Yap, PaDEL-Descriptor: An Open Source Software to Calculate Molecular Descriptors and Fingerprints, J. Comp. Chem., (2010):**32**: 1466-1474.

6. J.-L. Faulon, D.P. Visco Jr., R.S. Pophale, The signature molecular descriptor. 1. Extended valence sequences vs. topological indices in QSAR and QSPR studies, J. Chem. Inf. Comput. Sci. **43** (2003) 707–720.
7. J.-L. Faulon, C.J. Churchwell, J.D.P. Visco, The signature molecular descriptor. 2. Enumerating molecules from their extended valence sequence, J. Chem. Inf. Comput. Sci. **43** (2003) 721–734.
8. L. Breiman, J. H. Friedman, R. A. Olshen, C.J. Stone, „Classification and Regression Trees", CRC, Boca Raton, 1998.
9. T. Hastie, R. Tibshirani, J. Friedman, The Elements of Statistical Learning, Data Mining, Inference and Prediction, Springer Verlag, Second Edition, New York, 2009.
10. T. Chen, T. He and M. Benesty (2016). xgboost: Extreme Gradient Boosting. R package version 0.4-3. https://CRAN.R-project.org/package=xgboost
11. R Development Core Team, , R: A language and environment for statistical computing. R, Vienna, Austria,www.R-project.org 2015.

444791 Dissolution and Precipitation of Minerals Under Different Alkali Concentrations in a Reactive CO_2 Batch System

Abstract:

The primary focus of this research is to study the dissolution and precipitation of minerals and salts from Berea sandstone and brine samples taken from the Great Salt Lake in a reactive CO_2 batch system. Various experiments were constructed to measure dissolution and precipitation reaction rates under different alkali concentrations. Changing the alkali concentrations of the Great Salt Lake brine samples produced changes in dissolution and precipitation of minerals from the Berea sandstone and the Great Salt Lake brine samples. Data from ICP-MS was used to quantify mineralogical changes in the samples and Micro-CT was used to measure the porosity change in the Berea sandstone samples. The results from these analytical instruments showed patterns of mineralogical dissolution and precipitation in the post-experimental fluid that correlate to the change in alkali concentrations.

444825 Significance of Energy Intensity and GHG Emission Reduction Towards Sustainability of a Petrochemical Plant

Abstract:

Sustainability is an important pillar in today's petrochemical industry for improving process efficacy by reducing specific energy consumption, greenhouse gas (GHG) emissions, and wastewater. SABIC (Saudi Basic Industries Corporation) like other giant petrochemical companies has set stringent sustainability targets and investing huge resources to transform its plants from the present position to the first quartile in each category.

SABIC is the world largest producer of ethylene oxide and ethylene glycol. For each ton of glycol produced, 0.2 to 0.5 tons of CO_2 is emitted as GHG. The significance of sustainability "in general" while the methodology adopted to reduce energy intensity, carbon footprint, and wastewater "in particular", will be discussed thoroughly in our talk.

444827 Syngas to Ethylene Glycol: A Technology Overview

Abstract:

SABIC is the world largest MEG producer petrochemical company and most of its MEG is being produced through the hydrolysis of EO. Ethylene oxide (EO) is a simplest form of epoxides and due to its molecular structure, acts as a versatile intermediate for a variety of chemicals such as; monoethylene glycol (MEG), di- and triethylene glycols, ethoxylates, ethanol amines, and glycol ethers. The growing demand of MEG owing to its downstream utilization in polyester industry has prompted the research of an efficient, cost effective, and environmentally benign EO production process.

The major feedstock for MEG are ethylene and syngas. The heavy dependence of MEG producers on ethylene and its scarcity in some parts of the world promoted the need of alternate feedstock for highly demanded MEG. The technological developments utilizing syngas as a feedstock will be discussed along with the commercial analysis and future growth in EO/EG business.

444834 Stability and Rheology of Emulsified Acid in Waste Oil: Effect of Emulsifier Concentration, Mixing Speed, Temperature and Droplet Size

Abstract:

Acid in oil emulsion is used in the industry for stimulation purposes. In particular, HCl as the acid phase and diesel as the oil. Emulsifying HCl has numerous benefits over regular HCl. Perhaps the main reason is due to deeper penetration. Apart from that, less corrosion damages are caused since the external phase is hydrocarbon. Several studies showed the success of replacing diesel as an external phase with other hydrocarbon oil, such as crude oil and xylene.

This work utilizes the extra hydrocarbon left unused –or sometimes dumped- from refineries. This unused hydrocarbon is referred to as waste oil. First the chemical composition of waste oil was studied using HPLC, SARA fractionation as well as FTIR. This helped in selecting the proper emulsifier agent for the waste oil. The HCl-waste-oil emulsion was prepared using 15wt% HCl with a ratio of 70:30 of acid-to-oil. Extensive work was done to ensure reproducibility of data. This study was done to experimentally investigate the effects of distinct important variables on thermal stability and apparent viscosity of the HCl-waste-oil emulsion. These variables are; emulsifier concentration, mixing speed and the addition of corrosion inhibitor. All experiments were conducted at a high temperature 120˚C. To assess the quality of the HCl-waste-oil emulsion, droplet size was measured as a function of temperature (up to 100 ˚C).

The results showed that the HCl-waste-oil emulsion is a shear-thinning fluid. Power-law model was applicable seamlessly to all of the apparent viscosity data for all of measured temperatures. The optimum conditions were found to be; 0.5vol% and 600 rpm for emulsifier concentration and mixing speed, respectively. Apart from that, the emulsions prepared with corrosion inhibitor were less stable compared to the ones without it. However, the addition of corrosion inhibitor tended to generate monodispersed emulsion but bigger size droplets for temperatures 25˚C and 50˚C and polydispersed droplets for higher temperatures. Droplet size measurements showed an average size of 1.47 to 3.09 μm. Overall, the lab results showed a promising potential for the HCl-waste-oil emulsion to be used in the field.

This work takes into account; reducing the budget of acidizing job and at the same time environmental concerns. It makes use of the unwanted hydrocarbon from refiners. And, it saves the environment by using the waste oil instead of dumping it.

444994 Multiscale Simulation of Proteins Motion in the Microchannel

Abstract:

Protein adsorption is an important phenomenon, which plays a prominent role in chemical, industrial and medical areas. In this work, the dynamic behavior of the egg-white lysozyme has been investigated in a micro-channel with a multiscale modeling approach. The Molecular Dynamic simulation, free energy computation and Langevion Dynamics (LD) simulations have been combined to study lysozyme adsorption process on Au (111) surface. The gold surface has strong attraction to proteins. The protein-surface adsorption or desorption free energies is calculated. From the case studies of LD simulations, the surface has effects on the protein dynamic adsorption process. By the comparison of LD, MD and the experiments, the time step is concluded as an important factor to study protein adsorption behavior at the water-surface interface. Fine time step at the scale of 1 picosecond is necessary to provide enough accuracy to resolve proteins' dynamics trajectories.

445004 Grand Challenges in Future Energy and Chemicals: Climate Change and Sustainability As Drivers for New Technology and Innovation

Abstract:

Global demand for energy will double by 2050, driven by increasing population and affluence especially in emerging economies. While fossil fuels will continue to underpin the energy supply for the near future, concern over impact on climate change continues to grow, driving interest in expanded use of renewables and development of carbon capture and sequestration or re-use solutions. In addition to CO_2, water use and air emissions are critical issues to be addressed in future and current energy supply, as is energy storage for effective use of renewables. This presentation will provide an overview of what industry is doing today to address these issues, including the factors and metrics used in decision making.

445029 Electrospun Nanocomposite Fibers for the Removal of Heavy Metals in Water

Abstract:

Uniform electrospun nanofibers with small diameters were successfully prepared by systematically optimizing the electrospinning parameters including polymer concentration, salt loading, applied voltage, tip-to-collector distance, and flow rate. These parameters are shown to affect the fiber morphology and the addition of iron (III) nitrate salt was found to be able to reduce the fiber size. Due to the conductive nature of iron salt in polymer solution, even a relative lower voltage of 6 kV could generate a stable fluid jet and yield the nanofibers. The thermal properties of the composite fibers were investigated by thermal gravimetric analysis (TGA) and differential scanning calorimetry (DSC). Rheological studies were also conducted. The composite nanofibers were further carbonized to carbon nanofibers. Applications of the nanofibers in environment remediation have shown promise.

445030 Heavy Metal Detection in Treated Water Via a Fabric Nanocomposite Sensor

Abstract:

We present the conception of an electrochemical sensor for on-site/portable water analysis. This flexible sensor was developed to quantify the amount of heavy metals (in real-time), alleviating the wait time, large sample size, possible contamination, and expensive analytical equipment associated with current procedures for water analysis. This sensor is composed of a multiwalled carbon nanotube (MWNT) functionalized polymer mat, for a flexible yet conductive sensor. The MWNTs are then functionalized with a cyclo-oligomeric calixarene, which is shown to selectively form a supramolecular complex with metal ions. Upon complex formation, the charge carriers are drawn away from the carbon layer and the detection is prevalent at appropriate levels. The optimization of MWNT and calixarene functionalization, as well as the selectivity and sensitivity of the sensor will be discussed.

447119 Development R&D of Manoir Industries and Emphasis on Coating

Abstract:

The Senior VP Technical Support & Projects will present the development R&D of Manoir Industries and especially the coating (it is our lattest development) as well the services we can provide such as metallurgy expertise, advice, convector cleaning, site welding, welding supervision in order to increase the furnace productivity. So it is 60% R&D and 40% services. Manoir Industries manufactures cast components for the petrochemical market and develops high temperature alloys such as the Manaurites.

444478 Biogas Reforming with Addition of Carbon Dioxide to Syngas over Ni-Based Bimodal Pore Catalyst

Abstract:

Methane dry reforming of (MDR) has been widely studied due to its important applications in producing of syngas ($CO + H_2$), removal of two greenhouse gases, and upgrading of biogas (mainly composed of CH_2 and CO_2) into value added chemicals. Compared to the noble metal catalysts, nickel-based catalysts are suitable for industrial scaled-up MDR due to their high activity and low cost. However, nickel-based catalysts have been encountering a significant challenge that is the active metal sintering and coke deposition, leading to the catalyst deactivation and then the poor stability. Nevertheless, it has been reported that the deactivation of nickel catalysts can be suppressed by adding promoters like strong Lewis bases with the enhancement of chemisorb CO_2, such as MgO and CaO, and like lanthanide elements with the capacity of oxygen storage and release, such as CeO_2 and La_2O_3. Al_2O_3 is a commonly applied catalyst carrier own to its good pore size dispersion, high specific surface area and high mechanic strength.

Herein, we report that the bimodal NiCeMgAl catalysts with different NiO-loading were synthesized by the refluxed co-precipitation method and were evaluated for MDR performance by using real biogas from anaerobic digestion of food wastes. The sample containing 15wt% NiO-loading (Ni_{15}CeMgAl) was found to be active enough at 750 °C with a high CH_4 conversion of around 90%. Ni_{15}CeMgAl catalysts kept their bimodal porosity after reduction and MDR process. The evolution of the Ni_{15}CeMgAl catalyst before and after MDR process was investigated by SEM, TEM, XRD, and TGA techniques. The correlation between the structure evolution and catalytic performance change was also discussed.

444551 Continuously Adjustable, Molecular-Sieving "Gate" on 5A Zeolite for Distinguishing Small Organic Molecules By Size

Abstract:

Zeolites/molecular sieves with uniform, molecular-sized pores are important for many adsorption-based separation processes. Pore size gaps, however, exist in the current zeolite family. This leads to a great challenge of separating molecules with size differences at ~0.01 nm level. Here, we report a novel concept, pore misalignment, to form a continuously adjustable, molecular-sieving "gate" at the 5A zeolite pore entrance without sacrificing the internal capacity. Misalignment of the micropores of the alumina coating with the 5A zeolite pores was related with and facilely adjusted by the coating thickness. For the first time, organic molecules with sub-0.01 nm size differences were effectively distinguished via appropriate misalignment. This novel concept may have great potential to fill the pore size gaps of the zeolite family and realize size-selective adsorption separation.

16AIChE
Spring Meeting, Houston, TX

The Principle of Intrinsic Reality Solving the World's Woes through Chemical Engineering and Economics

Jed Moyano Bellen, SCPM, PMP
Hyundai Engineering & Construction Co. Ltd.
9 Temasek Blvd., #27-02 Suntec Tower 2, Singapore 038989
jed.bellen@hdec.co.kr

Prepared for Presentation at
American Institute of Chemical Engineers
2016 Spring Meeting
11th Global Congress on Process Safety
Houston, Texas
April 10-14, 2016

UNPUBLISHED

The Principle of Intrinsic Reality
Solving the World's Woes through Chemical Engineering and Economics

Jed Moyano Bellen, SCPM, PMP
Hyundai Engineering & Construction Co. Ltd.
9 Temasek Blvd., #27-02 Suntec Tower 2, Singapore 038989
jed.bellen@hdec.co.kr

Keywords: Principle of Intrinsic Reality, Chemical Engineering, Economics, Sustainability

Abstract

Human beings in profound rationality always seek for explanations of the phenomena around them. To understand these phenomena, human beings have developed systematized body of knowledge as a practical way of dealing with life's challenges. Among the systematized body of knowledge developed are chemical engineering and economics.

An avant-garde way of looking at each of these disciplines leads to the Principle of Intrinsic Reality. This principle relates the fundamental laws of the two disciplines to glean insights on the constraints imposed by conservation laws on economic systems at both the macro and micro levels.

It can be argued that if mass and energy in the universe are constant, then logically diminishing returns and marginal utility are bound to be observed. This logic is quite evident in a microorganism culture which follows the Monod growth model.

It can also be thought that conservation laws imply scarcity of resources. As such, it enjoins everyone to find optimal and efficient use of resources in the context of trade-offs, production decisions and sustainability. It even calls for the development of new technologies to produce more outputs from less inputs (i.e. shifting of production possibility frontier to the right). In this way, the Principle of Intrinsic Reality leads to the imperative for sustainable optimization.

Furthermore, it is deemed that the conservation laws can explain the phenomenon called "polarity of wealth." As mass and energy when harnessed create wealth, then the constancy of mass and energy implies boundary for wealth creation. In effect, wealth can neither be created nor destroyed but can only shift from one entity to another thereby creating polarity in wealth distribution. This assertion can be the key towards more equitable restructuring of economic systems to close income gap and thereby result to stable and resilient populace.

In conclusion, this paper envisions for a paradigm shift from traditional money-centric view of chemical engineering and economics to a synergistic convergence of disciplines which can powerfully tackle the world's woes and thereby effect sustainability.

1. The World in which We Live Today

The past few years saw unprecedented global challenges which have put into question existing economic models and scientific paradigms. Among these challenges are the dystopian global economy, climate change, environmental degradation and terrorism. The magnitude of these challenges has grown to a profound level that newer ways of thinking and perspectives become necessary. With that end in mind, this paper attempts to look at the fundamental laws of chemical engineering and economics in new ways to recognize our intrinsic reality from which these pressing challenges can be tackled at large.

2. The Principle of Intrinsic Reality

It is an observable fact that advancement in science increasingly blurs boundaries of various disciplines in tackling proactively the pressing global issues or problems. Within this purview, two seemingly diverging disciplines namely chemical engineering and economics can be taken together to glean insights on our intrinsic reality and what it means for our economic and ecological systems on a larger scale.

In the author's contention, intrinsic reality is grounded on the fundamental laws of chemical engineering and economics. These laws are the Conservation Laws (Chemical Engineering) and Law of Diminishing Returns (Economics). It can be argued that the Law of Diminishing Returns is observable because of the constraints imposed by the Conservation Laws. As such, this leads to the Principle of Intrinsic Reality which can be stated thus;

> *In any system, both at the macro- and micro-levels, the limiting factor for growth is the constancy of potential inputs.*

Potential inputs mean not the actual inputs to processes but the constant material or energy available which can be harnessed to create wealth. The reasoning or proof of this principle is presented in subsequent discussions.

2.1 Proof Based on the Monod Growth Model

The best illustration to understand the Principle of Intrinsic Reality is the Monod Growth Model. Monod in his 1949 seminal paper presented an empirical relation between exponential growth rate and concentration of essential nutrient. The data was fitted into hyperbolic equation:

$$R = R_K\left(\frac{C}{C_1 + C}\right)$$

In this equation, R is the growth rate during the exponential phase; R_K is the rate limit for increasing concentrations of C; C is the concentration of the nutrient; and C_1 is the concentration of nutrient at which the rate is half the maximum.

Figure 1: Graphical Representation of Monod Growth Model

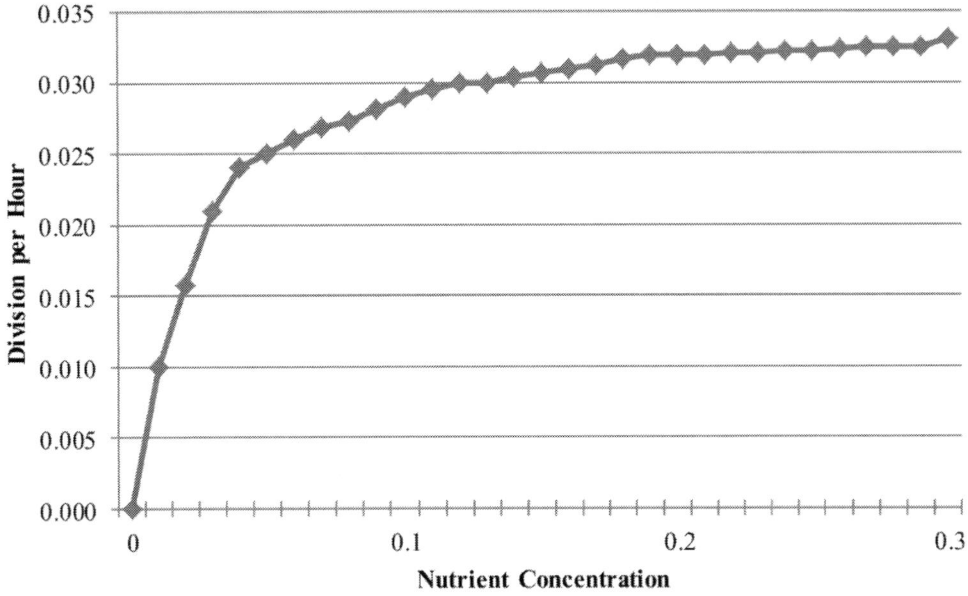

Monod (1949) further pointed out that the limiting factors commonly found are exhaustion of nutrients, accumulation of toxic metabolic products and changes in ion equilibrium.

The foregoing discussion has been observed at the micro-level with the bacterial culture. At the macro-level (i.e. the world at large), analogous phenomena are also observed to occur because of the same constraint imposed by the constant quantity of resources (i.e. material and energy). One example of these phenomena is that human beings cannot grow indefinitely. If they do, then how can the world sustain that growth with limited available resources (e.g. foods) to support it?

Another analogous phenomenon is in population growth. Thomas Malthus (1789) predicted that when population growth exceeds agricultural production, the population level is forced to return to subsistence-level conditions. This phenomenon as Malthus (1789) believed is due to the restriction of available resources supporting population growth, which others have claimed to be exploding as the earth becomes over populated.

In a surprising manner, a similar phenomenon can be observed in production processes in which the increase in a single factor of production assuming all other factors constant, results not to more marginal output but to a reduced one. This was attributed by classical economists, Thomas Malthus and David Ricardo, to the decreasing quality of inputs while neoclassical economists attributed this to the disruption in the entire production process as additional units of input are added to a fixed amount of capital. But in view of the Principle of Intrinsic Reality, the phenomenon is viewed intrinsic to the production system in which case infinite amounts of mass or energy cannot be processed with perfect efficiency. Hence, the law of conservation of mass and energy comes into the fore.

It can be deduced from the discussion that there is an underpinning intrinsic reality to the more obvious phenomena in the world from the micro-scale (bacterial growth) to the macro-scale (human growth and production processes).

2.2 *Proof Based on Diminishing Marginal Utility*

Another way to prove the Principle of Intrinsic Reality is through the diminishing marginal utility observable among humans and perhaps other organisms as well. Marginal utility, as defined in economics, is the gain from an increase or the loss from a decrease in the consumption of goods or services. The diminishing marginal utility is an observable phenomenon in which the utility or satisfaction an individual gains diminishes as he or she consumes more of the same unit of goods or services. Graphically, this can be represented as in Figure 2.

Figure 2: Graphical Representation of Diminishing Marginal Utility

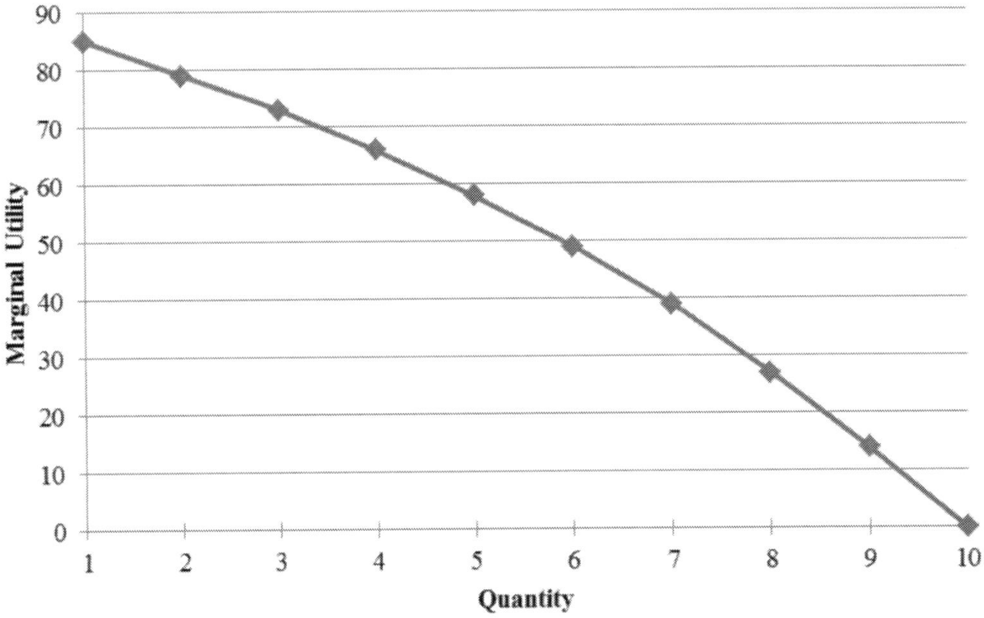

It can be deduced from this phenomenon and in support of the discussion in section 2.1 that human beings cannot grow indefinitely due to the fact that in our intrinsic reality, there is always a limit so as the constancy of available resources (i.e. mass and energy) can support our existence.

This means that in view of our intrinsic reality, we cannot have infinite greed otherwise an upsurge of unintended consequences are bound to happen (i.e. environmental degradation, terrorism and eventually demise of humanity).

Therefore, recognizing our intrinsic reality is the first step towards solving the world's woes and thereby resulting to sustainability and security.

2.3 Expanding the Production Possibility Curve through R&D

The caveat here is that the constant mass or energy (i.e. the limits) leads to trade-offs. Trade-offs is encountered when personal or business decisions are made. In making these decisions, one has to ensure that he or she becomes better off with his or her chosen option. We also realize in our life that we cannot get all that we want and we have to make trade-offs so long as it places us in a better or preferred position. This is more of a philosophical discourse, but it goes to show that the concepts of chemical engineering can be as useful in life and in business as it is in chemical processing.

To tackle these trade-offs, the Production Possibility Frontier in economics can be used. In view of the Principle of Intrinsic Reality where growth is limited by the fixed amount of material and energy, the Production Possibility Frontier needs to shift to the right as shown in Figure 3.

Figure 3. Graphical Representation of Expanding the Production Possibility Curve

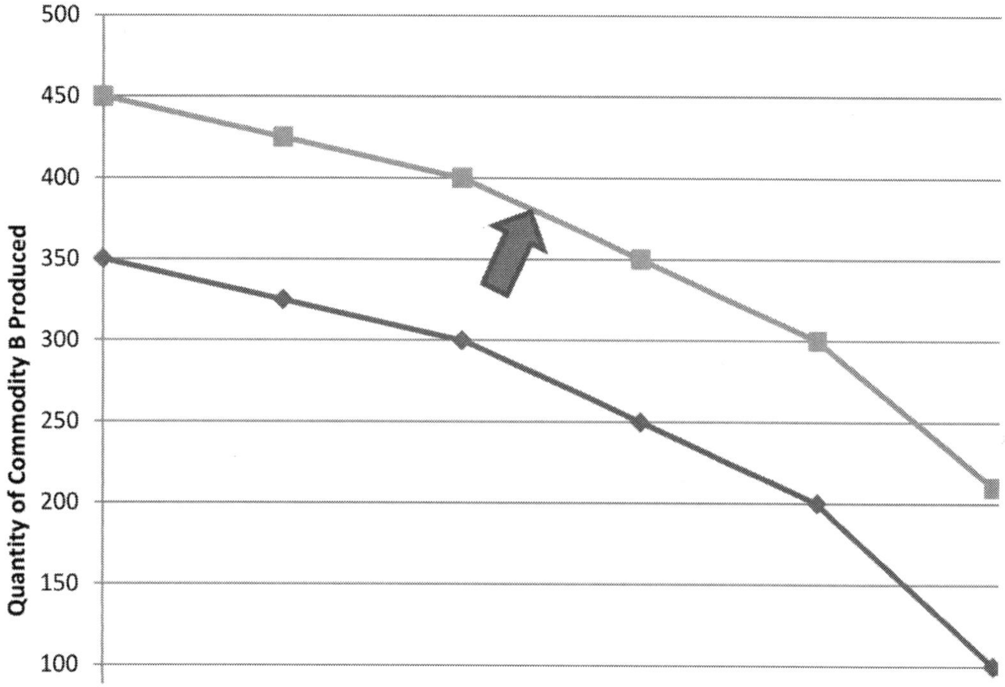

This shift has an implication on the practice of chemical engineering as Research and Development (R&D) is needed to achieve the said shift. R&D enables the economy to produce more of the two commodities (i.e. Commodity A and Commodity B) resulting in economic growth. The key challenge here to Chemical Engineering, particularly on Chemical Engineering R&D, is to look at the bigger picture where economic growth emanating from R&D output is pivoted on sustainability; considering its impact to the environment and the society at large.

3. Paradigm Shift in the Practice of Chemical Engineering

"Money" or "Profit Maximization" has been at the center of the practice of chemical engineering (Smith 2005). Various mathematical optimization methods are used to design systems, processes or components of processes with the end in mind of optimizing profits. This mental model however, will no longer work in our situation today as we face economic, ecological, social and spiritual crisis (Scharmer 2010).

To address this, Scharmer (2010) pointed out seven acupuncture points to shifting capitalism into a regenerative ecosystem economy. Within this purview, the practice of chemical engineering has a greater role to assume. For example, Scharmer (2010) suggested that the design of all production and consumption cycle be completely earth-to-earth, i.e. without necessitating for landfill and that it has to co-evolve with the natural ecosystem. In this regards, the way we do optimization should be revamped into sustainable optimization. By sustainable optimization, it means optimization of the entire system and adding in environmental as well as societal objectives to the equation.

In this definition, two things need to be considered. The first one is systems thinking which emphasizes the understanding of the entire system so the interaction among components will be known and unintended consequences mitigated (Senge 2006). This is a holistic approach to problem solving and finds enormous use in process design.

It should be noted however that the construction of processing plant, whether it is a greenfield or a brownfield project, is rooted on projected market demand. In some instances, economic volatility necessitates companies to experiment with multiple pathways to survive. In this circumstance and within the purview of systems thinking, design thinking comes into the fore. Design thinking is a collective principle emphasizing empathy for users, doing prototyping and tolerance for failure (Kolko 2015). This principle finds enormous use in chemical engineering designs such as in product development, process development, systems improvement and the like. This is also useful for simplifying or humanizing chemical engineering designs with the end in mind of contributing to sustainability. It also has to be noted that of paramount importance in using this tool is imagining the future with the view in mind of status quo and the likely outcome for sustainable development.

The second consideration is the inclusion of environmental and societal objective functions to the objective functions (economic criteria) of profit, cost, energy and yield (Edgar, et al. 2001) in optimization. The environmental objectives can be the impact or risk to environment and the societal objectives can be the impact or risk to society or common good.

These are just few examples of the needed paradigm shift in chemical engineering practice in consideration of the principle of intrinsic reality. But there can be more. A picture of the envisioned new paradigm for the practice of chemical engineering is shown in Figure 4.

Figure 4. New Paradigm in the Practice of Chemical Engineering

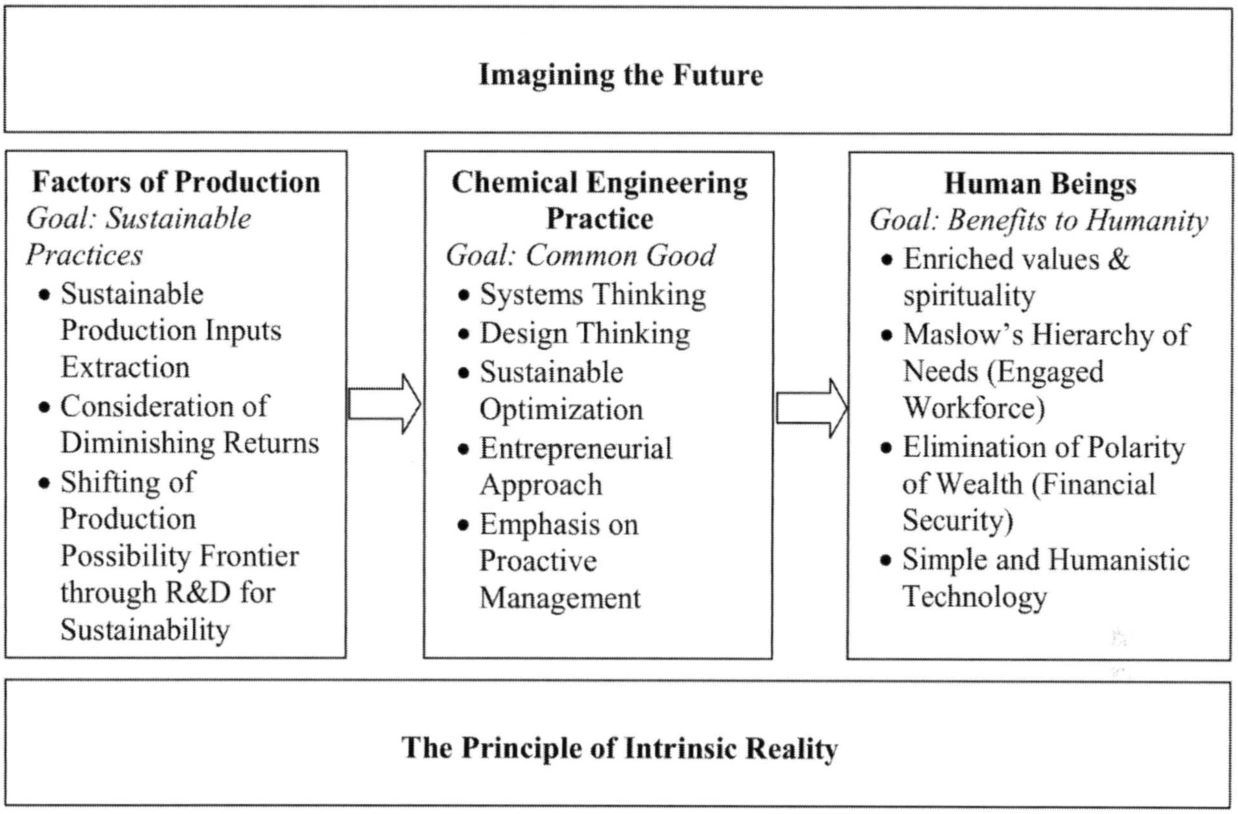

It should be realized that at the core of our economic systems is chemical engineering which can drive the economy into positive direction with the least unintended consequences. Related to that is capitalism in which upgrading of its operating systems can affect the practice of chemical engineering to a substantial degree. A discussion of this is provided in the following section.

4. Wealth Creation and Its Imperative for Sustainability

The development of this paper was inspired by the book "Freakonomics" in which the authors Steven D. Levitt and Steven J. Dubner (2005) asserted that economics can explain almost everything. It came to mind that if economics as a basic science can explain almost everything, then why not chemical engineering which was founded on chemistry, the central science?

Reflecting on it deeper, while explanations clear our doubts, we need more than that so we can tackle the world's woes. As such, this author thought that perhaps combining the two disciplines of chemical engineering and economics can be a potent force towards this end. How this can possibly be done is discussed in this section.

The starting point here is a review of the economic history of the world. In his book "A Farewell to Alms", Gregory Clark (2007) reviewed the economic history of the world in which he divided it into two periods; the period before 1800 and the period after. This is graphically represented in Figure 5.

Figure 5. Shares of Major Economies in World GDP through Time

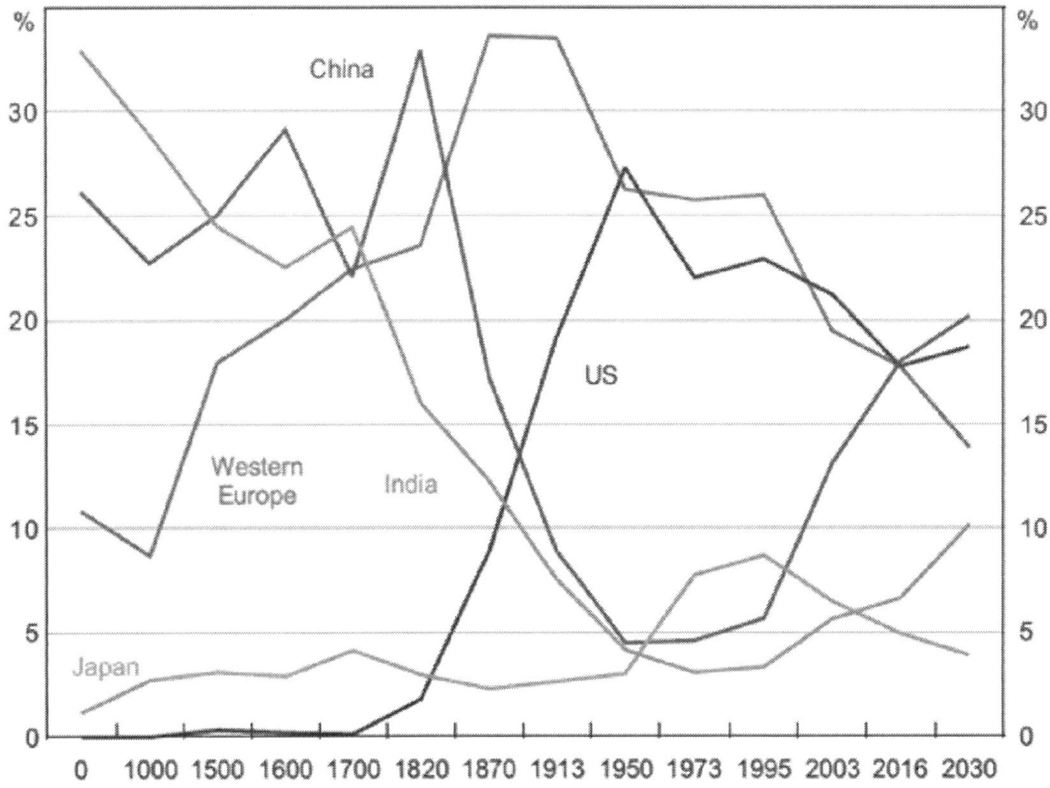

Note: Based on purchasing power parity (PPP) valuation of country GDP

Sources: International Monetary Fund (IMF), *World Economic Outlook* database, April 2011; Maddison (2006)

Thompson (2012) interpreted this graph by stating that "everything to the left of 1800 is an approximation of population distribution around the world and everything to the right of 1800 is a demonstration of productivity divergences around the world." In other words, everything to the left before Industrial Revolution depicts economic power by the sheer size of population and everything to the right after the Industrial Revolution shows the effect of mass production on country's economic output.

Clark (2007) on the other hand called the era before the industrial revolution, i.e. ca. 1800, as Malthusian era in which income per person varied across societies and epochs but growth is not observed. As Clark (2007) explained, this was due to a mechanism called Malthusian trap in which short-term increase in income through advancement in technology is dragged down by population growth. He further explicitly stated:

> "Economics as a discipline arose in the dying decades of the Malthusian era. Classical economics was a brilliantly successful description of this world. But the torrent of goods unleashed by the Industrial Revolution not only created extremes of wealth and poverty across nations, it also undermined the ability of economic theory to explain these differences."

In view of this statement, Figure 5 is reinterpreted in this paper. The contention of this author is that *polarity of wealth* can be gleaned from the graph. *Polarity of wealth* means the uneven concentration of wealth in the global economic system as some parts of the world are richer than others. The polarity can also be observed to shift as some richer parts of the world decline in economic power and some poorer parts of the world rise in economic wealth.

A historical example can be gleaned from Figure 5. India and China up until 1500 CE had 50 and 60 percent share of the world's economy, respectively. In 1 CE, China had a population of almost 60 million, while the area of the present day United States had a population of 680,000. In the late 18th century, the Industrial Revolution rendered countries with large population, just countries with the largest populations.

Looking at North America, it took them 1800 years to overtake China's economic output. However, by 1950, the US economic output was three times as great as China, even though its population is three times less. Comparing China with UK in 1913, the UK had a population of 45 million and China had 437 million but the economic output was almost the same. Analyzing what happened in Europe and later in the United States, Europe was on the lead even before 1800 (Nye 2012).

In 2010 according to data, China overtook Japan as the second biggest economy, though the numbers may not be strictly comparable. Japan's data are seasonally adjusted while those for China were not. Angus Maddison suggested that China and India were the biggest economies in the world for most of the past 2000 years. That is why economists find it a mystery why they fell so far behind and now they are flourishing ('China's economy' 2010). This may not be mysterious considering the *polarity of wealth* as was mentioned earlier and that this polarity can be argued to emanate from the Principle of Intrinsic Reality.

The reasoning here is that the world has constant mass and energy. Assuming that all these mass and energy can be harnessed to create wealth, it could mean that there is a limit for wealth creation. Theoretically, the limit can be expanded indefinitely through R&D but experience demonstrates that this was not achieved yet or getting anywhere near being achieved. As such, the constant mass and energy vis-à-vis the limit for wealth creation can be thought of as the reason for the polarity of wealth, i.e. wealth can only shift from one entity to another.

The implication of this assertion can be demonstrated by the fact that the 1% richest have accumulated more wealth than the rest of the world combined (Oxfam International 2016). The world has never seen 100% of its population to be wealthy. As such, time and again, equitable distribution of wealth is called for and mechanisms to achieve such_should be made a priority in government policies.

In a more concrete way, Scharmer (2010) suggested measures to address this growing inequality which is consistent with the mindset espoused by the Principle of Intrinsic Reality. These measures are as follows:

- Creation of a New Class of Common Property Rights

The creation of common property rights is intended to address the cost to society of private economic undertakings. For example, a private company is extracting natural resources from a certain place. While the private company can profit from that endeavor, the community in that place might suffer the undesirable consequences (e.g. flooding due to forest denudation). To ensure sustainability and institutionalize the voice of future generations in this case, a trust can be established which will own the rights and distribute funds (e.g. from cap-and-trade system for carbon emissions) to citizens (one share per person), communities and the government for investment in public infrastructure.

The trust is an effective means because it does not function like a company (driven by profits) or government (driven by special interest groups) but that it is accountable for long-term sustainability of the specific commons it manage for future generations.

- Creation of a Human Right to Basic Income

One way to ensure equitable distribution of wealth is to create a human right to basic income for every human being. This right when combined with free or affordable access to healthcare and education would create a playing field where everyone has a chance to pursue aspirations and dreams. In other words, it enables individuals to put their real creativity for the benefit of larger community. Another intention of this measure is to free up paid labor (a form of dependence) so that human beings can enhance their creativity and entrepreneurial skills.

- Inclusive Capitalism

To eradicate the *polarity of wealth*, capitalism should be made inclusive so that money and capital flows are directed to all sectors of the economic system. Currently, our money system leads to accumulation of oversupply of money and capital in high financial but low environmental and social gain areas. At the same time, there is an undersupply of money and capital in areas that address societal and community needs. It implies that there is too much money in areas where it is not needed but not enough on areas where it is needed. To address this imbalance, it is recommended to redesign the system so that the current extraction-based financial system is replaced by institutions which can support the real economy and thereby serve the ecology and community.

It should be noted that shifting of wealth is also observed in various geographical regions. When this occurs, geo-political instability can result. This goes to show that if we want lasting stability and security within the limits of our intrinsic reality, then we should not tolerate unabated greed. Otherwise, *polarity of wealth* will continue to persist and we cannot realize our full potential as individuals and collaborators for common good.

5. Way Forward...

Recognizing our intrinsic reality leads us to greater understanding of our very nature bounded by the constancy of mass and energy. It shed light on why growth stabilizes after addition of more inputs. Within this purview, trade-offs are also encountered which necessitates for R&D. R&D initiatives however should be underpinned on sustainability.

Looking at the practice of chemical engineering from the lens of intrinsic reality, the approach should be futuristic while considering relevant aspects related to factors of production and human beings.

The bigger picture shows that Engineers particularly Chemical Engineers are major players in driving the real economy. As Scharmer (2010) aptly puts it, "The essence of technology and knowledge creation, and the ultimate driving force of the economy, is the same: it is the human capacity to create." As such, if sustainability is to be achieved, Chemical Engineers have a greater responsibility to innovate for regenerative and social technologies. In this way, we can stem the polarity of wealth by providing affordable technology to enable disadvantaged individuals enhance their creativity, innovation and entrepreneurship. This can help them partake in the benefits of inclusive capitalism.

As demonstrated in this paper, the synergy emanating from the convergence of Chemical Engineering and Economics can powerfully tackle the world's woes. As such, there is a good reason for chemical engineers and economists alike to collaborate particularly on sustainability initiatives.

6. References

[1] 'China's economy overtakes Japan's in real terms' 2010, *The Economist*, 16 August, viewed 5 March 2016, http://www.economist.com/node/16834943

[2] Clark, G 2007, *A Farewell to Alms: A Brief Economic History of the World,* Princeton University Press, Princeton, NJ.

[3] Edgar, T, Himmelblau D & Lasdon, L 2001, *Optimization of Chemical Processes*, McGraw-Hill, New York.

[4] Huang, Y & Wang, B 2011, 'From the Asian Miracle to an Asian Century? Economic Transformation in the 2000s and Prospects for the 2010s', in H Gerard & J Kearns (eds), *Proceedings of the Reserve Bank of Australia annual conference*, Reserve Bank of Australia, Sydney, viewed 30 January 2016,

http://www.rba.gov.au/publications/confs/2011/huang-wang.html

[5] Kolko, J 2015, 'Design Thinking Comes of Age', *Harvard Business Review*, pp. 66-71, viewed 24 February 2016, https://hbr.org/2015/09/design-thinking-comes-of-age

[6] K.N.C. 2012, 'Mis-charting economic history: more 2,000 years in a single graphic', *The Economist blog*, web log post, 20 June, viewed 29 February 2016,

http://www.economist.com/blogs/graphicdetail/2012/06/mis-charting-economic-history

[7] Levitt, S & Dubner, S 2005, *Freakonomics: A Rogue Economist Explores the Hidden Side of Everything*, William Morrow, New York, NY.

[8] Malthus, T 1798, *An Essay on the Principle of Population*, J. Johnson in St. Paul's Church-yard, London.

[9] Maslow, A 1943, 'A Theory of Human Motivation', *Psychological Review*, vol. 50, pp. 370-396, viewed 24 February 2016, http://psychclassics.yorku.ca/Maslow/motivation.htm

[10] Monod, J 1949, 'The Growth of Bacterial Cultures', *Annual Review of Microbiology*, vol. 3, pp. 371-394, viewed 2 February 2016,

http://www.annualreviews.org/doi/abs/10.1146/annurev.mi.03.100149.002103

[11] Nye, J 2012, 'Fascinating new graph shows the 'economic history of the world since Jesus'', *MailOnline*, 23 June, viewed 15 November 2015,

http://www.dailymail.co.uk/news/article-2163610/Fascinating-new-graph-shows-economic-history-world-Jesus.html

[12] Oxfam International 2016, *An economy for the 1%, how privilege and power in the economy drive extreme inequality and how this can be stopped*, Oxford, viewed 5 March 2016,

https://www.oxfam.org/sites/www.oxfam.org/files/file_attachments/bp210-economy-one-percent-tax-havens-180116-en_0.pdf

[13] Samuelson, P & Nordhaus W 2009, *Economics*, McGraw-Hill Higher Education, New York, NY.

[14] Scharmer, O 2010, 'Seven Acupuncture Points for Shifting Capitalism to Create a Regenerative Ecosystem Economy', *Oxford Leadership Journal*, vol. 1, pp. 1-21, viewed 24 February 2016, http://www.oxfordleadership.com/journal/vol1_issue3/scharmer.pdf

[15] Senge, P 2006, *The Fifth Discipline: The Art & Practice of the Learning Organization*, Doubleday/Currency, New York.

[16] Smith, R 2005, *Chemical Process Design and Integration*, John Wiley & Sons Ltd., West Sussex, England.

[17] Thompson, D 2012, 'The economic history of the last 2,000 years in 1 little graph', *The Atlantic blog*, web log post, 19 June, viewed 29 February 2016, http://www.theatlantic.com/business/archive/2012/06/the-economic-history-of-the-last-2-000-years-in-1-little-graph/258676/

Acknowledgment

The author would like to thank Hyundai Engineering and Construction Co. Ltd. for the support in this endeavor. The author is also thankful to Ms. Sofia Ayu for meticulously editing this manuscript and Mr. Marcial N. Barroga for the poster design.

Azita Ahmadzadeh,
Paolo Palmas
Mike Sandacz

Computational Fluid Dynamic (CFD) Modeling Improves FCC Technology and Profitability

Honeywell
UOP

April 12th, 2016 | AIChE 2016 National Spring Meeting and 12th Global Congress on Process Safety | Houston, Texas

1

Outline

- Objective
- Introduction
- CFD modeling Setup
- Results
  - Comparing the UOP cyclone cluster design with conventional VSS™ configuration

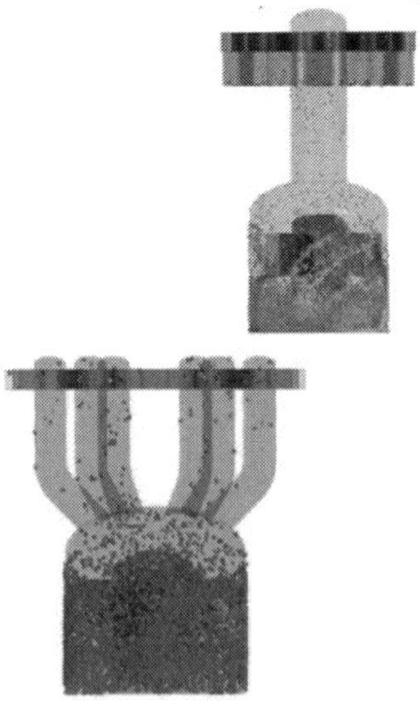

Honeywell
UOP

2

Objective

- Using CFD for equipment design & development
  - Reactor & regenerator design of Fluidized Catalytic Cracking (FCC) unit
  - Key challenge: how to optimize the size of the reactor vessel, which is determined by number and size of the cyclones
- The objectives of UOP's new FCC reactor design
  - Reduce the reactor size
  - Maintain the disengaging efficiency
  - Maintain the overall performance
  - Maintain equal and symmetric loading to all the cyclones

UOP 7331D-2

3

Introduction

FCC process is primary conversion unit to produce motor fuels and light olefins from

- Straight-run atmospheric gas oils
- Vacuum gas oils
- Certain atmospheric residues and
- Heavy stocks recovered from other refinery operations

UOP has dramatically increased the value of FCC technology

UOP continues to make advances and improvements in:

- Feed distribution system
- Riser termination devices and
- Spent catalyst stripping

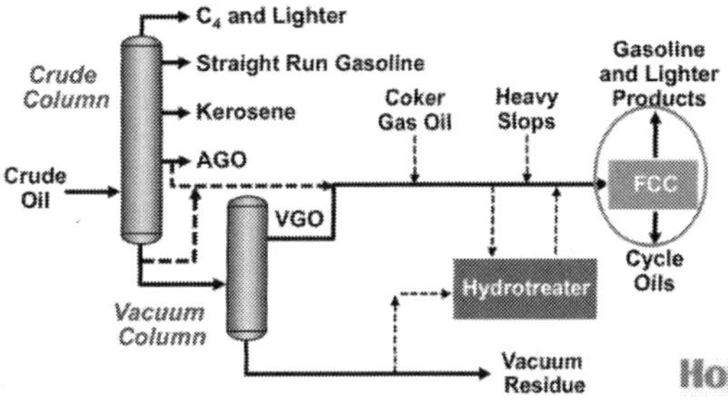

Honeywell UOP

UOP 7331D-3

Introduction

- Cracking reactions occur as the vaporized gas oil feed and catalyst flow up the riser.
- The catalyst is then regenerated to remove the coke deposition.
- Entrained catalyst is recovered through the cyclones in the upper section of the FCC unit.
- The cyclones are designed that gas velocity entering the cyclones is less than the critical velocity in which excessive erosion would occur.
- When increasing the through put, catalyst loading and gas velocity increase, resulting in larger units which add significant design challenges to design an FCC cyclone system that is efficient, reliable and cost effective.

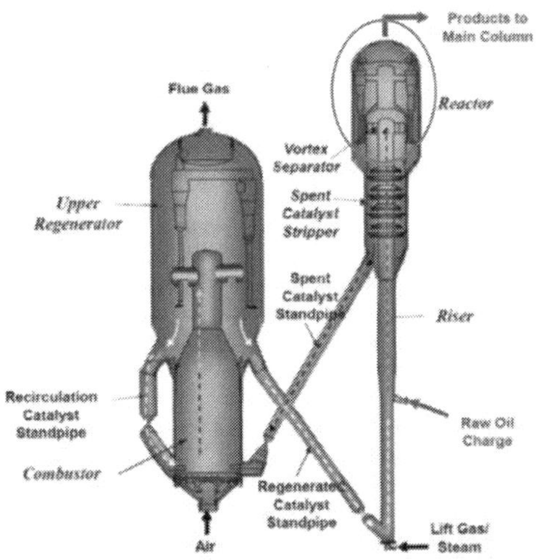

Honeywell UOP

VSS[SM] Vortex Chamber Flow

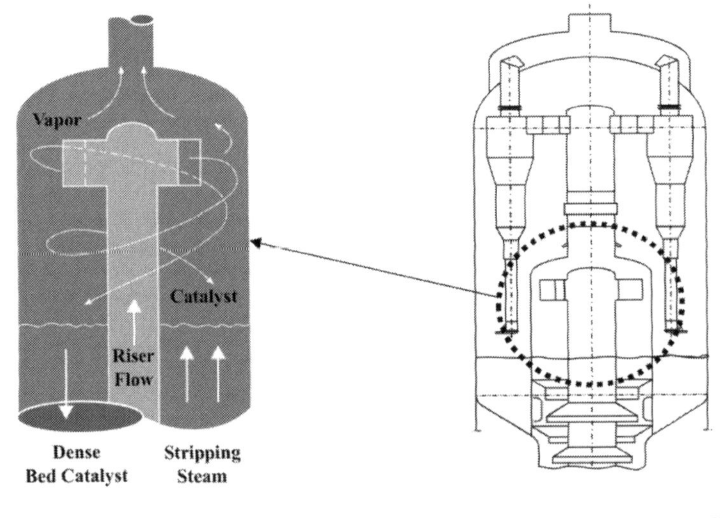

Honeywell UOP

FCC Reactor – World Class Size Unit
Typical VSS™ Configuration

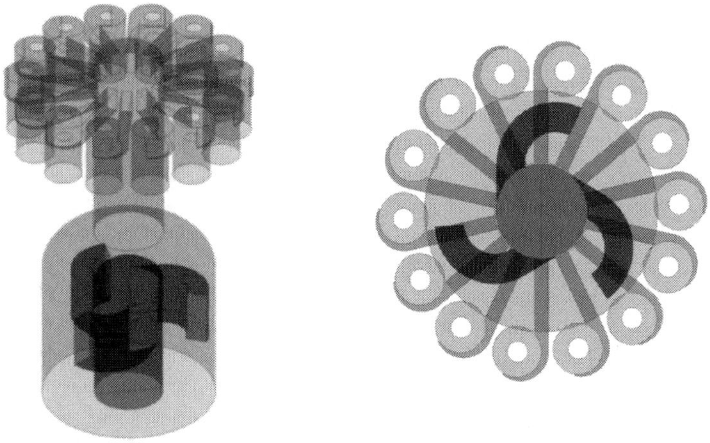

The number and size of cyclone set the diameter of the reactor

Honeywell UOP

FCC Reactors – Growing Pains

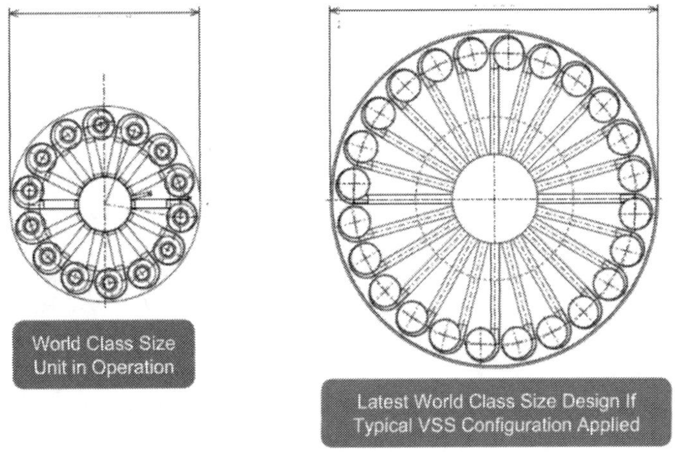

Following the traditional reactor design approach and cyclone configuration, the reactor diameter would be increased significantly.

Honeywell UOP

FCC Reactor – Decreasing the diameter

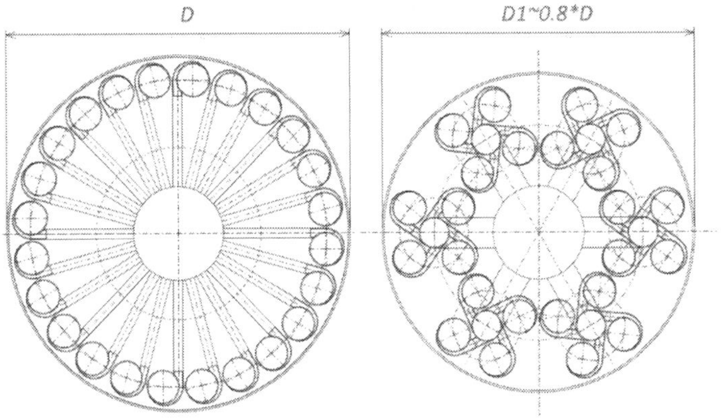

UOP cyclone cluster design solution

- *Split into multiple VSS outlet pipes and maintain equal cyclone distribution*
- *Cluster the cyclones to the respective outlet pipe with equal length cross-over ducts*
- *Reduced **CAPEX***

Honeywell UOP

CFD Modeling Set Up

- Disengaging Arm, for a better flow distribution into the cyclone clusters
- 3D Steady State
- Gas/Solid Flow
- Isothermal Flow
- Turbulent Flow (using RSM, Reynolds Stress Model)
- Incompressible Flow
- NO Reaction
- DPM (Discrete Phase Model)
- Compared with Conventional VSS Unit

Honeywell UOP

Results

& Discussion

Honeywell UOP

Post Processing Locations

Cyclone Cluster Design | *Conventional VSS Unit*

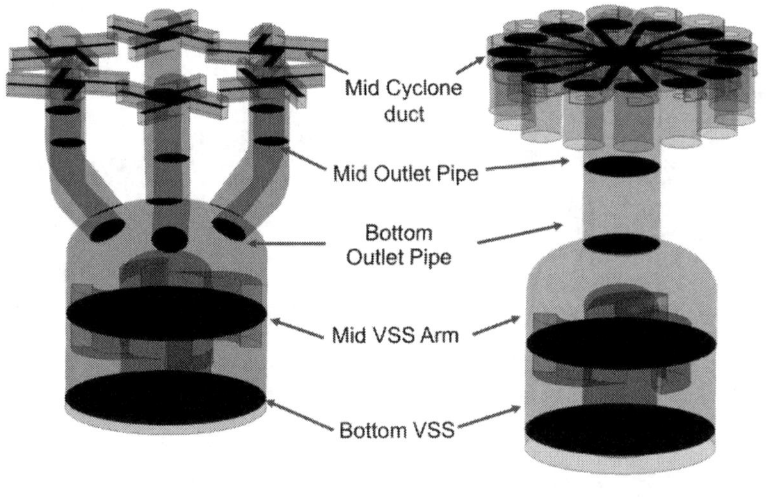

Honeywell UOP

Velocity Vector Plot at Bottom VSS

Cyclone Cluster Design

Gas Velocity Vector Plot at Bottom VSS

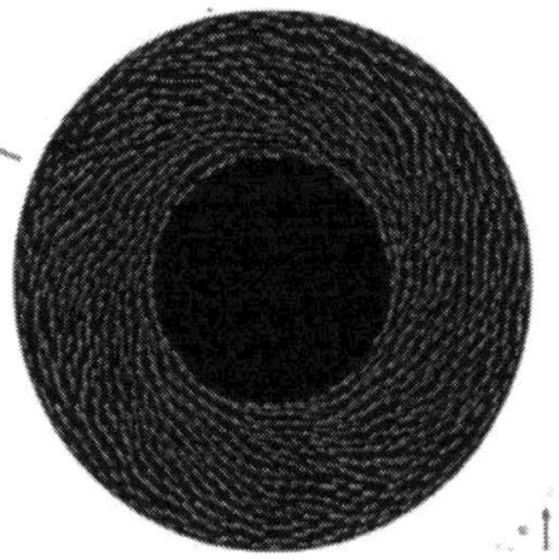

Conventional VSS Unit

Gas Velocity Vector Plot at Bottom VSS

- The Velocity Profile of the New FCC unit at the Bottom of the VSS is the same as Existing unit.
- The gas flow has rotational pattern at the bottom of the chamber, where the particles are leaving.

Honeywell
UOP

Velocity Profile – Mid VSS

Cyclone Cluster Design

Gas Velocity Vector Plot at Mid VSS Arm

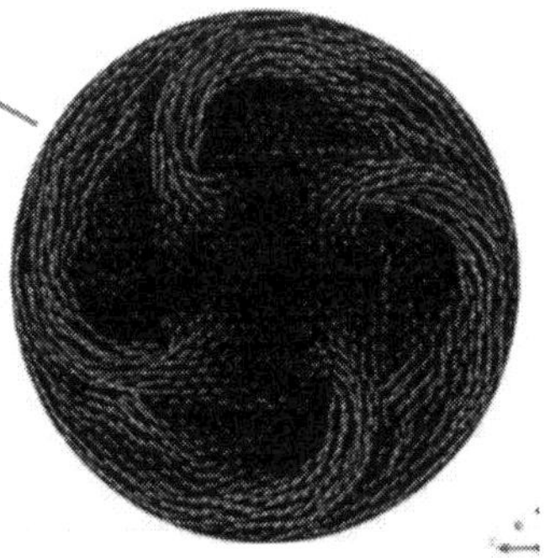

Conventional VSS Unit

Gas Velocity Vector Plot at Mid VSS Arm

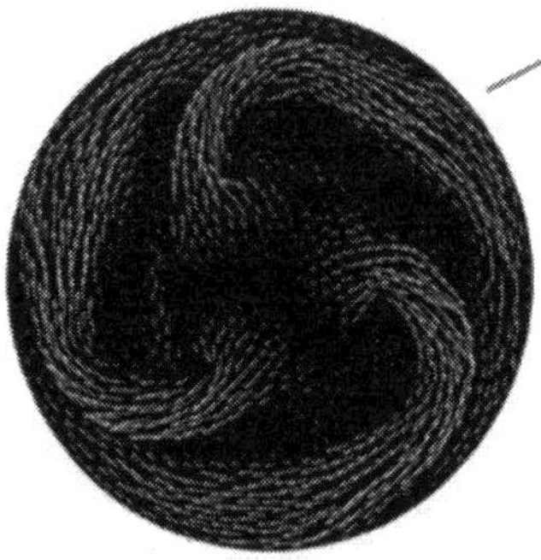

- The catalysts have high rotational velocity leaving the VSS arm and moving toward the chamber wall and move downward.

Honeywell
UOP

14

Velocity Profile - Chamber Pipe Entrance

Cyclone Cluster Design

Gas Velocity Vector Plot at Entrance of Chamber Pipes

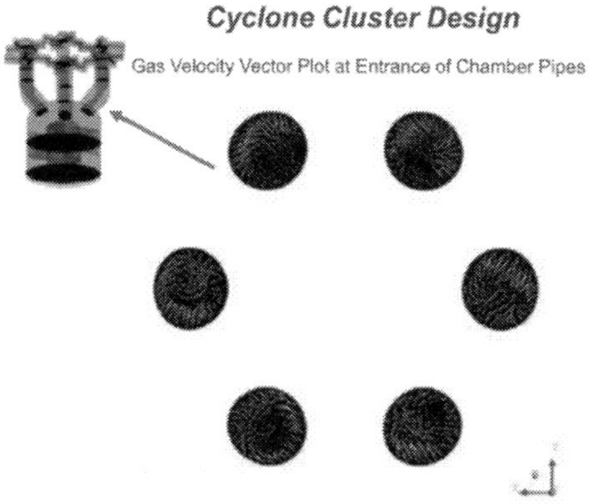

Conventional VSS Unit

Gas Velocity Vector Plot at Bottom Riser

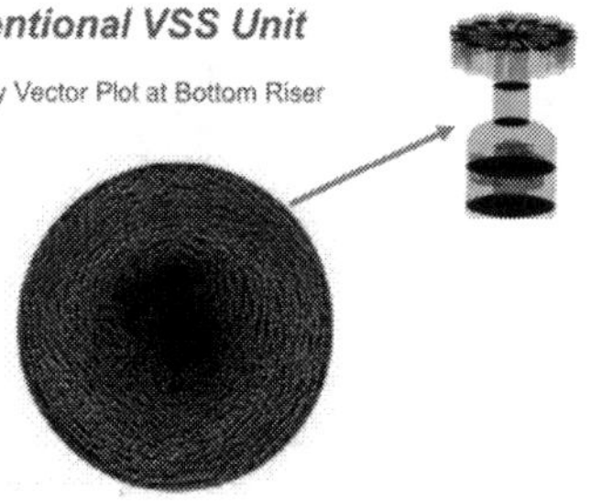

- Gas has a rotational pattern entering the chamber pipes

15

Velocity Profile - Mid Chamber Pipe & Mid Outlet Pipe

Cyclone Cluster Design

Gas Velocity Vector Plot at Mid Chamber Pipes

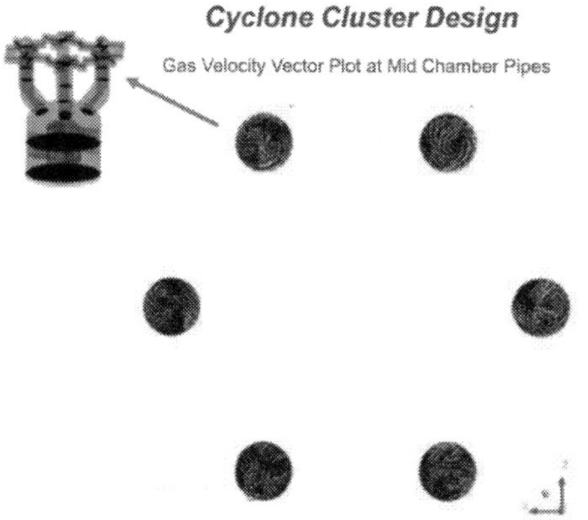

Conventional VSS Unit

Gas Velocity Vector Plot at Mid Riser

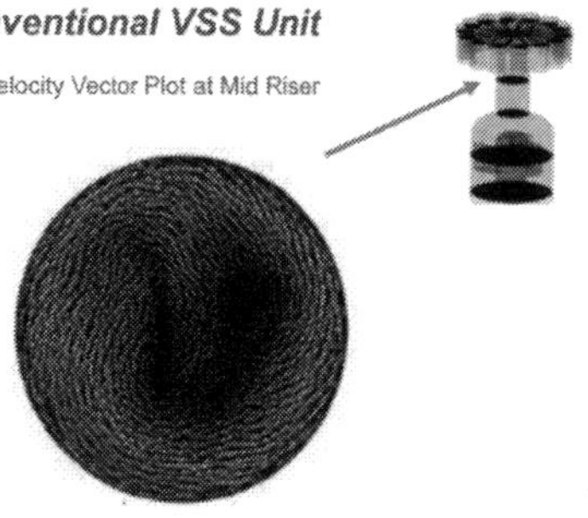

- Gas has a rotational pattern inside the chamber pipes of the Cyclone Cluster Design, the same flow pattern as in Conventional VSS Unit.

Velocity Profile – Mid Cyclone duct

Cyclone Cluster Design

Gas Velocity Vector Plot at Mid Cyclone

Conventional VSS Unit

Gas Velocity Vector Plot at Mid Cyclone

- Gas flow pattern inside the Cyclone Cluster Design ducts have the same pattern as Conventional VSS Unit.

Honeywell
UOP

Velocity Profile – Vertical Plane

Cyclone Cluster Design

Gas Velocity Vector Plot at XZ Plane

Conventional VSS Unit

Gas Velocity Vector Plot at XZ Plane

- In Cyclone Cluster Design, same as Conventional VSS Unit, there is a downward gas flow at the chamber wall, where it cause the catalysts to move downward.

Honeywell
UOP

Catalyst Concentration Profile

Cyclone Cluster Design

Catalyst Concentration Distribution

Conventional VSS Unit

Catalyst Concentration Distribution

- The Catalyst particles are moving downward by the wall and only a few particles leaving the chamber and going up to the outlet pipe(s).
- *The efficiency of the Cyclone Cluster Design is the same as Conventional VSS Unit.*

Honeywell UOP

UOP 73310-18

Residence Time Distribution

The residence time distribution in the Cyclone Cluster Design unit is the same as Conventional VSS Unit.

E Curve

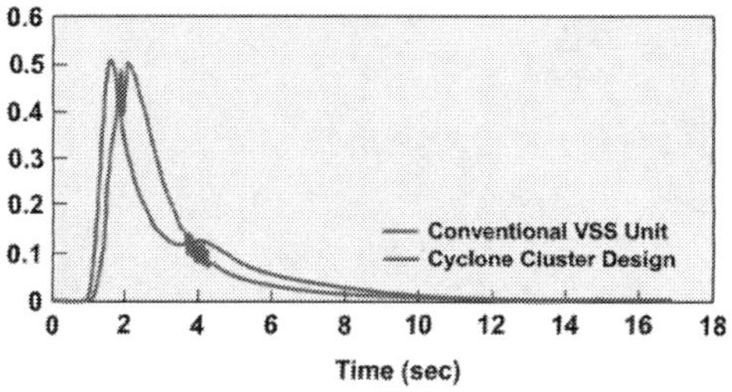

F Curve

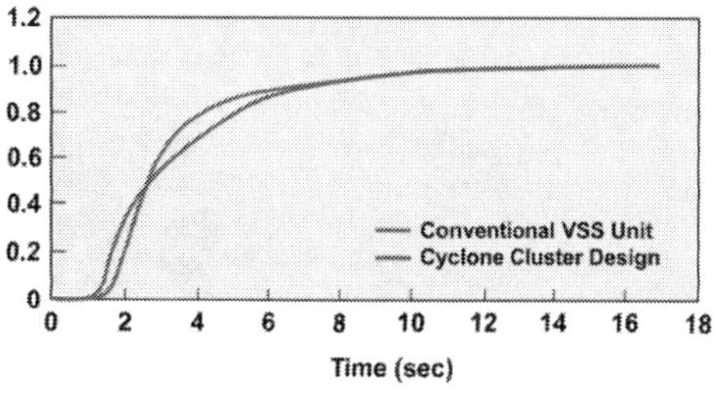

Honeywell UOP

UOP 73310-19

20

Conclusion

CFD modeling proved that this innovative approach achieved:

- Same flow pattern
- Same efficiency
- Same residence time

CFD Modeling allows for decreased vessel diameters which provides for a higher throughput at a lower capital & installed cost

Honeywell UOP

21

Honeywell UOP

444221 CFD Simulation of an Industrial High-Efficiency FCC Regenerator

Abstract:

A three-dimensional simulation of the gas-particle flow behavior, heat transfer, and coke combustion inside commercial high-efficiency FCC regenerator that includes a combustor and a regenerator vessel was carried out. The simulation has several goals: to evaluate the behavior under present conditions of operation, identify and propose solutions to operating troubles. The kinetic for coke combustion was simulated in MATLAB to understand its performance. The CFD simulation based on the Eulerian-Eulerian model for coke combustion was conducted using a commercial CFD package Fluent 15.0, that uses the finite volume method to solve the partial differential equations of continuity, momentum, energy and species transport by discretizing the equations over finite volumes. The Phase Coupled SIMPLE algorithm was used to adjust the pressure and velocities after each iteration when solve the gas-solid momentum equations. For simulate the combustor, a CFD simulation of the gas distributor was needed in order to define boundary conditions of the gas inlet for the combustor. The results obtained of the CFD simulation of the combustor shows lower concertation of the solid particles in this zone of the regenerator. Homogeneous mixture between solid particle and gas in the regeneration process is critical since most of the catalyst in the bed has very little coke on it, about 1 %w. At the same time, the CO content in the flue gases is high due inefficient mix between the air and the particles. Afterburning results when the CO content in the flue gases mix with the high oxygen content gases in the regenerator vessel. Afterburning causes thermal deactivation of the catalyst FCC and damage on cyclons due to hot spots. To correct these problems, different designs of the spent catalyst distributor was evaluated. The results obtained with the new designs of the distributor showed decreased of the afterburning phenomenon in the regenerator vessel.

April 12, 2016 AIChE Spring Meeting Presentation
Paper 66C

MODELING OF THE FIRST OF A KIND TECHNOLOGY FOR THE PROCESS DESIGN PACKAGES

MINGHUA YE (Ph.D.)
BRIAN PENG (Ph.D.)
And
SERGIO MOHEDAS

CHEMISTRY THAT MATTERS™

PROCESS DESIGN PACKAGE (PDP)

- **Purpose**: Develop and document technology information in sufficient details to allow an engineering, procurement, and construction (EPC)company to prepare the Front End Engineering Design (FEED) package for plant construction
- **Focus on:** ISBL design of all priori technology and licensed process units required for plant construction and operation to produce specified product and to achieve expected performance
- **Content:** All necessary process & technology documents

No. 1

PDP AND FEED DOCUMENTS

- ✓ Generate Plant-Wide H&MBs
- ✓ Generate Plant-Wide Utilities
  - Steam Consumption
  - Cooling Water Consumption
  - Coolant System
  - Electrical Power Consumption
- ✓ PFD (Major stream #, Operation conditions)
- ✓ P&IDs
- ✓ Equipment Process Datasheets & Specifications
- ✓ Piping, Instrument, and PSV Datasheets

No. 2

PROCESS AND DESIGN SOFTWARE

- ➢ **AspenPlus**
- ➢ **Hysys**
- ➢ **CHEMCAD**
- ➢ **PRO/II, Provision**
- ➢ **In-House Proprietary Design Tools**
- ➢ **Other special simulation software (CFD, Chemikin, etc,)**

No. 3

REQUIRED INFORMATION FROM R&D OR TECHNOLOGY DEVELOPER

sabic

- **Reaction:**
  Process chemistry
  Reaction Kinetics

- **Thermodynamic Package:**
  VLE, VLLE, LLE parameters for the process

- **Proprietary Calculations or Properties**

No. 4

SELECTION OF PROPERTY PREDICTION METHOD FOR SIMULATION

sabic

- **Chose the most suitable model / thermos method**
  Recommendation:
  1. Eric Carlson: "Don't gamble with physical properties for simulations", Chem. Eng. Prog., October 1996, 35-46
  2. Software recommendations (AspenPlus, Pro/II, etc)
  3. Bob Seader's Recommendation (University of Utah)

- **Verify the model calculated property typically VLE, LLE, VLLE with experimental data** (from lab, literature, encyclopedia, commercial databanks, etc.)

- **Estimate the parameters that unavailable from the simulation software**

- **Verify the theoretical stages of separation unit (absorption, distillation, extraction, etc.) vs. actual number of trays or the packing height used in commercial unit**

- **Run several scenarios to test if the simulated results are consistent with plant operating records**

No. 5

EXAMPLE: AZEOYTOPE VLE FOR HCl-H_2O – SELECT SUITABLE ACTIVITY MODEL

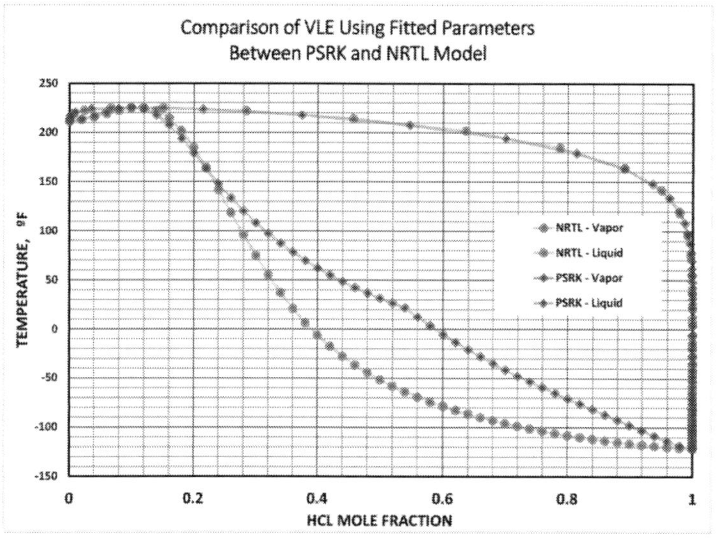

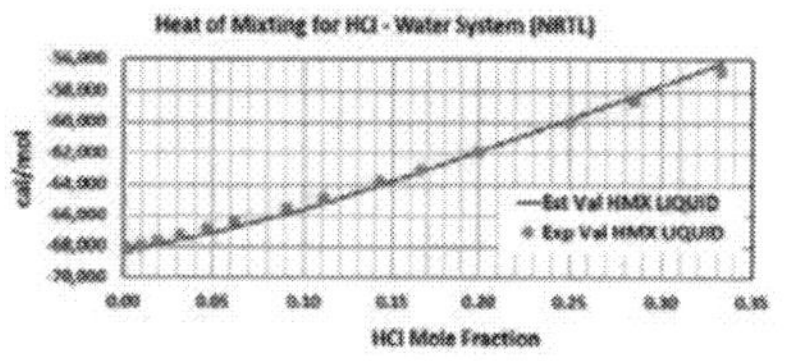

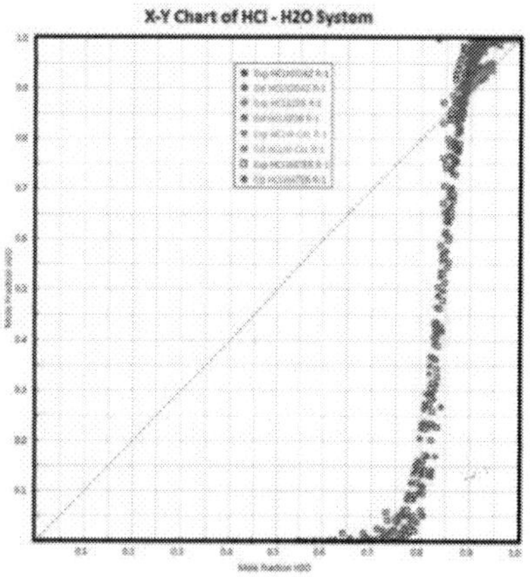

No. 6

CHEMICAL REACTION DATA FROM TECHNOLOGY DEVELOPER

- **Obtain reaction kinetics from the technology development group / literature**
- **Estimate kinetic model and parameters**
- **Implement the kinetic model into simulation software (subroutines, calculators)**
- **Verify simulated reactor results vs lab test, pilot plant data, and/or commercial unit data**
  - **Reactor type, size, and heat exchange**
  - **Catalyst amount required for SOR, NOR, and EOR**
  - **Yield for various operating conditions**

No. 7

EQUIPMENT DESIGN MUST MEET STANDARDS AND CODES

- **Except special proprietary equipment, the common equipment design must meet standards and code that used in industry**
- **The simulation result should provide sufficient data for process and mechanical design (listed in equipment datasheet).**
  - ✓ Capacity (flow rate)
  - ✓ Operating conditions (control case, design case, T, P)
  - ✓ Physical properties (density ρ; surface tension σ; thermal conductivity k; latent heat λ; critical property Tc, Pc; Vapor pressure of liquid at the operating temperature P_b^o;etc.)
- **Extra simulations need be performed for safety device calculations. PSV sizing requires property at relief conditions, for example:**

  $P_{max\ accumulated} = 1.10\ P_{max\ set}$ for single PSV, non-fire case
  $P_{max\ accumulated} = 1.16\ P_{max\ set}$ for multiple PSV, non-fire case
  $P_{max\ accumulated} = 1.21\ P_{max\ set}$ for single PSV, fire case

No. 8

LIST OF SOME STANDARDS AND CODES

- **API** (American Petroleum Institute)
  - ✓ API 521 Guide for Pressure-Relieving and Depressuring System
  - ✓ API 610 Centrifugal Pumps
  - ✓ API 660 Shell and Tube Heat Exchanger
  - ✓ TEMA Exchanger Selection
- **ANSI/ISA** (American National Standard Institute / International Society of Automation)
  - ✓ ANSI S5.1-1984 (R 1992): Instrumentation Symbols and Identification Process Design Base and Criteria
- **ASME** (American Society of Mechanical Engineers)
  - ✓ ASME B31.3 Process Piping Guide
  - ✓ ASME B73 Centrifugal Pump Datasheet
  - ✓ ASME Section I and VIII Pressure Vessel design

No. 9

EXAMPLE: SIMPLIFIED PROCESS FLOW DIAGRAM

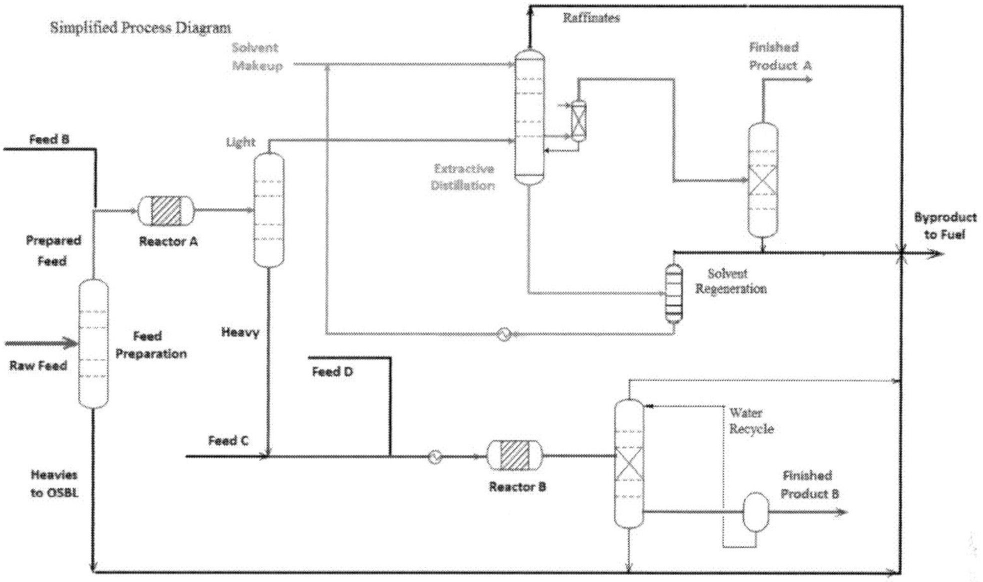

No. 10

CONSTRUCT SIMULATION MODEL ACCORDING TO PLANT PFD

- **Simulation Diagram must be consistent with Process Flow Diagram**
- **Simulation gives details for each unit and equipment**
- **Pseudo streams and equipment may be used to reflect the process and obtain stream flow rate and physical properties**
- **Understand the software limitations, Some features are not available in Pro/II and AspenPlus**
- **Make pseudo unit based on unit operation principals to simulate the technical data for equipment design**

No. 11

EXAMPLE: SIMULATION FLOW DIAGRAM OF A PLANT

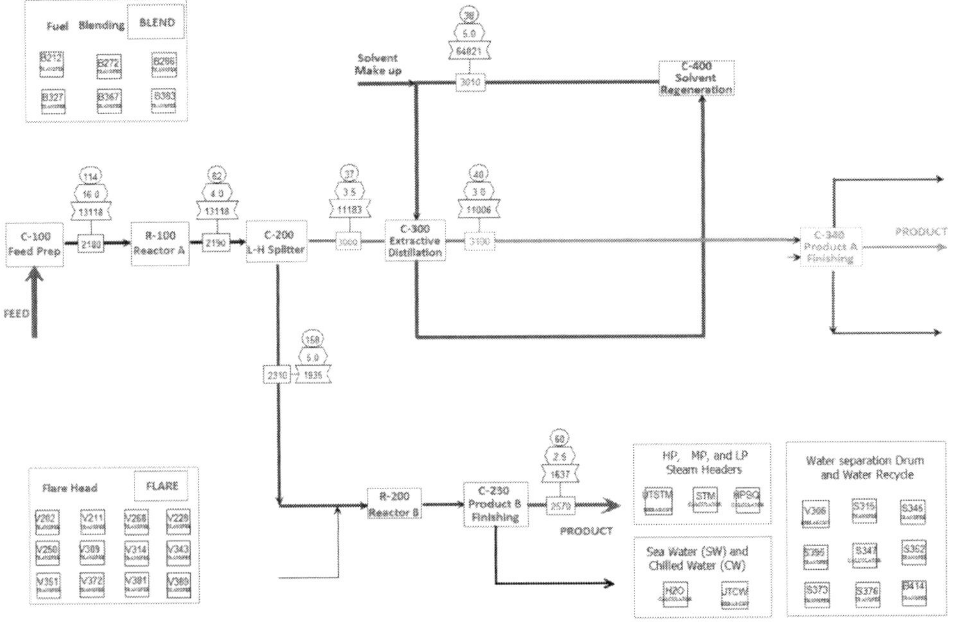

No. 12

EXAMPLE: PFD OF A DISTILLATION UNIT

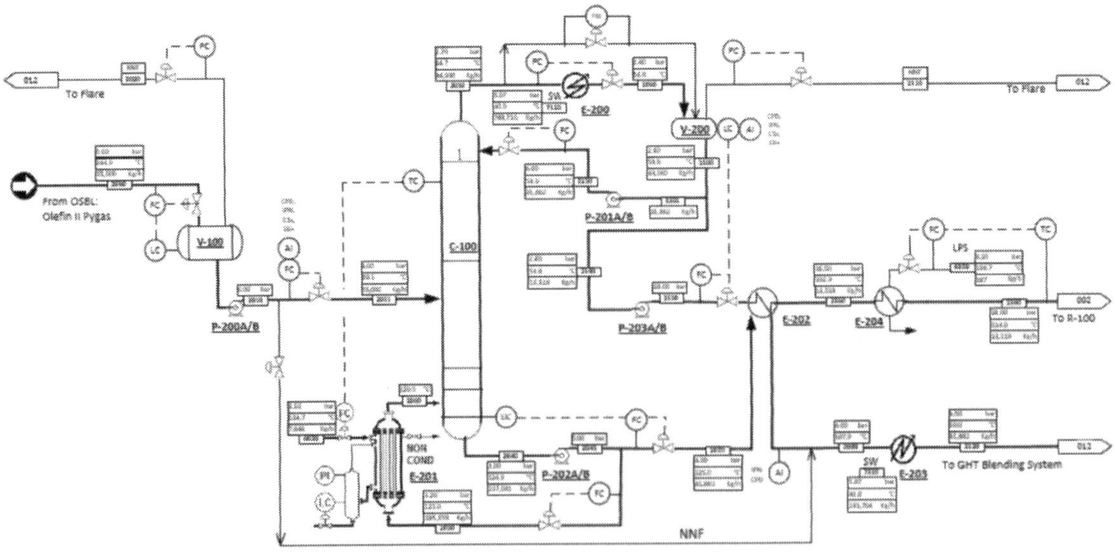

No. 13

EXAMPLE: DETAILED MODELLING THAT CONSISTENT TO PFD

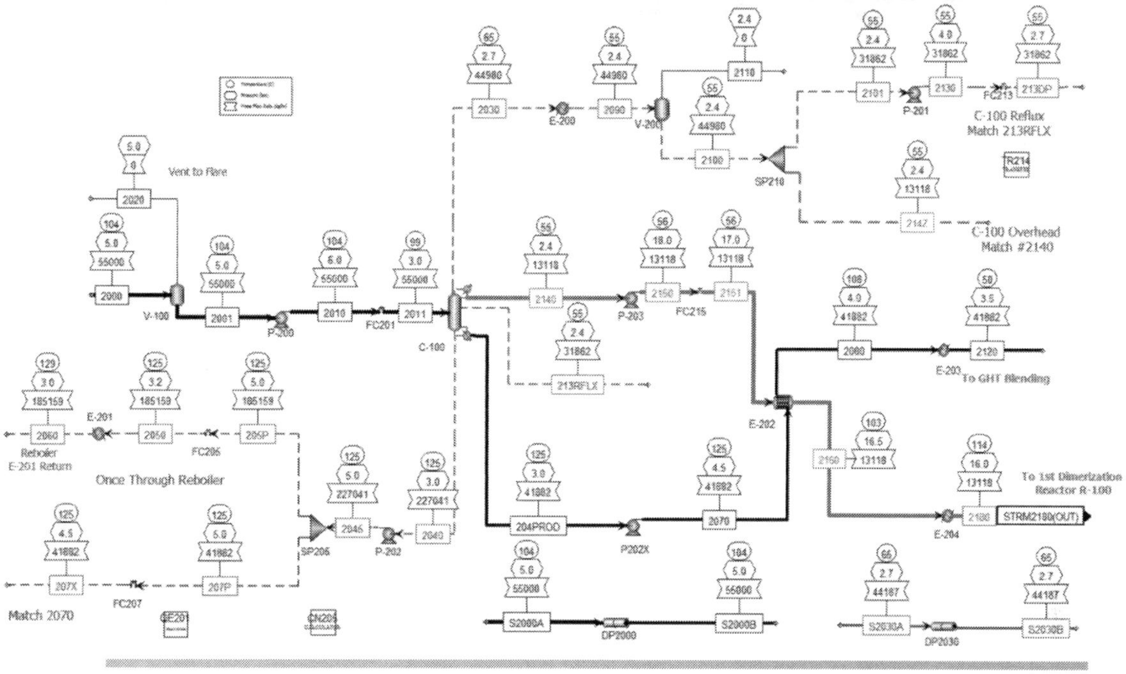

No. 14

HOW TO GENERATED PROCESS DATA FOR A THERMOSIPHON / ONCE-THROUGH REBOILER

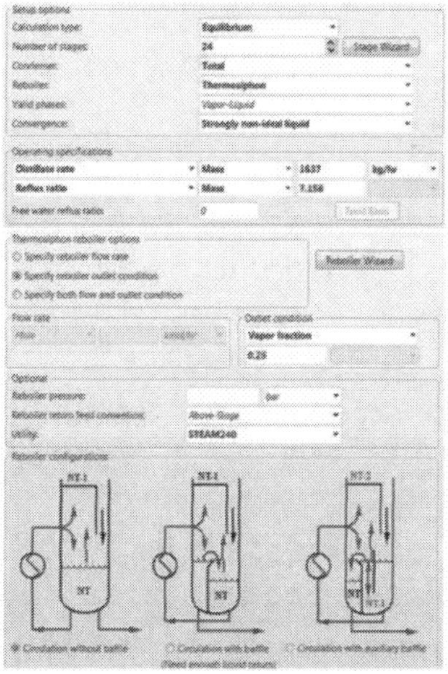

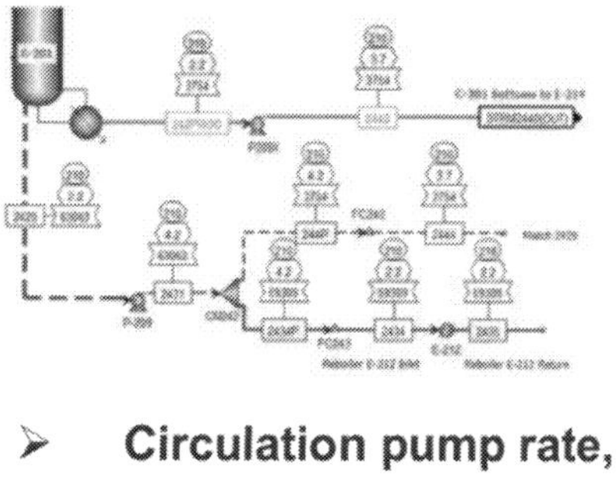

- Circulation pump rate, inlet / discharge pressure, electrical power required
- Control manfolder to divide bottoms product and the stream to thermosiphon reboiler
- Detailed stream property od reboiler inlet / outlet for piping and nozzle sizing

No. 15

HOW TO GENERATED PROCESS DATA FOR A THERMOSIPHON / ONCE-THROUGH REBOILER

Aspen Simulated Column Profile (Thermosiphon Reboiler)

Stage	T	P	Heat duty	Liquid from (Mole)	Vapor from (Mole)	Liquid feed (Mole)	Vapor feed (Mole)	Liquid product (Mole)	Vapor product (Mole)
	C	bar	MW	kmol/hr	kmol/hr	kmol/hr	kmol/hr	kmol/hr	kmol/hr
1	59.1	1.70	-1.45	199.1	0.0	0	0	24.41	0
2	66.3	1.72	0	131.2	199.1	0.00235	0	0	0
3	117.4	1.74	0	92.4	155.7	0	0	0	0
4	169.1	1.77	0	107.8	117.0	0	0	0	0
5	181.8	1.79	0	114.4	132.4	0	0	0	0
6	185.4	1.81	0	114.2	139.0	0	9.282	0	0
7	188.2	1.83	0	147.4	129.5	33.461	0	0	0
8	188.8	1.85	0	147.8	129.2	0	0	0	0
20	200.6	2.11	0	161.7	137.1	0	0	0	0
21	202.2	2.13	0	162.3	137.6	0	0	0	0
22	204.5	2.16	0	161.5	137.5	0	0	0	0
23	208.6	2.18	0	158.5	136.2	0	116.09	0	0
24	209.9	2.20	1.44	493.7	16.5	0	0	29.39	116.09

Aspen Thermosiphon Reboiler Heat Curves

Point No.	Heat duty	P	T	Vapor fraction
	MW	bar	C	
1	0	2.2	209.9	0
2	0.131	2.2	211.2	0.0181
3	0.262	2.2	212.3	0.0378
4	0.393	2.2	213.2	0.0589
5	0.524	2.2	214.0	0.0810
6	0.655	2.2	214.7	0.1039
7	0.785	2.2	215.3	0.1274
8	0.916	2.2	215.9	0.1514
9	1.047	2.2	216.4	0.1757
10	1.178	2.2	216.8	0.2003
11	1.309	2.2	217.2	0.2251
12	1.440	2.2	217.7	0.25

Thermosiphon Reboiler Data from Aspen Column Profiles

Results	Rate		T	P
	kmol/h	kg/h	C	bara
Liquid from column sump	493.7	82289	209.85	2.20
Liquid bottoms product	29.4	63456	209.85	2.20
Liquid to the reboiler	464.3	18832	209.85	2.20
Vapor Return to Reboiler	116.09	1967	Not shown	
Vapor fraction (mole)	0.25	= 116.09 / 464.3		

Pseudo Stream and Unit

Results	Rate		T	P
	kmol/h	kg/h	C	bara
Liquid from column sump	493.7	82289	209.9	2.2
Liquid bottoms product	29.4	63456	209.9	2.2
Liquid to the reboiler	464.3	18832	209.9	2.2
Vapor Return to Reboiler	116.1	1967	217.7	2.2
Vapor fraction (mole)	0.2			

HOW TO GENERATED PROCESS DATA FOR A KETTLE REBOILER

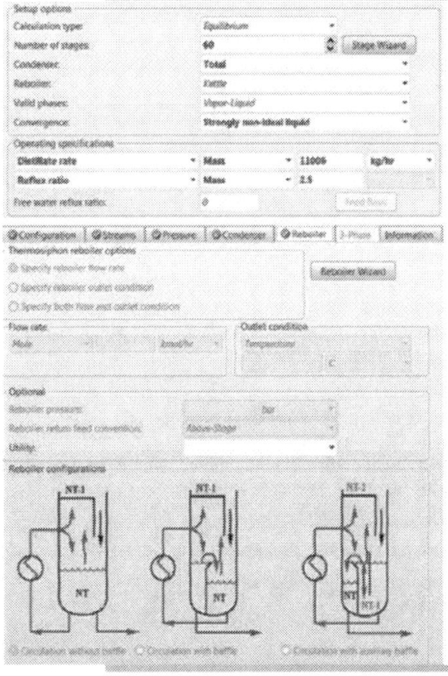

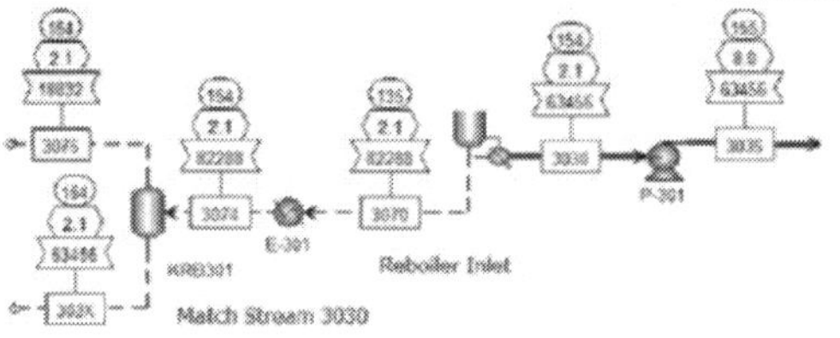

- The kettle reboiler performs one theoretical stage
- Liquid product is taken from the reboiler liquid outlet nozzle, not from column bottom (sump)
- Detailed stream property of reboiler liquid inlet / vapor outlet / Liquid Outlet for piping and nozzle sizing

HOW TO GENERATED PROCESS DATA FOR A KETTLE REBOILER

Aspen Simulated Column Profile (Kettle Reboiler)									
Stage	T	P	Heat duty	Liquid from (Mole)	Vapor from (Mole)	Liquid feed (Mole)	Vapor feed (Mole)	Liquid product (Mole)	Vapor product (Mole)
	C	bar	MW	kmol/hr	kmol/hr	kmol/hr	kmol/hr	kmol/hr	kmol/hr
1	39.6	1.70	-4.34	593.2	0.0	0	0	169.48	0
2	49.1	2.00	0	438.0	593.2	0.00000	0	0	0
3	56.0	2.00	0	1495.6	607.4	887.032	0	0	0
4	59.6	2.00	0	1512.2	778.1	0	0	0	0
39	82.8	2.06	0	1634.7	908.4	0	0	0	0
40	81.1	2.07	0	1780.3	917.2	162.012	0	0	0
45	128.6	2.07	0	1630.0	787.8	2.896	0	0	28.5618
46	131.4	2.08	0	1623.6	776.4	0	0	0	0
48	133.2	2.08	0	1619.3	767.0	0	0	0	0
57	133.8	2.09	0	1619.2	765.6	0	0	0	0
58	133.9	2.10	0	1618.3	765.7	0	0	0	0
59	135.2	2.10	0	1602.0	764.8	0	0	0	0
60	154.5	2.10	9.66	853.5	748.4	0	0	853.55	0.00

Aspen Kettle Reboiler Heat Curves				
Point No.	Heat duty	P	T	Vapor fraction
	MW	bar	C	
1	0	2.1	135.2	0
2	0.002	2.1	135.2	0.0000
3	0.879	2.1	136.1	0.0451
4	1.757	2.1	137.0	0.0899
5	2.636	2.1	138.0	0.1344
6	3.514	2.1	139.2	0.1784
7	4.393	2.1	140.6	0.2218
8	5.272	2.1	142.1	0.2647
9	6.150	2.1	144.0	0.3069
10	7.029	2.1	146.1	0.3483
11	7.907	2.1	148.5	0.3889
12	8.786	2.1	151.3	0.42852
13	9.665	2.1	154.5	**0.46718**

Kettle Reboiler Data from Aspen Column Profiles				
Results	Rate		T	P
	kmol/h	kg/h	C	bara
Liquid from column sump	1602.0	82289	135.2	2.10
Liquid to the reboiler	1602.0	82290	135.2	2.10
Vapor Return to Reboiler	748.41	18832	154.5	2.10
Liquid bottoms product	853.5	63456	154.5	2.10
Vapor fraction (mole)	0.467	**= 748.41 / 1602.0**		

Pseudo stream and Unit				
Results	Rate		T	P
	kmol/h	kg/h	C	bara
Liquid from column sump	1602.0	82289	135.2	2.10
Liquid to the reboiler	1602.0	82290	135.2	2.10
Vapor Return to Reboiler	748.4	18832	154.5	2.10
Liquid bottoms product	853.5	63456	154.5	2.10
Vapor fraction (mole)	0.467			

No. 18

SUMMARY

- **Be clear what data are required for develop a PDP and FEED**
- **Use commercial process and design software**
- **Obtain technology information from R&D or technology developer**
- **Select property prediction method for simulation**
- **Comply with Standards and Codes**
- **Verify thermodynamic parameters and reaction kinetics from various resources**
- **Construct simulation model according to plant PFD**
- **Detailed modelling to satisfied PDS demand**

No. 19

443615 Save Energy without Investing Capital

Abstract:

It is hard to get excited about saving energy during periods when energy prices are low, because the return on investment for energy efficiency projects is poor. However, not all energy savings require capital investment. Moreover, there are often additional benefits from improving energy efficiency, most notably increased plant throughput and reduced greenhouse gas emissions.

This paper examines various ways of saving energy in refineries and chemical plants without investing capital dollars, and the additional benefits that accrue from the efficiency improvements that are obtained. The discussion is illustrated with examples from recent projects around the world.

Heat Transfer Enhancement in Heat Exchangers to Reduce Plant Energy Demand

Tom Higley
Cal Gavin Limited
Cal Gavin Ltd, Minerva Mill Innovation Centre, Station Road, Alcester, Warwickshire, B49 5ET
tom.higley@calgavin.com

Prepared for Presentation at
American Institute of Chemical Engineers
2016 Spring Meeting
11th Global Congress on Process Safety
Houston, Texas
April 10-14, 2016

UNPUBLISHED

Heat Transfer Enhancement in Heat Exchangers to Reduce Plant Energy Demand

Tom Higley
Cal Gavin Limited
Cal Gavin Ltd, Minerva Mill Innovation Centre, Station Road, Alcester, Warwickshire, B49 5ET, UK
tom.higley@calgavin.com

Keywords: heat exchangers, heat transfer, energy consumption, low capital costs

Abstract

In today's market the pressure on process plants to maintain production rates whilst keeping within allocated budgets is increasing. The long term trend identifies energy consumption as a major cost source meaning that energy reduction strategies can be employed to improve plant economics. One form of this energy is heat input which is transferred around the plant using heat exchangers. If the efficiency of such heat interchangers can be improved then energy consumption can be reduced with an associated benefit to the economics of the process for the same throughput.

Utilizing heat transfer enhancement technologies is a powerful option to achieve reduction of energy demands. In comparison to the alternative of revamping or increasing plot space, the capital cost has shown to be significantly lower whilst achieving the same energy targets.

1. Introduction

In process plants the main source of energy consumption is through heat transfer. Whether it's heating or cooling of gases, liquids or 2 phase applications, heat transfer is needed to drive processes, either through recycled heat or provided through utilities.

The global market has increased competition in the process industries which has put plants under increasing pressure to improve throughput but maintain low production costs, i.e. becoming more efficient. Energy consumption has been given significant focus in processing due to its price volatility and environmental concerns. Therefore options to reduce energy consumption are a key tool for future competitiveness.

With new plant designs there is usually some flexibility built into the design to incorporate room for modifications or increases in production rates. Therefore heat exchangers are not necessarily designed on the limit with regards to the duties they can process compared to the process duties required. However judging the amount of additional margin to incorporate is difficult as it can depend on numerous factors, one being the service of the exchanger and also the fouling characteristics. In some cases where an exchanger is oversized, theoretically giving more heat transfer area and therefore more margin to process larger duties, the flow velocities are reduced such that the rate of fouling is increased, resulting in a 'bottlenecked' exchanger due to fouling. This scenario is counter intuitive as the prime aim of oversizing the exchanger has in fact accomplished the opposite and reduced the flexibility in the design in terms of processing duties. In order to cope with underperformance in these exchangers plant operations relies more heavily on utilities resulting in greater heat input and hence lower efficiency. Therefore when designing new exchangers for plants, particular care needs to be taken by not assuming more heat transfer surface area is the correct action needed for every service of exchanger.

For both new and existing plants there are traditional methods to analyze energy consumption which for new plants can optimize the overall layout. In addition to these methods there are technologies available that can either contribute to these traditional methods or resolve energy consumption problems where other methods don't.

2. Established Energy Saving Techniques

2.1 Pinch Analysis

For decades in the Process Industry there has been an established methodology for reviewing energy consumption for processes that have took different forms but fundamentally are based on the original approach of Pinch Analysis. Naturally this approach has developed to incorporate economic impacts, but overall this approach is defined as follows…
The methodology for minimising energy consumption of processes by calculating thermodynamically feasible energy targets (or minimum energy consumption) and achieving them by optimising heat recovery systems, energy supply methods and process operating conditions also known as process integration.

The historical fluctuations in the price of energy has at times focussed attention on energy saving. This naturally has steered the development of systematic process design methodologies for targeting and design of industrial processes which requires minimum energy consumptions. Pinch Analysis or Heat Integration emerged from early 1980 as a powerful tool to guide engineers to understand about how energy is utilised in the process context, and how much energy can be saved (e.g. for the technology state-of-the-art overview compendium see [4], [3]). This design approach, based on graphical concept of "composite curves", has been effectively applied to wide range of industrial sectors. This is mainly due to the design methodology being able to provide minimum energy targets as well as to achieve the targets through the design in an integrated and systematic manner.

The success of pinch technology comes from:

i) Pinch Analysis being able to screen objectives (e.g. minimum utility consumptions) before the detailed design.
ii) The engineer having full control of the decision making for the design of energy systems, using effective graphical representation of energy flows [7].

In order to deal with complex interactions existing in energy recovery systems, automated design methods using mathematical optimisation techniques have been widely developed during the last three decades. To mention just few: Kemp[2] providing overview of practical application to design of HENs (Heat Exchanger Networks), Ponce-Ortega [5] describing simultaneous retrofit and heat integration of chemical processes, suggesting HEN retrofit method using stage-model mathematical programming and commercial software for the design of heat recovery systems, e.g. Sprint © software, [8] or its commercial counterpart HEAT-int [1] and its industrial applications have been observed. The improved design methodologies exploit not only energy targeting but addresses the trade-off between area targeting and economics, (i.e. capital and total cost targeting) and complex network design [9].

2.1.1 Limitation of Pinch Analysis

When using Pinch Analysis, the scale of the analysis can cover a unit of a plant, or for the whole plant. Within this analysis hot and cold streams and the heat exchangers located on the streams are analyzed which form a heat exchanger network (HEN) of streams. For some plants these HEN's can vary in complexity and make it almost impossible to calculate using various algorithms without using Pinch Analysis. So fundamentally it has been found that this methodology can cater for such complexities as Mixed Integer Non Linear Programming. This programming however does need sophisticated software in order to optimize and generate an overall solution.

Significant developments have been observed in the design of energy recovery systems so far, however, there still exist considerable gaps and limitations for securing energy-efficient processing in manufacturing industries.

The main limitation of Pinch Analysis is that with powerful software suites processing the methodology, the proposed outcome may identify the optimum network but the practical implementation on a plant can be challenging. The idealised network for reduced energy consumption and changes needed are often not achievable within a given Return On Investment (ROI) for a plant, the capital costs can exceed plant budgets. This may occur because of limited space on a plant or process streams are established and require significant changes to pipe work and control systems if the scheme is implemented. This is a limitation that tends to occur with existing plants, for new plants Pinch Analysis is a definite approach to take in order to design an energy optimised process from the start.

In cases where Pinch Analysis is applicable on an existing plant, the assumptions made within the models do not necessarily take everything in to account. The models within this analysis do not take into account fouling characteristics or underperforming exchangers, and operational issues create a difference between the theory and the reality. Operational issues such as maldistribution of fluid flow through exchangers, available plot space on site, mechanical problems such as vibration, corrosion/erosion of static equipment and control issues are all drivers for a more audited approach to individual heat exchangers in order to increase energy efficiency.

2.2 Heat Exchanger enhancement

Focusing on individual or groups of heat exchangers naturally brings attention to operational problems that are occurring. In cases where operational issues impact the performance; heat exchangers tend to operate at full capacity with associated high energy consumption.
For new plants, the benefits of advanced heat exchanger technology are maximized if consideration is given during the conceptual process design phase. It is only at this stage that the engineer can manipulate the fundamental variables of the design to take full advantage of the range of equipment available. The later in the design process these options are considered, the greater the tendency for the rework cost to reduce the net cost benefits.

But for existing plant, consideration will first be directed at the debottlenecking of an individual heat exchanger. The purpose of such a project is usually to achieve higher throughput through increased heat transfer where heat transfer enhancement is an option. If this can be completed without the need for replacing pumps or the necessity for large pipework modifications, payback time is reduced.
Heat Transfer enhancement enables the increase of heat transfer for an exchanger such that it can cope with larger duties without the significant relative high capital costs traditionally incurred in adding primary surface area. Since various enhancements available work on different principles they are however not universally applicable (i.e. different enhancements will be suited to different process fluids, temperatures and operating flows).
Enhancement can be used on the shell side or tube side of the exchangers. It works either via promoting turbulence to increase heat transfer or extending the heat transfer surface area to achieve the same purpose.

2.2.1 Impacts of increasing heat transfer on a process

With an option of heat exchanger enhancement being made available without the need for significant plant changes, high capital costs immediately are eliminated when looking to improve performance.

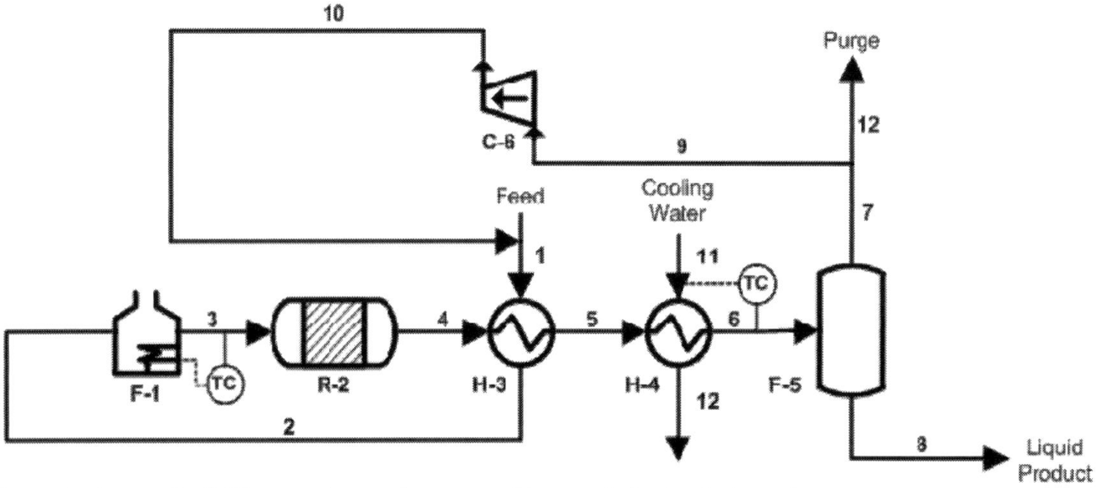

Figure 1: shows a feed/effluent exchanger, incorporated in a catalytic chemical reaction process.

Fig 1 shows a catalytic reactor that is a form of a plug flow reactor involving complicated treatment. The rate of this reaction inside the reactor is proportional to the amount of catalysts the reagents contact, as well as the concentration of the reactants. Another important factor is the inlet temperature of the feed (reagents); this temperature needs to be equal to the activation temperature for the reaction to occur effectively.

In order to maintain this inlet feed temperature a heat exchanger, known as a feed/effluent exchanger, followed by a fired heater preheats the feed to the required inlet temperature. Once the reaction takes place the effluent from the reactor enters the feed/effluent exchanger heating the feed entering the exchanger on the opposite side. The effluent stream is then further cooled by an effluent cooler before it reaches a separator where the liquid product is separated. The vapour phase is then recycled via a gas compressor and back in to the feed stream to be processed again. Any excess product in the vapour phase that cannot be compressed is purged.

In order to identify opportunities to reduce energy we need to determine where heat is input and which operations can affect this. For this particular process, energy is consumed by the fired heater in preheating the feed stream. The gas compressor will also consume energy in compressing the gas to recover the pressure loss from the feed stage to the point it reaches the compressor.

In addition to the reactor, the feed/effluent exchanger is also preheating the feed before the fired heater and is therefore ideally placed to affect the fired heater load.

If there is further heat recovery made by the feed/effluent exchanger the potential effects could be:

- Reduction in energy needed for the fired heater and as a result reduction in fuel consumed by the fired heater.
- With a higher inlet temperature sustained through increased heat recovery, this gives margin when the catalyst degrades and will sustain the product quality. With increased product quality, less product will be found in the vapour phase which reduces the amount being flared off in the purge stream.
- With less product in the vapour phase the energy needed to compress vapour will be reduced, reducing the energy consumption of the gas compressor.

The improved energy efficiency of the process associated with efficient heat transfer showcases the benefit of heat exchanger enhancement.

3. Heat Transfer Enhancement Technologies

On the market there are heat transfer enhancement technologies available that can be applied to both sides of a heat exchanger.
Traditionally increasing heat transfer surface area through extended surfaces has been a chosen route to pursue in enhancing the heat transfer. But apart from secondary surfaces different forms of exchanger components can be effective. A variety of baffles on the shell side can aid in improving heat transfer by changing flow and creating turbulence. Helical and EMBaffle-style baffles will in many circumstances perform better than standard conventional baffles. They can offer better heat transfer / pressure drop performance by changing the flow on the shell side. Each type of baffle creates the enhancement needed in different ways.

EM Baffle improves the heat transfer longitudinally along the tube creating turbulence around the tubes with every EM Baffle the fluid meets. With Longitudinal flow induced this enables shell side high flows to occur as the baffles create lower pressure drop compared to conventional baffles. Helical Baffles work on the principle of creating a helical flow on the shell side around the tube bundle, thus creating turbulence on the outside of the tubes improving the heat transfer. The advantage of the helical flow reduces the amount of dead flow zones usually found in conventional designs and also creates a beneficial heat transfer vs. pressure drop ratio.

For tube side heat exchanger enhancement technology, these technologies promote turbulence in different ways. In some cases creating turbulence incurred has demonstrated an increase of the heat transfer coefficient by as much as 14 times on the tube side, though the benefit varies with different flow regimes. In laminar to transitional flow regimes wire matrix turbulators (i.e. hiTRAN thermal systems) and twisted tapes have be shown to yield the best improvements

Twisted tapes work by inducing a swirl flow velocity which in turn creates shear at the tube wall creating turbulence, for this enhancement there needs to be a sufficient velocity in order for the swirl flow to be created. Cal Gavin's proprietary technology, hiTRAN thermal systems, physically disrupts the fluid flow creating radial mixing of the fluid at the tube wall as well as mixing the fluid with the bulk flow too.

Each type of technology has shown to be effective in increasing heat transfer and benefit in heat recovery.

3.1 hiTRAN Thermal Systems

For fluid flow that is operating in the laminar to transitional flow regime, transferring heat to or from a fluid flowing through a tube has limitations. The problem lies at the tube wall, where friction and viscous forces within the fluid combine to produce a thin layer of fluid that is scarcely moving called the boundary layer. The mechanism which carries heat through this "boundary layer" is conduction, which takes place only slowly for fluids with low thermal conductivity – such as oil.
It is when this boundary layer is disrupted that the heat transfer can be increased which is achievable with hiTRAN thermal systems.

3.1.1 Fundamentals of hiTRAN thermal systems

hiTRAN thermal systems solve the problem of poor heat transfer by increasing the velocity close to the wall, eliminating the troublesome boundary layer. As Figure 2: shows, under laminar flow conditions streams of blue and red dye cling to the wall of an empty tube (A), but quickly mix into the bulk fluid as soon as they encounter the hiTRAN thermal system (B). This mixing occurs due to the shear stress induced by the wire loops of the hiTRAN thermal system at the tube wall. This shear stress then creates forced turbulence reducing the impact of conduction. As the boundary layer is significantly reduced, this changes the heat transfer from conduction to convection. As convection transfers heat at a faster rate compared to conduction then this can greatly improve the tube side heat transfer. In cases where the heat transfer is limited on the tube side, i.e. (the heat transfer coefficient on the tube side is lower to that of the shell side) then the increase in the tube side heat transfer favourably increases the overall heat transfer of the heat exchanger.
With a higher rate of heat transfer the heat flux increase, which increases efficiency and reduces energy demand on the process system.

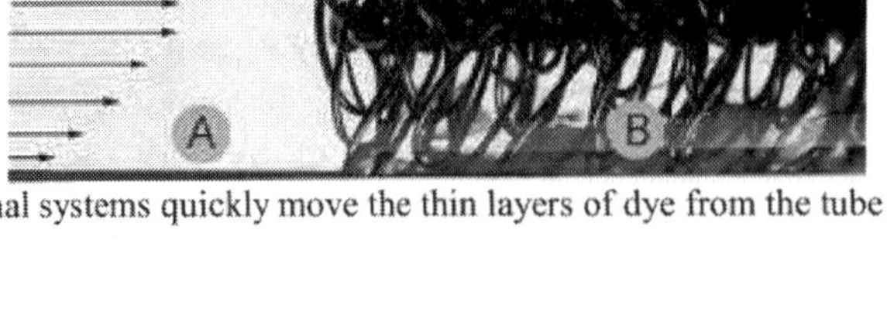

Figure 2: hiTRAN thermal systems quickly move the thin layers of dye from the tube wall into the bulk fluid.

4. Heat transfer enhanced case studies

4.1 Enhancement of a feed/effluent Texas Tower

Similarly to section 2.2.1, where the operation of a feed/effluent exchanger was given, a case study based on such a process is shown. At Cal Gavin Ltd there was an opportunity to reduce the overall energy consumption of a process by improving heat recovery on a feed/effluent heat exchanger.

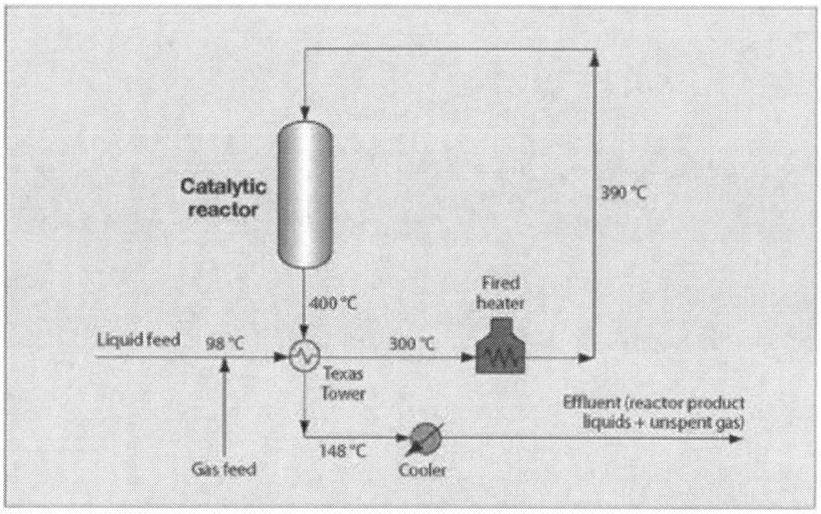

Figure 3: PFD showing feed/effluent Texas Tower

Shown in Fig 3, the product stream leaving the reactor consists of a mixture of unreacted feed gases and desired products. This is fed to the shell side of the Texas Tower, where it enters at the top and flows downwards. As heat is transferred to the feed stream, the product stream is first desuperheated and then partially condensed. It is then fed to a series of coolers before the desired liquid products are separated from the unreacted gases.
Most catalytic reactors are subject to decay of the catalyst. A common means of maintaining the production rate as the catalyst decays is to increase the temperature of the reaction. Eventually, the operation will reach either the maximum reactor temperature or the maximum firing capacity of the fired heater and the plant will have to be shut down in order to renew or regenerate the catalyst. The load on the fired heater increases across the operating period, so therefore the duty is ever increasing

The obvious way in which the feed/ effluent exchanger contributes to the economics of plant operation is by reducing energy consumption explained in section 2.2.1. However, while the saving in energy costs accumulated over the plant's operating period is substantial, this is not the only benefit. For a fired heater of fixed capacity, the higher the level of heat recovery in the feed/effluent exchanger, the longer the plant's operating period. This has two major cost advantages: the plant has a higher capacity; and the plant produces more products for a given batch of catalyst.

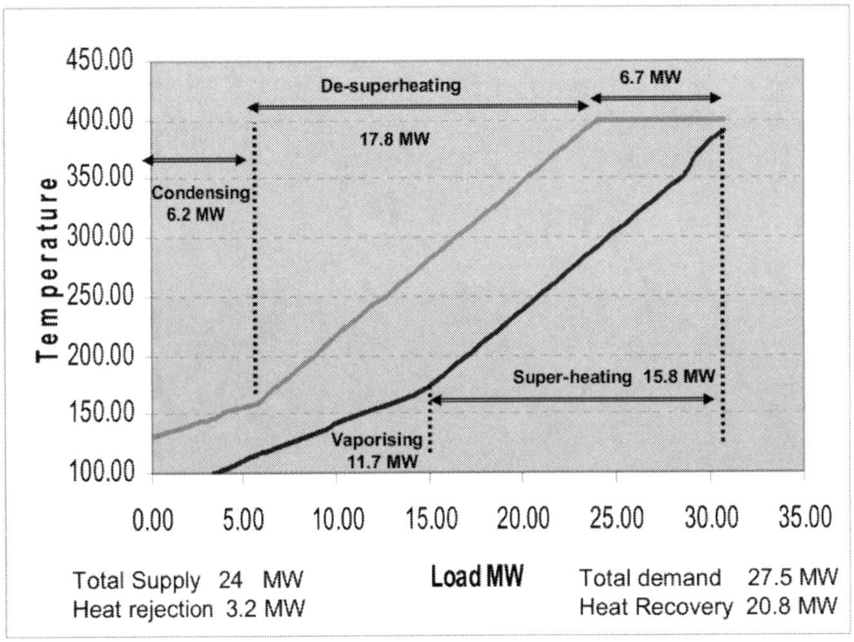

Figure 4: Heat recovery around reactor

The thermodynamics of heat recovery in a feed/effluent exchanger can be determined by superimposing curves showing the heat demand (reactor feed) and the heat release (reactor effluent) for the case described above as functions of temperature (see Figure 4).

The curves have been superimposed at a minimum temperature approach of 50°C. The distance the demand curve (the lower curve) extends beyond the release curve at the hot end of the scale shows the demand placed upon the fired heater. In this example, the reactor feed is heated to a temperature of 300°C, the feed has to enter the reactor at a temperature of 390°C and the load on the fired heater is 6.7 MW.

The total demand of the feed stream is 27.5 MW and consists of 11.7 MW associated with vapourisation of the feed stream and 15.8 MW for superheating the stream to 390°C. Given that 6.7 MW is provided by the fired heater, heat recovery provides 20.8 MW.

The heat released by the reactor product as it is cooled to 130°C is 24.0 MW. This figure comprises 17.8 MW for de-superheating the gas and 6.2 MW associated with condensation of product. The amount of heat being absorbed in the cooler positioned after the feed/effluent exchanger is 3.2 MW. Hence, the 6.2 MW associated with condensation is divided between the feed/effluent exchanger (3 MW) and the cooler (3.2 MW).

In the laminar and transitional flow regions hiTRAN thermal systems are the optimum enhancement technology to use. Not only can hiTRAN be used in the liquid phase, it can also be used in gas flows at quite high Reynolds numbers.

A project in which hiTRAN thermal systems were used involves the Ruhr Oel plant in Gelsenkirchen, Germany. The plant had been operating successfully for 30 years, but had reached maximum capacity. The operator wanted to increase throughput by 17.5%.

The existing feed effluent exchanger matched feed and effluent associated with an Isomar Process catalytic reactor for paraxylene production. The exchanger comprised 1734 tubes of 20 mm outside diameter and 12 m length in a 1360 mm diameter shell. This gave a heat transfer surface area of 1256 m^2.

At the established throughput, the feed stream exited the exchanger at a temperature of 327°C. Simulation of the performance of the Texas Tower at increased throughput indicated that the temperature at which the feed left the unit would fall to 310°C. This meant that at end-of-run conditions the load on the fired heater would be above its capacity. Consequently, the operator had either to improve heat recovery or renew the catalyst more frequently. The last option would defeat the object of the revamp, because it would result in a significant increase in operating costs associated with catalyst renewal, which would in turn lead to reduced plant capacity because of increased outage time.

The operator was left with three options: replace the fired heater and burn more fuel; replace the Texas Tower with a new, larger unit; or improve the performance of the Texas Tower. Replacement of the fired heater would involve significant capital cost, increased energy consumption and increased carbon emissions. Replacement of the existing tower with a larger unit would incur the capital expenditure associated with the purchase of a new, large unit and could also attract significant civil engineering costs associated with the strengthening of existing support structures to accommodate the additional weight of the tower. The operator opted to examine the benefits of using heat transfer enhancement.

Given the flow regime and the availability of pressure drop, it was clear that hiTRAN would provide a means of debottlenecking the unit. It was assumed that the fouling levels encountered in the revamped unit would be similar to those estimated for the current unit. With an increase in heat flux at the tube wall, the vapour film sometimes created through boiling is broken up and can be shown in this case study, favourably influencing the reduction of potential fouling. However the potential benefits were ignored in the exchanger analysis, so the results estimated were deemed to be conservative.

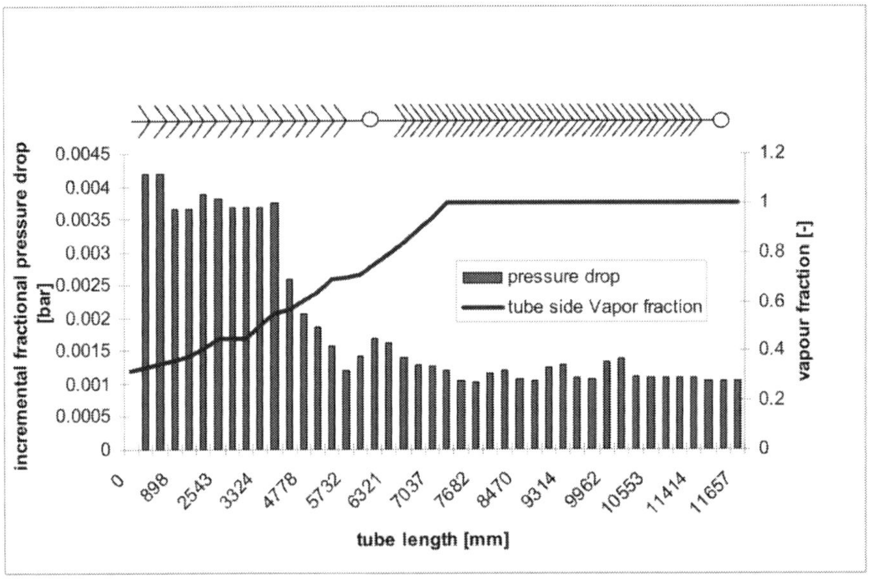

Figure 5: Incremental frictional pressure drop and vapour fraction as a function of tube length

Simulation of the tube-side behaviour of the unit showed, for about half of the tube length, the presence of a boiling two-phase mixture. After this region, the fluid is superheated before entering the fired heater. Installing a bespoke hiTRAN giving lower pressure drop (Figure 5) into the flow boiling region of the tube resulted in a modest increase in the tube-side heat transfer coefficient from 740–825 W/m^{2}K.

Placement of a different hiTRAN into the upper section of the tube (where single-phase heat transfer is occurring) resulted in the overall heat transfer coefficient being increased from 273–405 W/m^{2}K, with a 48% increase. The pressure drop was still well within the allowable value of 110 kPa. The geometries chosen for the hiTRAN were designed for an effective heat transfer to pressure drop performance.

The heat recovery was, increased, such that the required duty of the fired heater was within its capability and subsequently feedback suggests that the unit is performing well.

Heat transfer in the superheating region increased by 48%, while the reduction in fired heater duty delivered energy savings of about 0.8 MW, which represents a reduced CO_2 output of about 1700 t/y. Increased heat recovery led to energy cost savings, which paid for the revamp within 18 months. But more importantly large-scale investment in new equipment was avoided.

4.2 Enhancement Raffinate Air Cooler design

A plant in Russia, the OJSC Rosneft Refinery, was looking to design new air coolers for their raffinate cooling in their Lubricant production process. This cooling was either needed for storing the Lube oil at a safe temperature or similarly transporting the Lube oil at a safe temperature.

Due to the high throughput from production the customer knew that at the initial design phase high capital costs would be incurred. With the requirement to keep operating costs down, this would in turn would be achieved by keeping energy consumption down.

The air cooler duty was 2.9MW and designing based on conventional empty tube bundles, showed the capital costs exceeded the profitable investment for the plant. Also the plot space would be too large to fit within the available space for the existing plant.

Table 1. Heat Transfer enhanced design comparison with conventional design

Non-enhanced design	**hiTRAN enhanced design**	**Comparison savings**
5 Bays (2 bundles per bay)	2 Bays (2 bundles per bay)	**3 Bays**
10 Tube bundles	4 Tube bundles	**6 Tube bundles**
18125m^2 heat transfer surface area	7255m^2 heat transfer surface area	**10870m^2 heat transfer surface area**
3 working fans per bay	2 out of 3 working fans per bay	**1 working fan per bay**
15 Fans	4 fans	**11 Fans in total**
Power consumption per fans 10.2kW	Power consumption per fan 5.75kW	**4.45kW Power consumption per fan**
Total Energy consumption 153kW	Total energy consumption 23kW	**130kW energy consumption saving**
US$706800 Capital	US$285700 Capital	**US$421100 Capital saving**
US$101792 per annum Operating cost	US$24890 per annum Operating cost	**US$76902 Operating cost saving per annum**

Without enhancement, the design to process 2.9MW of cooling was producing a plot space of 18125m^2 of area needed and energy consumption excluding the pumping costs of 153kW. The associated cost of this energy consumption taking into account the initial capital cost was ~ US$808592 for the first year. This suggested there was scope to improve upon the competitiveness of the unit of the plant.

The next approach was to analyse utilizing exchanger enhancement in order to reduce energy demands. As the heat transfer was controlled on the tube side this lent the application to use tube side enhancement using hiTRAN thermal systems. The total energy saving achieved by using a hiTRAN design was 130kW. This translated to an overall economic saving of US$498002 for the first year.

This design example of using heat exchanger enhancement shows that if the reduction of energy consumption is explored, as in this case by 85%, significant economic savings are available. This case also achieved release of extra plot space on the plant as well as maintaining the original civils and minimal piping modifications.

Conclusion

The importance of energy efficiency on plant economics has been highlighted and various strategies for reducing plant energy consumption discussed. The option of heat exchanger enhancement is assessed in detail and various options have been addressed. In particular the use of hiTRAN has been highlighted and case studies show that it has been used to reduce energy input as well, as highlighting additional potential benefits such as increased throughput, longer runtimes or reduced capital costs as appropriate. This makes the case for the consideration of enhancement technologies to reduce plant costs to maintain competitiveness.

References

[1] HEAT-int (2013) Process Integration Ltd. www.processint.com/chemical-industrial-software/heat-int . Accessed 13 March 2013

[2] Kemp, I. C., 2007 (authors of the first edition Linnhoff, B., Townsend, D. W., Boland, D., Hewitt, G. F., Thomas, B. E. A., Guy, A. R. and Marsland, R.) Pinch analysis and process integration. A User Guide on Process Integration for Efficient Use of Energy. Amsterdam, Elsevier, the Netherlands.

[3] Klemeš J (ed) (2013) Handbook of Process Integration (PI). Woodhead Publishing Limited, Cambridge, UK, Philadelphia, USA

[4] Linnhoff B, Townsend DW, Boland D, Hewitt GF, Thomas BEA, Guy AR and Marsland RH (1982) User Guide on Process Integration for the Efficient Use of Energy, IChemE, Rugby, UK

[5] Ponce-Ortega JM, Jimenez-Gutierrez A, Grossmann IE (2008) Simultaneous retrofit and heat integration of chemical processes. Industrial and Engineering Chemistry Research, 47, 5512–5528

[6] Siemanond K and Kosol S (2012) Heat Exchanger Network Retrofit by Pinch Design Method using Stage-Model Mathematical Programming. Chemical Engineering Transactions, 29, 367–372, DOI: 10.3303/CET1229062

[7] Smith R (2005) Chemical Process Design and Integration, John Wiley & Sons Ltd, West Sussex, UK

[8] Sprint © software (2013) CPI, The University of Manchester

[9] Vengateson U, (2009) Design of multiple shell and tube heat exchangers in series: E shell and F shell. Chemical Engineering & Design 88: (5–6),725–736

442489 Inefficient Operation Leads to Higher Capacity

Abstract:

Conventional wisdom always talks about efficient operation. We conduct training schools and classes to teach the operators. However when it comes to practice, we find that typically operators are able to increase the capacity of their fired heaters by operating inefficiently. We have recently observed this situation in several heaters especially coker heaters where they could run their heaters at higher capacity and for longer time by increasing the excess air. We will review several cases of different fired heaters where you can increase capacity by lowering efficiency.

445187 Naphtha Processing Solutions for North American Crude Blends

Abstract:

Unconventional crudes being produced in the United States, such as shale oil, yield significantly more naphtha than conventional imported crude blends. As a result, refiners need solutions for processing naphtha to profit from these local feedstocks. This paper will discuss a broad range of options for increased naphtha processing with reforming and isomerization technologies.

444983 Preflash Options for Light Tight Oil Processing

Abstract:

As light tight oil continues to increase in availability, refiners are challenged with how best to process this unique type of crude oil. With a much higher amount of naphtha and lighter materials in tight oil, the preflash step takes on increasing significance in the crude unit. This presentation will compare two types of preflash – a simple drum with vapor further refined in the crude tower, and a full preflash fractionator, where on-spec light naphtha products can be made and bypass the unit's crude tower. The choice for an individual refinery is dependent on many factors. Advantages and disadvantages of each will be explored, and factors to consider in making a selection will be presented. Burns & McDonnell has significant experience with preflash designs in crude units and specific examples will be given to provide a clear understanding.

444679 *UOP RZ Platforming^{TM} Technology - Maximizing Profitability from Tight Oil*

Abstract:

UOP's RZ Platforming^{TM} Technology - Maximizing Profitability from Tight Oil

Presentation Highlights

The recent shale oil/gas phenomenon in the US is leading to an unprecedented surplus of light naphtha. The consequential price pressure is now paving the way to the resurgence of the aromatic derivatives industry in the US.

The dramatic growth exhibited by the Eagle Ford formation (EF), rich in C6s/C7s, is presenting unparalleled challenges and opportunities alike for the US refining industry.

The extraordinary high yields, RVP and low octane of the EF light naphtha require significant reconfigurations, revamps of the naphtha complex and ultimately, suboptimal sale of excess naphtha.

The route of Zeolitic-based catalytic reforming technology, such as UOP RZ Platforming offers the opportunity to unlock the full EF potential by producing high-value benzene and toluene from cost-advantaged C6s and C7s.

This presentation summarizes the outcome of a configuration study conducted for a generic USGC refinery faced with the challenge of maximizing EF processing and, ultimately, its profitability. The case study analyses the re-configuration options to allow maximum EF processing vis-à-vis the economic environment for light naphtha.

444753 Aspen Model for Propane/Propylene Separation By Pressure Swing Adsorption Using Carbon Molecular Sieves

Abstract:

Cryogenic distillation is the current method for separating propylene/propane mixtures. It is a difficult separation requiring high reflux ratios, high energy input, and tall columns, because of the small difference in the volatilities of propylene and propane. Adsorption based processes have been explored in recent years as potential replacement of or augment to the incumbent cryogenic distillation process to reduce the energy intensity or debottleneck existing assets.

Dow has developed new carbon molecular sieves (CMS) that show a high separation factor for propane and propylene. This presentation shows the development of a PSA model using ASPEN ADSORPTION™. The adsorption isotherms of propylene and propane on the CMS were measured at different temperatures that cover the commercial operating range. A temperature dependent, four-parameter Langmuir model was selected to fit the single component isotherm data and the model parameters were obtained by using the Aspen ADSORPTION™ equilibrium parameter estimation tool. Different multicomponent isotherm methods including the Ideal Adsorbed Solution Theory were explored. The heat of adsorption of each component was determined from the isotherm models according to the Clausius – Clapeyron equation. The mass transfer coefficients of propylene and propane were estimated using Aspen ADSORPTION™ simulations by fitting breakthrough curves at different flow rates. Models for a simple Skarsktrom cycle as well as alternate cycle configurations from literature were set up using ASPEN. Model predictions are compared to PSA experimental results. Propylene purity and recovery are the key output variables. The validated PSA model can then be used for future process optimization and scale-up.

444768 Delivering Real Value with Process Simulation

Abstract:

The usage of Process Simulation technology has grown greatly over the past decade. This presentation will focus on how current technology can be used to return real and tangible benefits in several phases of project development. We will review the "lifecycle simulation" concept and explore some ways that a well implemented lifecycle solution can deliver value. Case studies will be presented that show how process simulation can be used across the lifecycle, providing benefits in cost reduction, improved safety and confidence in the design and lower overall costs.

444565 Equation Oriented Approach for Solving Complex Flowsheets

Abstract:

Equation Oriented (EO) approach is a powerful tool that can be used for simulating complex flowsheets with many recycle streams and significant heat integration. EO approach has been successfully used for solving plant operational issues, evaluating plant revamps and developing conceptual process designs. The use of this approach will be illustrated for several industrial applications. These examples will show case different objectives: simulation, optimization, and data reconciliation.

Analysis of Energy Consumption in Industrial Distillation Columns

Akshaya Sundar
Research Scholar
Indian Institute of Technology, Madras
akshayas27@gmail.com

Dr A Kannan
Professor
Indian Institute of Technology, Madras

Prepared for Presentation at
American Institute of Chemical Engineers
2016 Spring Meeting
11th Global Congress on Process Safety
Houston, Texas
April 10-14, 2016

UNPUBLISHED

Analysis of Energy Consumption in Industrial Distillation Columns

Akshaya Sundar
Research Scholar
Indian Institute of Technology, Madras
akshayas27@gmail.com

Dr A Kannan
Professor
Indian Institute of Technology, Madras

Keywords: Energy, Distillation, CGCC profiles, Murphree efficiency, Column Targeting

Abstract

Distillation is the commonly applied separation process in the chemical process industry as heat is used as the non-contaminating separating agent. However, the process is also inefficient and hence leads to considerable energy consumption. Methods have been proposed in literature for quantifying the energy inefficiencies for ideal columns. Suitable recommendations based on Column Grand Composite Method (CGCC) analysis such as reduced reflux ratio, feed preheating and use of side condensers/reboilers have been suggested to improve the efficiency after assuming the vapor and liquid exiting the stages to be in equilibrium. However, industrial columns are rarely ideal and hence the conclusions based on ideal column assumption may not work well. To overcome this limitation in the analysis, either the rigorous rate based approach or the simpler efficiency approach may be adopted. In the present work, we have carried out the energy utilization analysis for ideal distillation columns handling non-azeotropic and azeotropic systems. Considerable differences were noticed between the CGCC curves of ideal and non-ideal columns. The assumption of leaving liquid and vapor streams in a distillation column are in equilibrium with each other cannot be incorporated in case of real industrial columns. CGCC curves will be dissimilar for ideal and non-ideal columns. Based on this we have shown that the recommendations for improved energy utilization from ideal columns do not work well for actual columns. Column modifications recommended by thermodynamic analysis uses the available CGCC curves and hence the differences will lead to poor retrofits when accounting for real industrial columns. The rate based model calculations are difficult and require many parameters such as multi component transport coefficients, tray dimensions and holdup etc. Hence, the relatively simpler efficiency approach was adopted and these were estimated independent of the rigorous rate based method. The methodology based on Murphree efficiencies was developed to carry out the CGCC calculations. The recommendations from the simplified model based on these efficiencies were in close agreement with those made from the rate based model. Incorporating this methodology in actual practice will considerably reduce the energy consumption from popular separation processes.

1. Introduction

Distillation is a popular separation process for multi component separation regardless of thermodynamic inefficiencies and high energy consumption. Under a high energy crisis scenario and lack of an equally effective energy source, the most practical way to reduce energy consumption is the intelligent use of existing processes. Minimizing use of energy consumption may be done through effective heat integration and thermodynamic analysis.
The stage-enthalpy and temperature-enthalpy profiles of Column Grand Composite Curves (CGCC) represent the theoretical minimum cooling and heating requirements over the stages or the temperature range. The compositions of liquid and vapor streams emerging from the same stage are the equilibrium compositions at that stage temperature. Suitable recommendations for the column based on Column Grand Composite Method (CGCC) analysis may be made for improving energy. This method is generally called column targeting. On analysis, column modifications such as reducing the reflux ratio, feed preheating and use of side condensers/reboilers can be been to improve the efficiency. CGCC analysis had become more popular methodology because of its reliability.

2. Methodology

In this paper we are presenting a methodology to analyze the energy consumption of the real column using the stage-enthalpy CGCC column profiles obtained from any converged simulations from the Aspen plus simulator. The benzene-toluene distillation system was taken for our study. Rate Based converged simulations from the aspen was taken as the real column performances [1]. The profiles obtained from the simulator are also cross validated with hand calculations by the methodology developed by Dhole and Linhoff [2] and from the paper of S Bandyopadhyay, Ranjan K Malik and Uday V Shenoy [3]. First we basically tried to model an equivalent equilibrium column analogous to the real column and studied and compared the CGCC profiles. We tried to incorporate the similar modifications and suggestion made to the equivalent equilibrium column to the real column. In parallel we also developed a tray efficiency incorporated equilibrium column and performed the same analysis. The CGCC profiles of the three models are studied in detail. Based on the obtained column profiles column targeting options are found out. The study conducted helped to identify the closest comparable model with that of a real column

The specifications of the benzene-toluene column used here to demonstrate the CGCC validation are listed in Table. 1

Table 1: Column specifications

Feed Flow rate [kmol/hr]	100	Column diameter [m]	1.34
Temperature [C]	35	No of stages	20
Pressure [atm]	1	Feed stage	10
Distillate rate [kmol/hr]	50	Bubble cap tray	

3. Results and Discussion

The stage-enthalpy and temperature-enthalpy profiles of Column Grand Composite Curves (CGCC) are the representation of theoretical heating or cooling requirements over the stages or temperature range. The preliminary work was to validate the HGCC data obtained from a converged simulation results from Aspen with stage wise material and energy balance calculations. The same was validated using both the Top-down and Bottom-up approaches. The results are shown in figure 1

Figure 1: Stage-Enthalpy profile

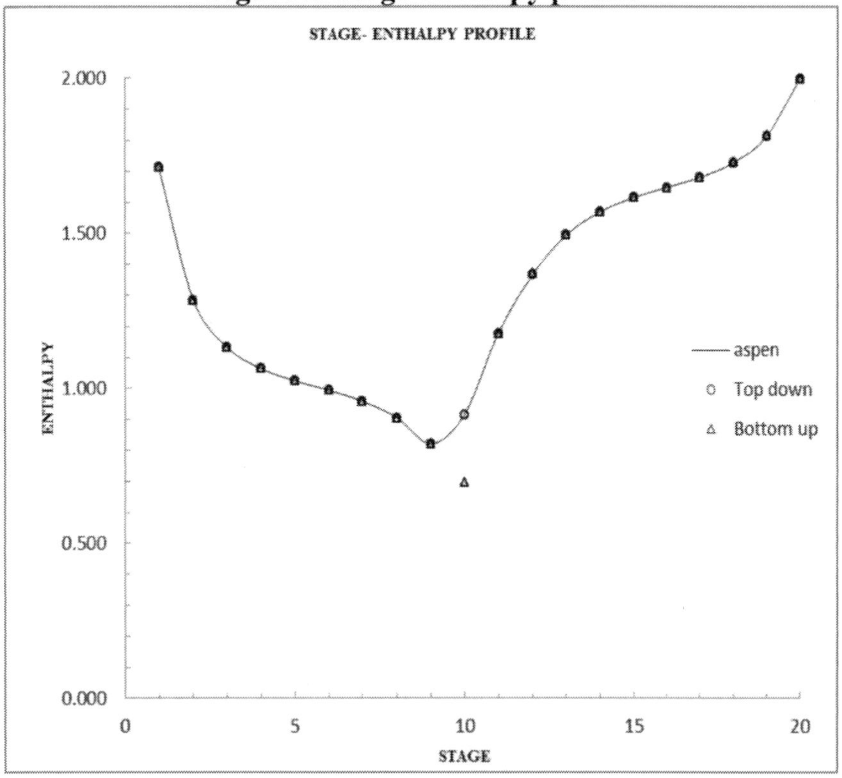

In the rate based model RBM which operates more like a real column, the assumption of leaving streams in equilibrium cannot be incorporated for the calculations of CGCC profiles. Hence we proposed a methodology to find an equivalent equilibrium based column EQC analogous to the existing real column. The CGCC profiles of the equivalent column are used for the energy analysis and suitable modifications are made. These modifications were then tried to map on the real column.

The equivalent equilibrium model EQC was developed on the basis of product purity. This was done by setting up an ideal column for the same product purity that will be obtained from a rate based model RBM with identical column conditions and specifications. The closest column so chosen was by various random trials for same column conditions of RBM. The specifications of the equivalent column so obtained are tabulated in Table 2

Table 2: Column specifications of the Equivalent Column

Specifications	RB Column	Equivalent column
Feed Flow rate [kmol/hr]	100	100
Temperature [C]	35	35
Pressure [atm]	1	1
Column diameter [m]	1.34	1.34
No of stages	20	11
Feed stage	10	4
Distillate purity	0.954	0.952

The equivalency between the models was made sure by comparing various column temperature profiles, stage wise liquid and vapor enthalpies and product compositions. These were not exactly matching as expected. CGCC profiles obtained from EQC model were analyzed and modifications made were mapped to the real column. Considerable differences were noticed between the CGCC curves of non-ideal and equivalent ideal column. Thus the targeting results from EQC model if mapped will give unclear results even if the overall energy was found to be reduced by the modification. Figure 2 shows the differences in CGCC profiles of these two models.

Figure 2: Stage-Enthalpy profiles of EQC and RBM before and after column targeting

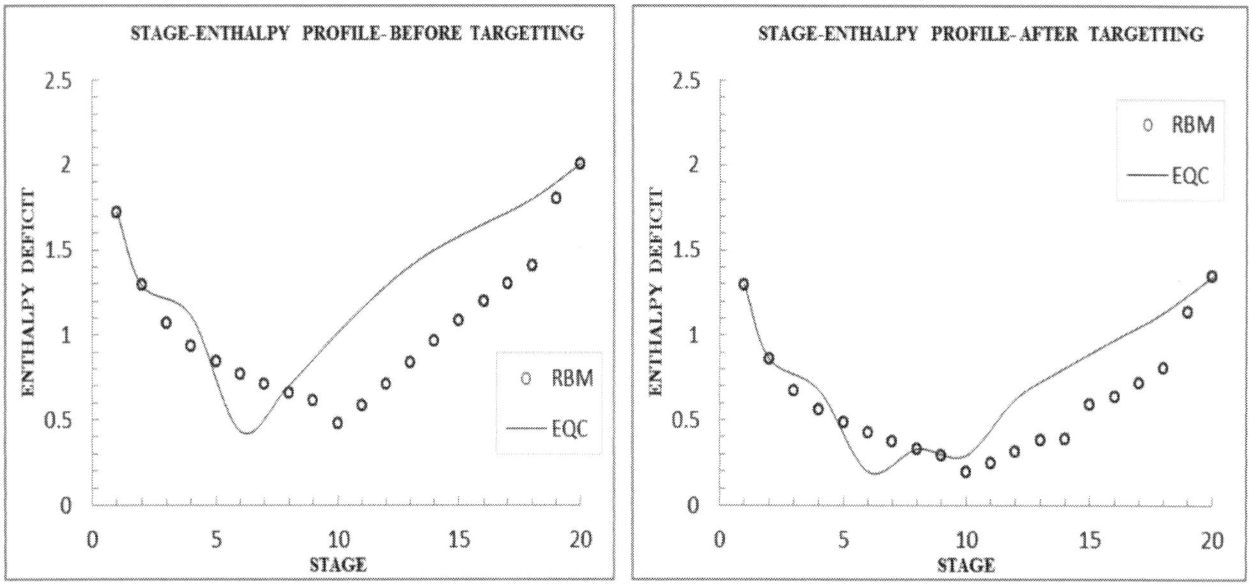

This variations may also have occurred possibly because of the lack of fit of equivalent model and mapping errors. Better equivalent model can be predicted by making this comparison tight. In our work we developed another model which is based on the Murphree tray efficiency TEM and compared with real column. TEM model is developed by incorporating stage efficiencies to an equilibirum column operating under the same column conditions and specifications. The stage efficiencies incorporated in our work was taken from the RBM model. The conventional CGCC analysis and column targeting is performed. Column Grand Composite Method (CGCC) analysis such as reduced reflux ratio, feed preheating and use of side condensers/reboilers have been suggested to improve the efficiency after assuming the vapor and liquid exiting the stages to be in equilibrium. Temperature profiles, stage wise enthalpy profiles and composition profiles before and after column targeting are also examined and found to have close match. Figure 3 shows the column profiles of RBM and TEM models before targeting.

Figure 3: Column profiles of TEM and RBM before column targeting

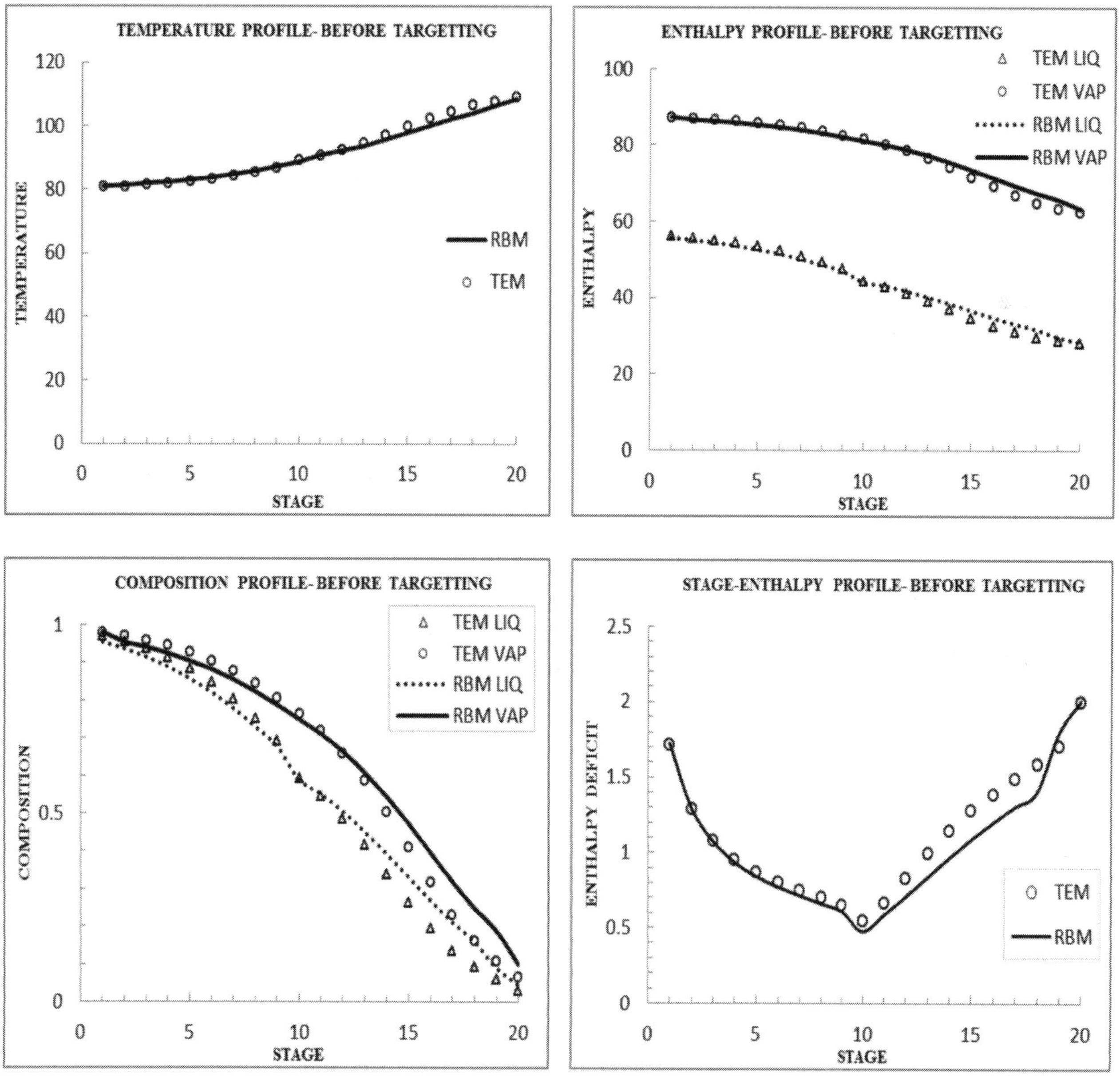

Both the models were comparable on this basis. Column targeting performed in this work was by reducing reflux ratio along with feed preheating and introducing a side reboiler by keeping the number of stages and feed stage location constant. This modifications had considerably reduced both condenser and reboiler load along with product purity. The work was intended to validate the equivalency between the two models and therefore product purity was not strictly maintained. Figure 4 shows the results after targeting column. The column temperature, composition and enthalpies were also found to match even after column targeting.

Figure 4: Column profiles of TEM and RBM after column targeting

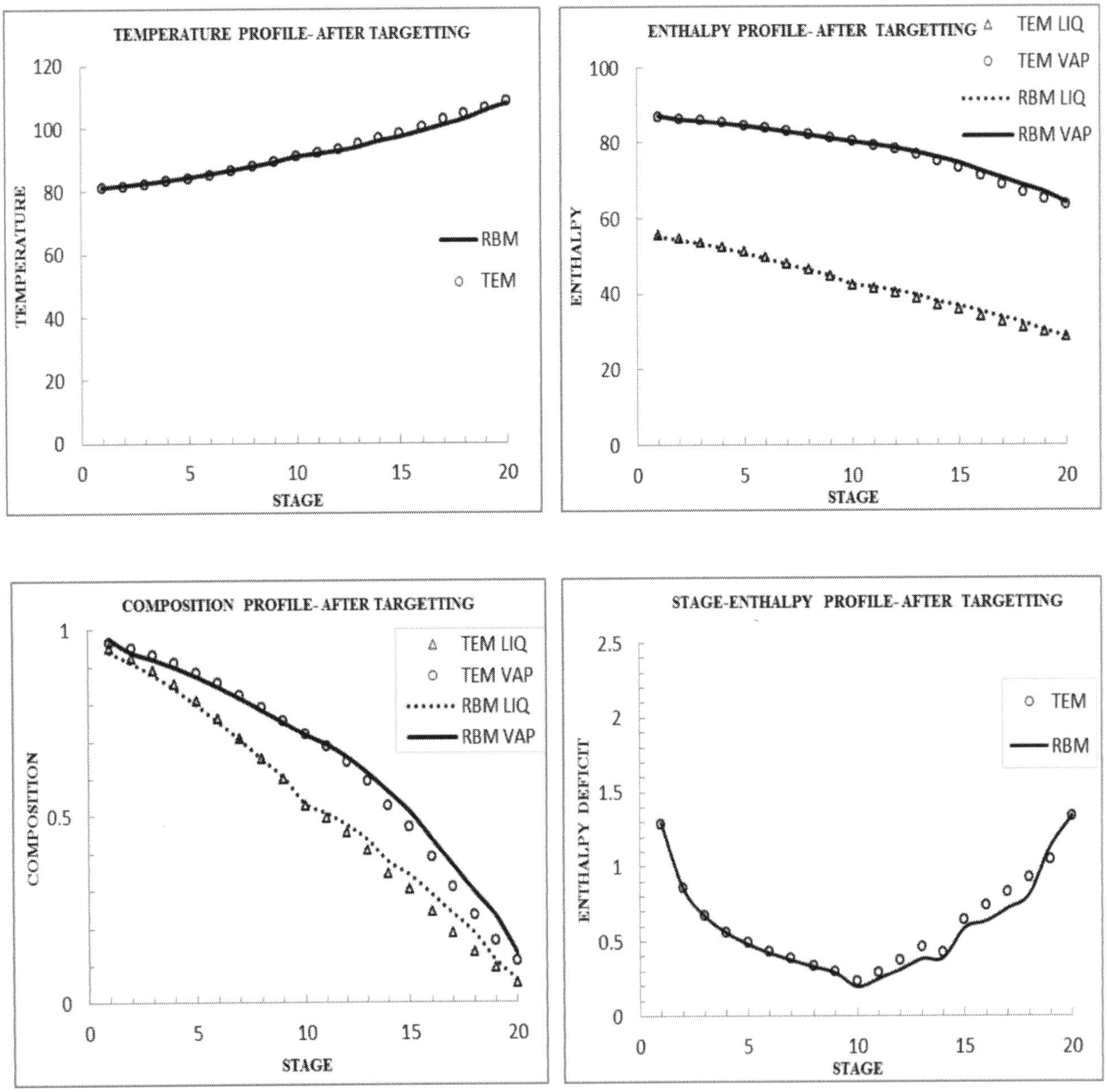

On analysing the stage enthalpy profiles for TEM and RBM before and after targeting the column overall column energy reduction can be noticed easily. Figure 5 shows the considerable reduction in heat after the modifications made to the column for both the models.

Figure 5: Stage Vs Enthalpy of TEM and RBM before and after column targeting

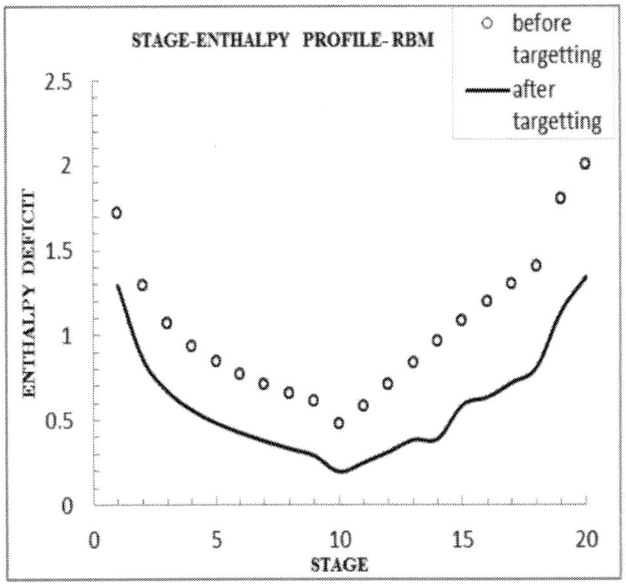

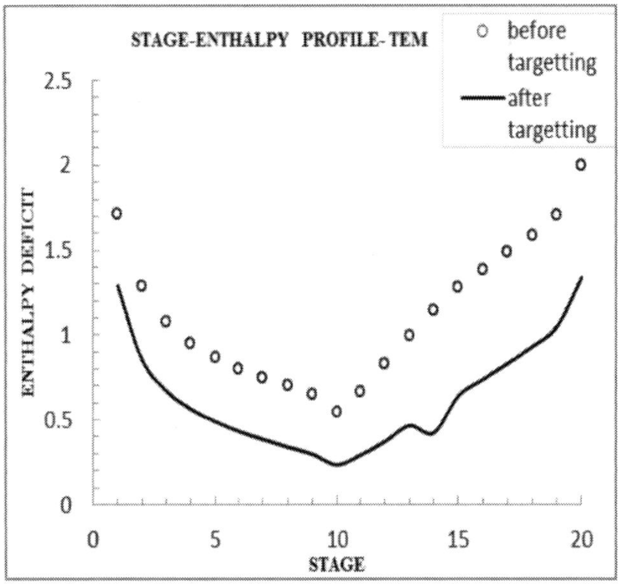

Same work has been extended to a 10 staged azeotropic Ethanol-Water system and carried out the the similar procedure for analysis. The column profiles and CGCC profiles were found matching before the column targeting. Hence the methodology can be effectively extend to the system for further energy efficient modifications. Figure 6 shows the profiles obtained for the azeotropic system before column targeting for both TEM and RBM models

Figure 5: Stage Vs Enthalpy of TEM and RBM before targeting for azeotropic system

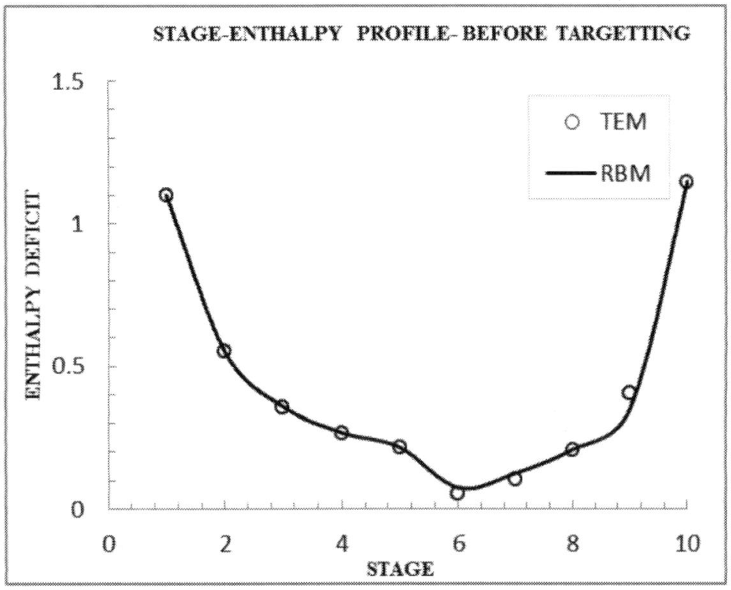

4. Conclusion

Energy analysis of real column can be performed by a converged rate based model in aspen simulator. Column Grand Composite Curve or CGCC analysis is a reliable method for making appropriate column modifications for improving energy utilization. CGCC calculations are basically derived formulas from material and energy balance equations. CGCC curves were found to be dissimilar for ideal and non-ideal columns. This is because the assumption of leaving liquid and vapor streams in a distillation column are in equilibrium with each other cannot be incorporated in case of real industrial columns. Equilibrium columns are always easier for analysis. In our work we incorporated murphree tray efficiencies in an equilibrium column having same column specifications and conditions. The CGCC analysis was performed on both this columns. The column profiles of both the models are all found to be similar before and after the column targeting. The recommendations from the simplified model based on efficiencies were in close agreement with those made from the rate based model. The methodology was found effective when the same procedure was extended to an azeotropic system. Incorporating this methodology in actual practice will considerably reduce the energy consumption from popular separation processes.

5. References

[1] Debashish Panda, Kannan A, Equilibrium and Rate Based Simulation of MTBE Reactive Distillation Column, International Scholarly and Scientific Research & Innovation, Vol:8, No:10, 2014

[2] V.R.Dholle & B.Linnhoff, Distillation column Targets, Computers chem.Engng.,vol 17, 549-560, 1992

[3] Santanu Bandyopadhyay, Ranjan K Malik and Uday V Shenoy ,Temperature- Enthalpy curve for energy targeting of distillation Columns, Computers chem.Engng.,vol 22, 1733-1744, 1998

444907 Heat Exchanger Selection Improves the Performance of Refinery Preheat Trains: Industrial Case Studies

Abstract:

This paper shows how the advanced heat exchanger network simulation software SmartPM® allows refinery engineers to better understand and manage heat exchanger fouling behaviour. In addition to guiding on the most profitable cleaning schedules, the predictive analytics in SmartPM® can assess the benefits of using HELIXCHANGER® heat exchangers to extend run lengths between cleans in refinery heat exchanger trains.

Refinery case studies of crude preheat and vacuum distillation heat exchanger trains show how HELIXCHANGER® shells can improve fouling performance among other benefits. These benefits are being realized as HELIXCHANGERS® are increasingly adopted by leading engineering designers where SmartPM® provides the framework for their decisions.

443270 Optimization of Refinery Configuration Processes for Maximizing Gasoline Based on Crude Oil Characterization

Abstract:

Refineries today have become increasingly more efficient and efforts across the globe are being made to invest in better performing technologies and processes that save energy and decrease each refineries environmental impact whilst meeting increasingly more stringent fuel product specifications. The demand for high value petroleum products is increasing, while the demand for low value products such as fuel oil and residue based products is decreasing. Therefore, maximizing of liquid products yield from various processes and valorization residues is of immediate attention to refineries.

In this work, two refinery configurations are investigated for upgrading projects to increase gasoline production for local market demand. Different alternatives for the upgrading can be tackled. Either direct upgrading of the atmospheric residue, or first subject the atmospheric residue to vacuum distillation then upgrade the vacuum residue and vacuum gas oil to more valuable and lighter products.

This work provides the techno-economic feasibility study based on characterization of different crude oils and optimum selection of the conversion process units to produce more gasoline to cover the local consumption. Selection of optimum conversion process is based on some physical properties (conradson carbon residue, and metal content (i.e. vanadium and nickel) of atmospheric residue for sixteen crude oil samples. Obtained results show that, the scenario which included residue fluidized catalytic cracking (RFCC) has shown the optimum in terms of both maximum gasoline and less capital cost compared with configuration that included the delayed coking process.

444224 Benefits of Crude Pretreatment for Unconventional, Heavy Crude Processing

Abstract:

Heavy crudes represent a significant proportion of today's global crude supply, and forecasts indicate the barrels of available heavy crude will continue to increase in the decades ahead. As refiners increase the percentage of these cost-advantaged feedstocks in their crude mix, they also increase the potential for downstream consequences, including waste water treatment plant excursions, catalyst poisoning, corrosion, and increased energy costs. The first defense against these many risks is generally the desalter; however, as a standalone process, desalting is limited in its ability to mitigate the downstream impacts of heavy crudes. Baker Hughes Crude Oil Management™ philosophy considers the desalter as one part of a much larger operation that encompasses the entire feed delivery and handling system. A central feature of Crude Oil Management is crude oil pretreatment: the application of proprietary low-temperature demulsifier technology in the crude delivery and storage facilities to improve feedstock quality, optimize desalter operations, and minimize the downstream impacts of heavy crudes. Case histories are presented, including a multi-year study of tank farm operations, detailing the benefits of crude oil pretreatment strategies.

442623 Calcium Removal from Heavy Crude Oil Using Aqueous Solution Containing a Calcium Removal Agent

Abstract:

Due to the growth limitation of light crude oil production, refinery companies are showing interest in heavy crude oil containing the high contents of S, Ca, organic acids, and metals. The market price of heavy crude oil is much lower than that of light crude oil and the discount rate depend on the impurity contents (averagely 10 ~ 25%). Thus, as increasing the handling capacity of heavy crude oil, economic feasibility improved.

Ca is one of the major impurities in heavy oils, occurring several problems during the refining processes such as fouling in pipe systems and heat exchangers and the deactivation of catalyst. Calcium naphthenate (CaNA), combined with the Ca cation and the anion of naphthenic acids, is difficult to remove because of complex and various molecular structures of naphthenic acids.

In this study, calcium removal from heavy crude oil was conducted using aqueous solution containing a calcium removal agent. The effect of operating variables on the Ca removal efficiency was experimentally studied. Batch type of electric coalescer was employed for the demulsification of the oil samples. The Ca concentration in the oil samples were analysed using inductively coupled plasma (ICP).

16AIChE
Spring Meeting, Houston, TX

The use of hiTRAN wire matrix technology for the size reduction of heavy crude exchangers in separation processes

Peter Drögemüller, Cal Gavin LTD, Minerva Mill Technology Centre, Alcester, UK
peter.droegemueller@calgavin.com

William Osley, Cal Gavin LTD, Minerva Mill Technology Centre, Alcester, UK
william.osley@calgavin.com

Luis Quillares Vargas, Tecnopinch Integral Solutions, México. D.F., Mexico
luis.quillares@tecnopinch.com.mx

Prepared for Presentation at
American Institute of Chemical Engineers
2016 Spring Meeting
11th Global Congress on Process Safety
Houston, Texas
April 10-14, 2016

UNPUBLISHED

The use of hiTRAN wire matrix technology for the size reduction of heavy crude exchangers in separation processes

Peter Drögemüller, Cal Gavin LTD, Alcester, UK

William Osley, Cal Gavin LTD, Alcester, UK

Luis Quillares Vargas, Tecnopinch Integral Solutions, México. D.F., Mexico

peter.droegemueller@calgavin.com

Keywords: hiTRAN, heavy crude, dehydration, heat transfer

Abstract

The flow characteristics of heavy crude oils make them difficult to process with conventional refining equipment. Heating and cooling capacity is limited by the laminar flow behaviour of such crudes. The flow patterns are complex and poorly understood. Heat transfer can be of several magnitudes lower than under turbulent flow conditions encountered processing light crudes. This results in large and expensive equipment. This paper explains the underlying heat transfer mechanisms for such flows, employing CFD simulations backed up by experimental research. To overcome such limitations hiTRAN wire matrix tube side enhancement technology can be applied and is evaluated here.

1 Introduction

The hydrodynamic of flow in tubes equipped with hiTRAN technology is changed, leading to much improved heat transfer compared to pure laminar flow. Enhancement levels up to 16 times the rate of heat transfer for similar flow velocities are measured. This results in corresponding smaller equipment size, with substantial savings in exchanger and piping costs. In offshore applications, where equipment size and weight are the constraining factors, this technology can offer substantial benefits when processing viscous fluids. One Installation case from the Gulf of Mexico is presented in detail. In this application heavy crude with water content is heated for further processing. Heat transfer is limited by the viscous nature of the heated and the cooled fluid, resulting in large and ineffective heat transfer equipment. By applying tube side enhancement the original plain empty tube design was reduced from **20** to **8** exchanger shells, with substantial impact on the CAPEX costs for exchanger and piping. For this particular case the hydrodynamics and heat transfer is explained in detail for the original and the final hiTRAN enhanced design.

2 Separation process for heavy crude oils

Heavy-oil development is becoming increasingly important because of the continuous decline in conventional oil production. There is no common definition of what constitutes heavy crude oil. The term is relative in different regions. In the North Sea heavy crudes is typically between 15°API – 20°API, whereas in Canada and Venezuela heavy crudes are as low as 8°API(1). The dynamic viscosity for heavy crude oils range from about 50mPas to more than several Pas for low API crudes under ambient conditions. As the name suggests heavy crudes are often characterised by densities similar to water densities and higher solid loadings compared to conventional crude oils. This poses challenges for the separation process of the oil /water. For the sizing of the separation units one of the main variables is the separation speed from water droplets in a crude oil continuous phase. The speed can be determined by Stoke's Law or its derivatives [1]:

$$v_s = \frac{d^2\,(\rho_W - \rho_O)}{18 \cdot \eta} \cdot g \qquad (1)$$

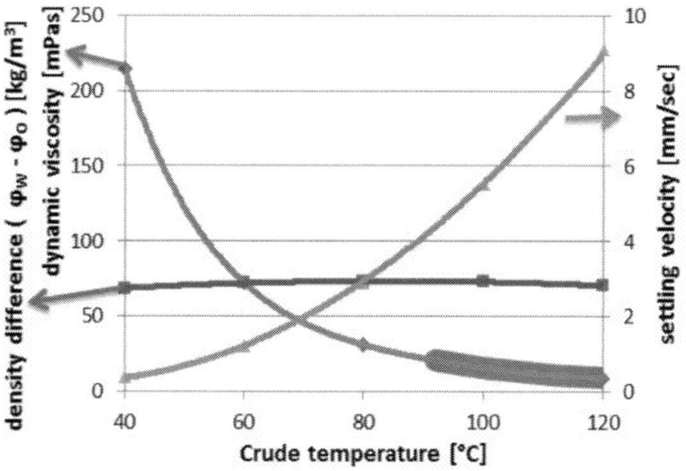

***Figure 1**: Viscosity temperature function of 18API crude and Density difference between 1% salinity Brine and 18API crude as function of temperature. Resulting settling velocity.*

In **Figure 1** calculation results from equation (1) for the settling velocity (v_s) of a 1.5mm diameter water droplet (1% salinity) in API 18 crude oil are shown. The settling velocity is mainly influenced by the viscosity of the continuous oil phase. For that reason the sizing of oil/water separators are often based on a target viscosity around 5mPas to 20mPas as indicated with the bold grey line in **Figure 1**. To achieve this viscosity the oil / water emulsion in the example case has to be heated to about 90°C. It is evident that the required heat input varies with the API grade of the processed crude. In general low API crudes require higher operation temperatures of the separators. Due to the often high mass flows the required heat duty to achieve the required operation temperatures can be significant. For that reason large dehydration units are in general designed with heat recovery after the dehydration process.

In **Figure 2** the different sources for heat recovery are shown. Heat is recovered from the dry crude oil, which usually accounts for the largest mass flow fraction from the separation process. In processes where large amounts of water are separated heat recovery from the produced water can also be an option for a more efficient overall process. The required inlet temperature to the separator is then reached with the use of a fired heater or an indirect heating system with heat transfer fluid (HTF). These heat exchangers are characterised by the large surface area required to heat laminar flow fluids.

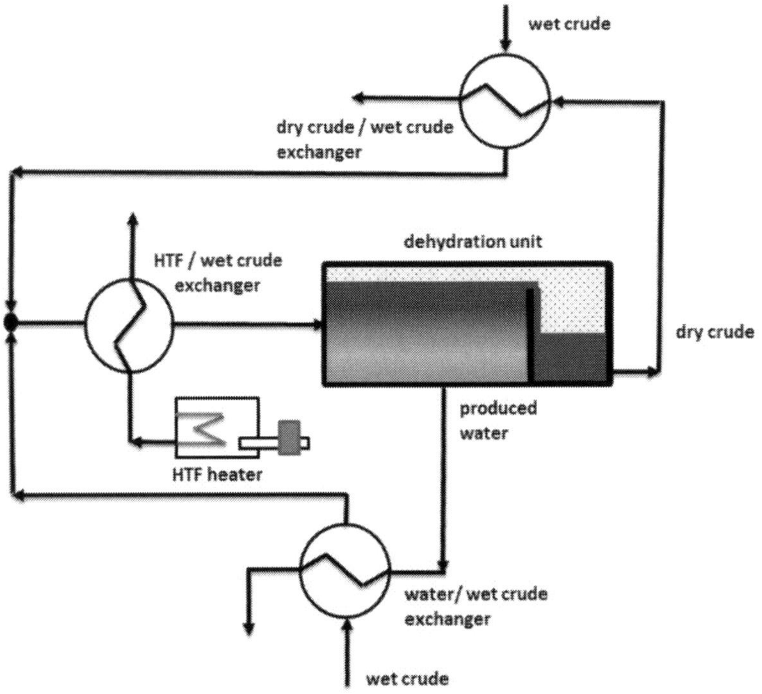

Figure 2*: Flow sheet of heat recovery in a oil/water separation unit*

3 Characteristics for heavy crudes feed / effluent exchangers

The required heating duty for the wet crude oil can be expressed with:

$$Q = m \cdot cp \cdot (T_{out} - T_{in}) \tag{2}$$

From equation (2) and the observations in section 2 it is evident that with higher required outlet temperature the duty increases with reduced API of the heavy crude. For the dehydration process the heavy crude **including** the containing water has to be heated. Common duties (Q) of such units are large in the range from 5 to 30MW. The heat transfer area A is directly linked to the required pre-heat duty:

$$A = \frac{Q}{U \cdot EMTD} \tag{3}$$

In order to achieve high heat recovery rates these types of exchangers are designed for small EMTD's between the heated wet crude and the effluent. Large duties with a small temperature approach require large surface areas as evident from equation (3). The only variable to reduce the exchanger size is the overall coefficient U, which is defined as follows:

$$\frac{1}{U} = \frac{1}{h_i} + \frac{1}{h_o} + foulingresistance + wallresistance \tag{4}$$

The overall coefficient U will be lower than the smallest coefficient in equation (4). In a dry / wet heavy crude exchanger with the fluid viscosities outlined in section 2 the lowest coefficient will be found on the tube side. Here the exchanger operates under laminar flow conditions, opening up the possibility for tube side enhancement.

3.1 Temperature cross

In order to recover as much heat as possible from the hot streams exiting the dehydration unit the outlet temperature of the wet crude has to be higher than the outlet temperature of the effluent (dry crude or produced water). For the heat exchanger design this *temperature cross* has to be taken into account. Under such operation in a single shell&tube exchanger multi-passing tube side flow leads to reverse internal heat flow in the exchanger. For that reason one option is to split the multi-pass exchanger into several shells in series and avoid the temperature cross in each shell. Alternatively a shell&tube exchanger operating counter-current with a single tube pass can operate under temperature cross conditions. Single pass designs often lead to low tube side flow velocities, this is especially true for exchangers with viscous flows and therefore poor heat transfer on the shell side since for those designs large heat transfer areas and therefore increased cross-sectional flow areas are required, resulting in even lower flow velocities.

Here hiTRAN offers an option with substantial benefits, due to the low pressure drop usually encountered in single pass design hiTRAN can be employed without exceeding the allowable pressure drop. With this the full benefit of heat transfer increase is taken.

4 Heat transfer and flow pattern in viscous tube side flow

When considering heat transfer in pipe flow one has to differentiate between turbulent, transitional and laminar flow conditions. In case the heavy crude is placed on the tube side the heat transfer equipment will operate under laminar flow conditions. Those are defined by a Reynolds number below ~2300:

4.1 Empty tube heat transfer

In **Figure 3** where the dimensionless heat transfer parameter inside a tube is shown on a logarithmic scale, it is evident that the heat transfer rates in laminar flow are in general more than one order of magnitude lower than in turbulent flow.

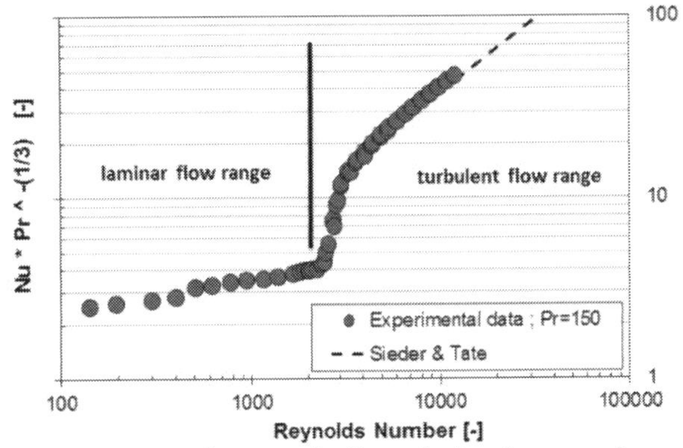

Figure 3: Measured empty tube dimensionless heat transfer as function of Reynolds and theory [2]

In laminar flow the increase of heat transfer with Reynolds number is only modest; typically a doubling of the velocity gives just a 10% to 20% increase in the heat transfer rate. For that reason using multiple tube passes to increase the heat transfer coefficient is much less effective than in turbulent flow, where the heat transfer almost doubles with a doubling of the fluid velocity. It is also evident from the graph that once the transition to turbulent flow is reached, the increase in heat transfer is very sudden and steeper compared to the laminar and even fully developed turbulent region and therefore extremely sensitive to a change in flow or physical properties. In laminar flow the heat transfer can be dominated by either forced or mixed convection. In mixed convection flow conditions, a secondary flow profile caused by density differences is superimposed on the forced velocity profile in the flow direction. The dominant mechanism depends on the conditions and physical properties of the fluid being heated or cooled. The effect was first quantitatively examined by Colburn [3]. Under these conditions it is very important to predict the heat transfer coefficient accurately, because a low value may become the controlling resistance in the design. Calculations are more complex than under turbulent conditions.

4.2 hiTRAN Wire Matrix heat transfer

hiTRAN wire matrix technology alters hydrodynamic and therefore also the heat transfer in laminar flow. The change in hydrodynamic can be visualised as shown in **Figure 4**. In the area of plain empty tube laminar flow (A) injected red dye is not mixed with the bulk flow.

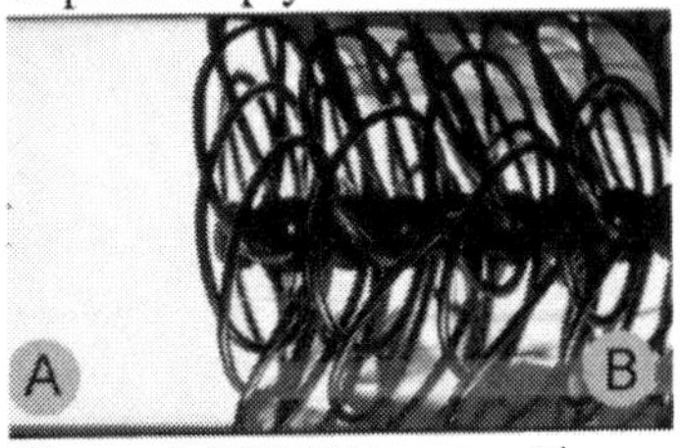

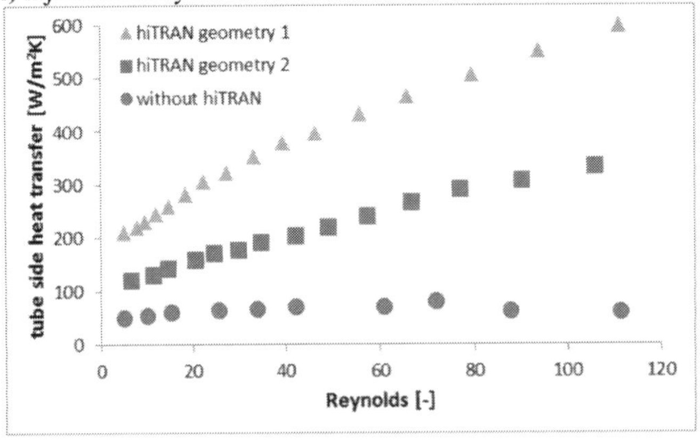

Figure 4: hiTRAN inserts with disrupted flow; measured tube side heat transfer as function of Reynolds for two hiTRAN geometries and empty tube. (η~80cP ; Pr ~ 1200)

Once the dye hits the insert wire the coloured fluid is deflected towards the bulk flow (B) indicating a much improved mixing. The better mixing results in a considerable improved tube side heat transfer.

Heavy crude heat exchanger with the wet crude on the tube side, often operate in a Reynolds range well below 100. In **Figure 4** measured heat transfer coefficient for an empty tube and two different hiTRAN geometries are shown. Heat transfer and pressure drop characteristics of the elements can be changed by varying geometrical parameters such as matrix density and wire sizes. Correlations for all Element variations are implemented in hiTRAN.SP a free software tool which integrates with HTRI and Aspen software [4]. What can be noted is the very low tube side heat transfer for the **empty** tube operation with just about ~50W/m^{2}K. Measurements show that the value remains almost constant with some scatter in the readings. hiTRAN improves the heat transfer in this Reynolds range by more than 10 times and even for single digit Reynolds numbers considerable improvement can be achieved. This translates directly into size reduction as shown later.

5 Exchanger design

In this section the three different exchanger options are discussed

5.1 Wet / dry crude exchanger

In applications with moderate to low water content most of the heat is recovered using the sensible heat from the dry crude leaving the dehydration unit. The viscosity of the wet crude will be considerably higher compared to the dry crude. Wet crude typically behaves as an emulsion. where the viscosity will be substantially higher than the viscosity of the oil phase alone. Since the average temperature of the wet crude is also lower the wet crude has a higher viscosity than the dry.

5.1.1 Fluid allocation

In conventional shell&tube heat exchanger design it is common to allocate the more viscous fluid to the shell side of the exchanger. The combination of parallel and cross flow improves the heat transfer to the viscous fluid. Compared to the heat transfer in not enhanced laminar tube side flow the heat transfer is considerably higher. Another important aspect for the fluid allocation is often the fouling behaviour of the different fluids

5.1.2 Dry crude tube side

This is the preferred option for enhanced hiTRAN design. Good exchanger design aims to balance the heat transfer on the tube side and the shell side as much as possible. Calculation show that this can be often achieved when placing the wet heavy crude on the shell side and the dry crude on the tube side. Dry crude has in general a lower viscosity compared to the oil / water emulsion of the wet crude oil. For that reason typical tube side Reynolds numbers for heavy **dry** crude viscosities are in the range from about 200 to 2000. In this Reynolds range hiTRAN offers the best potential for heat transfer enhancement. Heat transfer measurements of two different hiTRAN geometries are shown in **Figure 5**. It is evident that in this Reynolds range the heat transfer can increase up to a factor of 16 compared to a design without enhancement. An example case below shows the potential for enhancement.

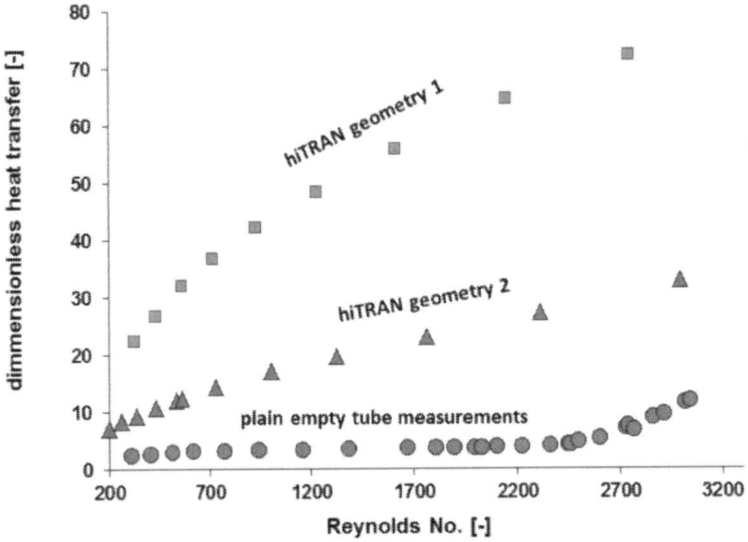

***Figure 5**: Dimensionless empty tube heat transfer compared with hiTRAN enhanced heat transfer for two different insert geometries in relevant dry crude range*

Case study

In this case study the size reduction when applying hiTRAN technology are shown. The application is in operation for wet crude preheat train in the Dos Bocas terminal, located in the Gulf of Mexico. In this train, a cold stream of wet crude oil is heated from 33.6°C to 39.3°C. The heat is recovered from a hot stream of desalted dry crude oil available at 59.5°C inlet temperature. Due to the high mass flow of 577kg/sec per train the required energy is about 13MW in total. The wet heavy crude has an inlet viscosity of 77cP, whereas the dry crude viscosity is about 20cP at inlet condition. Initial designs without heat transfer enhancement calculated very large equipment size. After consultation with Cal Gavin Ltd the unit was re-designed with hiTRAN technology implemented and commissioned in 2009.

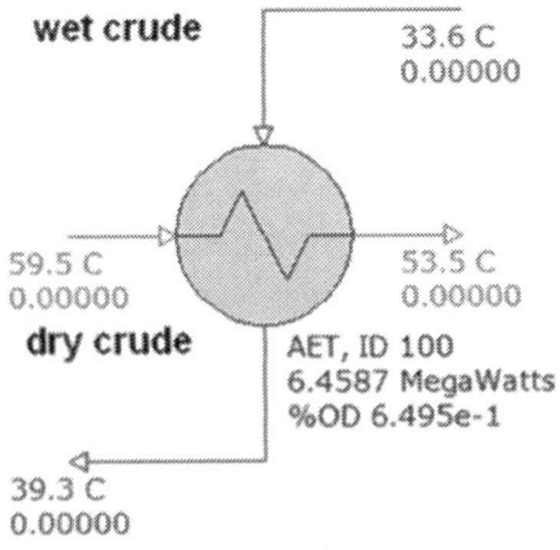

***Figure 6**: Tube sheet after installation of hiTRAN; temperatures in the train (HTRI output)*

Heat transfer calculation show typical problems when designing heavy crude wet/dry heat exchangers. Due to the high viscosity, even the dry crude will operate under laminar flow conditions when placed on the tube side, with very low heat transfer coefficients as indicated in **Figure 3**: Measured empty tube dimensionless heat transfer as function of Reynolds. In this example resulting in a tube side coefficient of 64W/m^{2}K, this leads to a very low overall coefficient, which determines the size of the exchanger according to equation (3). When applying hiTRAN, the tube side heat transfer increases by the factor of about 10 for this case. With the much increased tube side coefficient it was possible to reduce the size of the unit from 20 shells to 8 shells. The pressure drop was kept within the allowed pressure drop of 100kPa by reducing the number of passes and therefore the flow velocity. The size reduction had considerable impact on CAPEX cost for heat exchanger but also for the associated piping costs. HTRI exchanger optimizer [5] was used in order to get an indication about the cost savings possible shown. The results are summarized in **Table 1**.

***Table 1**: Comparison empty tube and hiTRAN design, calculation with HTRI Xchanger suite [6] (empty pipe) and hiTRAN.SP plug-in (hiTRAN design) [4]*

	Empty tube	**hiTRAN design**
No of shells	**20 shells**;2 trains (2 ser. ; 5 par.)	**8 shells**; 2 trains (1 ser. ; 4 par.)
Bundle		
Tubes [-]	1550 25.4mm x 6.1m	1366 25.4mm x 6.1m
Passes [-]	4	2
Total area [m^2]	14453	3267
Exchanger Performance		
Tube side Δp [kPa]	109	100
Shell side HTC [W/m^{2}k]	388	463
Tube side HTC [W/m^{2}k]	**64**	**641**
Tube side Reynolds [-]	~800	~540
OHTC [W/m^{2}k]	45	208
Exchanger duty [MW]	13	13
Flow path [m] / time [s.]	48 / 56	8 / 13
Exchanger / piping costs [$] with installation [5]	**9200000**	**4600000**

Another aspect is the change in hydrodynamics in this assembly. The flow path was reduced considerably from 48m to 8m by the exchanger size reduction, this impacted on the residence time of the crude in the exchanger as shown in **Table 1**. The other major difference in flow behaviour can be seen when applying CFD in order to interrogate the temperature planes

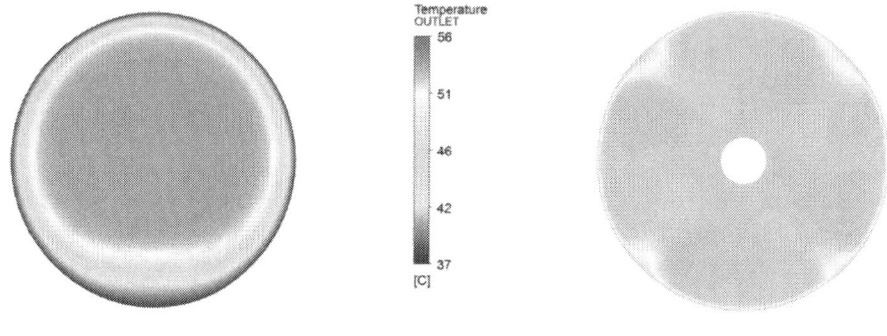

Figure 7: Dry crude outlet temperatures of case study in tube planes; left in empty tube; right in hiTRAN enhanced exchanger.

In **Figure 7** the temperature distribution in the outlet plane of the initial plain empty tube design and the subsequently used hiTRAN design are compared. For both designs the outlet temperature is 53.5°C Due to the mixed convection effect in laminar flow empty tube heat exchangers the temperature within the tube becomes stratified. It is notable that the fluid temperature towards the inner wall of the tube approaches the temperature of the wall. This effect is more pronounced towards the bottom of the tube. The core of the fluid is hardly affected by heat transfer. In the hiTRAN case the temperature is almost evenly distributed with just a small temperature boundary layer at the wall. Reduced hold-up due to size reduction and shorter residence time at the inner wall due to better mixing. This results in a reduced the potential for fouling.

5.1.3 Wet crude tube side

Under certain circumstances it is beneficial to allocate the wet crude to the tube side. One of the reasons can be the high corrosion potential for wet crude with saline water. The viscosities of the crude / water emulsion are higher compared to the dry crude the temperature level of this process stream is also lower. For that reason it is not uncommon to encounter tube side Reynolds numbers in the single or double digit range for this type of exchanger. Under those flow conditions the plain empty tube heat transfer is very low but as seen in **Figure 4** heat transfer improvements up to factor 5 are possible even for single digit Reynolds numbers when hiTRAN enhancement is applied. Feasibility studies [7] for Venezuelan low API diluted heavy crude oil with wet crude viscosities in the range of 600cP and Reynolds numbers in the single figure range (4 to 16) show for a 15MW unit a potential size reduction to ¼ of the original empty tube design size.

5.2 Produced water exchangers / wet crude exchangers

As shown in **Figure 2** heat can also be recovered from the produced water leaving the dehydration unit. Whether this is a valuable option depends very much on the water cut and the amount of heat possible to recover. Due to the high heat capacity of water even moderate flows do offer potential to improve process efficiency. In those exchangers the wet crude is in general placed on the shell side. One reason being the higher heat transfer on the shell side for viscous liquids and also the highly corrosive nature of the saline water. The low overall coefficient due to the viscous wet crude will result in large heat transfer areas. As outlined in section 3.1 in feed effluent exchangers with a temperature cross a single pass arrangement is generally the most efficient option. The combination of large cross sectional area, single pass and often lower flow rates compared to the shell side will lead to low tube side velocities. Even for water with a low dynamic viscosity it is common to encounter laminar or transitional flow behaviour on the tube side. The typical Reynolds range would be 800 to 4000. Under flow conditions in the transition region the heat transfer is very sensitive to any change in flow rate (Reynolds number) as seen in **Figure 3** and very difficult to predict. In case the velocity of the water drops further and the flow becomes laminar then natural convection effects become dominant and lead to thermal stratification.

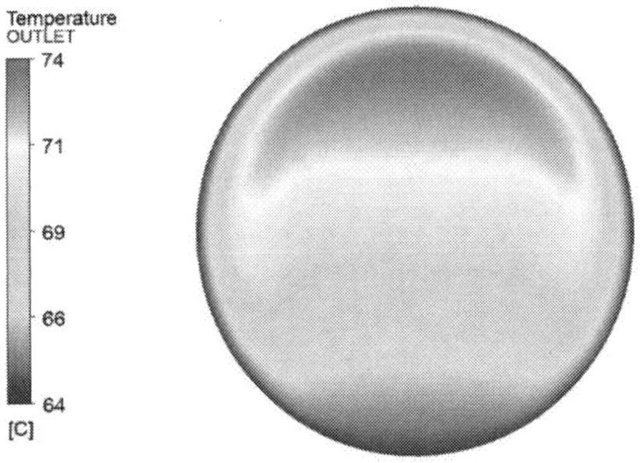

***Figure 8**: CFD simulation of thermal stratification*

This can be seen in **Figure 8**, where the temperature distribution in the outlet plane of a 12m long tube in a wet / produced water exchanger is shown. The temperature difference in the plane is more than 10°C. At the bottom of the tube the blue colour indicates that the produced water temperature drops to the same temperature level as the wall temperature. This corresponds to a loss in driving temperature difference between wall and liquid reducing the effective heat transfer area. Dooley [8] investigated those effects, and identified them as a local temperature pinch. They are reported as HTRI warnings [6]. hiTRAN wire matrix technology alters those conditions in multiple ways:

- Much improved heat transfer up to 16 time compared to the empty tube under such flow conditions
- No sudden increase in heat transfer at transition from laminar to turbulent flow.
- Removal of the local internal temperature pinch.
- Higher tube side pressure drop, which improves the fluid distribution at bundle inlet

5.3 Shell side design

The main focus in this paper was given to the tube side heat transfer enhancement. Often in those scenarios the controlling heat transfer resistance switches after enhancement from tube side to shell side. As in the Case study summarized in **Table 1** the heat transfer increases for the hiTRAN option from 64W/m^{2}K to 641 W/m^{2}K, the shell side coefficient changes due to the smaller bundle size from 388 W/m^{2}K to 463 W/m^{2}K. In this case it changes from heavily tube side controlled to shell side controlled. This highlights the possibility for further enhancement with shell side enhancement technologies. Due to the viscous nature of the fluid HELIXCHANGER (r) Heat Exchangers with a helical baffle design can bring considerable benefit when applied in conjunction with hiTRAN technology [9]. Apart from increased shell side coefficients, one of the main advantages of such a design is the reduction of stagnant shell side zones, therefore mitigating the fouling behaviour.

6 Conclusion

In the paper it has been demonstrated that the use of hiTRAN Wire Matrix technology can be used for substantial CAPEX saving when applied in shell & tube heat exchangers for heavy crude dehydration processes. Apart from the space reduction it is evident that the use also offers operation benefits in terms smaller inventories and reduced residence time in the exchanger. It reduces the risk of internal tube side temperature pinches when designing under laminar flow conditions.

7 Symbols

7.1 Abbreviations

CFD	**C**omputational **F**luid **D**ynamic
EMTD	**E**ffective **M**ean **T**emperature **D**ifference [°C]

7.2 Symbols

A	[m^2]	area
cp	[J/kg K]	specific heat
d	[m]	diameter
h	[W/m^2K]	heat transfer
v	[m/sec]	velocity
$\dot{m}$	[kg/sec]	mass flow
Δp	[Pa]	pressure drop
Q	[W]	duty
T	[°C]	temperature
U	[W/m^2K]	Overall heat transfer

7.3 Greek Symbols

ρ	[kg/m^3]	density
η	[Pas]	dynamic viscosity

7.4 Subscripts

i	inside
in	inlet
out	outlet
o	outside
O	oil
s	settling
W	water
dro	droplet
dro	droplet
dro	droplet

7.5 Dimensionless numbers

Nu	Nusselt number
Re	Pramdtl number
Re	Reynolds number

8 References

[1] E. Sellman, G.S. Sams, and S.P.K. Mandawalkar, "Use of Advanced Electrostatic Fields for Improved Dehydration and Dasalting of Heavy Crude Oil and DilBit," in *World Heavy Oil Congress [WHOC12]*, Aberdeen, 2012, p. 16.

[2] E.N. Sieder and G.E. Tate, "Heat Transfer and Pressure Drop of Liquids in Tubes," *Ind. Eng. Chem.*, vol. 28, pp. 1429 - 1435, Dec. 1936.

[3] A. P. Colburn, "A method of correlating forced convection heat transfer data and a comparison with fluid friction," *Trans. Amer. Inst. Chem. Eng.*, vol. 29, pp. 174 - 210, 1933.

[4] P. Droegemueller and C. Mann. (2009) hiTRAN.SP Software. [Online]. http://www.calgavin.com/software-downloads/

[5] HTRI. (2015) Exchanger Optimizer. Software. [Online]. www.exchangeroptimizer.com

[6] Heat Transfer Research Inc. (2015) HTRI Xchanger suite 7.2. Software. [Online]. www.HTRI.net

[7] M. Baritto, P. Drögemüller, W. Osley, and L. Quillares Vargas, "THE USE OF HITRAN WIRE MATRIX ELEMENTS IN HEAVY CRUDE OIL PROCESSING," in *South American Oil and Gas Congress*, Maracaibo, 27 –30 October 2015.

[8] J.D. Dooley, "CFD Study of Diabatic Single-Phase Laminar Liquid Flow in Horizontal Tubes," HTRI, HTRI Research Report S-ST-1-13, 2010.

[9] M. J. Gough et al., "Dual Enhanced Heat Exchangers using Cal Gavin hiTRAN® Wire Matrix Tube Inserts along with the ABB-LGHT HELIXCHANGER® Technology," in *HEE ASIA*, Kuala Lumpur, 2007.

444377 CFD Simulation of a Coke Drum for the Evaluation of Coke Deposition on the Outlet Pipe of the Reactor (Vapor Line)

Abstract:

A CFD simulation of the top of a coke drum reactor and the outlet pipe of the reactor (vapor line) was developed for the evaluation of the coke deposition on the vapor line. The coke drum is the main part of the delayed coking process. One of the main bottlenecks in the coke drum is the fouling of the vapor line due to coke formation because it could cause plugging and other inefficiencies in the process.

The CFD simulation was implemented in the CFD commercial package Fluent 16.0 using a $k-\varepsilon$ turbulent model. ANSYS ICEM CFD pre-processing software was used to discretize the 3D geometry. A mesh was created using mainly tetrahedral elements with approximately 750000 elements. The mesh in the vapor line was refined close to the wall to obtain a value of Y+ tat would best capture the hydrodynamics as coke deposition is a surface phenomenon.

For the evaluation of the coke deposition on the vapor line, a four lump mechanism with three reactions was obtained from experimental data from literature related to the thermal cracking and delayed coking processes. The four lumps are Vacuum Residue Crude (VRC), gases, distillates and coke. From the kinetic mechanism and the hydrodynamic of the reactor gas phase, an initial approximation about the plugging time of the vapor line was made. As two weeks for 30% tube blockage. The coke formation from condensation was not taken into account because it is not possible to obtain condensation under specific operational conditions (temperature and pressure), composition and characteristics of the components in the vapor line. New geometries, configurations and operational conditions are proposed to reduce the amount of coke formed in the vapor line.

Corresponding author: amolinao@unal.edu.co

444968 Multi-Scale Modeling of Fired Process Heaters in Chemical Process Industry

Abstract:

Comprehensive design and analysis of a fired process heater entails the study of the air duct system for combustion air flow regulation, burner and tube placement design on the firebox side to achieve the required heat flux distributions along the tubes, and an adequate description of the process side physics to be able to predict the right conversion rates. Computational Fluid Dynamics (CFD) is a cost effective alternative to time-consuming design of experiments approach to study burner performance, identify potential issues related to tube life, meet emissions standards, and help maintain product quality.

Since the complete systems is made up of several sub-systems each of which could influence the overall performance, a multi-scale modeling approach is employed in this paper where, CFD is used to model each of component parts in a typical fired process heater. First, the air delivery system to the burner is studied and optimized. Next, the burner used on the firebox side is modeled and optimized to meet flame length and emission requirements. The heat from the firebox side is used to heat the process fluid flowing in the tubes. The heat transfer to the process side and the subsequent process side reactions are modeled using a co-simulation approach and the resulting product yield is monitored for a given firing rate and process fluid flow rate. Depending on the application, the tube side physics could involve reactions taking place in a packed bed system as in the case of steam methane reformers. In such a situation, the heat transfer in packed beds depends on particle loading and particle shapes. A methodology to study the pressure drop and evaluate heat transfer in packed bed systems by explicitly modeling the packing in tubes is also presented.

442166 Simulation of Process Heater Burners Utilizing Multiphysics

Abstract:

Process heater burner design has traditionally relied on calculations based around abstract representations of the underlying geometry and physical testing. Advances in Computational Fluid Dynamics (CFD) workflows allow for co-simulation of the fluid and solid domain of specific burner geometry with very limited abstraction.

Six Sigma design methodologies are used in conjunction with modern simulation tools to produce designs that both meet the burner specification limits and improve equipment reliability. A development case study is presented which utilizes simulation tools and physical testing as well as a comparison of the results.

445225 Integration of Divided Wall Columns in a Shale Oil Processing Unit

Abstract:

Use of Divided Wall Column (DWC) in petroleum processing has significantly increased in the past decade. Savings of 20-30% in capital and energy have been reported. This paper discusses the application of DWC in a Shale Oil Processing unit. Impact of utility constraints on the economics for application of DWC is discussed. The paper will also discuss how Pinch Technology and Process Integration techniques can be used to integrate the DWC with the rest of the process and the insights that can be developed to optimize the process configuration.

443937 Dividing-Wall Column Screening Guidelines and Applications

Abstract:

A dividing-wall column (DWC) is a distillation column with a vertical partition that can be implemented as a potentially attractive alternative to conventional multi-column or sidecut column arrangements; use of a DWC may enable capital cost and energy savings of around 30% or even higher to be achieved. Rules-of-thumb involving process conditions like feed composition and relative volatility are often cited for consideration when trying to determine if a DWC might make sense (or not) in a given flowscheme. The intent of this work is to present and review these guidelines in the context of refining and petrochemical processes and obtain an updated perspective that will allow for more detailed guidance to be provided for future screening evaluations.

444354 New Superstructure-Based Optimization for Process Intensification

Abstract:

Process intensification enables "multi-tasking" by combining reaction, separation and other operations in a single unit, and leads to a substantially smaller, cleaner, safer, and more energy-efficient technology [1-3]. However, there is a lack of a systematic method for the identification of intensification opportunities and the optimal synthesis of intensified processes at the flowsheet level. Traditional process synthesis and integration methods have focused on classical unit operations, where units are integrated but they are not merged to perform multiple operations *in situ*. Though research on process intensification has brought about many promising technologies including reactive distillation, divided wall columns, membrane reactors and task integrated columns, current methods are often suboptimal and consider only a few of all plausible alternatives. While recent phenomena-based design methods [4-6] incorporate process intensification, they are difficult to solve and are based on heuristics and/or decomposition approaches leading to suboptimal solution. Instead of considering classical unit-operation based superstructure representation of a flowsheet, we propose a new flowsheet based on building blocks in which each block represents a unit use of materials performing a specific function (reaction, separation, storage, etc.). An assembly of blocks of the same material would obtain a classical unit operation, while blocks with different materials or functionalities would result in intensified unit operation. This allows a systematic identification, representation and generation of intensification alternatives at the flowsheet level without exhaustive enumeration. We pose the optimal process intensification problem as a mixed-integer nonlinear optimization (MINLP) problem which aims to synthesize a process with intensified units by minimizing or maximizing a process metric given the feed and product specifications, feed and product prices, material properties and bounds on flow rates. In this talk, we will present the overall process intensification method with specific applications.

References

[1] Reay, D.; Ramshaw, C.; Harvey, A. Process intensification: engineering for efficiency, sustainability and flexibility. Waltham, MA: Butterworth-Heinemann, 2013.

[2] Stankiwicz, A.I. , Moulijn, J.A.,. Process intensification: transforming chemical engineering. Chem. Eng. Process, January, 2000, 22-34.

[3] Baldea, M. From process integration to process intensification. *Computers & Chemical Engineering* 2015, Doi.10.1016/j.compchemeng.2015.03.011.

[4] Arizmendi-Sanchez, J. A., and Sharratt, P. Phenomena-based modularisation of chemical process models to approach intensive options. *Chemical Engineering Journal* 2008,135, 83-94.

[5] Lutze P, Babi DK, Woodley JM, Gani R. Phenomena based methodology forprocess synthesis incorporating process intensification. *Ind Eng Chem Res* 2013, 52(22):7127–44.

[6] Babi, D. K., Holtbruegge, J., Lutze, P., G_orak, A., Woodley, J. M., and Gani, R. Sustainable process Synthesis–intensification. *Comput. Chem. Eng.* 2015, .doi:10.1016/j. compchemeng.2015.04.030.

442487 Feasibility Study for Hydrogen Sulfide (H_2S) Optical Gas Imaging

Abstract:

Infrared (IR) cameras have been used extensively to detect hydrocarbon leaks from process equipment for both process safety and environmental applications. Recent advancements in the field have allowed for continuous and autonomous detection as well as quantification of hydrocarbon emissions. While these advances have led to new applications for hydrocarbon leak detection, optical gas imaging to date has not be utilized for detection of H_2S leaks. H_2S is a very toxic chemical commonly found in oil and gas streams and some process streams. Due to the toxic nature of the gas, it is highly desirable to remotely image and detect the presence of H_2S. Unfortunately, due to its IR spectral characteristics there has been no demonstrated method to achieve IR imaging of H_2S plumes.

A series of tests were conducted to evaluate the feasibility of imaging H_2S and to determine the sensitivity of an infrared imager in terms of ppm-m. Various spectral regions were examined for suitability of H_2S imaging. The results of the test show that it is feasible to image H_2S in the SWIR region. The results of these tests and the corresponding sensitivity of an H_2S imager are presented, along with potential applications for this technology.

12TH GLOBAL CONGRESS ON PROCESS SAFETY

Design Considerations for Oily Water Lift Stations (OWLS)

William J Celenza
KBR Inc
Houston, TX
William.celenza@kbr.com

Janani Sankar, AIChE Member

Prepared for Presentation at
American Institute of Chemical Engineers
2016 Spring Meeting
12th Global Congress on Process Safety
Houston, Texas
April 11-23, 2016

Design Considerations for Oily Water Lift Stations (OWLS)

William J Celenza
KBR Inc
Houston, TX
William.celenza@kbr.com

Janani Sankar
AIChE Member

Keywords: Oily Water Sewer, Pump Station, Sump, Gravity Drain

Abstract

Oily water, a mixture of hydrocarbon residues and water, is contaminated water released during normal operations at an oil refinery. Process wastewater flows are collected by an oily water or industrial sewer system that can be combined with area drains to collect potentially contaminated surface runoff from the on plot process paved areas as well as areas where spillages and wash water may occur. Surface runoff includes precipitation and fire water which is drained to catch basins, or area drains, through a network of gravity drainage piping and manholes. A clean storm water system collects precipitation from unpaved areas, roofs, roads and other areas not likely to be subject to contamination for direct offsite discharge. An oily water system needs to safely transfer collected flows to treatment or storage facilities to prevent contamination from impacting human health or the environment.

Refineries that segregate the oily water collection system from storm or fire water flows may drain runoff by gravity directly to one influent lift station at the entrance to the treatment plant, without using any intermediate lift stations. However, with combined systems the reliance on gravity conveyance can lead to excessive excavation depths and increased sewer construction complexity. The use of lift stations strategically placed throughout a combined collection network can simplify underground sewer piping networks, accommodate high water tables, and provide functional isolation for complexes or units within a refinery.

The design of an Oily Water Lift Station (OWLS) is distinct from that of municipal or highway related storm water lift stations due mainly to the industrial nature of the inflows, which typically are non-flammable but can include limited amounts of flammable and combustible hydrocarbons. The potential for explosive atmospheres in oily water lift stations, typically sealed and vented for VOC emission compliance, presents what is to some an unacceptable risk and has led to a design philosophy of minimizing or even eliminating the use of lift stations in this service, regardless of the increase in installation complexity. Also system flexibility can be increased and environmental risk associated with operating an Oily Water Drainage System (OWDS) can be better managed by including OWLS. Even though relying on lift stations presumes lower system reliability, appropriate pump selection, use of an emergency power supply, and automatic control and alarm systems have been successfully used to reduce this risk.

This paper describes methods for managing risks associated with operating OWLS through design considerations and related operational and maintenance procedures. Compliance with current US codes and standards for fire, electrical, and flammable vapor emission issues are discussed for gravity drain systems with lift stations, as well as the use of complementary closed drain pressurized collection systems, chemical sewers, and dedicated collection systems for constituents not compatible with atmospheric gravity drain systems. However, specialized systems such as Benzene Waste Operations NESHAPs compliant collection networks are not included in this discussion.

1 Background and Purpose

Modern refineries segregate direct contact process wastewater, such as desalting washes, process condensates and sour water, as well as solvent discharges in dedicated collection systems. Segregation allows for pre-treatment near the point of generation and potential cascading or reuse onsite with the aim of reducing end-of-pipe treatment loads. Wastewater is then typically pumped directly to the Wastewater Treatment Plant (WWTP) headworks. Also, flammable, toxic, corrosive, reactive or high pour-point liquid materials are segregated into closed drain systems (CDS), consisting of dedicated piping and vessels connected to selected drains that collect these liquid types for recovery or safe disposal. Hydrocarbons from tank drains, solvents or sour water from sulfur recovery or tail gas treatment units, and amines from acid gas removal units are examples of process streams which should use CDS. Typically only residuals or cleaning washes from these operations would be drained to a gravity Oily Water Drainage Systems (OWDS).

Oil-free wastewater such as boiler blowdown, cooling system blowdown, and demineralizer waste if not needed or suitable for reuse is typically collected then pumped around the headworks and primary treatment section of the WWTP as shown in Figure 1.

Not many refineries have segregated OWDS and process area storm water drainage systems possibly due to local regulations. But if segregated systems exist, the water is directed to separate treatment systems and the storm water system is likely to follow first flush principles subject to local regulation.

The combined OWDS will collect potentially contaminated storm water runoff from the on plot process paved areas and any areas where spillages and wash water may occur. This system will also collect and drain all fire water. Collecting these flows ensures that any contaminated runoff will not be discharged to the environment, or infiltrate into the groundwater. The areas being drained to the OWDS system are usually concrete paved.

Sanitary sewer systems should be entirely segregated from an OWDS. Sanitary wastewater may eventually be commingled with the oily water if the mixing may prove beneficial for the treatment of the combined wastewater at the WWTP.

As schematically shown in Figure 1, only some wastewater generated from refining operations is conveyed to the site WWTP via an underground gravity OWDS. The wastewater streams presented in Table 1 when only nonvolatile, nonflammable, and non-toxic components are present are typically managed in a refinery gravity OWDS. Since these streams are either continuously or potentially oil contaminated, treatment through the primary treatment section of a refinery WWTP is always recommended. Also as indicated in Table 1, the process area storm water is collected through the OWDS as a combined system. When compared to the typical influent for a refinery API Separator treatment system, the OWDS pollutant levels would generally be lower in chemical oxygen demand (COD), phenols, sulfides and ammonia but higher in oil & grease and total suspended solids (TSS). The API separator influent includes the OWDS combined with other wastewater streams at the WWTP headworks.

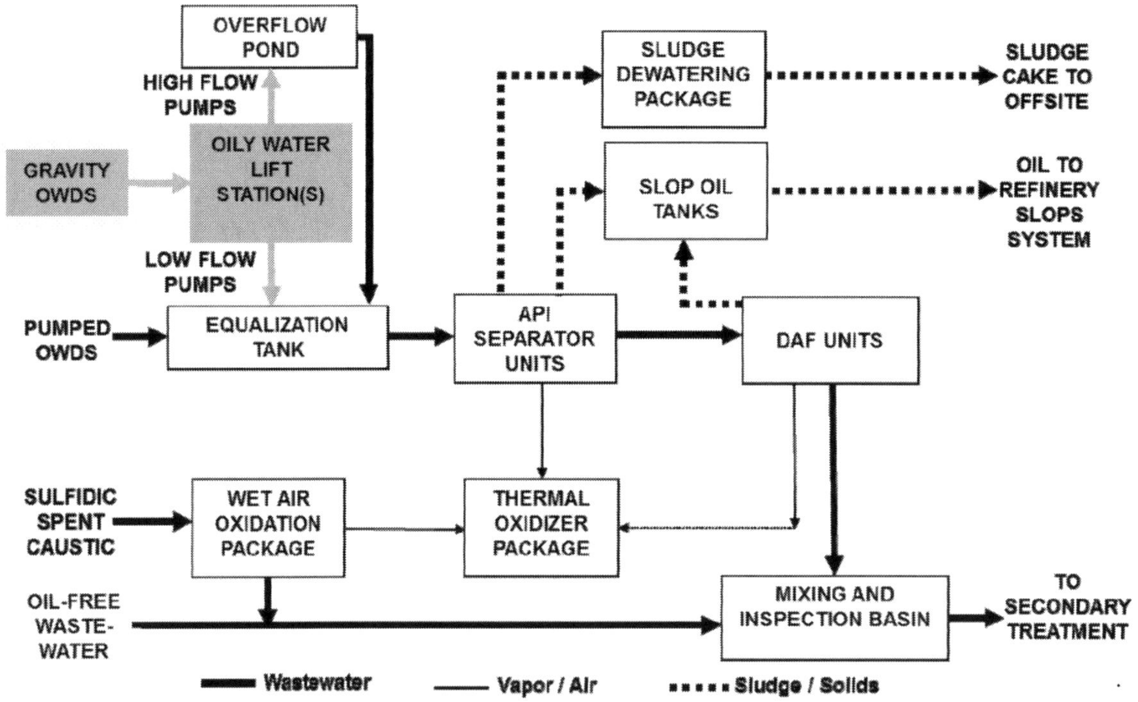

Figure 1 Refinery Wastewater Collection and Primary Treatment System

An OWLS, as part of the OWDS, receives flows that are incidental in nature and thus highly variable in flowrate as well as pollutant loading. In addition to storm water and fire water events, incidents such as hydrocarbon spills, hose station wash-downs, testing of fire water sprays, process vessel washes, pump baseplate working loses, etc. need to be anticipated.

Table 1 Wastewater typically managed in a refinery gravity OWDS (derived from US EPA 1980)

Wastewater Source	Typical Conventional Pollutant Levels (mg/l)						
	Oil & Grease (1)	COD	TSS (2)	Phenolic Cmpnds	Ammonia Nitrogen (3)	Sulfides (3)	pH
Hydrocarbon Equipment Drains (tanks, vessels, etc.) for dewatering, cleaning, hydrotesting, etc.	< 1,000	< 300	< 1,500	< 5	Trace	<10	5 to 9
Pump Station Drains (casing, base plate, etc.)							
Utility Drains (sample points, level gauges, instruments, etc.)							
(Un)loading Station Drains							
Pipe Trench Drains							
Area Drains (Wash, Storm, Fire Water)							

Notes:

(1) Oily Water is a mixture of non-flammable hydrocarbon residues and water, released during normal operations and is not expected to create hazardous conditions if discharged directly to a gravity OWDS.
(2) Can include fine dust/carbonate type solids that have been observed to produce a sticky paste when mixed with water and oils at some sites.
(3) For instance, Sour Water which contains more than 10 mg/l of hydrogen sulfide with other components such as ammonia, phenols, and benzene would not be permitted as discharge to an OWDS.

2 OWDS SYSTEM DESCRIPTION

The basic types of gravity OWDS are "dry" or "flooded". Dry systems consist of free-draining sewer networks interconnecting catch basins and manholes. A typical catch basin or manhole has a sand trap and a liquid seal as shown in Figure 2. The sewers free drain to an OWLS that transfers the wastewater to a site WWTP for treatment or temporary storage.

Flooded systems maintain continuous wastewater gravity flow within the sewer piping and drain directly over a weir located at the entrance to an OWLS or to an API Separator at a WWTP. The system is flooded by the weir at these locations; and the catch basin and manhole components are the "dry-box" type with sewer inlet and outlet lines located flush with the flat bottom (Mason 1984). Water entering the OWDS will cause the level to rise and water will eventually flow into the OWLS by overflowing the weir. The weir needs to be designed to manage the peak instantaneous flow. Flooded systems may be impractical in locations sufficiently dusty where excessive solids may cause clogging (Lees 2012a, Mason 1984).

Compared to a flooded OWDS, a dry OWDS requires deeper excavation to accommodate free draining sewer lines and operation of the OWLS nearly empty. Thus a dry OWDS typically requires more intermediate OWLS for the same topographic conditions, such as available terrain slope, and water table elevation.

Per the OWDS in Figure 3, raw wastewater streams normally are introduced into the collection system through individual or area drains, which can be open to the atmosphere or sealed to prevent wastewater contact with the atmosphere. Individual drains may be dedicated to a single source or piece of equipment.

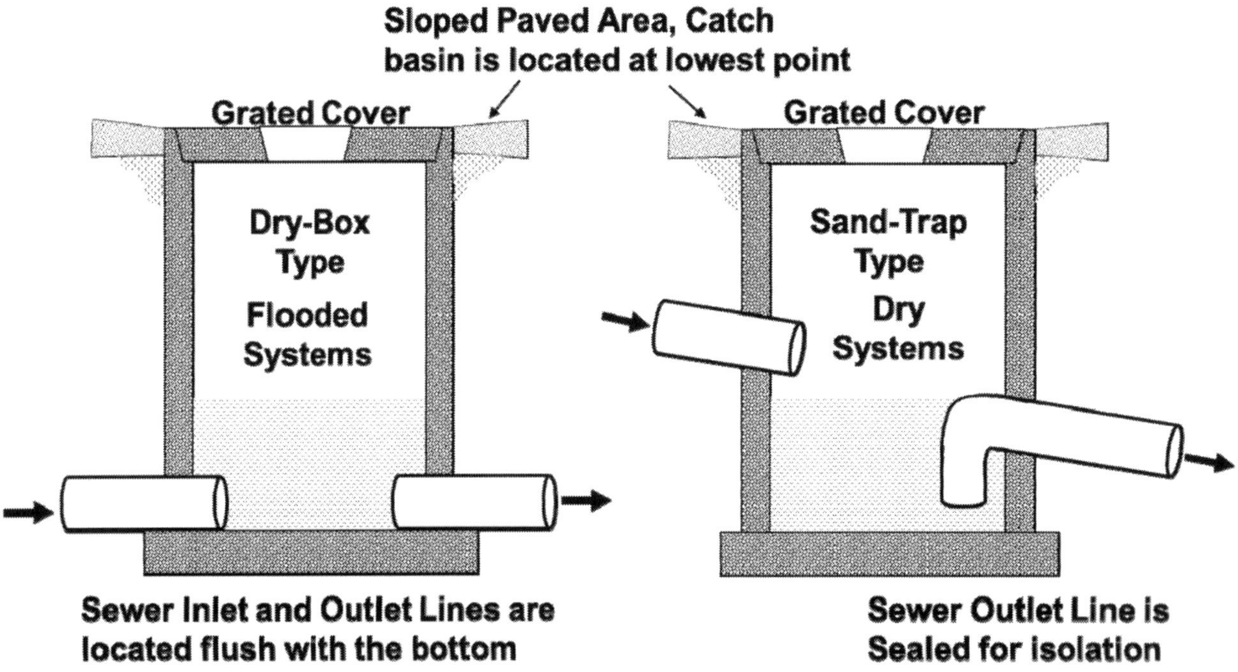

Figure 2 Dry-box "Flooded" vs. Sand-Trap "Dry" Type Catch Basins

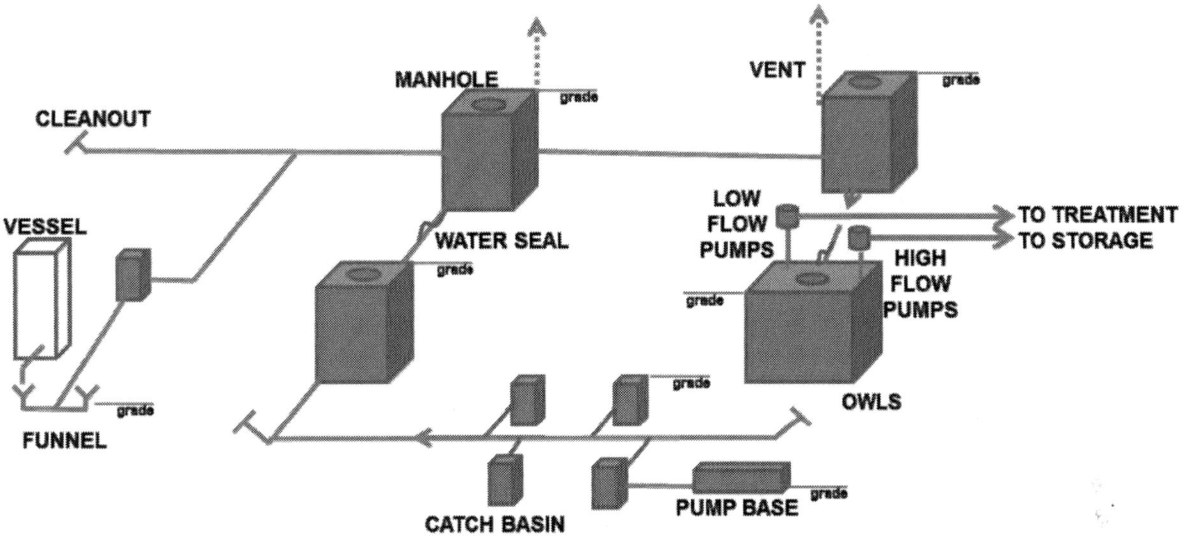

Figure 3 OWDS, Local Oily Wastewater Lift Station (OWLS) System

Drain Pipe / Funnels are outlet pipes that collect oily water from points such as pump bases, drain points, or vessel drains and discharge through a catch basin or manhole into the OWDS.

Area drains, or catch basins, serve several sources and are located centrally among the sources or pieces of equipment that they serve. Catch basins have a grate on top, an integral sediment trap and outlet liquid seal for dry systems and also serve as a collection drain for surface flows. Individual catch basin drainage areas are typically limited to approximately 5,000 square feet to ensure rapid surface runoff and to isolate groups of equipment for fire protection

Manholes in sewer lines permit service, inspection, and cleaning of a line. They may be located where sewer lines intersect or where there is a significant change in direction, grade, or sewer line diameter. Manholes include an outlet liquid seal for dry systems.

An OWLS, as part of the OWDS, is usually the last collection unit before transfer for treatment or storage, accepting wastewater from one or several sewer lines. An OWLS is a retaining structure, usually under atmospheric pressure, located below the normal grade of the gravity OWDS network. The collected wastewater is normally pumped to the WWTP.

An OWLS is sized to manage the larger of the following design flow rate cases:

- Normal process operation flows plus the peak fire water flow from the local risk area; or

- Cumulative storm water flow, typically based on a 5-year or 10-year return frequency, to prevent flooding that would disrupt operations or prevent a safe shutdown.

In some regions, the allowance for fire water may be more than the volume allowed for a 5-year or even 10-year storm water event; while in other regions the storm water event sizes the system. In those cases where fire water dominates, the normal process flows would be added to the storm water for the sizing analysis. This analysis also depends on the size of the collection area, since storm water flows are proportional to the ground surface area while fire water flows are not.

An OWDS with a network of intermediate OWLS is shown in Figure 4. The OWLS pumps typically would discharge into common headers, with one header dedicated to convey the wastewater to the WWTP for treatment and another header to storage.

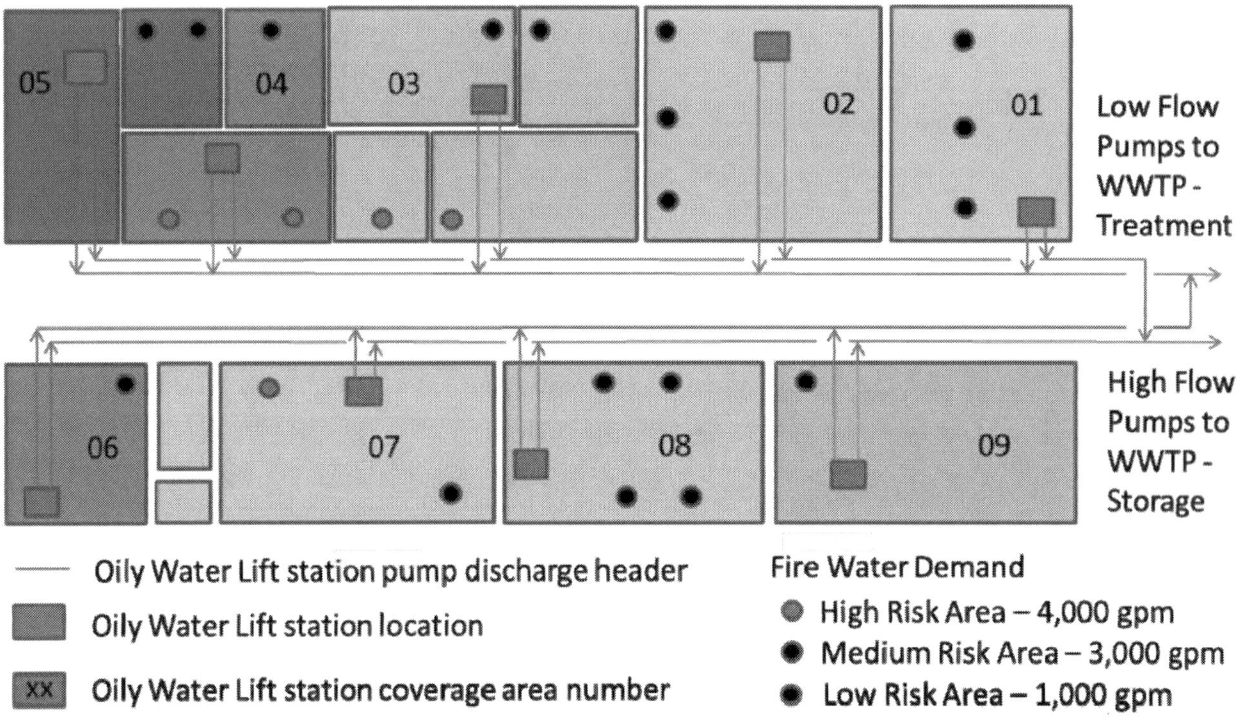

Figure 4 OWDS With A Network of Intermediate OWLS

Depending on the difference between normal and peak or high inflows, an OWLS may have one or two sets of pumps. If the high flow is less than 10% more than normal flow, a single set of pumps sized for the high flow condition can be used efficiently. If these flows vary by more than 10%, two sets of pumps, one for the normal and one for the high flow condition may be more efficient. For a two pump set system, a single or dual pumping chamber OWLS may be considered. Figures 5 and 6 show plan and section views of a dual chambered OWLS. Normal flows are managed in a smaller chamber with the larger chamber of the same OWLS reserved for the high flows.

All OWDS wastewater initially enters the low flow or process flow chamber and the high flow or wet weather chamber receives overflow from the low flow chamber. The minimum required working volumes of the chambers are related to their pump capacities, with the low flow pumps typically around 50 gpm or less and the high flow pumps 500 gpm or more each. However as indicated in Figure 6, the low flow chamber must also accommodate an underflow baffle and overflow weir rated at the design inlet flow. These weir dimensions often dictate the footprint shape, and often result in a working volume larger than the required minimum.

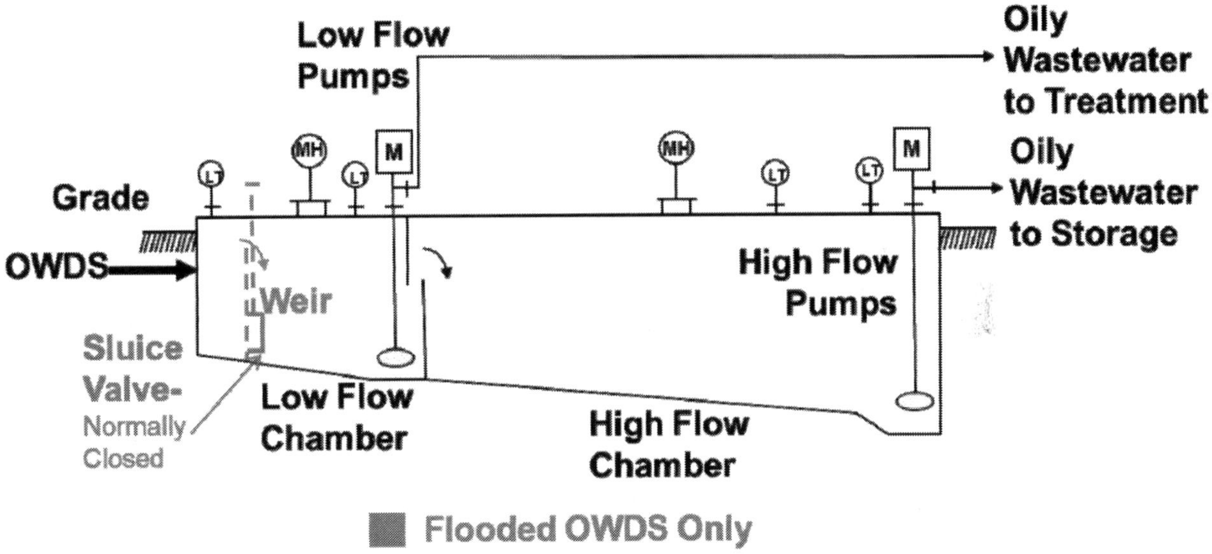

Figure 5 Dual Chamber Oily Water Lift Station (OWLS) – Section A-A

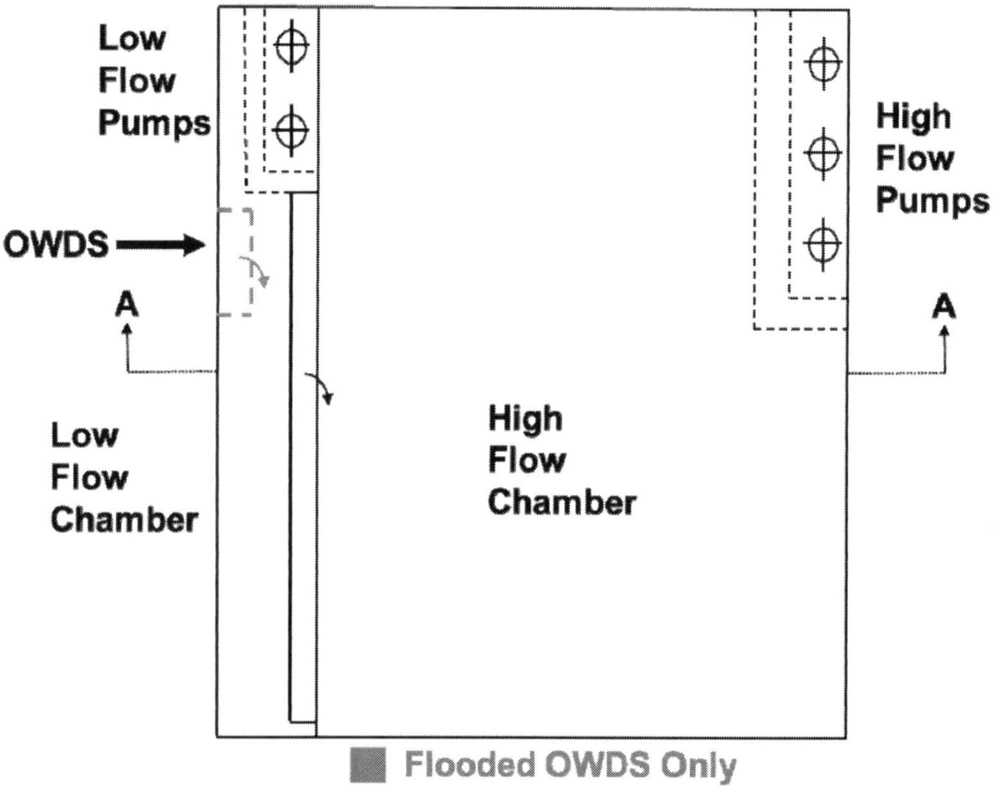

Figure 6 Dual Chamber Oily Water Lift Station (OWLS) – Plan View

3 SAFETY CONSIDERATIONS

A majority of safety incidents occur during startup or in auxiliary areas such as transport, storage, waste management, and recovery of raw materials or products (Lamberti 2011). Flammable liquids in wastewater discharged to an OWDS can cause fires / explosions in collection systems-particularly near the point of discharge, where high temperatures promote evaporation into a relatively fixed volume of air and form vapors that are trapped within the collection system. As the heated gases expand within the confined space, they create underground pressure seeking an opportunity to vent. Due to the vapor space confinement within OWDS and OWLS, when a flammable vapor cloud is ignited, the combustion probably will give rise to an overpressure resulting in a vapor cloud explosion (VCE) rather than a fire (Lees 2012b). This pressure will blow off at the weakest point, such as a manhole cover, a liquid seal trap, and / or an OWLS cover. Once dislodged, more oxygen may be introduced triggering additional explosions. A VCE can be ignited by electric sparks, friction, surfaces such as manhole covers heated by the sun, or heat generated by chemical reactions. Often workers use metal tools that can accidentally strike against a concrete surface or manhole cover creating sparks. (US EPA, 1992). Also, potential pump-related ignition sources include defective pumps, dry running pumps and faulty cables on submersible pumps (Bloch 2015).

Twenty years of historical data from a refinery and terminal facilities operator recorded a total of three dozen OWLS incidents in dry type systems (confidential 2008). Forty percent of the incidents involved a fire / explosion, while the rest were overflows. The main contributors to the incidents were:

- Presence of flammable liquids that formed an explosive vapor mixture – fire incident,
- Malfunction of level instrumentation - overflow,
- Failure of pumps - overflow.

Table 2 summarizes methods for managing risks associated with operating the components of an OWDS, preventing system failures, and increasing reliability.

Table 2 Methods for Managing Risks Associated with oily water lift stations

Design Considerations	
Electrical Area Classification	OWDS are classified as Class I, Zone 0 locations for manholes, catch basins, OWLS, underground piping, and vents. Zone 0 is defined as an area where an explosive gas mixture is continuously present or expected to be present 1,000 or more hours/year or 10% of the time (API RP 505, reaffirmed 2013).
Seals / Isolation	All flows into a dry OWDS enter through sealed catch basins or through a sealed connection to a manhole. All sealed connections, including sections of a drainage line located between liquid seals, from upstream of the first sealed manhole in an OWDS to downstream of the farthest catch basin, need to be vented to direct the release of any flammable or hazardous gases and to prevent a vacuum which could drain out the water seal connection. A water seal typically consists of an elbow or a tee with an outlet extending downward to provide a minimum of a 6 inch seal.
	In exceptionally hot and dry locations the seals in dry OWDS may require water addition to avoid evaporation; and a flooded OWDS may require a water purge to maintain a level of working liquid (Mason 1984).
	The use of a network of independent OWLS permits the isolation of individual process areas, units or complexes during site maintenance shutdowns. Whereas a site-wide sewer network increases the risk of a fire spreading and may lead to effluent stagnation during emergency situations.
Managing	Closed Drain Systems (CDS) are needed to manage flammable liquids and toxic materials from process water draw-offs, hydrocarbon contaminated

Flammable Liquids and Corrosive or Toxic Materials	cooling water, de-inventorying equipment drainage, and streams which could otherwise create hazardous conditions if they were discharged directly to OWDS. Even though no continuous flammable liquid source connections should exist, flammable liquid hydrocarbon content less than 50 mg/l may be considered for drainage. A combustible liquid at a temperature within 15 F (8 C) of its flashpoint would be considered flammable.
	Examples of toxic materials include: spent caustic, benzene, phenol, sour water that contains H2S (>10 mg/l) and ammonia (> 0 mg/l). Corrosive materials have a pH < 3 (acid) or pH > 12 (base).
Venting	Since flammable gases may accumulate in an OWDS, manhole covers and OWLS covers should be sealed or gasketed and vented to a safe location away from any potential ignition sources.
	End-of-Line Deflagration Flame Arresters may be used on vent lines to prevent flames from entering an OWLS, such as if a flammable gas mixture flowing from the vents were to encounter an ignition source.
Piping line depths	The OWDS piping in onplot unit areas is located under reinforced concrete slabs. Thus the configuration of the standard catch basin or funnel / drain pipe assembly vs. Minimum Cover Over Top of Pipe typically sets the minimum depths. Since a connection at the extremity of an OWDS sets the network depth, drain lines at these locations may be embedded in concrete to limit depth if the required soil-bearing loads are high.
	In a dry OWDS the slope for a free draining line is usually 0.6% to 0.8%. A flooded OWDS line slope would only require a nominal 0.1% slope (Mason 1984).
	For a dry OWDS, all gravity flow lines entering an OWLS must be above the maximum water level that will be maintained during operation. This design requirement ensures that the line will not become overfilled and not work properly. A 6-inch elevation difference is typically required to not impede incoming flow and cause a backflow and accumulation of sediment (confidential report 2008).
	Typically, when gravity OWDS lines are installed in trenches deeper than 20 feet, the installation cost increases significantly due to more complex excavation and trench shoring techniques (US EPA 2000a)
Siting	When locating OWLS, consideration for the distance to major process equipment should be considered. Offset, such as 150 to 250 ft. from process equipment, 30 to 70 ft. from occupied buildings, and 10 ft. from pipeways; may limit options and even lead to consideration of off unit plot locations.
	If a high water table or large distances are an impediment to the design of an OWLS, even after considering converting from a dry OWDS to a flooded and / or segregation of the OWDS from the storm / fire water, then additional OWLS may need to be installed.
	The number of local OWLS is determined by the following:

	• Type of OWDS, dry vs. flooded. • Site topography, available elevation change. • Minimum gravity flow line velocity, • Maximum invert depth to prevent the OWLS bottom elevation from being located below the water table, • Prevention of gravity flow lines from crossing large underground site utility lines, • Prohibition of locating near process equipment, inhabited buildings, drainage swales, diked areas, or designated pipeline corridors. • Avoidance of the pump lift station head from exceeding 20 ft. to allow the use of self-priming pumps.
	For OWLS vent locations, a safe location should be considered as no closer than 50 ft. horizontally from a furnace or other ignition source; and if within 100 ft. consider ending the vent line at least 50 ft. above grade.
Downstream impact	As the number of OWLS pumping to an API Separator increases, the turbulence impact from on-off pumping on the oil / water separation process should be considered. Either installation of a break tank with equalization or ensuring that the API Separator design is robust enough to manage the potential fluctuations and turbulence is suggested.
	Due to potential disruption of the pump(s), such as by mechanical or power supply reasons, the WWTP design may need to include quick operational turn down.
Pump Selection	Due to a Class 1 Zone 0 Electrical Area Classification, pumps with submersible motors would not be permitted. Bloch (2015) further discusses pump types for Zone 0 service.
	Several types of centrifugal pumps are frequently used in OWLS applications. Straight-flow centrifugal pumps are suitable for low-flow / high head conditions, angle-flow are appropriate for low to moderate heads, and mixed flow for large flows at low head (US EPA 2000c). Concerns for causing emulsification that would impact downstream treatment may also guide pump selection.
	Self-priming pumps have been found to be more reliable in OWLS service (confidential report 2008). If the required pump flow / discharge pressure is within the operating range of a self-priming pump, and the OWLS depth does not exceed a 20 ft. suction lift then the use of a self-priming pump may be an option. Use of a piping foot valve or anti-siphon valve would be recommended in this application. Or an option would be a vertically suspended line shaft type pump.
	US EPA (2000b and c) states that the number of pumps should be chosen to correspond as nearly as possible to flow fluctuations. In OWLS with peak inflows less than 700 gpm, two pumps are customarily installed with each

	able to meet the peak influent rate, subject to flow variability concerns. For instance, if the peak flow is less than 10% more than normal flow, a single set of pumps sized for the peak flow condition can be used efficiently. For larger OWLS or OWLS with greater variability, strategies include: installing pumps of different sizes, using parallel pumps, trimming impellers, installing variable speed drives or two-speed motors in order to prevent frequent starting / stopping and to not require excessive OWLS storage capacity. However for OWLS with rapid flow increase or high head losses, a single peak pump may be favored because two pumps operating together yield only slightly higher flows than one pump. Parallel peak pumping is favored for large OWLS with relatively flat system head curves, where multiple pumps deliver substantially more flow and provide more flexibility.
	Ursino (2014) states that adding a redundant pumping element in standby is less effective than increasing the number of pumping units that may function in parallel. Thus a larger number of smaller online pumps may prove more reliable than a less number of larger pumps, despite provisions for standby operation.
	Overall pump efficiency depends on the type of pump, the control system, solids loading, and the fluctuation of influent. Well-designed pump systems may operate at up to 75 to 85 percent efficiency most of the time (US EPA 2000c); however, discharge header conditions may limit efficiencies to closer to 50 percent at times (US EPA 2000a). Also, the mechanical efficiency of the pump impeller would normally decrease with increased solids handling ability, in relation to the solid size to be passed.
	Installing an inlet OWLS screen with means to clean it, will provide for filtering out incoming debris to protect the pumps from blockage.
Power Supply	Including separate transformer feeds to all pump systems can help to ensure a continuous source of power is provided. Emergency generators whether onsite standby or portable with quick connection can also be used as a backup power source. However, in the case of a power failure, options to de-inventory an OWLS would be: a) Use of Vacuum Trucks with suction hose access through a sump cover manway and, b) Use of portable diesel pumps with suction hose access through a sump cover manway or by way of dedicated connections on pump piping lines.
Instrumentation and Controls	Instrumentation to ensure overflows or running pumps dry does not occur is recommended; however, faulty or complex instrumentation should not hold up pumps from automatically operating when needed for instances such as removing fire water. The confidential report (2008) states: 1. Provide two radar level transmitters (LTs) for redundant level measurement in the OWLS. Level indication shall be in the control room. LTs should be configured to fail low such that pumps will not start on

	transmitter failures. LTs are preferred over switches because switches are blind devices that can fail in an undetected manner. Typically one LT should be designed to control the sequenced start-stop of pumps. Each LT should indicate its level on a DCS display in the control room and be provided with low and high level alarms. The two associated LTs should be compared and alarm on deviation, to provide more confidence that the level signal is true. 2. Maintain the water seal within an OWLS by having the incoming flow line termination point at 6 inches below the pump shut off level. Also the OWLS high-high level shall be reset below the incoming flow line. 3. A flow or pressure transmitter should be provided on each pump to trip on low flow. An orifice based flowmeter is recommended, but a pressure transmitter may be considered if the pump discharge pressure will reliably drop to a known pressure on loss of flow. Flow transmitters (FTs) are preferred over switches because switches are blind devices that can fail in an undetected manner. FTs should trip its associated pump on low flow and be configured to fail low. Low flow tripping should be active when the pump is on automatic or manual operation. However, consider low flow alarming only for pumps likely to be used for fire water service, to allow for enhanced availability. 4. Local controls should be located at a safe distance from the edge of the OWLS – 10 ft. away or more - to keep operators away from potential explosion exposure. Consider a local analog flow indicator from each pump FT and a local analog level indicator for each sump. Each pump should also have a local hand-auto-off (HOA) switch and push button start. However, since operating the pumps in manual mode increases the risk of pumps dry running, this switch should be locked or car-sealed in auto mode. Also consider providing the control room DCS with an indication of HOA switch position. 5 The control room DCS should be provided with a running and stopped indicator for each pump. Also consider DCS logic that will monitor both the OWLS level and pump flow / running conditions, so to provide an alarm if a pump is running with acceptable flow but the level is not moving.
<u>Operational and Maintenance Procedures</u>	
<u>Spills</u>	Operating procedures will need to stress the importance of detecting and isolating oil spills.
	The dual chamber version of the OWLS contains an overflow / underflow weir system to segregate free oil in the inlet portion of the system. However, an automated oil removal system is usually not installed. As a result, oil will collect in this first chamber and needs to be removed

	manually. The main function of the underflow / overflow weir system is to avoid oil carryover into the second or high flow chamber since only the low flow or dry chamber discharges directly to the WWTP's oil removal system.
	During vessel turn-around after depressurization, equipment inventory is drained into a CDS. Only residual inventories of flammable and toxic liquids, less than 26 gallons from equipment should be drained to an OWDS under special procedures. Flushing the sewer with water for dilution after draining until the percent lower explosive limit (%LEL), determined by lowering an LEL tester sample hose near the liquid level, is reduced to < 10% is suggested. Before draining any small amounts of flammable liquids or hazardous materials to the OWDs, also consider: (a) wind direction, (b) positioning of any ignition source beyond 50 ft., (c) H2S or ammonia content, and (d) use of proper personal protective equipment (PPE) including breathing apparatus.
<u>Removal of Surface Oil</u>	For lighter hydrocarbons, the insoluble oil tends to float on the surface. When the OWLS pumps down, the oils often are retained when the low level is reached. Special pump designs are available to allow for removal of much of this oil, and / or vacuum truck access can be provided to remove surface oil.
<u>By-pass for maintenance</u>	Facilities for bypassing an OWLS should be considered, such as emergency pump connections on the underground gravity inflow line and on the outlet pump discharge line. In addition, provisions for placement of an isolation sluice gate on underground inflow lines to assist in a temporary pump-around may be practical.
<u>Pump Failure</u>	Confidential report (2008) states that OWLS pump failures are mainly due to: • Dry running: from dirt plugging the pump strainers or faulty level instruments. • Lack of lubrication: poor dry packing design or poor maintenance practices. • Vertical pump application: more prone to failure than horizontal self-priming pumps because of their suspended long shaft, which is more likely to wear against pump bushings and packing if dirt or vibration increases. Also more difficult to detect problems as components are below the surface and not easily accessible. • Lack of effective preventative maintenance (PM): even if an OWLS pump PM program exists; confirm that all pumps are covered.
	Maintenance goal should be to have Mean Time Between Failures (MTBF) for pumps above two years.
<u>Temperature</u>	From a materials of construction consideration, the temperature of any drainage should not be higher than the temperature rating of the drainage piping. Hot steam condensate is not recommended to be discharged to the

	OWDS.
	In the case that Potentially Contaminated Condensate (PCC) has been detected from a leaking exchanger, cooling would be required before any PCC would be approved for draining to an OWDS subject to flammability concerns. Collecting PCC in a CDS for pumping directly to a WWTP based dedicated cooler should be considered. Cooling to at least 140F is recommended in either case. For operational input: 1. In the event of any contamination of steam condensate with hydrocarbons, identify the source of contamination and isolate for correction. 2. If for any reason, hydrocarbon-contaminated hot steam condensate is directed to OWDS, then it must be cooled by dilution with utility or fire water.
Solids	Recognize that EPA HAZARDOUS WASTE CODE F037 pertains to Petroleum refinery primary oil/water/solids separation sludge –including any conveyances; sumps; and storm water units receiving dry weather flow. Also, in some regions fine dust/carbonate which produces a sticky paste when mixed with water and oils can be expected in an OWDS. Any sand or grit that collects in process areas and gets washed into an OWDS may mix with hydrocarbons and create sticky oily solids. One pound of dry solids can create ten pounds of oily sludge (IPIECA 2010). Solids issues favor the use of non-clog pumps, typically open impeller trash pumps, since they have been demonstrated to be reliable for handling oily wastewater and provide protection against oily waste causing pump failure.
Firewater	The OWDS function to remove fire water is essential to firefighting activities, so OWLS pumps must be set-up to operate when required. The primary concern is to contain firewater runoff onsite. Firewater within unit battery limits should gravity drain through the OWDS and collect in a local OWLS. Both the OWDS and OWLS in a combined system should be designed for firewater loads. For dual chamber OWLS, the excess inflow will overflow to the high flow chamber. When the level in the high flow chamber reaches the high flow pump start level, the high flow pumps will start and the low flow pumps stop. However, an operator can shut off the low flow pump and eventually rely on only the high flow pumps without loss of containment. The high flow pumps will discharge to storage, thus isolating the collected firewater from the WWTP treatment system. Note, the low flow pumps discharge directly to the WWTP treatment system but their relatively small flow rate will limit the impact of firewater on the WWTP. Once collected in storage, firewater can be analyzed and even pre-treated before discharging through the WWTP for treatment at a controlled rate to limit potential impact. Any significant free oil layer should be manually skimmed off before discharging back to the WWTP.

	The first flush approach while applicable to storm water, may not necessarily apply to firewater runoff which can remain contaminated throughout a fire fighting event. Therefore, the disposition of any collected firewater runoff will have to be determined on a case-by-case basis, and should be expected to depend on the specific firefighting scenario. Firewater can carry environmental contaminants, with a scenario including flammable and/or combustible materials, such as petroleum products, foam solution containing perfluorooctane sulphonates, dissolved and particulate materials from combustion processes, materials generated through quenching, as well as heavily contaminated solids and grit. Since the hazards and associated risks, such as unknown health effects and the impact on a biological treatment system are difficult to estimate in practice, both the liquids and solids should be considered as potentially hazardous wastes. Also, if the analyses indicate that the detained solids / sludge in an OWDS / OWLS after a fire event may not be recommended for treatment at the onsite WWTP then arrangements for offsite treatment and/or disposal via licensed waste contractors may be necessary. Note, even if the firewater runoff liquid is determined acceptable for the onsite WWTP, the solids may require a different disposal route.

4 CONCLUDING REMARKS

The management of oily water involves managing process safety hazards. Since all hydrocarbons can form an explosive atmosphere under the right temperature and pressure conditions; methods for mitigating and controlling risks associated with operating OWLS should be firmly established.

Safety hazards associated with OWLS can be mitigated through proper design, permits and procedures. The "safety culture" must extend to control auxiliary areas such as wastewater management systems; and should consider the OWLS as a hydrocarbon process facility in regards to HAZOPs and process safety analysis. The most labor-intensive task for OWLS is routine preventive maintenance (US EPA 2000c); however regardless of the excellence of a maintenance program, equipment use causes wear and ultimate failure or breakage of parts should be accounted for in planning.

5 REFERENCES

API RP 505, (Reaffirmed 2013), Recommended Practice for Classification of Locations for Electrical Installations at Petroleum Facilities Classified as Class I, Zone 0, Zone I, and Zone 2.

Bloch, H.P. (August 2015). Consider Zone 0 vertical pumps for slop pits. Hydrocarbon Processing. Vol. 94 No. 8.

Confidential Report, (2008).

IPIECA (2010), Petroleum refining water / wastewater use and management. Operations Best Practice Series. Refinery Water Management Task Force. Storm water and sewerage section.

Lamberti, P.G., JL Woodward. (March 2011). Learning from Events: Major Process Safety Incidents from Wastewater in Process Chemical Plants, Or Take Care of Your Wastewater or It Will Take Care of You. Process Safety Progress Vol. 30, No. 1.

Lees, F.P. (2012). Lees' Loss Prevention in the Process Industries (4th Ed) Elsevier. (a)Chapter 10 – Plant Siting and Layout, Section 10.17 Drain Systems.

(b)Chapter 17—Explosion, Section 17.28 Vapor Cloud Explosions

Mason, G.S., C. Arnold. (1984) Contain Liquid Spills and Improve Safety with a Flooded Storm Water Sewer. Chemical Engineering V 91 (Sept 17th), pp 105-109.

Ursino, N. (2014), P. Salandin. Combined effects of mechanical availability and storage capacity in pumping station risk assessment. Procedia Engineering V 70, pp 1679-1687.

US EPA, (July 1980) Treatability Manual Volume II Industrial Descriptions, Petroleum Refining Category Section14, EPA-600/8-80-042b

US EPA, (June 1992) Guidance to Protect POTW Workers From Toxic And Reactive Gases and Vapors, Section 3.1. EPA 812-B-92-001.

US EPA, (September 2000a) Wastewater Technology Fact Sheet – Sewers, Force Main, EPA 832-F-00-071.

US EPA, (September 2000b) Collections Systems Technology Fact Sheet – Sewers, Lift Station, EPA 832-F-00-073.

US EPA, (September 2000c) Wastewater Technology Fact Sheet – In-Plant Pump Stations, EPA 832-F-00-069.

12TH GLOBAL CONGRESS ON PROCESS SAFETY

Critical Considerations in the Pressure Relief Analysis of Refinery Sour Water Systems

James (Jim) R. Bridgett
james.bridgett@siemens.com

Robert (Bob) Siml
robert.siml@siemens.com

Gabriel Gandhi Martiniano Ribeiro de Andrade
andrade.gabriel@siemens.com

Siemens Energy, Inc.
4615 Southwest Freeway, Suite 900,
Houston TX 77027

Prepared for Presentation at
American Institute of Chemical Engineers
2016 Spring Meeting
12th Global Congress on Process Safety
Houston, Texas
April 10-14, 2016

UNPUBLISHED

Critical Considerations in the Pressure Relief Analysis of Refinery Sour Water Systems

James (Jim) R. Bridgett
james.bridgett@siemens.com

Robert (Bob) Siml
robert.siml@siemens.com

Gabriel Gandhi Martiniano Ribeiro de Andrade
andrade.gabriel@siemens.com

Siemens Energy, Inc.
4615 Southwest Freeway, Suite 900,
Houston TX 77027

Keywords: safety, pressure relief, two-phase flow, inherently safer design, sour water

Abstract:

Sour water systems are an integral part of all refineries. They are designed to collect sour water (i.e., water containing hydrogen sulfide, other sulfur compounds, and trace amounts of hydrocarbons) from numerous sources for treatment prior to disposal. Due to the complexity of these systems, the pressure relief analysis is inherently difficult. This paper discusses critical considerations in the pressure relief analysis, including the impact of light hydrocarbons entering the sour water system, design flaws that set traps for operators, deficiencies in operating procedures, and some critical considerations for the pressure relief analysis. Also presented will be potential mitigations for existing systems and recommended best practices for the design of new systems.

1. Introduction

Sour water collection and treatment systems are an integral part of all refineries. Although not a core production unit, the proper function of the system is important to the safe, reliable economical operation of a refinery. While it might be tempting to assume that non-core production units such as sour water systems are low risk and therefore don't require the same level of overpressure protection analysis and review as other parts of the refinery, the work recently done by Siemens to support the study of a process upset in a sour water system strongly indicates that is not the case. Since sour water systems inherently contain numerous high to low pressure transitions, the overpressure protection analysis is much more complex and difficult than might first appear. This paper will discuss in detail the lessons learned from our recent work and how those lessons can be used to strengthen the overpressure protection of sour water and similar systems. Improving the overpressure protection of these systems is a key part of reducing risks associated with operating these types of systems to as low as reasonably practicable (ALARP).

2. Initiating Event

A typical refinery sour water system consists of multiple sources, including streams that are pumped or pressured in on level control, as well as streams that are manually drained.

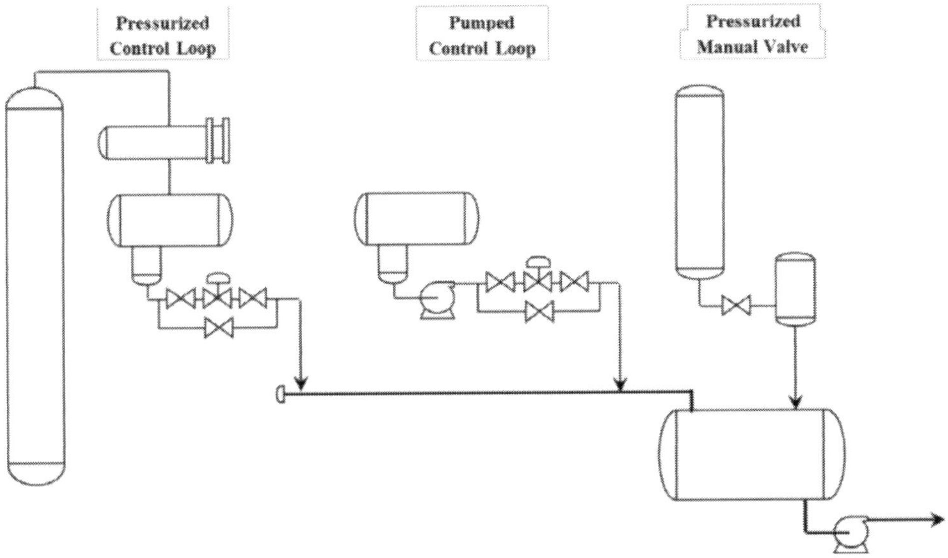

Due to a process upset, the feed to the column contained much more water than normal. Water was entering the overhead accumulator and separating out in the water boot faster than the interface level control could remove the water to the sour water system. Therefore the outside operator was asked to open the bypass valve around the interface level control valve, to reduce the water level.

It appears that the operator opened the bypass valve gradually, in steps, over a few minutes. The water level in the boot decreased in corresponding steps, and then sharply dropped to zero. At the same time the pressure in the sour water drum increased rapidly. Over about 30 seconds, the pressure increased from the normal operating pressure of approximately 10 psig to at least 55 psig. The actual pressure the vessel reached is not known because 55 psig was top of scale for the pressure transmitter on the sour water drum. As the pressure in the drum spiked, both the pressure control vent valve and the relief valve on the sour water drum opened as designed, releasing vapors and/or gases to the flare.

The pressure transmitter remained "pegged" at the maximum reading of 55 psig for another 30 seconds. The pressure decreased to approximately 15 psig over the next 4 minutes. As the pressure in the system dropped, the relief valve closed. Over the next 30 minutes or so, the pressure gradually dropped from 15 psig and stabilized at the normal system pressure of 10 psig. As the pressure dropped, the vent valve returned to normal operating position.

In response to the loss of interface level and the pressure increase at the sour water drum, the operator immediately closed the level control bypass valve. The interface level was re-established in 10 to 15 minutes.

3. Preliminary Review

The above information is based on both process data as captured by the data historian and information from operations. Siemens first conducted a preliminary evaluation to validate the process data and time line as described above. This evaluation was intended to answer the following questions:

a. What was the root cause of the pressure rise in the sour water system?
b. How much liquid hydrocarbon, if any, entered the sour water system?
c. What maximum pressure did the sour water system reach?
d. Did the time line make sense based on estimated flow rates through the bypass valve, etc.?

Since these questions can be answered using standard and well understood calculation techniques, detailed calculations will not be presented in this paper. Instead, the focus will be on the results and conclusions.

3.1 Root Cause of the Sour Water System Pressure Rise

Three potential root causes of the pressure spike in the sour water drum were considered:

a. *Release of trace amounts of hydrocarbons, hydrogen sulfide and other dissolved gases from the additional sour water flow that occurred when the level control bypass valve was opened.*

   Based on the dimensions of the water boot, the maximum amount of sour water that could have been drained to the system was about 15 gallons (approximately 120 pounds). This volume of water is simply too small, and would have contained too little dissolved hydrocarbons, hydrogen sulfide, etc. to contribute significantly to the pressure rise in the sour water drum.

   If the additional sour water was the only source of light gases and vapors into the system, this analysis indicates that the process vent valve would have had more than sufficient capacity to handle the event; as a result, the pressure would not have reached the relief valve set point.

b. *High pressure hydrocarbon vapors from the overhead accumulator (i.e. "gas blow-by" scenario).*

   For a gas blow-by type scenario to occur, all of the liquid in the overhead accumulator would have to have been (drained) transferred to the sour water drum. Since the hydrocarbon level in the overhead accumulator did not drop to zero at any time during this upset, this ruled out the gas blow-by scenario as the root cause of the pressure spike in the sour water drum.

In addition, even though there was more than enough hydrocarbon in the overhead accumulator, the high level alarm in the sour water drum never activated, no frost line occurred, and the interface level was quickly reestablished.

c. *Liquid hydrocarbon from the overhead accumulator flashing as it entered sour water drum.*

Primarily by a process of elimination, the root cause of the pressure spike in the sour water system was determined to be a moderate amount of liquid hydrocarbon entering the sour water drum (see 3.2 for details). Due to the difference in pressure between the overhead accumulator and the sour water drum, the hydrocarbon flashes as it enters the sour water system. The process vent valve could not respond fast enough and/or did not have enough capacity by itself to vent the vapor volume generated to the flare, and as a result, the pressure in the system increased until the relief valve opened.

Additional evidence that supports this conclusion includes:

- Little to no change in the hydrocarbon level in the overhead accumulator during this process upset.
- The relatively short duration of the maximum pressure reading and the relatively quick return of the system to normal pressure.
- Interface water level quickly being reestablished.

3.2 How much, if any, liquid hydrocarbon entered the sour water system

Based on the available empty volume of the sour water drum and the documented pressure changes, this analysis indicates that about 100 pounds of hydrocarbon vapor was required to increase the pressure in the sour water drum from the normal 10 psig to 55 psig. This sets a lower limit on the amount of hydrocarbons that could have entered the sour water drum.

The total amount of hydrocarbon that entered the sour water drum, in addition to the amount required to pressure up the system, includes the following:

a. *Amount released to the flare through the process vent valve.*

   Based on the flow coefficient (C_V) of the control valve and estimated vessel pressures plus the established time line, the best estimate of the total amount of material released to the flare through the vent valve is about 300 pounds. This estimate includes approximately 200 pounds over the 5 minutes as the pressure spiked and then decreased to 15 psig and at least 100 pounds over the next 30 minutes as the pressure dropped from approximately 15 psig back to the normal 10 psig.

b. *Amount released to the flare through the relief valve*

   Since there was no indication that release was two phase or liquid, the flow through the relief valve was based on standard relief device vapor sizing equations using ASME certified parameters for the relief valve orifice area and coefficient of discharge. Based on the time line, the pressure relief device was open for less than 2 minutes. During this time, approximately 90 to 110 pounds of material was released through the relief valve to the flare.

Overall, it is estimated that approximately 500 pounds (110 gallons) of hydrocarbon drained from the overhead accumulator into the sour water drum during this upset.

3.3 What maximum pressure did the sour water drum reach?

Based on the recorded process data, it was known that the pressure in the sour water drum reached at least 55 psig. Since 55 psig was the maximum reading of the pressure transmitter, the actual pressure reached was not known. To address this question, a process simulator was used to plot the vapor pressure of the overhead accumulator hydrocarbon vs. temperature. Unfortunately, the temperature of the sour water drum was not available. Therefore, the temperature of the overhead accumulator, 85 °F, was used as the best available estimate of the temperature in the sour water drum. This temperature is consistent with cooling water temperatures expected during the time of year the upset occurred.

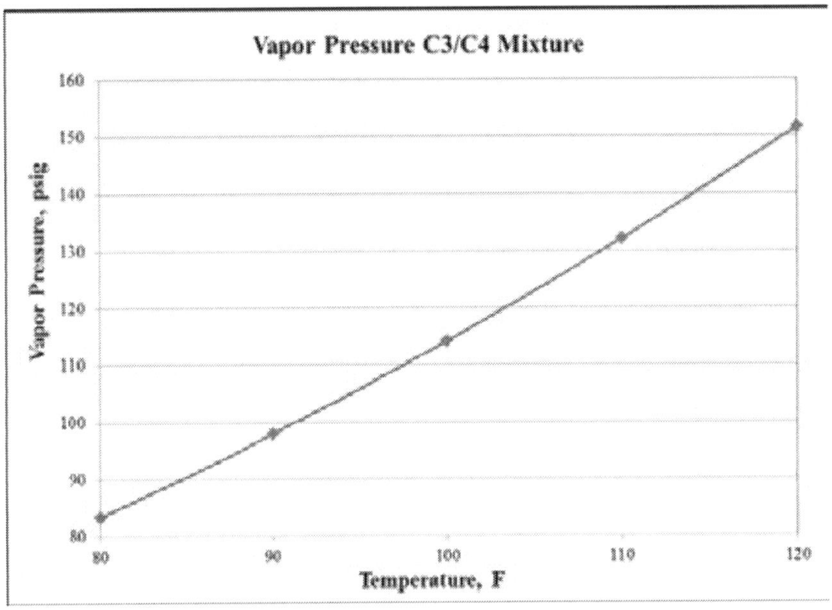

Using the estimated temperature of the sour water system, the maximum potential pressure in the sour water drum is approximately 90 psig when liquid hydrocarbon is present. Based on the capacities of the relief valve and process vent valve, the best estimate for the pressure in the sour water drum during this upset is 60 to 70 psig.

This pressure is just above the top of scale pressure for the pressure transmitter on the sour water drum and is an indication that the pressure transmitter was reasonably accurate.

3.4 Does the time line make sense based on estimated flow rates through the bypass valve, etc.?

As previously indicated, the estimated volume of sour water in the overhead accumulator boot was approximately 15 gallons. The interface control station consists of a 1 inch control valve and a 2 inch gate valve as the bypass. Although it is not unusual for gate valves to be used as a control valve bypass it is generally recognized that gate valves are not suitable for flow control. It is unusual to see an oversized gate valve as a bypass. Obviously, due to both size and flow characteristics, at the same pressure drop, the capacity of the gate valve is much greater than the control valve.

For sour water, the wide open capacity of the bypass was as high as 500 gpm based on the normal operating pressure in the overhead accumulator (185 psig) and the normal pressure in the sour water drum (10 psig). This estimate assumes little or no flashing occurs across the control valve.

For the hydrocarbon, the wide open capacity of the bypass was approximately 150 gpm based on the same pressure differential before the pressure in the sour water drum started to rise. This flow rate is much lower than the flow rate on water due to the lower density of the hydrocarbon and the flashing of the hydrocarbon liquid across the bypass.

In general, the estimated flow rates through the bypass valve are consistent with the rapid short duration of the pressure spike in the sour water drum.

Some time and effort was spent to investigate whether the flow characteristics of the gate valve also contributed to this process upset.

The number of turns from fully closed to fully open for a 2 inch gate valve varies between 11 and 13, depending on manufacturer. Definitive opening characteristics for gate valves were not readily available in the open literature, but it was generally noted that due to the geometry of the valve, the flow characteristics of gate valves are highly non-linear. There is some anecdotal evidence that gate valves may flow as much as 80% of capacity at only 20% open, but this is not well documented and should only be used with appropriate care and judgement as a relative indicator of the opening characteristics of gate valves.

While the evidence is not as definitive as desired it is doubtful that the bypass valve was ever fully open and was closed quickly once the interface level indication in the water boot went to zero (0) and the pressure spiked in the sour water drum. All indications are the 2 inch gate valve was grossly oversized and not suitable for control of flow out of the water boot.

4. Overpressure Protection Analysis

The following scenarios would typically be considered during the overpressure protection analysis of the sour water drum:

a. External fire
b. Blocked liquid outlet (overfilling)
c. Process vent control valve failing closed
d. Failure of automatic controls on each feed stream

Of these scenarios, a, b and c did not play a role in this process upset, and the analysis of these scenarios is fairly straightforward using standard overpressure protection assumptions and calculation methods appropriate to each type of scenario.

The failure of automatic controls cases were the primary focus for two reasons:

a. A failure of automatic controls and/or associated bypass played a central role in the process upset being studied.
b. Sour water systems inherently contain numerous high to low pressure interfaces at control valves or manual valves.

The failure of automatic controls scenarios typically include:

1. The control valve failing or being driven 100% open
2. The control valve bypass valve being 100% open
3. Some combination of control valve and bypass valve open, per client preference.

For all of these cases, the required relief rate calculation is based on:

a. Control valve flow coefficient (C_v) provided by the valve manufacturer
b. The pressure differential across the control valve. The typical assumption used in the calculation is normal upstream pressure minus downstream relief pressure.

When evaluating control valve and bypass scenarios the first step in the analysis is to determine if the liquid in the upstream system will fit in the downstream system.

If the liquid will fit, then the relief device on the downstream vessel should be sized for the vapor breakthrough or gas blow-by case.

If the liquid does not fit, then the relief device on the downstream vessel should be sized for the "liquid displacement" case. Liquid displacement is a phenomenon where vapor or two-phase material entering a liquid full system displaces the liquid. The relief device must be sized to handle the displaced liquid at the volumetric flow rate entering the system. Typically, sizing for liquid displacement requires a considerably larger relief device.

Liquid Displacement

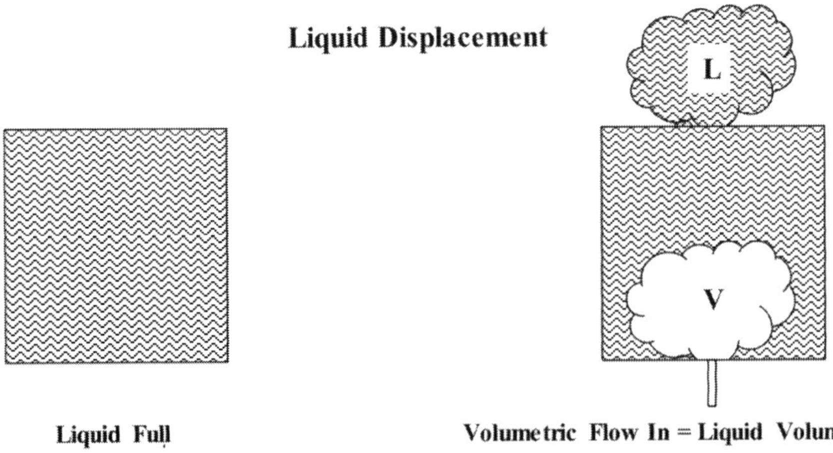

Liquid Full **Volumetric Flow In = Liquid Volume Flow Out**

In this particular case, the hydrocarbon is flashing at the lower pressure in the sour water system. Initially, the liquid will tend to separate out and accumulate in the drum. The flow through the relief device would only be vapor. Note that any liquid hydrocarbon that accumulates in the drum would continue to flash and generate additional vapor due to transfer of heat from the sour water. In general assuming all the liquid will vaporize is conservative. Evaluating the potential for liquid to collect and instantaneously generate a higher vapor rate due to mixing is very difficult to predict and is beyond the scope of this paper.

One of realizations of this analysis is as liquid separates out and begins to fill the drum, the space available for liquid-vapor disengagement is reduced. At some point, the vapor velocity in the drum will increase to the point that the liquid will not separate out and instead will be re-entrained out the relief device. This two phase release case can significantly impact the size of the relief valve. Therefore any process vessel with a source of two-phase flow must be evaluated for vapor-liquid disengagement and the relief system must be sized for two-phase flow if re-entrainment will tend to occur from the level rising.

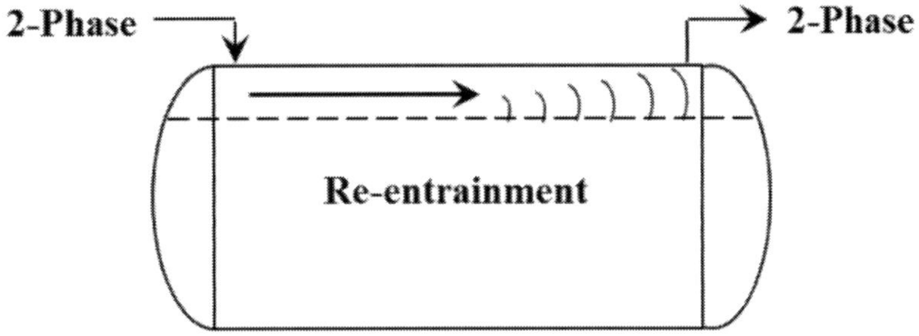

One of the most surprising realizations is liquid displacement will not occur due to the entry of two-phase flow when downstream vessel is not initially liquid full, even when there is enough liquid in the upstream system, because re-entrainment will occur first.

5. Conclusions

- Overpressure protection analysis of sour water systems in particular, and other similar collection systems in general, is not simple and straightforward. Particular attention must be paid to all high pressure to low pressure interfaces and ensure all credible scenarios have been identified.
- Liquid displacement will not occur when the flow entering is two phase if the vessel is not initially liquid full because re-entrainment will occur first.
- An oversized gate valve is not suitable for use as the bypass valve and can set a trap for operations.
- Any process upset is an opportunity for continuous improvement.

6. Recommendations

Corrective Actions for Existing Systems

- Check for oversized bypass valves:
  - Consider downsizing, installing a mechanical stop, or restricting orifice.
  - Consider replacing gate valves with globe valves.
- Review the overpressure protection analysis.
- Consider increasing the span of pressure transmitter.
- Consider recording process temperatures in the data historian

Inherently Safer Design for New Systems

- It's generally not practical to design the sour water system for max upstream pressure of every source. To the degree practical, design for the vapor pressure of the hydrocarbon at upper end of the sour water system operating temperature.
- In the design phase, balance the size of the water boot versus design of control and bypass valves to ensure the operator has adequate time to respond.
- Bypass valves for use in normal operation should generally be globe valves of similar size to the control valve.
- If oversized bypass valves are required for start-up, shut down, or other specific operating conditions, consider installation of a separate bypass valve that is car-sealed closed during normal operations.
- When reasonably practicable, the process vent valve should be sized to keep the relief valve from opening.

Disclaimer:

The information contained in this document represents the current view of the authors at the time of publication. Process safety management is complex and this document cannot embody all possible scenarios or solutions related to compliance. This document is for informational purposes only. Siemens makes no warranties, express or implied, in this paper.

For more information, please contact:

Siemens Energy, Inc.
Process Safety Consulting (SAF)
4615 SW Freeway, Suite 900
Houston, TX 77027
1 (800) 658-8809
+1 (713) 570-2940
Website: www.siemens.com/energy/oilandgasconsulting
Email: info.ogconsulting.energy@siemens.com

12TH GLOBAL CONGRESS ON PROCESS SAFETY

Practical Application of Advancing Process Safety Tools for the Petroleum Refining and Petrochemical Industries

Rodney Reibold
Marathon Petroleum Company
539 S. Main, Findlay, OH 45840
rrreibold@marathonpetroleum.com

Prepared for Presentation at
American Institute of Chemical Engineers
2016 Spring Meeting
12th Global Congress on Process Safety
Houston, Texas
April 11-13, 2016

UNPUBLISHED

Practical Application of Advancing Process Safety Tools for the Petroleum Refining and Petrochemical Industries

Rodney Reibold
Marathon Petroleum Company
539 S. Main, Findlay, OH 45840
rrreibold@marathonpetroleum.com

Keywords: AFPM, API, Advancing Process Safety, APS, Hazard Identification, Practices Sharing, Event Sharing, API 754, Human Reliability, Process Safety Site Assessments

Abstract

Over the past 5 years, the American Fuel & Petrochemical Manufacturers (AFPM) and American Petroleum Institute (API) have developed process safety tools for use by the refining and petrochemical industries to assist in improving process safety across the entire industry. The suite of programs are called Advancing Process Safety. This paper describes what industry is seeing in the way of process safety performance, the tools and programs available to these industries to help them improve performance, and practical ways companies are using the tools to improve performance.

1. Background and Development of Advancing Process Safety Programs

The petroleum refining and petrochemical manufacturing industry is committed to protecting the health and safety of its workers, contractors, neighbors, and the environment in which it operates. For decades, industry has worked together to develop safe work practices and standards aimed at reducing risks of major industry hazards and preventing process safety incidents. Despite these efforts, process safety incidents occur. As a result, industry, government, non-governmental organizations (NGO's) and the media have increased focus and attention on process safety.

In 2010, a group of refining and petrochemical industry executives belonging to the then NPRA (National Petrochemical and Refiners Association) now AFPM, tasked process safety professionals within their companies to develop a working group to explore ways to collectively improve process safety performance across the refining and petrochemical industries. The goal was to develop programs that would create a significant improvement in process safety performance. Shortly after work began, API was asked to join the effort to leverage the strengths and resources of both organizations to successfully develop and implement the programs. Over a two year period, the group developed a suite of six programs currently being utilized by the petroleum refining and petrochemical industries. These programs are Process Safety Metrics, Process Safety Event Sharing, Hazard Identification and Practices Sharing Program, Process Safety Regional Networks, Process Safety Site Assessment, and Focused Improvement. Together these programs are called the Advancing Process Safety (APS) program.

A key objective of the APS program is to develop tools and information, aimed at improving the industry's process safety performance, and to make these tools and information freely available to member companies. The APS program is able to use data streams unique to the program and specific to our industry to target areas for improvement.

This paper introduces and illustrates some of the tools and information available to member companies and how these tools and information can be used to raise process safety awareness and improve process safety performance.

2. Advancing Process Safety Data Collection and Analysis

AFPM and API members provide the following data on an annual basis to APS to be analyzed by industry experts:

- Facts regarding Process Safety Events (PSEs) per *ANSI/API RP 754 Process Safety Indicators for the Refining and Petrochemical Industries,*
- Tier 1 and 2 PSE summary reports to the Event Sharing database, and
- Results and observations from the API Process Safety Site Assessments.[1][2]

The following sections describe the data collected, the analysis process, and the resulting observations for industry improvement. This data analysis occurs on an annual basis. At the time of this paper, the most recent full year of data was 2014 process safety data, which was reported and analyzed in 2015. For the purposes of this paper, this will be referred to as the 2015 data analysis.

2.1 Data Analysis

Each group, Process Safety Metrics, Event Sharing, and Site Assessment individually analyze their data and present their observations during the annual Deep Dive Analysis meeting. The meeting participants consist of over forty process safety experts from member companies.

The results of the analyses are combined into one Deep Dive analysis to provide insight into opportunities for improvement that will be discussed later. It should be noted that data from each program (Process Safety Metrics, Event Sharing Database, and Site Assessments) are discrete data sets, and therefore some observations might vary from data set to data set. Trends observed in the 2015 data analysis are discussed in the following sections.

2.1.1 API 754 Process Safety Metric Data

The experts analyzed the 2014 Process Safety Metric data submitted by 59 companies representing 199 sites and were able to make several observations.[3] The data consisted of 921 events: 268 Tier 1 PSEs and 653 Tier 2 PSEs.

Figure 1 represents the four Tiers of performance indicators, as explained in API RP 754. For more information, please refer to Appendix A for API RP 754 Tier 1 and 2 definitions.

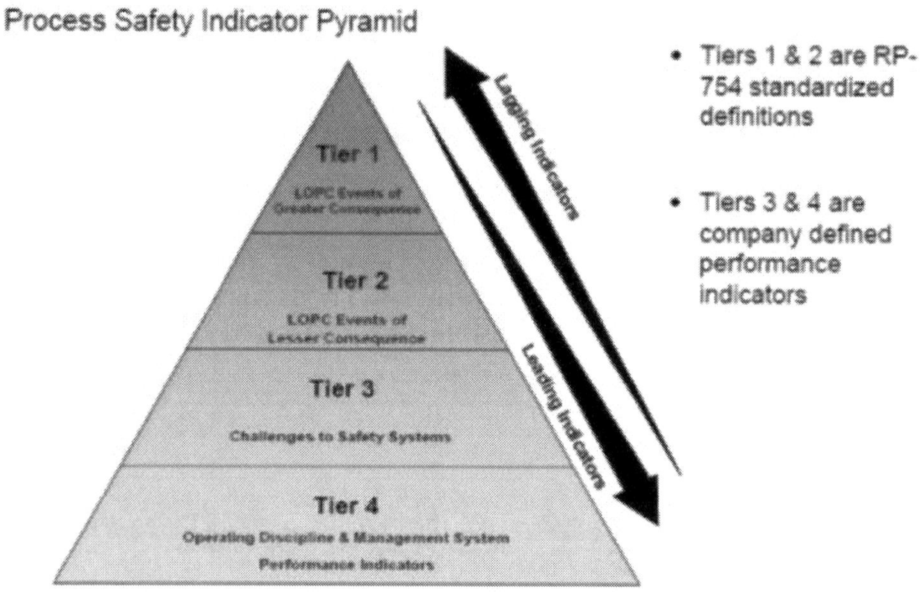

Figure 1. API RP 754 Process Safety Indicator Pyramid

Basic Observations:

- Mode of Operations – 70% of reported incidents occurred during Normal Operation with a majority of those events occurring during Steady State (72%). The next largest category of releases occuring during Normal Operation occurred with Loading and Unloading acvtivities (10%).
- Point of Release – Loss of Primary Containment (LOPC) from Piping Systems and Tanks accounted for 63% of the incidents reported (Piping Systems 49% and Tanks 14%). Figure 2 illustrates that the largest subset of releases from Piping Systems were caused by Piping and valve leaks.

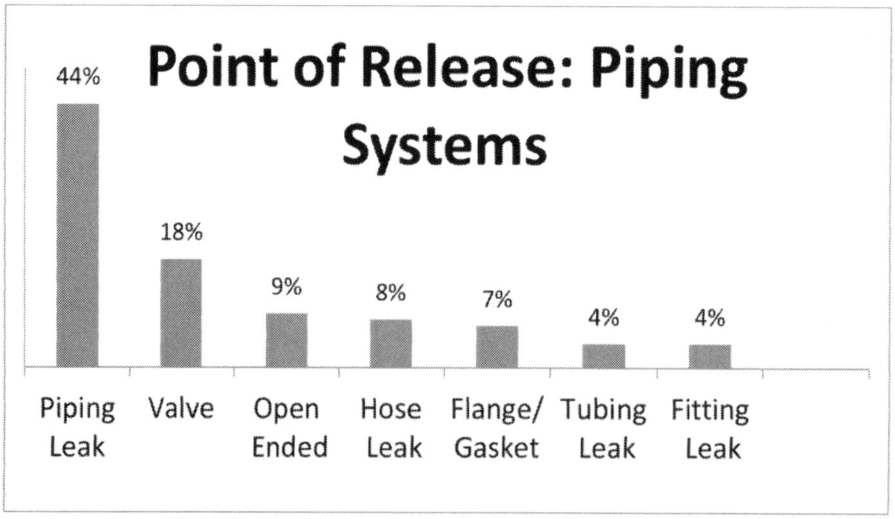

Figure 2. Points of Release from Piping System PSEs

- Piping Leaks – The top contributing factors to piping leaks were fixed equipment inspection, equipment reliability and design. Fixed equipment inspection issues are related to programs for detecting and mitigating metallurgical issues such as corrosion or cracking, before piping failure.
- Valves – Observations from incidents that involved releases from valves indicated that a majority of events were related to "Valves Left Open" and a majority of those events occurred during "normal operation". See Figure 3.

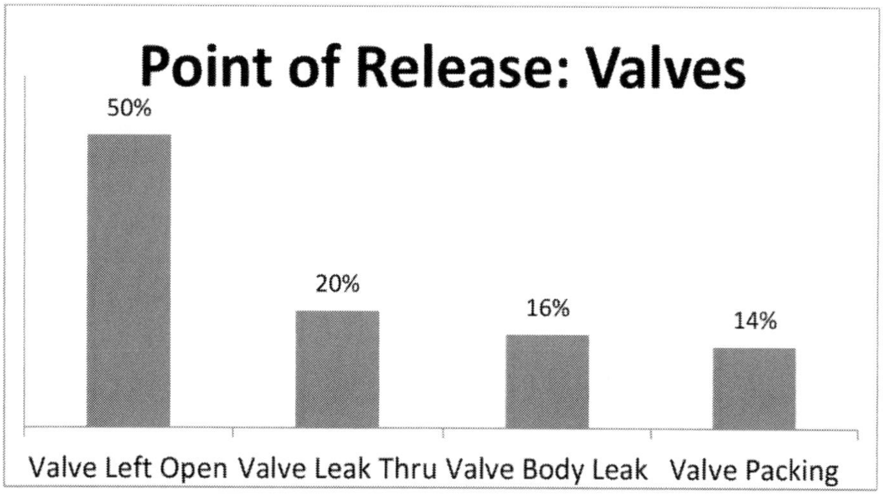

Figure 3. Valve Related PSE Subcategories

- Release from Atmospheric Tanks – Atmospheric tanks (including floating roof tanks) were the second largest point of release locations for PSE Tier 1 and 2 events after Piping Systems. A majority of the events occurred in the Tank Farm followed by the Sulfur Recovery unit. The largest group of those events were considered "overfill" events. Human factors and equipment reliability were the top causes for tank related events and human factors and procedures were the main causes contributing to the overfilling of atmospheric tanks. Figure 4 illustrates the types of releases from tanks.

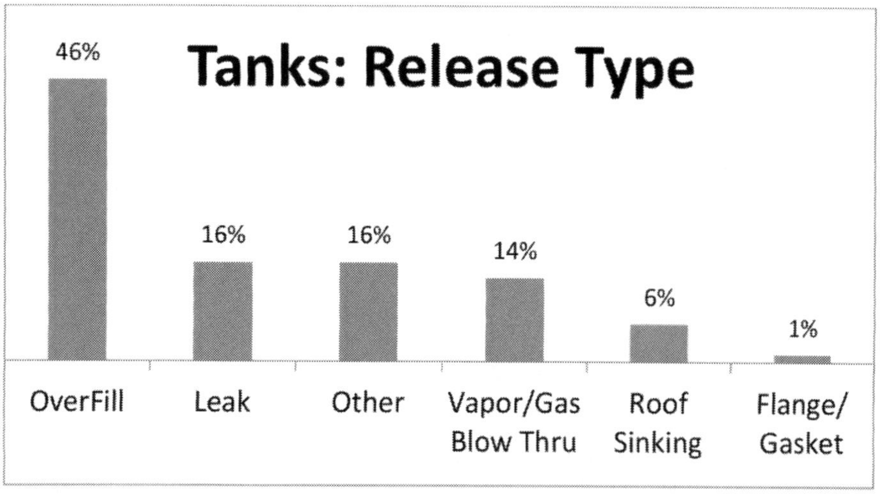

Figure 4. Release Type for Tank PSEs

- Causes – One of the limitations of API RP 754 statistics is that the data does not contain the cause of the incident as the member company determined during their incident investigations. The information provided under API RP 754 includes a synopsis of each process safety event that occurred. The team analyzing this data was able to identify some apparent causes for about 50% of the data, by reading the event description that was included. Although the analysis was based on the limited sample size, the team concluded that human factors, equipment reliability and inadequate fixed equipment inspection were the primary apparent causes of the events. Figure 5 shows the apparent causes of the PSEs, derived from the 2015 API RP 754 data analysis.

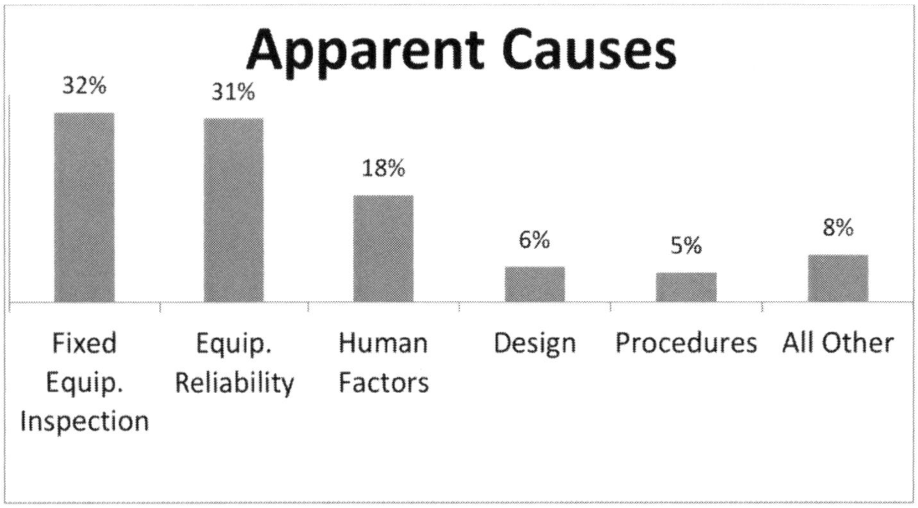

Figure 5. Overall Apparent Causes of PSEs

2.1.2 Event Sharing Database

The Event Sharing Database is a smaller subset of the API RP 754 data and provides more descriptors of the event, causes as determined by the member company incident investigations, and lessons learned.[4] At the time of the 2015 data analysis, approximately 428 process safety events were reviewed from the AFPM Event Sharing Database. Eighty percent of those events were API Tier 1 events that occurred from 1989 to 2014. Some general themes observed from the deep dive analysis include LOPC as a result of valves being left open or inadvertently opened, loss of primary containment originating from small bore piping and/or tubing (2" or less in diameter), and corrosion related LOPC events. Figure 6, illustrates the proportion of these Human Reliability events and how they break down into sub-categories.

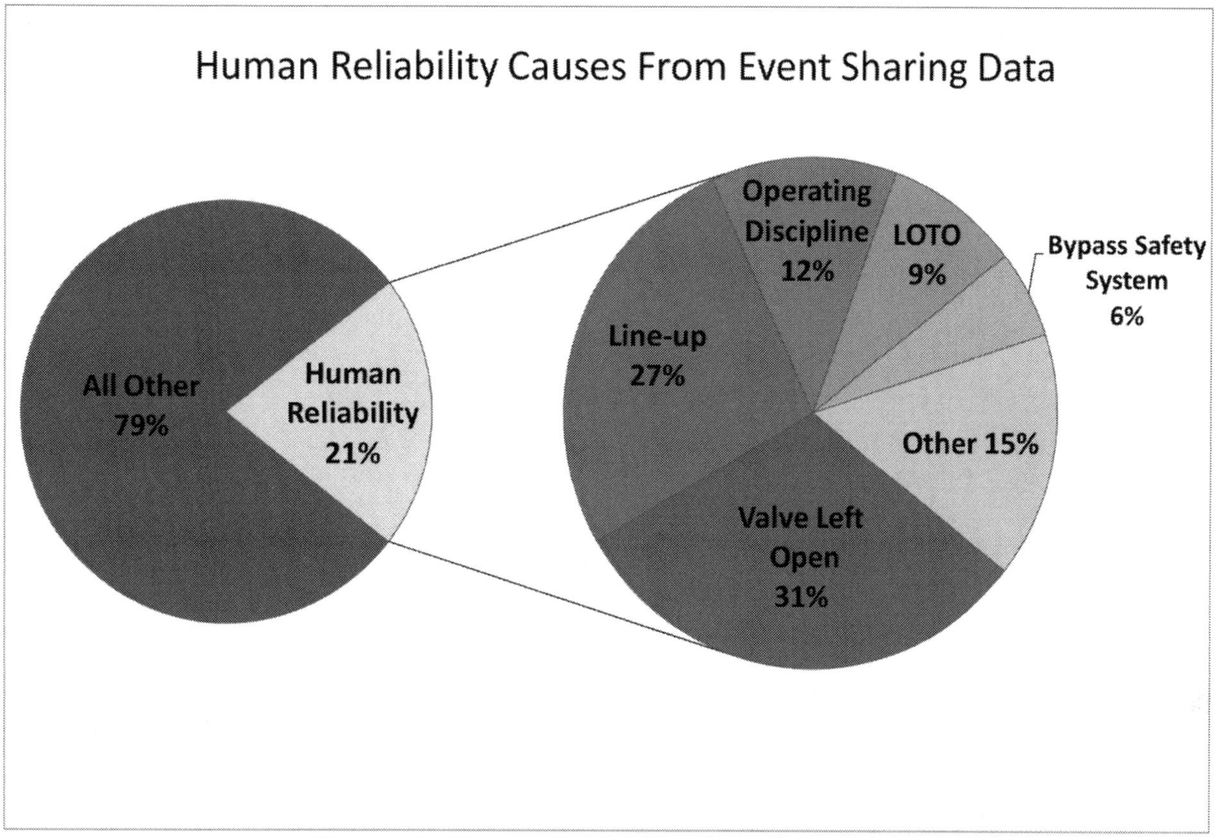

Figure 6. Human Reliability Sub-Categories

Open Valves – Twenty-one percent of the events from the database were identified as being caused by human reliability problems, including valves that were inadvertently left open to atmosphere or closed valves that were inadvertently opened. Nearly 1/3 of this subset involved bleeders, vents, and drains with 40% of the "open valve" events occurring during the startup of equipment.

Operating procedures and human factors (i.e., operating discipline) were identified in the database as the two most prevalent causes for these types of events. "Procedures" were often named as a cause because they did not mention the valve(s) in question or the specific task(s) being performed. In only a few cases were the procedures found to be unclear or just not followed.

Small Bore Piping – LOPC events involving small diameter piping continue to be prevalent in the 2015 data analysis with at least 19% of the reported events being related to piping or tubing 2" in diameter or less.

Piping Corrosion – Events associated with corrosion represented about 19% of events in the database. Corrosion continues to relate to a significant number of PSEs.

Winterization – Freeze-related events trended much higher in 2014 than historical rates. In 2014, 15% of events entered were associated with freeze-ups compared to 6% in the entire database.

2.1.3 Site Assessment Program

Observations from the Site Assessment program are shown in Figures 7 and 8.[2]

Basic observations:

- The lowest scores came from Mechanical Integrity (MI) and Process Hazards Analysis (PHA).
- When looking at the protocol categories, there appeared to be a wide distribution of scores for topics like shift turnover, operator rounds, and facility siting which indicates that some sites are doing those things very well and that others have an opportunity to learn from those sites.

A note about assessment scores: Higher is better, but the scoring isn't quite the same as a college exam. In the Site Assessment world, a score of 75% represents strong performance that meets the requirements of the assessment protocol. To achieve a 100% score, a program must show it meets or exceeds the requirements of the protocol, show a cycle of continuous improvement, must demonstrate sustainability, and show evidence of being embraced by the workforce.

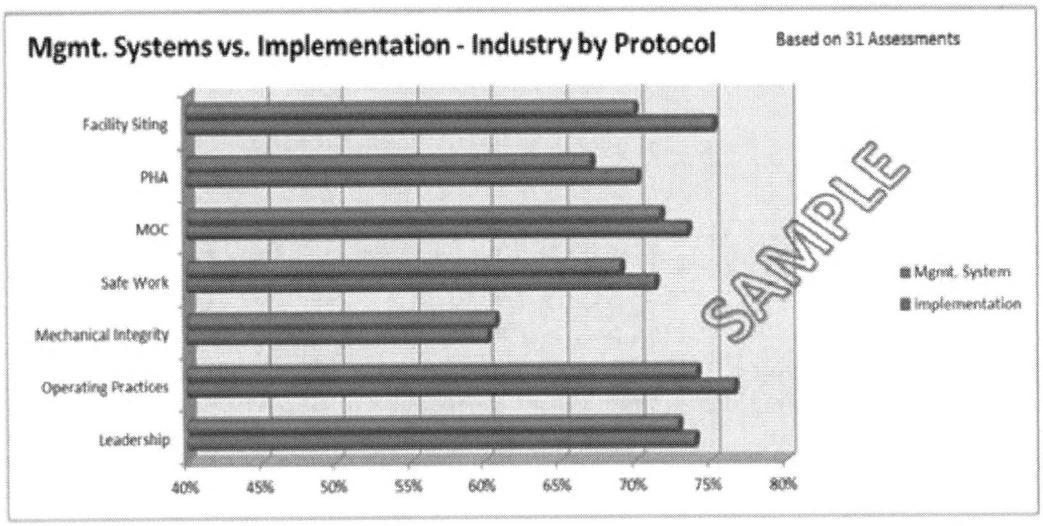

Figure 7. Management System versus Implementation by Protocol

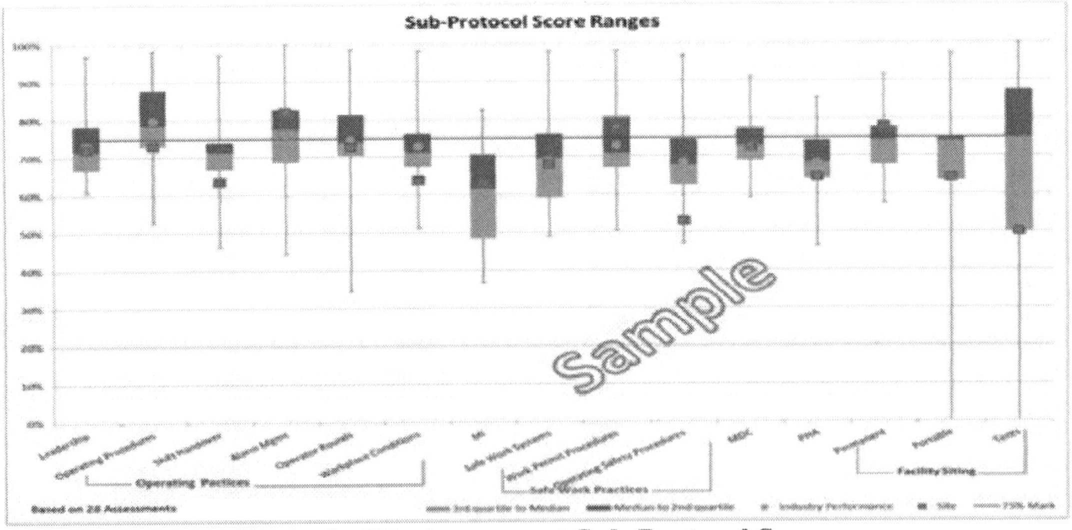

Figure 8. Site Assessment Sub-Protocol Scores

2.2 Using the Data to Improve

Does the data above look familiar? Are you experiencing similar types of events at your facility? It is likely that similar types of incidents ARE occurring at your processing facilities. The APS data indicate that these common themes are occurring over a broad cross-section of the industry; small plants and large plants, refineries and chemical plants, domestic and international. It is incumbent upon each company to collect, analyze, and make improvements based on their own data. The foundation of APS is to collect, analyze, and develop tools for improvement based on the industry's data.

The Advancing Process Safety program has used the data for industry improvement initiatives and to develop many tools for member company sites to use to improve their own performance. The following are a few actions that have been pursued based on the annual deep dive analysis:

- Developed and proposed a number of revisions to strengthen API 570 (Piping inspection code) and API RP 574 (piping inspection practices) in the areas of small bore piping.[5][6] The main idea is to increase awareness of small bore piping issues as part of a company's routine piping inspections procedures.
- Developed a number of Hazard Identification documents to address hazards associated with equipment, or work practices on equipment, uncovered in the data analysis. Examples include topics such as Piping Dead-legs, Winterization, and Small Bore Piping.[7]
- Developed a number of Hazard Identification documents and Shared Practices to address common personnel issues uncovered in the data analysis. Examples include topics such as Shift Change issues, Operator Communications, Operator Line-ups, and Operator Evaluation Rounds.[7]
- Developed a program called "Walk the Line" to address common Human Reliability issues. This includes prepackaged training materials, prepared tool box meeting outlines (scripts), informational newsletters, and videos.[8]

All of these materials and much more are available to member companies for use in their process safety improvement journey. The following section describes how information and tools developed by the APS Program can be used to improve your process safety performance.

3. Advancing Process Safety Tools and Ways to Use Them

The following sections outline the individual APS programs, tools, and information as well as how some companies are using them to improve their process safety performance. They are organized according the APS program where the information or tools are developed, and are freely available to member companies. It should be noted that this paper only discusses a few of the ways these tools can be used. In practice, there are many more ways that each of the tools discussed below can be used. The products described in these sections can be found on the AFPM Safety Portal website.[9]

Figure 9. AFPM Safety Portal

3.1 API 754 Process Safety Metrics

The purpose of API RP 754 is to identify and track leading and lagging process safety indicators.[1] Knowledge of these indicators and taking appropriate action based on those indicators reduces the overall risk of major hazards in industry. On an annual basis AFPM and API publish a Process Safety Event Report and data analysis tool for both Tier 1and Tier 2 events.

The following illustrates a few practical examples of how APS member companies use Process Safety Metrics data.[3]

3.1.1 Benchmarking Against Industry

Figure 10 shows how benchmarking can be used by a company to compare its process safety performance to industry performance. Smaller companies, without a central corporate Process Safety Management (PSM) organization, may also use this tool to compare their sites to each other without need for a centralized compilation of data. Performance can be compared based on multiple criteria including the type of facility, size of facility, exposure hours, type of incident, incident year, or for specific process units. Comparisons can be done by both PSE count and PSE rate. In this example, a fictional company has four sites and is comparing its number of Tier 1 Process Safety Events over a three year period to other refining facilities of any size. The metric database is an extremely flexible benchmarking tool. By making simple choices from drop down menus, similar comparisons can be quickly created based on a wide variety of criteria. Figure 11 shows a comparison of Site D's performance on its fluid catalytic cracking (FCC) unit compared to other small facility FCC units. Figure 11 also shows that 44 of the FCC units at small facilities went the entire period without a Tier 1 or Tier 2 process safety event.

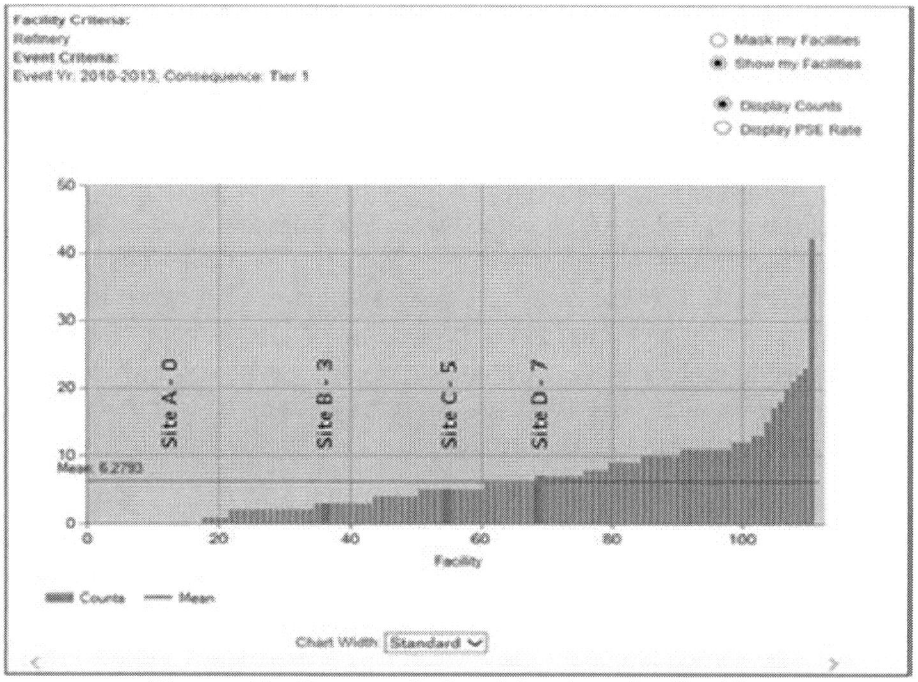

Figure 10. Process Safety Metrics – By Facility

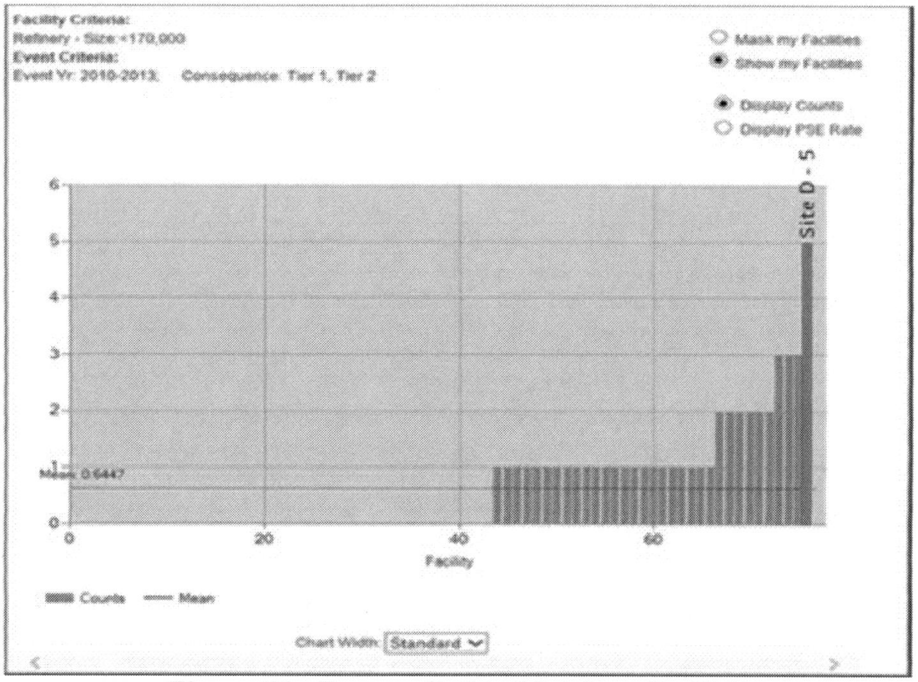

Figure 11. Process Safety Metrics – By Unit

- Goal Setting – Companies set process safety performance goals for their sites. This tool allows a company to identify what level of performance they will need to target to be an industry leader.

- Learning – The benchmarking tool can be used to identify successful practices within a company so that they can be shared across their sites. For instance in the example in Figure 10, Site D can use the tool to identify that Site A has implemented successful PSM programs and practices. Site D can then engage Site A to understand these programs and practices to implement at their own site.

- Prioritization of Resources – The benchmarking tool can be used to identify where resources need to be focused to improve PSM performance. For instance in the example in Figure 11, Site D would be able to identify that the large majority of their PSEs are occurring in their FCC unit, and that they are the worst performer in FCC process safety among similar sized sites. With this information, the site can target improvement efforts at the FCC. The site may also use this tool to identify which of their sister sites have strong FCC process safety performance, and direct resources to learning from the other sites and implementing existing successful practices to improve performance.

3.2 Event Sharing

The Event Sharing Database is a smaller subset of the API RP 754 Process Safety Metrics data and provides additional descriptors of the event, causes, and lessons learned. Where the main intent of API RP 754 data collection is **benchmarking**, the main intent of the APS Event Sharing program is **learning**.[4] Data submitted to APS includes a description of the event and equipment involved, as well as the cause(s) and learning(s) identified by root cause analysis performed by the member company. Individuals have the opportunity to search the database for specific information or export the database in a variety of forms. Key information about the event, cause(s), and learning(s) are clearly laid out, as is other information not shown here. See Figure 12. Other tools derived from the database are Safety Bulletins and the Annual Learnings Report.

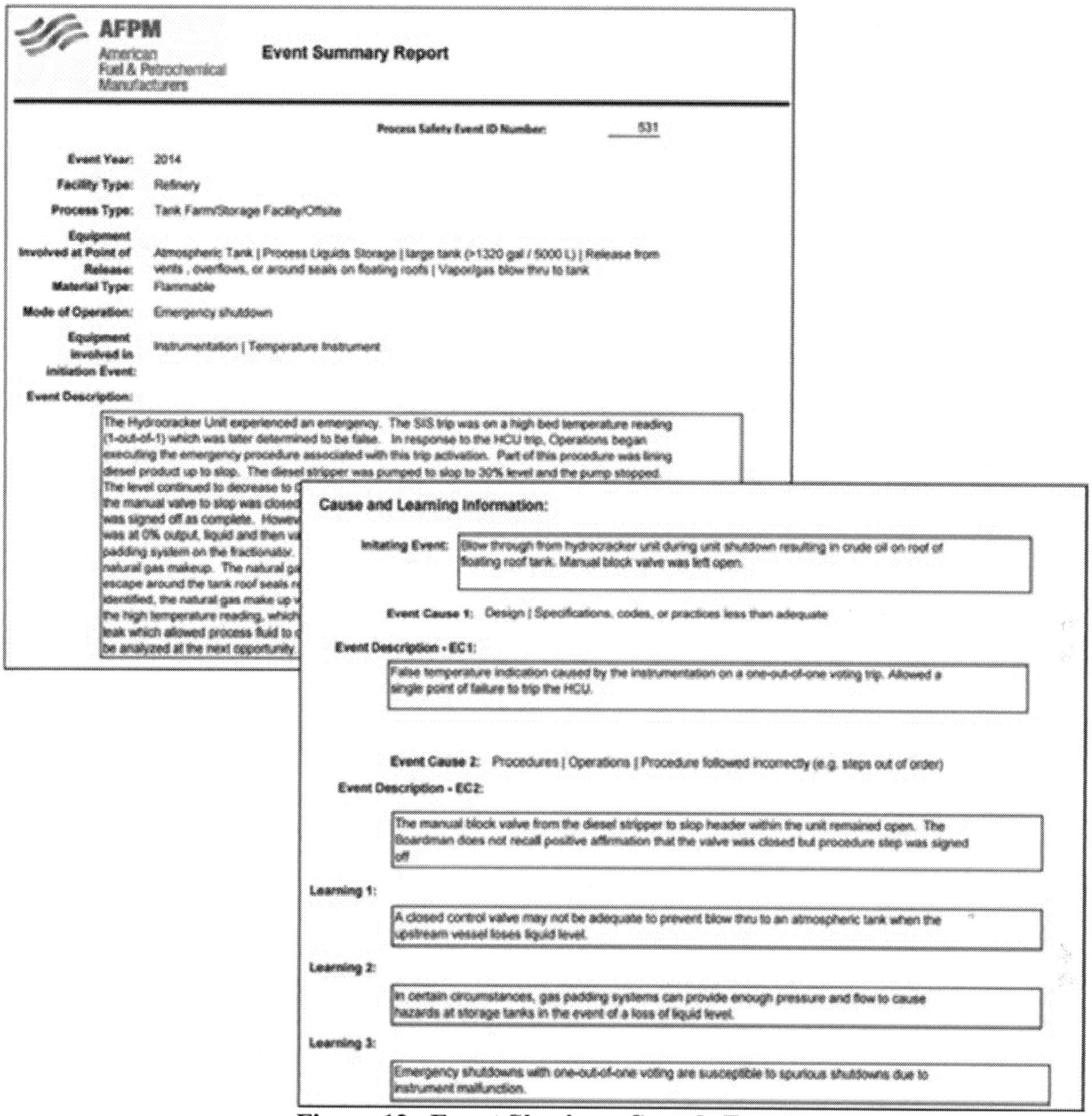

AFPM
American
Fuel & Petrochemical
Manufacturers

Event Summary Report

Process Safety Event ID Number: 531

Event Year: 2014

Facility Type: Refinery

Process Type: Tank Farm/Storage Facility/Offsite

Equipment Involved at Point of Release: Atmospheric Tank | Process Liquids Storage | large tank (>1320 gal / 5000 L) | Release from vents , overflows, or around seals on floating roofs | Vapor/gas blow thru to tank

Material Type: Flammable

Mode of Operation: Emergency shutdown

Equipment Involved in Initiation Event: Instrumentation | Temperature Instrument

Event Description:

The Hydrocracker Unit experienced an emergency. The SIS trip was on a high bed temperature reading (1-out-of-1) which was later determined to be false. In response to the HCU trip, Operations began executing the emergency procedure associated with this trip activation. Part of this procedure was lining diesel product up to slop. The diesel stripper was pumped to slop to 30% level and the pump stopped.
The level continued to decrease to
the manual valve to slop was closed
was signed off as complete. Howev
was at 0% output, liquid and then va
padding system on the fractionator.
natural gas makeup. The natural ga
escape around the tank roof seals n
identified, the natural gas make up v
the high temperature reading, which
leak which allowed process fluid to c
be analyzed at the next opportunity.

Cause and Learning Information:

Initiating Event: Blow through from hydrocracker unit during unit shutdown resulting in crude oil on roof of floating roof tank. Manual block valve was left open.

Event Cause 1: Design | Specifications, codes, or practices less than adequate

Event Description - EC1:

False temperature indication caused by the instrumentation on a one-out-of-one voting trip. Allowed a single point of failure to trip the HCU.

Event Cause 2: Procedures | Operations | Procedure followed incorrectly (e.g. steps out of order)

Event Description - EC2:

The manual block valve from the diesel stripper to slop header within the unit remained open. The Boardman does not recall positive affirmation that the valve was closed but procedure step was signed off

Learning 1:

A closed control valve may not be adequate to prevent blow thru to an atmospheric tank when the upstream vessel loses liquid level.

Learning 2:

In certain circumstances, gas padding systems can provide enough pressure and flow to cause hazards at storage tanks in the event of a loss of liquid level.

Learning 3:

Emergency shutdowns with one-out-of-one voting are susceptible to spurious shutdowns due to instrument malfunction.

Figure 12. Event Sharing – Sample Event

There are a number of ways companies use information gathered through Event Sharing submissions including:

- PHA Preparation and Conduct – One tenet of conducting a process hazard analysis is to identify and evaluate incident scenarios which could have catastrophic consequences. The Event Sharing Database is a convenient tool for supplementing a site's own incidents to ensure hazard scenarios from the broader industry are identified and evaluated.[4] The database can be queried to extract events related to the process unit in question. The query can be further refined to pinpoint a specific type of equipment (e.g. Fired Heater) or a mode of operation (e.g. Start-up). The events can be prescreened by the PHA leader and appropriate events selected for review by the PHA team. The team can consider the database events as well as the causes and learnings as they evaluate hazard scenarios, causes, and safeguards.

- Procedure and Safe Work Practice Development – A company or facility can use the database when developing procedures or practices. The database can be queried to find events that relate to a specific unit and further refined to find events where procedures or safe work practices were listed as a cause. The company or site can then use the information in these events to ensure they have adequate measures incorporated into their procedures or practices to prevent a similar incident from happening.

- Mode of Operation Scenarios – In preparation for starting up a unit after turnaround, an operations foreman can query the database to find events on the same type of unit which occurred during start-up. The foreman can use the learnings from the event(s) to ensure the same incident does not happen during the pending start-up. For example: Many events during start-up are related to bleeder valves left open when feed is started to the unit. The Foreman could use this information to verify that a unit walk down occurs prior to start-up which looks for open bleeder valves.

- Design Reviews – When designing a new unit or making modifications to an existing unit, a design engineer can use the database to find events on the same type of unit where Design was listed as a cause. The engineer can then take appropriate measures to ensure their design will not be susceptible to a similar event.

- Tool Box Topics – The Event Sharing database is an excellent source of events to use as tool box topics or "What-if" drills. An event for a pertinent unit or task can be selected for discussion of what happened, how to respond to a similar event, and how to prevent a similar event from occurring in the first place.

3.2.1 Process Safety Bulletins

There are times when it is important to highlight key and/or urgent information to companies when a high learning value event or trend has been identified. Each Bulletin contains detailed information about events, causes, and lessons. Process Safety Bulletins are located in the Safety Portal under the Reference tab. Figure 13 is a sampling of recent bulletins which were issued to member companies.

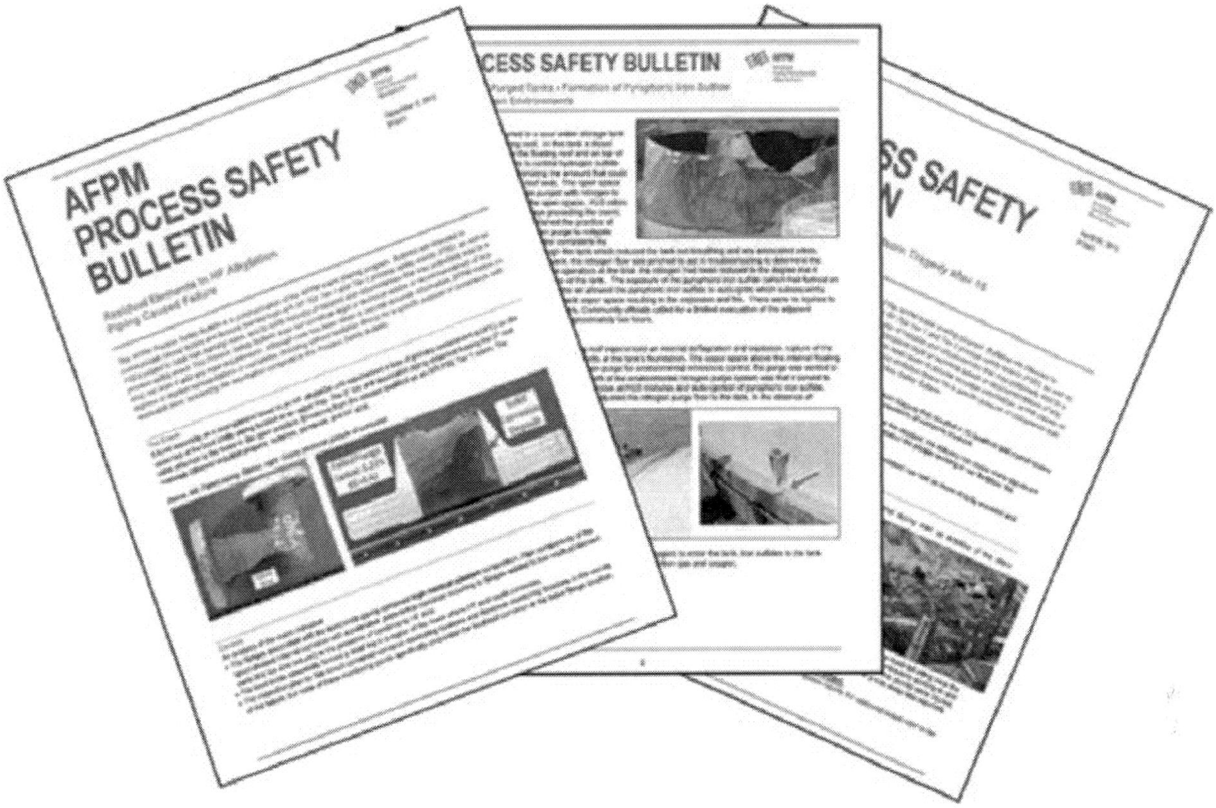

Figure 13. Process Safety Bulletin - Samples

Companies can use Process Safety Bulletins in a number of ways, including:

- Awareness – A company may use the bulletins to inform personnel of a serious incident that has occurred or a risk that exists, which could affect them.
- Standards, Specifications, and Practices – Subject matter experts can use the bulletins to evaluate the Standards, Specifications, and Practices for their company or facility to make sure that they have adequate requirements, features, and/or safeguards to prevent a similar event from happening at their company.
- PHA Preparation and Conduct – The process safety bulletins are appropriate for use by PHA teams to identify hazards and ensure adequate safeguards are in place.
- Leadership/Management Review – A facility leadership team can review the content of the bulletins to increase awareness of hazards and to determine if action at the facility is necessary to prevent future incidents. Any actions can be assigned and tracked to completion by the facility.
- Tool Box Topics – Like the Event Sharing database, the Process Safety Bulletins are an excellent resource to use as tool box topics or "What-if" drills. A pertinent bulletin can be selected for discussion of what happened, what the hazard is, how to respond to a similar event, or how to prevent a similar event from occurring at your site.

3.2.2 Annual Learnings Report

The Annual Learnings Report contains the results of an annual data analysis that identifies event trends in the industry. Figure 14 is a sampling of a recent Learnings Report.

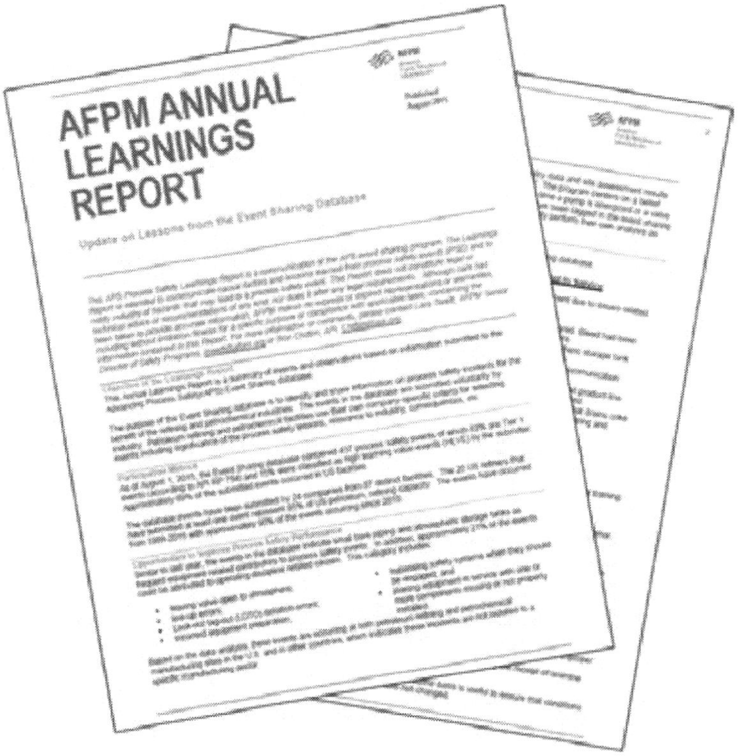

Figure 14. Event Sharing Annual Learnings Report

In addition to the same uses as outlined for the Process Safety Bulletins, the Learnings Report is also an opportunity for introspection and a chance to recognize "blind spots." A company can consider that "If this is an industry trend, is it also happening at our facilities?" Where appropriate, the company may choose to take measures or actions intended to prevent the events from happening at their facilities.

3.3 Hazard Identification

The AFPM Process Hazard Identification program supports the sharing of process safety hazards.[7] The program provides information and ready guides for potentially overlooked and not widely known process safety hazards in order to improve process safety awareness. The goal is to create tools that capture and share knowledge that could be used by other companies or sites when developing new process safety practices or improving existing ones. A sample Hazard ID document is shown in Figure 15. The Hazard ID documents are located on the AFPM Safety Portal.[7] Current Hazard ID documents are shown in Figure 16.

Each Hazard ID document contains sections with information on:

- Purpose and Use of the document,
- Scope,

- Examples of Inherent Concerns or Hazards,
- Detailed breakdown of potential causes for the hazards,
- A listing of available references and resources, and
- Links to known industry events related to the hazard.

Figure 15. Sample Hazard Identification Tool

Hazard ID documents can be used in a number of ways, including:

- PHA – The Hazard ID documents are useful during hazard reviews as a check to make sure that no credible hazards were missed or omitted by the review team. The Hazard ID documents can be prescreened by the PHA leader and appropriate documents selected for review by the PHA team. The team can consider these hazard identification tools as they evaluate hazard scenarios, to ensure no credible hazards are missed, and to verify they have adequate safeguards or controls to address the identified hazards. The References and Resources section can be used to help the team determine what actions to take if they feel the existing safeguards are not adequate.
- Pre-Job Hazard Assessments – A number of the Hazard ID documents are associated with recurring tasks in a Refining environment such as vacuum truck operation, sampling, lighting or tuning a fired heater, live flare work, etc. These

documents are great tools for a pre-job hazard review to facilitate discussion and calibrate a technician before they go do a task. They are brief, to the point, and include a robust listing of potential hazards. During the meeting the Hazard ID document can be used to discuss how each potential hazard will be mitigated during the upcoming task.

- Procedure and Safe Work Practice Development – A company or facility can use the Hazard ID tools when they are developing procedures or practices. The Hazard ID portal can be reviewed to identify documents that are applicable to procedures or practices that are being developed, or are up for their periodic review. The company or site can then use the information in these documents to ensure they have adequate measures incorporated into their procedures or practices to mitigate the identified hazards. Another way to use these tools would be to stay abreast of new Hazard ID documents as they are created, and proactively complete a gap assessment with your existing management systems to ensure all identified hazards are adequately addressed.
- Design Reviews – When designing a new unit or making modifications to an existing unit, a design engineer can review the Hazard ID tools during the design to ensure hazards are identified and mitigated.
- Training Development – The Hazard ID documents can be useful tools when preparing training materials. One key aspect of refinery training is to include an overview of the hazards of the process. These tools are a great way to make sure all of the bases are covered in this portion of the training.

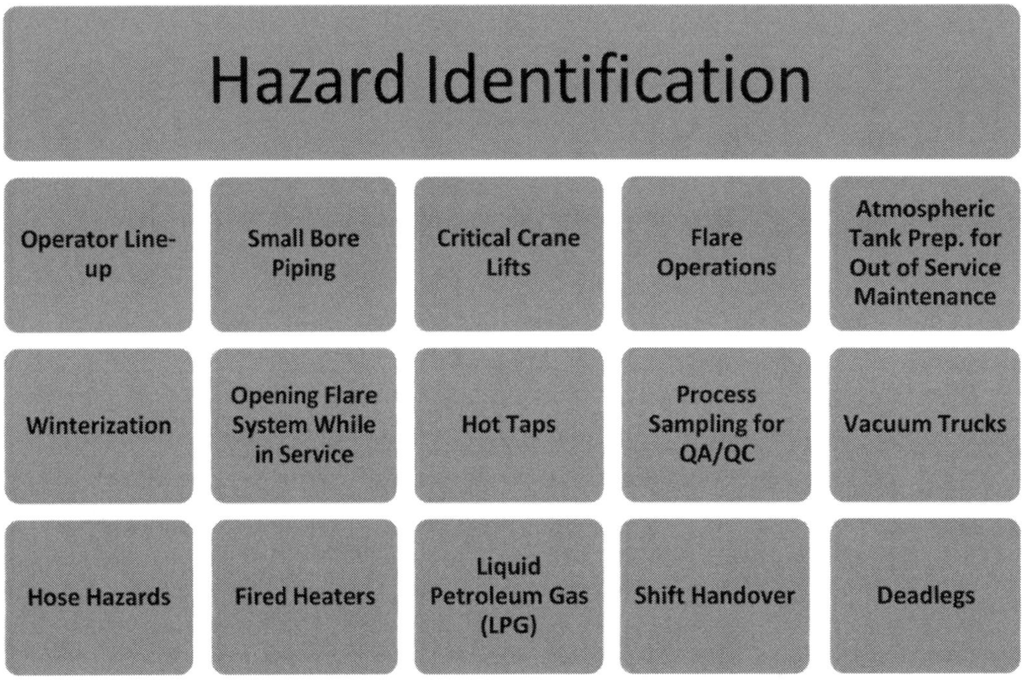

Figure 16. Currently Available Hazard ID Documents

3.4 *Practice Sharing Documents*

Practice Sharing Documents are designed to share information on process safety practices in order to help improve process safety performance and awareness throughout industry.[7] The goal is to capture and share knowledge that could be used by other companies or sites when developing new process safety practices or improving existing ones. The Practices being shared have been successfully used by an industry member, but this does not mean they are right for everyone, or that they will produce similar results at your site. Rather, they are an option to consider when implementing or adjusting programs and practices at a site. Figure 17 shows a sample Practice Sharing document.

Each Practice Sharing document contains sections with information on:

- Purpose and Use of the document,
- Scope,
- A description of the practice,
- Implementation suggestions, and
- References and Resources.

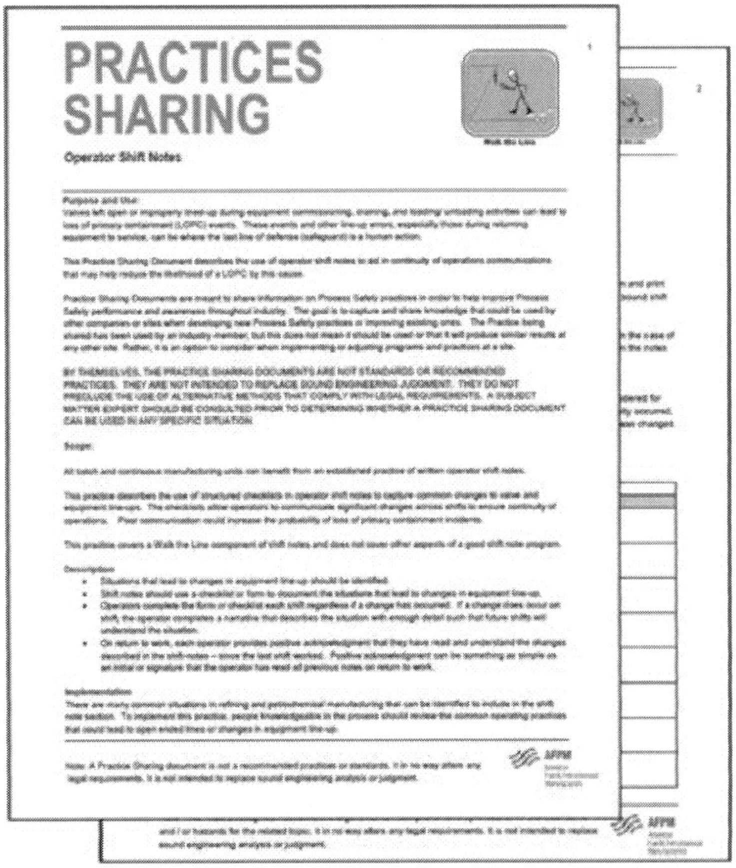

Figure 17. Sample Practices Sharing Tool

Practice Sharing Documents can be used in the following ways:

- Procedure and Safe Work Practice Development – A company or facility can use the Practice Sharing tools when they are developing or updating procedures or practices. A shared practice can be helpful when a facility is developing a new practice, by providing related concepts so that the facility is not starting from scratch. Similarly, the practice shared could be consulted by a site developing a related practice, to see if there are any aspects in the shared practice that the site might want to implement. Where a company or facility has an existing practice of their own, the shared practice can be useful during the periodic reviews of the existing practice to determine if there are any useful concepts to incorporate from the shared practice.
- Incident Investigation – When a site has completed a root cause investigation and is looking for ways to address a gap in site practices, a relevant shared practice could be a useful tool for developing a new site practice or modifying an existing one.

Practices Sharing is a new program developed in 2015. Currently there are nine Practices Sharing Documents available:

- Critical Bleeder Valves,
- Threaded Lanyard Plug,
- Operating Instructions,
- Operator Shift Notes,
- Shift Change Meeting,
- Shift Change Communications,
- Process Safety Field Audit,
- Operator Evaluation Rounds, and
- Informal Unit Walk-Through.

More practices will be identified and shared as time goes on. These practices can be located on the AFPM Safety Portal.

3.5 *Process Safety Site Assessments*

The Process Safety Site Assessment (PSSA) Program is a collaborative industry effort managed by API, focused on higher risk process safety activities in petroleum refining and petrochemical facilities.[2] This is **NOT** a run-of-the-mill compliance audit! This program primarily involves the assessment of a site's Process Safety systems by a small team of independent and credible third-party industry-qualified process safety assessors. Each assessor is an industry veteran subject matter expert with decades of experience. The expertise of the assessors is the strength of this program and what makes it so successful. Through the use of industry-developed protocols that span over 1,000 questions in total, the site assessments evaluate both the quality of the written programs and the effectiveness of field implementation in eight focus areas: Leadership, Management of Change, Mechanical Integrity, Safe Work Practices, Operating Practices, Facility Siting, Process Hazards Analysis, and HF Alkylation/API RP 751. More information about the Process Safety Site Assessment program is available HERE.

There are two primary ways to use the Site Assessment Program

Actual Assessment

When you have a Site Assessment performed at your site, you will get the benefit of seasoned veterans, looking for soft spots in your critical programs, using an industry developed protocol. The site assessment protocols are not just a check-the-box verification of compliance with the PSM regulations, but far more. The site assessment team evaluates your written programs and how your facilities implement them, using over 1000 very specific protocol questions. The resulting report provides a scoring which shows your performance in overall programs, sub-programs, and also drills down to very specific protocol provisions. The team also provides the facility management with a benchmarking report that compares your performance against the industry or against sister sites within your company. Figure 18 shows a sample page from an assessment protocol. One of the many benefits of a site assessment is having access to industry experts in each area; it's "like going to school." While they are on-site, they are also available to answer questions and engage in discussions on topics pertinent to the protocols.

Benefits of a Site Assessment Include:

- Identification of soft spots in your programs or their implementation.
- Suggestions for how to improve.
- Access to hundreds of years of combined expertise from the assessors. It's almost a training opportunity for your people.
- Obtaining a scored report and benchmarking against the industry.
- Having multiple sites assessed and use the scores to benchmark them against each other, to identify strong programs that could be shared from one site to another.
- Satisfies API 751 audit requirements with an additional audit.

Self-Assessment

Conduct a self-assessment, using subject matter professionals from your company. The same protocols used by the expert assessors are available, to member companies to use in self-assessments. These free tools are an excellent way to evaluate some of your site's critical programs. You won't get benchmarking, but you can score your own assessment to find areas for improvement. It's as simple as this: (1) APS member companies request the protocols from the API PSSA webpage, (2) put a team of your own employees together to apply the protocols to your site, (3) identify areas where improvements could be made in either your management systems or their implementation, and (4) go make the selected improvements.

Benefits of a Self-Assessment Include:

- The protocols alone can serve as roadmap for a successful program.
- Assessing your programs against the items suggested in the protocols can identify weak areas in your programs or their implementation.

- The protocols can also be used when developing new management systems as a check to ensure key aspects are covered in the program or practice being developed.
- Provide a standardized way to compare multiple sites within a company.

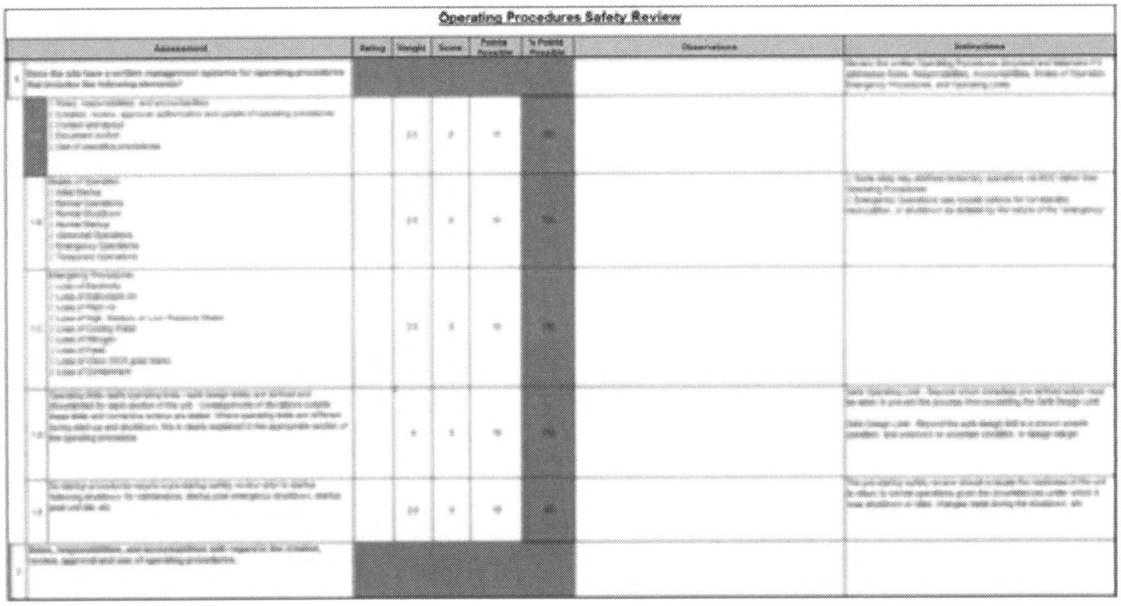
Operating Procedures Safety Review

Figure 18. Sample Site Assessment Protocol

3.6 *Human Reliability – "Walk the Line"*

Analysis completed by the APS Program has determined that approximately 20% of Tier 1 and Tier 2 process safety events are caused by human reliability issues. Types of events in this group include:

- Leaving a valve open to atmosphere,
- Valve line-up errors,
- Lock-out tag-out (LOTO) isolation errors,
- Incorrect equipment preparation,
- Bypassing safety systems when they should be engaged, and
- Placing equipment in service with one or more components missing or not properly installed.

"Walk the Line" is a program developed by the APS members to create a set of tools, free to member companies, to address the issues and cultures that lead to human reliability related events.[8] Walk the Line addresses the human reliability associated with conduct of operations and specifically, operating discipline by providing guidance for operations leadership and support to set the expectation to "walk the line" (to know exactly where a material will flow in a process). It reinforces the fundamental responsibility of operations to know with 100% certainty where a material will flow between any two points in a process. Walk the Line addresses the culture of operator line-up and provides tools that help the operator and front-line operations supervision become successful in meeting this responsibility.

The Walk the Line program is meant to generically address loss of primary containment in the refining and petrochemical industry caused by operator error when making a valve line-up to transfer process materials or operate a process. While there are many common causes and solutions, each company and site should have an understanding of their specific causes. Here is how you can use Walk the Line at your site or company:

- Set a Goal of Zero Walk the Line Incidents and commit to it.
- Analyze your LOPC Data.
- Identify Walk the Line Causes.
- Review and customize Walk the Line tools.
- Train employees using your site's customized Walk the Line tools.
- Repeat the discussions often in various meetings, tool box talks, and informal settings with process operations personnel.

The suite of Walk the Line tools are available to APS members HERE.

- Newsletters – The Walk the Line Newsletters are useful for increasing awareness about these Human Reliability issues, and for planning your sites effort to confront these causes and shift culture. The newsletters are suitable for top to bottom awareness and planning, from management level review to "boots on the ground" front line personnel.
- Tool Box Topics – The tool box topics offer short, scripted tool box meeting topics aimed at identifying common incident scenarios, engaging front line personnel in discussions, and providing suggested solutions to prevent incidents. There are 10 tool boxes related to Walk the Line topics, that are available for immediate use. While the tool box topics are fully complete, it is expected that each company or site might modify them to better align with their work practices or culture.
- Hazard ID Documents and Practices Sharing – A number of the tools being developed as part of the Walk the Line project are Hazard Identification Documents and Shared Practices. These have been previously discussed in this paper as has their suggested uses. Preparation of these documents for the Walk the Line project is on-going and these tools will be made available to APS member companies as they are completed. A development schedule is included on the Walk the Line page of the Safety Portal.
- Training Materials – Packages of "ready to go" training materials have been prepared and are available for off the shelf use. There are packages intended for both front line personnel and front line supervision.

4.0 Conclusion

The AFPM and API have been working together on an industry led initiative, Advancing Process Safety, with the overall objective of improving our process safety performance. Data obtained through the APS programs of Metric Reporting, Event Sharing, and Site Assessment identify high priority areas for improvement. Each area for improvement is being addressed through a APS work group, where tools and materials are developed and made available for AFPM and API member companies.

References

[1] API. "Process Safety Performance Indicators for the Refining and Petrochemical Industries," ANSI/API Recommended Practice 754, First Edition, April 2010.

[2] AFPM. AFPM Safety Portal Process Safety Site Assessment Programs, Available at http://safetyportal.afpm.org/SiteAssessment.aspx – Login credentials required. Accessed on February 10, 2016.

[3] AFPM. The AFPM Safety Portal Process Safety Metrics Database, Available at http://safetyportal.afpm.org/ProcessSafetymetrics-access.aspx – Login credentials required. Accessed on February 10, 2016.

[4] AFPM. The AFPM Safety Portal Event Sharing Database, Available at http://safetyportal.afpm.org/EventSharing-access2.aspx – Login credentials required. Accessed on February 10, 2016.

[5] API. "Piping Inspection Code: In-service Inspection, Rating, Repair, and Alteration of Piping Systems," API 570, Fourth Edition, February 2016.

[6] API. "Inspection Practices for Piping System Components," API Recommended Practice 574, Third Edition, November 2009.

[7] AFPM. AFPM Safety Portal Hazard Identification & Practice Sharing, Available at http://safetyportal.afpm.org/HazardIdentification-access.aspx – Login credentials required. Accessed on February 10, 2016.

[8] AFPM. AFPM Safety Portal – Walk the Line, Available at http://safetyportal.afpm.org/walktheline.aspx - Login credentials required. Accessed on February 10, 2016.

[9] AFPM. AFPM Safety Portal, Available at http://safetyportal.afpm.org/default.aspx – Login credentials required. Accessed on February 10, 2016.

[10] AFPM. AFPM Safety Portal – References, Available at http://safetyportal.afpm.org/References.aspx - Login credentials required. Accessed on February 10, 2016.

Appendix A: API RP 754 Tier 1 and Tier 2 Definitions

A Tier 1 or Tier 2 Process Safety Event begins with an unplanned or uncontrolled release of any material, including non-toxic and non-flammable materials from a process that results in one or more consequences described in the RP. The consequences that define a Tier 1 and Tier 2 process safety events are tabulated below.

Tier 1 Consequences	Tier 2 Consequences
An employee, contractor or subcontractor "days away from work" injury and/or fatality	An employee, contractor or subcontractor recordable injury
A hospital admission and/or fatality of a third-party	
An officially declared community evacuation or community shelter-in-place	
A fire or explosion resulting in greater than or equal to $25,000 of direct cost to the Company	A fire or explosion resulting in greater than or equal to $2,500 of direct cost to the Company
A pressure relief device (PRD) discharge to atmosphere whether directly or via a downstream destructive device that results in one or more of the following four consequences: ▪ liquid carryover; ▪ discharge to a potentially unsafe location; ▪ an on-site shelter-in-place; ▪ public protective measures (e.g. road closure); and, a PRD discharge quantity greater than the threshold quantities in Table 1 in any one-hour period.	A pressure relief device (PRD) discharge to atmosphere whether directly or via a downstream destructive device that results in one or more of the following four consequences: ▪ liquid carryover; ▪ discharge to a potentially unsafe location; ▪ an on-site shelter-in-place; ▪ public protective measures (e.g. road closure); and, a PRD discharge quantity greater than the threshold quantities in Table 2 in any one-hour period.
A release of material greater than the threshold quantities described in Table 1 of the standard in any one-hour period	A release of material greater than the threshold quantities described in Table 2 of the standard in any one-hour period

442500 Forge Bonding: A Safer Metal Joining Process

Abstract:

When planning in-service repairs or enhancements that involve the joining of metals; sparks, flames and even elevated temperatures are a concern for refinery and storage terminal owners and operators. Processes that generate uncontrolled elevated temperatures that exceed Auto Ignition Temperatures (AITs) published by American Petroleum Institute (API) or published Hot Surface Ignition Temperatures (HSITs) are often scrutinized and classified as "hot work" within these facilities. A metal bonding process has been developed that in inherently safe in these environments where flammable hydrocarbons are present.

Friction welding is a solid-state welding process that generates heat through mechanical friction between work pieces in relative motion to one another, with the addition of a lateral force to plastically displace and fuse the materials. Technically, because no melt occurs, friction welding is not actually a welding process in the traditional sense, but a forging technique. An advanced form of friction welding called forge bonding has been developed for portable entry into refineries, improving worksite safety by reducing the time at temperature and the hot surface area at the bonding site. In addition, forge bonding has a distinct advantage over welding in that it can join dissimilar metals that have largely different melting points.

An introduction to the forge bonding process followed by results from extensive ignition testing to determine the safety margin of multiple chemicals commonly found in refineries when in the presence of forge bonding, will be presented. The dependency of the ignition temperature of hydrocarbons on the hot surface area and the ignition lag time are discussed along with support from data from API 2216 and other well known publications. A presentation of the problem followed by an investigation of the forge bonding solution will prove that ignition of common hydrocarbons is not possible in the presence of the low energy levels created by the small hot surface area, short ignition lag time and moderate temperatures.

Real life implementations add further proof that forge bonding is an alternative to welding that is inherently safe in refineries, storage terminals and anywhere flammable hydrocarbons are present. This technology has far reaching implications for improved safety in other industries, such as pharmaceutical, pipelines, mining and marine to name just a few.

16AIChE
Spring Meeting, Houston, TX

Best Safety Practice for Refinery and Petrochemical Plants During the Major T&I (Turnaround & Inspection)

Ayman Natto
Safety Department, Technical Safety Section, Rabigh Refining and Petrochemical Company (Petro Rabigh),
Rabigh, Saudi Arabia
ayman.natto@petrorabigh.com

Prepared for Presentation at
American Institute of Chemical Engineers
2016 Spring Meeting
12th Global Congress on Process Safety
Houston, Texas
April 10-13, 2016

UNPUBLISHED

Safety Practice for Refinery and Petrochemical Plants During the Major T&I (Turnaround & Inspection)

Ayman Natto, Process Safety Engineer
Safety Department, Technical Safety Section, Rabigh Refining and Petrochemical Company (Petro Rabigh),
Rabigh, Saudi Arabia
ayman.natto@petrorabigh.com

Keywords: Safety Practice during T&I, Safety in Major shutdown, Process Safety roles in shutdown, Pre-Start up Safety Review, Management of Change, Mechanical Completion Certificate.

Abstract

Planned shutdowns or Turnaround and Inspections (T&I) are normal activities for gas and oil companies the world over, but Petro Rabigh carried out a major shutdown at its huge refinery and petrochemical complex consisting of Refinery, Monomer, Polymer and Utility plants. This paper is aimed at the best Safety Practice implemented during the T&I which was completed at the end of December 2015. It explains the Safety Program developed and implemented by Safety Department to execute the T&I safely. Planning was a key element of the T&I preparation. A dedicated taskforce was established for the T&I planning with representatives from Manufacturing, Maintenance, Engineering, Safety, and other involved departments. The taskforce addressed various safety programs including Work Permit System, Critical Lift Plan, Confined Space Entry, Process Hazard Analysis (PHA) etc. Moreover, critical jobs were identified and Job Safety Analysis (JSA), Management of Change (MOC), Mechanical Completion Certificate (MCC) and Pre-Startup Safety Review (PSSR) were conducted for each of these jobs. Special training and presentations were given to both Petro Rabigh and Contractor Employees on the lessons learned from T&I incidents experience worldwide. A huge contractor workforce and the massive scale of activities were managed safely with the above innovative implementation of the practice, Petro Rabigh would like to improve in the 2019 T&I with many of its new Phase II plants.

1. Introduction

Most oil and gas companies and organizations worldwide have periodic planned plant shutdowns known as Turnaround and Inspections (T&I). They covers activities such as inspection, overhaul, modification and the installation of new parts or equipment. Activities that cannot be performed when the plant is in normal operation are also carried out during the shutdown period.

Rabigh Refining and Petrochemical Company (Petro Rabigh) is located on a 3,000 acre site in Rabigh, on the Red Sea coast of western Saudi Arabia. Petro Rabigh - a joint venture between Saudi Aramco and Sumitomo Chemical Company, began operations in April 2009, with the capacity to produce 18.4 million tons of petroleum based products and 2.4 million tons of ethylene and propylene-based derivatives per annum.

This integrated Refinery and Petrochemical complex has a 400,000 barrels per day Refinery with Distillation, Hydrotreating, Hydrogen Production and Recovery plants, and petrochemical plants consisting of the world's largest High Olefin Fluidized Catalytic Cracking plant, Ethane Cracker plant, Alkylation & Isomerization Unit, Propylene Oxide, Mono Ethylene Glycol, Polyethylene, Polypropylene, High Density Polyethylene plants, etc. These are supported by a vast network of utilities, such as a storage tank farm and marine berthing facility for the receipt of crude oil/raw materials and dispatch of finished products. Moreover, Petro Rabigh is currently constructing many new production plants - tied-in with existing plants during the 2015 T&I - for its Phase II project, known as Petro Rabigh II.

The first T&I at Petro Rabigh was successfully completed in 2011 with about 7 million man-hours logged, over seven weeks. With proper planning, preparation and perfect execution, Petro Rabigh was able to exhibit an enviable safety record, viz., two lost-time injury incidents of contractor employees, three medical treatment cases and five first-aid injury incidents. A huge contractor workforce and the massive scale of activities were managed safely with the above innovative practices and by team work, cooperation and dedication from all. Petro Rabigh would like to continue these practices in the 2019 T&I with improvement in the field of Process Safety, in addition to the hundreds of tie-in points for the Petro Rabigh II.

2. T&I Preparation

This was a major complex-wide turnaround, as it involved all the plants and utilities in the Petro Rabigh Complex. More than one hundred CAPEX works and Warranty Claim works were also included in the T&I scope of turnaround, adding to its daunting magnitude.

Based on the work scope and schedule, the peak contractor manpower in the Petro Rabigh Complex during the turnaround was over 13,000. This was in addition to the regular company employees, nearly 4000. On average the daily man-hours logged was 165,000 per day during the twelve weeks long turnaround period. At the end of the turnaround, the total man-hours logged were more than 15 million.

Petro Rabigh developed a T&I link in the company main web page which showed the day to day status of all Petro Rabigh T&I activities. Safety was one the main activities and included data such as daily safety observations, daily safety performance dashboards, safety flyers, incident flyers, lessons learned, near misses and safety suggestions. This link was updated daily by the Petro Rabigh Safety Department below is the screen print from the web page of that time:

2.1. Contractor Safety Management

Planning is Good Business. A Good Plan ensures safe execution.

It was essential that all the contractor organizations were made sufficiently aware of Petro Rabigh' s minimum safety requirements. The following actions were implemented during the T&I:

No. of Contractor Companies	57
Approved Hazard Identification Plan (HIP)	66
Approved CVs Of the Safety Representative	181

2.1.1. Safety Orientation sessions

Petro Rabigh provided about 13,000 attendees for the Safety Orientation to all the contractor employees through a specially developed Safety Orientation video film. This film was produced in five different languages for the different nationalities who would be working in the Turnaround.

2.1.2. Turnaround Safety Forum Sessions

Turnaround Safety Forums for Contractors and Petro Rabigh staff were conducted before the Turnaround. The Forum included short presentations and discussions on the safe execution of work activities related to turnaround, lessons learned from the 2011 T&I and best industrial safety practices etc. to further enhance safety awareness and improve safety communication during the execution.

In addition, several sessions of Petro Rabigh Work Permit Receiver Training were provided to the selected, qualified Contractor Supervisors to understand the work permits system requirements and standards. This ensured alignment of safe work systems of the contractors with the Petro Rabigh Work Permit System.

2.1.3 Confined Space Standby Man Training

Thousands of confined spaces were opened during the T&I, with high risk especially in such a large refinery and petrochemical plant. Petro Rabigh trained over 1100 standby personnel to understand their roles and responsibilities upon confined space entry. Petro Rabigh issued a standby man certificate for the trained contractor employees to ensure the confined spaces were monitored and managed by a competent person.

2.2. Critical jobs

From among the listed jobs for the Turnaround, critical jobs were identified by a team of Petro Rabigh Operations, Maintenance, Engineering and Safety representatives and a JSA was conducted by the team for each such job. Similar review was done for all scaffold erection plans and critical lift plans.

Awareness Training was also given to Operations and Maintenance staff of Petro Rabigh covering JSA and Scaffold Safety.

2.2.1 Tools Inspection and Certifications

A dedicated team of Petro Rabigh Maintenance Department examined all tools and equipment of the contractors and issued approval sticker on only the properly functioning and safe tool / equipment. This prevented the contractors from using damaged tools and equipment in Petro Rabigh work site. Such inspection also covered Heavy Equipment – such as cranes, loaders, forklifts, low-bed trailers etc. through approved Third Party agencies.

Petro Rabigh established the certification requirement for the Scaffold Inspectors, Scaffold Supervisors and Crane Riggers.

All such measures in the planning stage collectively helped in managing the Turnaround safely.

3. T&I Safety Implementation and Monitoring

All the planning and preparation were converted into implementation, which ensured an excellent safety record during the Turnaround.

The Safety Department hired 30 contractor safety advisors to monitor the T&I in addition to dedicated Petro Rabigh safety Inspectors and Engineers who went around the contractor work areas to evaluate contractor safety performance, observing the activities, identifying unsafe actions, initiating talks with the workers and supervisors, establishing rapport, correcting unsafe practices at site and investigating incidents. Checklists were developed to record field observations.

The data provided by the Safety Inspectors was analyzed on a daily basis and transformed into a daily dashboard called the Turnaround Daily Safety Performance Dashboard. The dashboard indicated the Top Risk Category Items, the Top Three Unsafe Behaviors, the percentage of unsafe activities – areas-wise, etc.

Petro Rabigh Safety personnel conducted a total of 61,022 daily formal interactions with contractor employees and Petro Rabigh staff, in each area. Meetings were established with contractor management directly, to highlight important concerns to improve safety. Such meetings were held twice every week.

Daily Safety interaction meetings were held by Safety Department with contractor safety representatives on day shifts and night shifts.

The daily Tool Box Talks became very popular and served as a means of safety communication. Topics of the tool box talk sessions were selected based on the Top Risk Category of the dashboard for the previous day. This session was conducted in the beginning of each shift by each contractor.

Field safety observations and analysis were listed together and released in the form of Safety Flyers by Petro Rabigh. The lessons learned from the near miss and injury incidents were also released in the form of Safety Flyers.

3.1. Safety Observation Tours

Based on the statistics of observations shown in Table 1, "Sub-standard use of PPE" and "Unsafe Tools and Equipment" have appeared as our Top Risk Categories in turns. Top Three Unsafe Observations range from various categories related to "Housekeeping, Scaffold Inspection, Confined Space, Basic Safety Rules, and Barricading" etc.

The dashboard along with the Top Risk Category and Unsafe Behaviors generated constructive discussions in T&I meetings thereby providing focus on areas of concern for necessary improvement. Also, it included the percentage of observed people working safely, indicating the status of improvement in safety performance and reflecting the joint efforts of the whole organization.

The daily dashboard will continue to highlight the priority areas where everyone's efforts must be concentrated.

3.1.1 Safety Observation Statistics

Table 1. Safety Observation Statistics

Counselling meetings were conducted with low performing contractors top management to communicate safety concern and observations and action plan was demanded from contractors to reduce safety observations at site. Visible improvement was observed in contractor performance after these meetings.

Area	Total No. of Observations
Refinery	1439
Monomer	2781
Polymer	621
Utilities & Terminal	646
Shops and Laydown Yards	1560
Total	7047

3.1.2 Safety Flyers and Incident Alerts

In order to address Top Risk Categories and Unsafe Behaviors, safety flyers were developed and disseminated among all T&I concerned parties, including contractors.

The flyer information contents were discussed during Day and Night T&I meetings, and also during contractors' Toolbox Talks.

3.2. Near Miss & Safety Suggestions Reports

3.2.1 Lessons learned

Petro Rabigh encouraged contractors and employees to enter lessons learned whenever on the T&I web page visible to all Petro Rabigh employees to avoid recurrences and learn from it. About 550 lessons learned were received during the T&I which was already communicated and closed in the site.

3.2.2 Near Miss & Safety Suggestions

Petro Rabigh requested employees to submit near miss incidents and safety suggestions through the Petro Rabigh web page throughout the year, and this practice continued during the T&I as shown in Table 2.

Table 2. Near Miss & Safety Suggestion during the T&I

Total number of PRC submitted Safety Suggestions	01
Total number of PRC submitted Near Misses	10
Total number of Contractors submitted Safety Suggestions	06
Total number of Contractors submitted Near Misses	44

4. Process Safety through the T&I

Petro Rabigh had about 25 major modifications that were considered as a project which included designing, reviewing and constructing that had completed during the T&I in addition to the hundreds of tied-in points for the Phase II Project. The Safety Department had to ensure that all these projects complied with the process Safety regulation and standards. By implementing the Engineering and Process Safety elements smooth and timely transition from construction to start-up phase was achieved by implementing the Engineering and Process Safety elements smooth and timely transition from construction to start-up phase was achieved.

4.1 Management of Change (MOC)

Management of Change (MOC) is a proactive process with a two-fold objective: first, to challenge the need for change and to analyze its wider impacts on the business goals; and second to evaluate, authorize changes before they are made and ensure proper closure after the changes are completed. Its purpose is to establish the procedures necessary to ensure that the all the risks arising from proposed changes are decreased to an acceptable level. It is also an excellent communication tool to inform impacted groups of areas that a change is being considered and/or about to occur.

The core elements of the MOC process that are implemented in all modifications during the T&I:

- Agreement on the justification and reasons for the change by appropriate management,
- Risk analysis of the proposed change by a multi-disciplinary team of competent people who have required knowledge and experience, including vendors and specialists,
- Approval of complex changes or design deviations by engineering managers or technical authorities of sufficient experience and knowledge,
- Develop operating procedures to supplement or implement the change,
- Communicate the change to all concerned and train the workforce, those who are directly affected by the change,
- Conduct a pre-startup safety review to ensure system is ready for safe start, i.e. that all actions from the risk analysis are implemented, operating procedures are developed and staff are trained,
- Monitor the change during implementation period and capture any lesson learned for future reference, and
- Close out the change after specified time period otherwise initiate a permanent change process.

The MOC process is also critical in day to day operations in case of safe operating envelope excursion during process upset; it requires immediate action such as shut down or reporting of incident for further investigation if conditions were normalized quickly. Overall a robust MOC process and its effective implementation is an integral element of chemical process safety in order to prevent and control major incidents risks.

4.2 Mechanical Completion Certificate (MCC)

An important initiative during this T&I was the implementation of MCC through a Committee chaired by Proponent representative and comprised of Construction Agency, Inspection Organizations, Safety and other representatives as appropriate, depending upon the nature of the facility. The committee had to complete the following MCC steps: (1) accepting new facilities as mechanically complete, (2) commissioning and start-up, and (3) final performance acceptance and financial closeout. The Construction Agency was responsible for delivering a completed facility as defined in the project scope of work. After the facility was accepted as Mechanically Complete, the Commissioning, Start-Up and Performance Acceptance became the responsibility of the Proponent Organization with the Construction Agency providing necessary design and construction personnel, technicians and vendor representatives to assist in these activities.

For all facilities and especially where an existing facility was undergoing an upgrade or retrofit, both the Construction Agency and the Proponent were to assign personnel during the design phase to establish a Pre- Commissioning, Commissioning and Start-Up Plan.

4.3 Pre Startup Safety Review (PSSR)

Pre Startup Safety Review (PSSR) aids in the transfer of projects from the construction stage to the operations stage. It provides assurance that commissioning and start-ups can be carried out in a safe manner following the installation of new facilities. In particular, PSSR aims to confirm that what was built and installed conforms to the original design and standards, and that the hazards during start-up of the project have been properly addressed and controlled. Petro Rabigh dedicated a team to conduct about 25 PSSR for the projects taken up during the T&I, in all four Operation departments.

Nominated leaders from each Operation department led their respective PSSR teams (comprising of members from Operation, Engineering, Maintenance, Project Management Team, Safety and Environment) and started conducting PSSR for such modifications.

5. Conclusion

T&I execution is a major challenge that needs team work, cooperation, dedication, and flexibility in order to be successfully accomplished. With proper planning, implementation of safety programs, innovative practices and active involvement of contractors, Petro Rabigh completed the T&I scope safely, timely and reliably.

443166 Novel Dividing Wall Columns for Ternary and Quaternary Distillation

Abstract:

Dividing wall columns for fully thermally coupled distillation can provide significant cost benefits. For many applications, they are known to offer savings on capital and operating costs by as much as 30% compared to conventional sequences using direct/indirect splits. However, during operation, these benefits may not necessarily be realized because the vapor flows in the different sections of the DWCs are not conveniently controlled by external means. To solve the issue of the unregulated vapor split at the bottom of a vertical partition, we propose new, alternate DWCs for any thermally coupled configuration. The vapor flow in each section of the new DWCs can be regulated by means external to the column, thereby making optimal/near-optimal operation, a practical possibility. We will illustrate our method by presenting novel DWCs for 3-component and 4-component distillations.

443932 Bridging the Know-How Gap - Hydraulic Design of Packed Dividing Wall Columns

Abstract:

By virtue of its design a Dividing Wall Column (DWC) minimizes energy and equipment as well as plot area requirements of three and more component separations by distillation, which makes it to the most sustainable among distillation technologies. Although proven and well established in practice, and highly appealing regarding energy and capital saving potential, the number of industrial applications of DWCs is still limited. Indeed, as elaborated in greater details elsewhere, e.g. Jansen et al., Chemical Engineering, August 2014, 40-48, a DWC is an atypical distillation column and there are certain process and mechanical design and operation-related concerns and potential constraints that may discourage inexperienced practitioners and make them reluctant to consider design and implementation of a DWC in prospective applications.

In the meantime, many process engineers have developed skills and are capable of using available tools within commercial software packages to simulate performance, i.e. to establish reflux and stage requirements of a DWC. However, what is not readily available in commercial software packages is a proven DWC dimensioning method. Indeed, a reliable hydraulic design method is needed to translate performance simulation results into diameter(s) and height of the shell of a new DWC, or to examine appropriateness of an existing column shell in a retrofit case, including detailed internal layout and proper dimensions of normal and partitioned sections, depending on chosen type and size of structured packing and auxiliary equipment. This must be done with certain rigor if one wishes to arrive at a reliable basis for evaluation of cost-effectiveness of both new DWC designs and retrofits.

To provide basis for filling the existing know-how gap, the present paper addresses all relevant aspects of packed DWC dimensioning, using engineering tools available in public domain. Emphasis is on equipment choice and a predictive model that enables accurate estimation and proper balancing of the pressure drop in parallel sections. The latter being the key to successful design and operation of DWCs. Capability of quantification in this respect opens the door for exploration of possibilities for expansion of the application window of DWCs, both as a new design to enhance potential benefits and as an energy saving retrofit option.

Dividing Wall Column – Design for a Retrofit

Steven G. Schon, P.E.

Engineering Fellow

Arkema Inc.

AIChE Spring 2016
Houston, Texas
April 10 – 14, 2016

Overview

- **Existing products produced for many years at existing plant**
  - Campaign operation (2 products)
  - Derived from impure olefins; key impurities are isomers and heavy homologs
- **Vacuum distillation train used trayed columns**
  - Plant built prior to introduction of high efficiency structured packing
  - HETP of 12" tray spacing < HETP of earlier-generation random packing
  - Small-diameter columns (4 foot)
- **Market opportunity for higher-purity products**
  - New market
  - Higher added-value
  - Need to reduce heavy homologs to < 80 ppm (vs. 2000 ppm for standard grade)
- **Good process simulation model of distillation train**
  - T profiles & purity vs. reflux agree well with plant data

Product Distillation Train

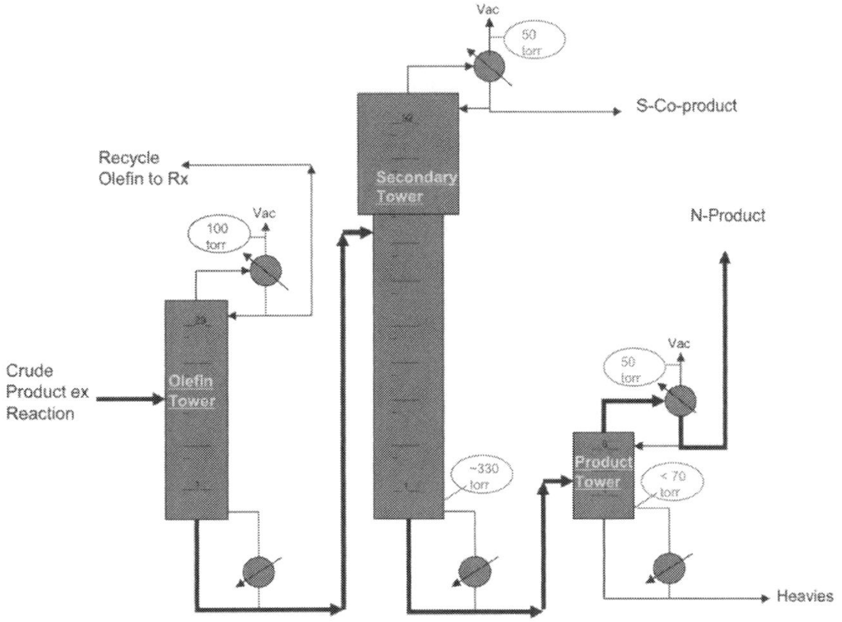

High-Purity Product Requirements

- **Existing 6-tray product tower insufficient for separation**
- **Proposed extending product tower shell and replacing trays with structured packing**
- **BUT, foundation, reboiler & condenser inadequate**
- **Revised proposal: New product tower**
  - Higher cost (new foundation, longer piping)
  - Longer lead time
  - Constructability concerns (congested area)

Conventional Train for High-Purity Product

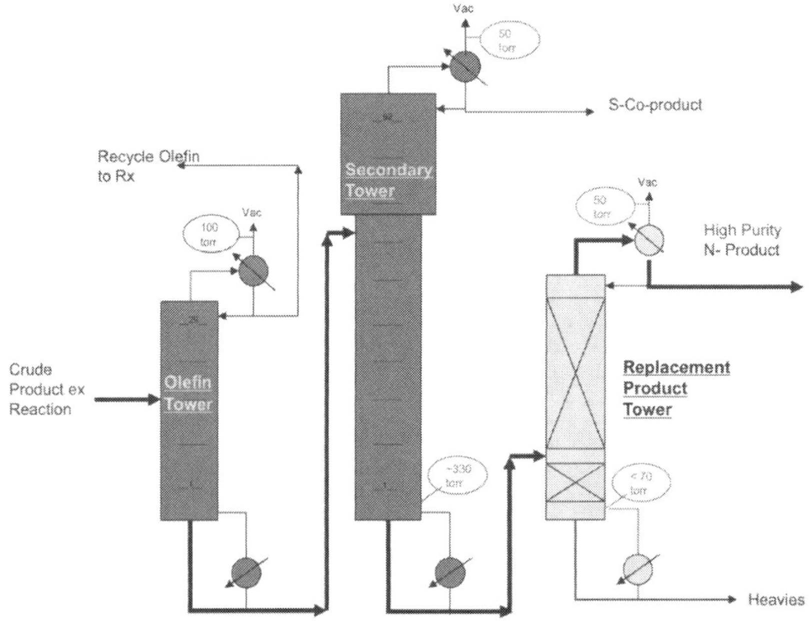

Other Considerations

- **Interest in debottlenecking entire production unit**
- **Olefin tower near flood limit, and reboiler near capacity**
- **Potential thermal degradation in olefin tower bottoms**
- **Consider revamping secondary tower with structured packing**
  - Lower HETP → more theoretical stages / easier separation
    - Sieve trays @ 70 % Efficiency, 12" spacing ~17 " HETP
    - 250 "Plus" style structured packing ~12" HETP
    - Well suited for high V/L → allows higher boil-up w/o vapor flood
  - Less dP → lower bottoms T → increase reboiler capacity; less degradation

Insight & Inspiration

- **Olefin tower has much higher capacity if packing vs. trays**
  - Vapor-flooding limited (vacuum service)
  - Low HETP / more theoretical stages with structured packing → reduce reflux
  - ~ 60 % flood with Plus-style structured packing
- **New product tower vapor rates < ½ that of olefin tower**
  - Equivalent to ~ 30 % of olefin tower flooding capacity
- **Packed olefin tower P profile close to that of new product tower**
- **Dividing Wall Column (DWC) does work of 2 towers in series**
  - If both conventional towers at comparable operating pressure
- **Can olefin tower be converted to DWC ?**
  - Use "excess" hydraulic capacity (with structured packing) for purification-side hydraulics
  - Remove both lights and heavies simultaneously in one column
  - Potentially reduce energy requirement → less boilup required
  - No new product tower → faster, cheaper project

Dividing Wall Column (DWC)

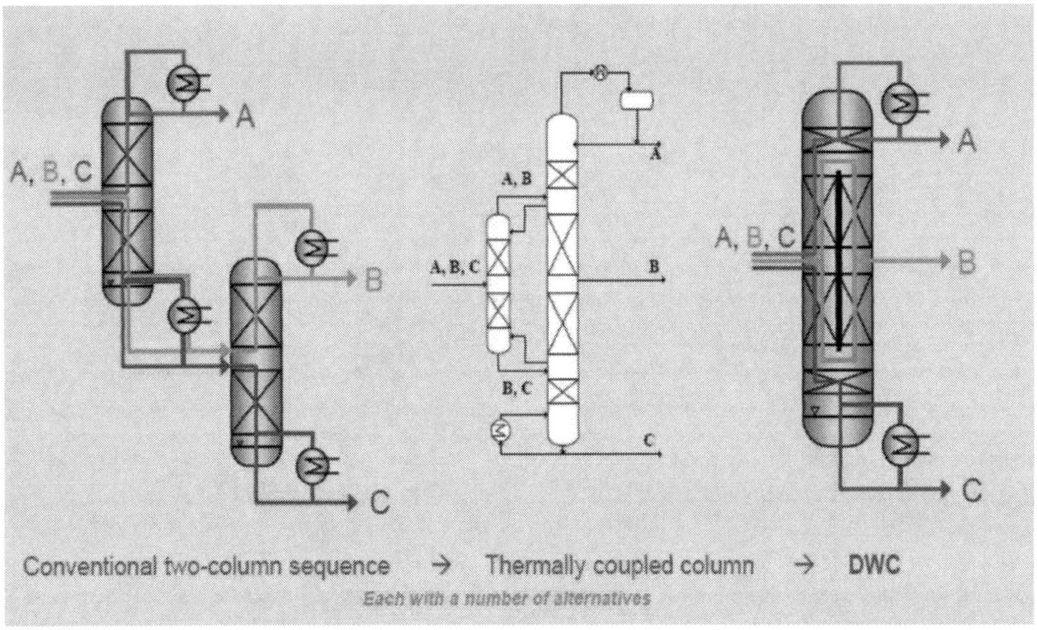

Courtesy of Julius Montz GmbH

DWC Retrofit Concept

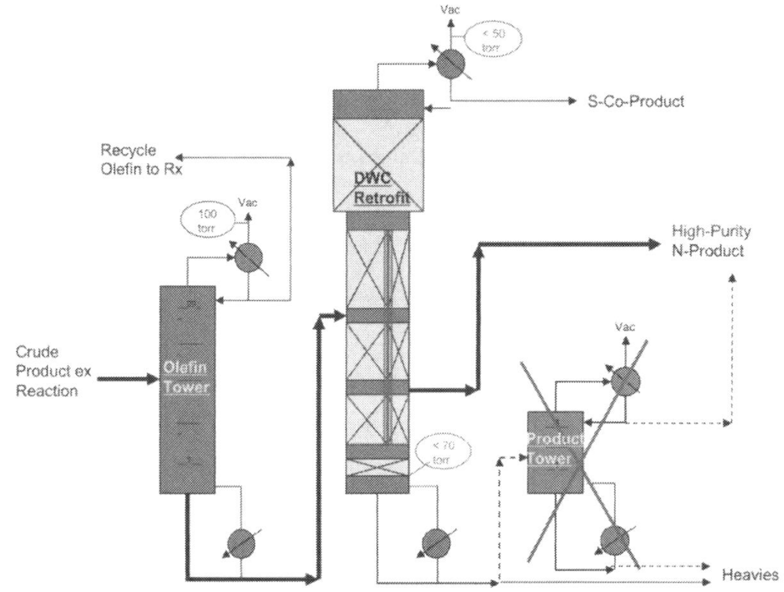

DWC Equivalency / Model

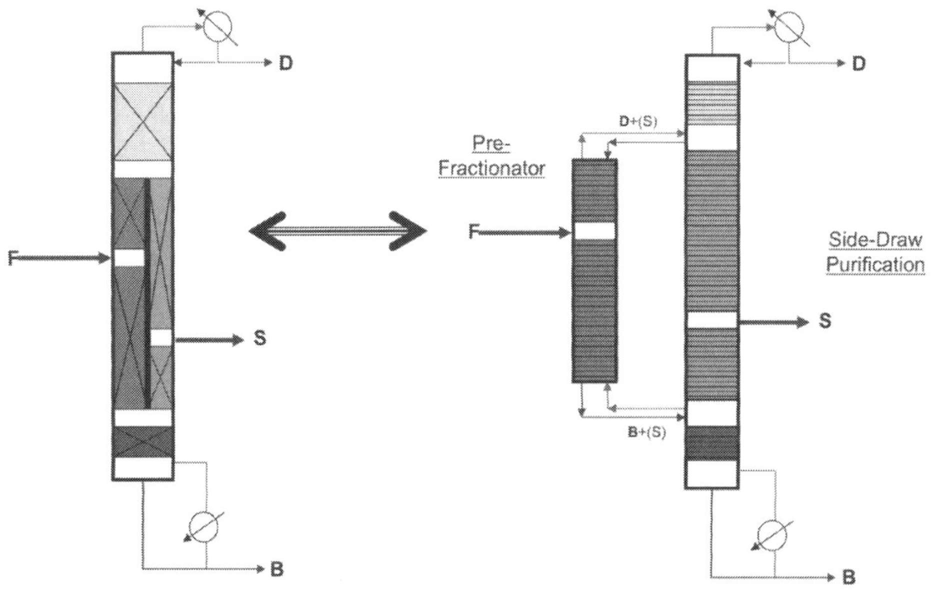

Aspen Simulation of DWC vs. Conventional

Conventional 2-Column Sequence

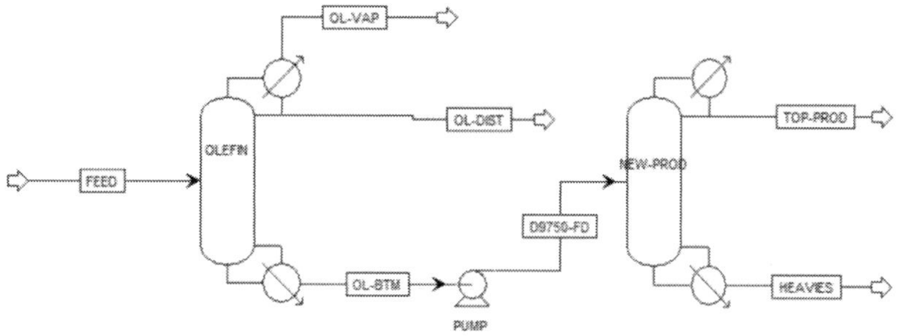

Dividing-Wall Sequence

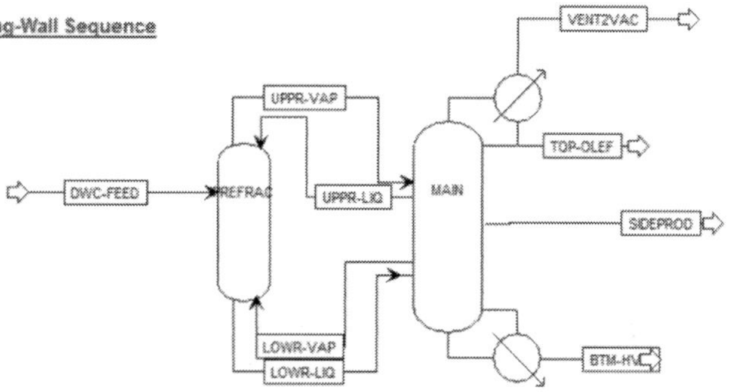

DWC Simulation Approach

- **Start with optimized conventional 2-column simulation**
- **Model DWC as prefractionator coupled to regular column**
  - Main column
    - Total stages = larger of 1st or 2nd conventional column stages
    - Upper connection stage = # stages below 1st conventional column rectifying stages
    - Lower connection stage = # stages above 2nd conventional column stripping stages
    - Side-draw = # 2nd conventional column rectifying below upper connection
  - Prefractionator
    - No condenser or reboiler
    - Same # stages as main column between upper & lower connections
    - Initialize top liquid & bottom vapor feeds based on corresponding conventional 2nd column stage flows & compositions
- **Activate dP calculations** (specify packing, diameter, HETP)
- **Adjust section diameters and prefractionator inlet flows until prefrac & corresponding main column section dPs match**
- **Optimize # stages & feed locations to minimize energy**

DWC Initialization

Conventional 2-Column Sequence

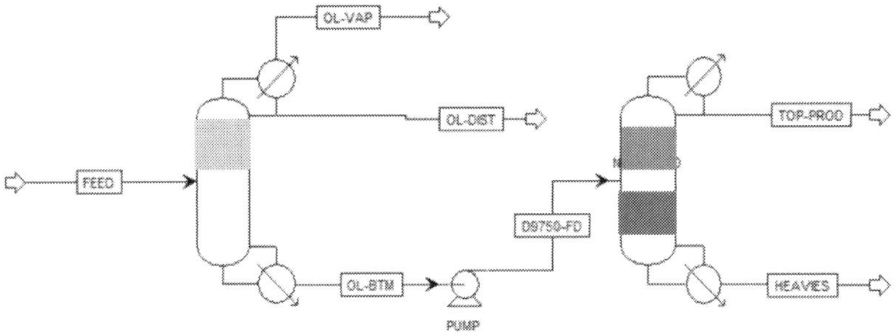

Dividing-Wall Sequence

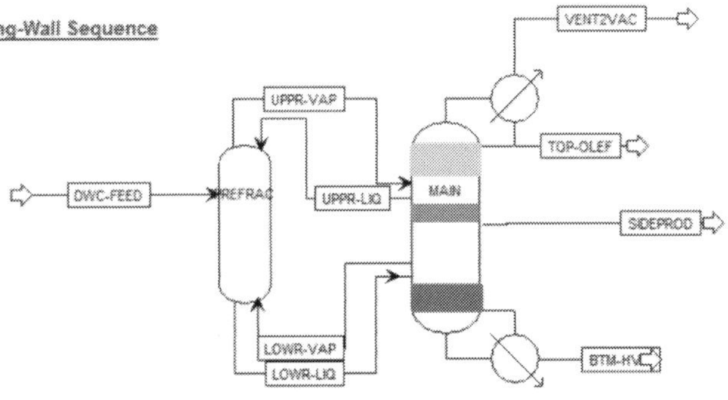

DWC Simulation Results

- **Revamping Olefin Tower as Packed DWC Appears Feasible**
  - Meets High-Purity Specs & Rates
- **73 Theoretical Stages**
  - Accounted for non-packed height "lost" for re-distributors
- **18 % Lower Reboiler Duty Than 2-Column Conventional**
- **5 Beds (3 Divided) of High-Efficiency Structured Packing**
  - Avg. HETP ~ 12" ; with distributors, fits in space of current 92 trays
  - Different packing to match section HETP & dP needs
  - < 80 % Flood in All Sections ➔ potentially higher capacity

DWC Zone Distribution

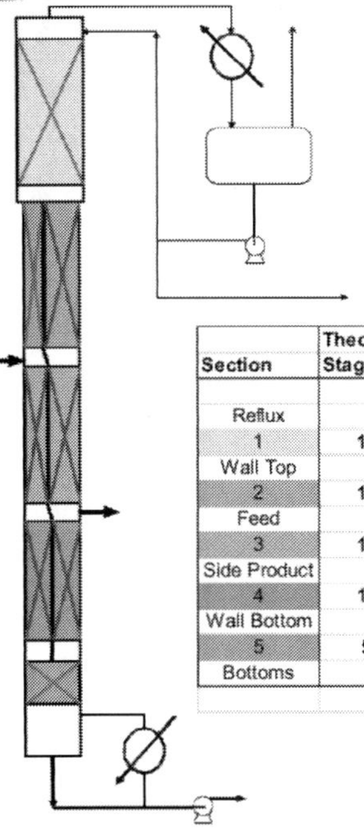

Section	Theor. Stages	Equivalent Diamter (feet) Pre-Frac	Common	Purif.	Overall	Packing (Mellapak Plus) Pre-Frac	Common	Purif.	Height (feet) Packing	Redistirb
Reflux										
1	18		4.5		4.5		452Y		18	
Wall Top			|				|			4
2	19	2.4	|	3.2	4.0	752Y	|	252Y	19	
Feed			|				|			4
3	16	2.6	|	3.0	4.0	252Y	|	252Y	16	
Side Product			|				|			4
4	15	2.8	|	2.8	4.0	252Y	|	252Y	15	
Wall Bottom			|				|			4
5	5		4.0		4.0		452Y		5	
Bottoms										
									89 total	

DWC Main Section Composition Profiles

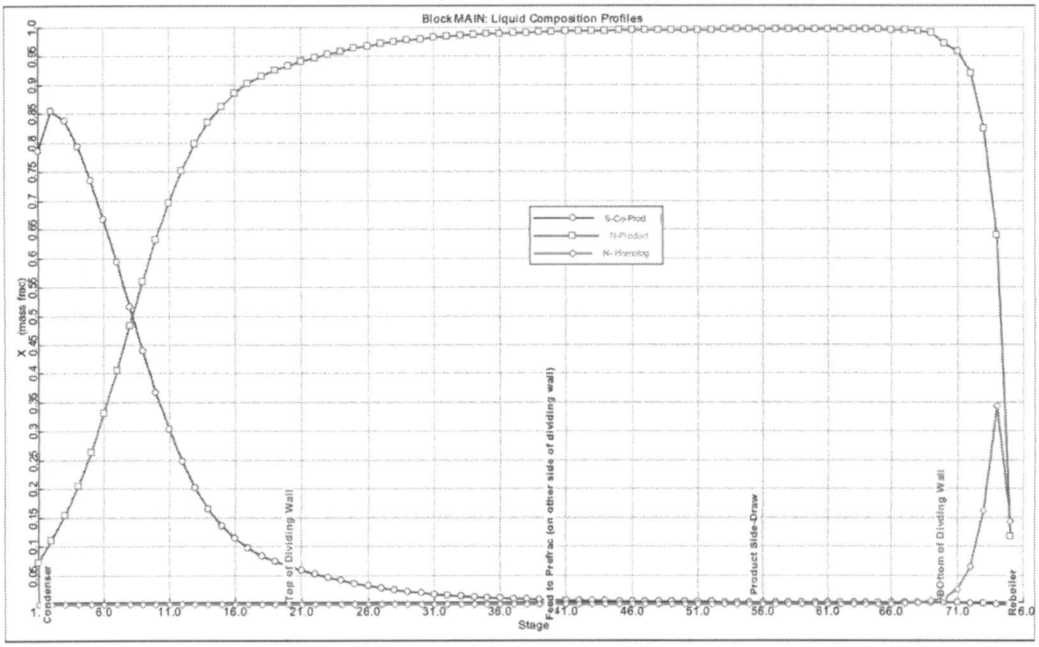

Predicted DWC Column Profiles

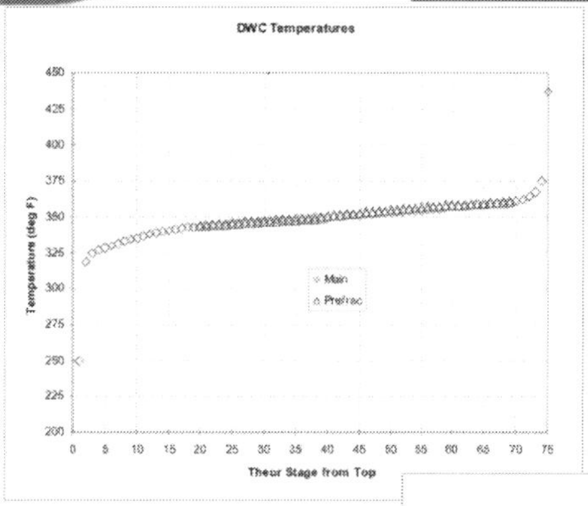

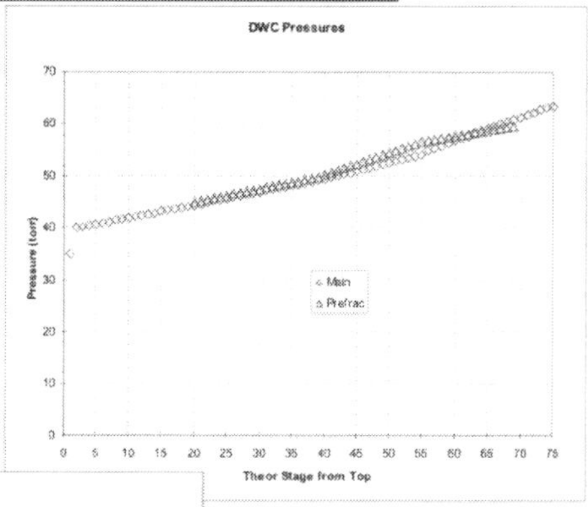

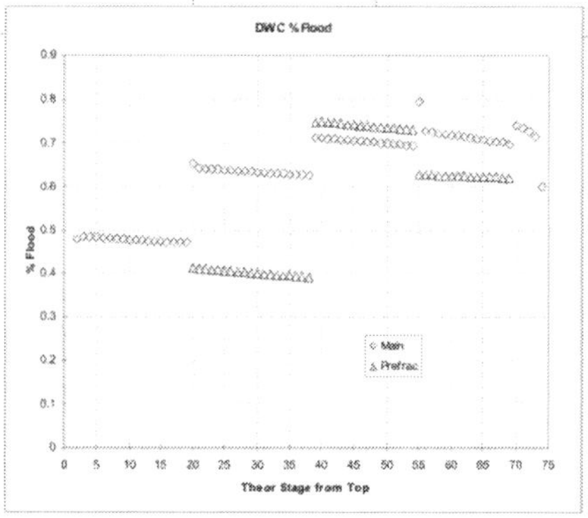

DWC Column Vendors' Assessments

- **Process excellent fit for DWC**
- **Un-equal sections / wall placement in small tower is manageable**
- **Options for DWC design**
  - Welded dividing wall
    - Small diameter requires multiple body flanges
    - Pre-install packing & distributors prior to setting column sections
    - Ensures no cross-wall leakage that can impact purity / efficiency
  - Non-fixed wall w/ wall wipers
    - No body flanges: Can do "replacement-in-kind" shell
    - Field-install wall, packing & distributors through manways after setting
    - Ensures good leveling of internals & liquid distribution
- **Other considerations**
  - New shell vs. retrofit
    - Removal of existing tray support rings labor-intensive
    - Cost
    - Down-time
  - More complicated control scheme
    - Recommend dynamic simulation & automated startup sequence

Preferred Option

- **Replacement shell with non-fixed wall**
  - "In-kind" replacement → easy permit approval
  - Cost comparable to grinding-off old tray support rings
  - Minimal down-time for column vessel replacement (setting & tie-ins)
  - Make use of existing infrastructure (foundations, piping, etc.)
  - Ensuring good liquid distribution outweighs cross-wall leakage concerns
  - No body flanges to leak (high T & vacuum service)
- **Additional studies to determine debottlenecking potential**
  - Column hydraulic limitations
  - Reboiler & Condenser capacities
  - Dynamic simulations by vendor / university partner
- **Keep old product column in place for other products; allow DWC to operate as (debottlenecked) conventional olefin tower if desired**
  - Option to feed both sides of wall & shut-off side-draw
  - Allay plant concerns (If it doesn't work, we won't be able to make current products)

Conclusions

- **Criteria when DWC retrofit may be appropriate**
  - Business driver for upgrading existing distillation train
  - At least 2 columns in conventional sequence at comparable pressure
  - Larger of existing columns has old / inefficient internals
    - Higher hydraulic capacity / more theoretical stages if modernized
  - Opportunities for energy savings
  - Reliable simulation of process
- **Practical considerations**
  - Mechanics / logistics of revamp; is a replacement shell justified?
  - Permit / regulatory implications
  - Additional instrumentation & controls needed: include in scope
  - Tricky hydraulics; leave final design to DWC experts

QUESTIONS ?

444452 The New Wave of Alkylation Activity in the U.S. - Economics and Options

Abstract:

There is a new wave of interest in alkylation projects currently in the U.S. Butanes are in cheap supply and octane value is high. Economics are driving projects towards alkylating available olefins streams as a way to bolster octane in the gasoline pool and to maximize profit. There are several alkylation projects that have been announced or are under study. These projects range from revamping existing units to constructing stand-alone grassroots facilities that are not located at a refinery, but near key NGL processing and storage facilities. This paper will review the economics driving this interest and provide an overview of the various alkylation unit configurations that are under study. Burns & McDonnell has significant experience with alkylation projects and specific examples will be given to provide a clear understanding.

444053 Recovery of Light Olefins and NGLs from Refinery Off-Gas

Abstract:

Abstract

Recovery of Light Olefins and NGL's from Refinery Off-Gas

Kathy Bigger*, Danielle Goldbeck

Linde Process Plants, Inc., Tulsa, USA

*corresponding author: kathy.bigger@lppusa.com

Refinery Off-Gas can contain products such as light olefins and NGL's. The Off-Gas streams from various refinery units are normally collected in a common system and burned as refinery fuel gas. Linde Process Plants, Inc. (LPP) has a proven technology for the recovery of valuable light olefins and NGL's from Refinery Off-Gas. Moderately pure hydrogen can also be recovered. High recoveries of ethylene and ethane (95+%) and very high recoveries of propylene, propane, and C4+ can also be achieved.

There are many technical challenges in designing a Refinery Off-Gas recovery process. In addition to the challenge of minimizing capital cost and achieving high energy efficiency, it is also essential to be able to clearly define the variations in Refinery Off-Gas composition and volumes that can occur due to changing conditions in the various Refinery units. It is imperative to design the Refinery Off-Gas recovery process to be robust enough to handle these variations. Accurately determining what contaminant species are present in the Refinery Off-Gas and what type of contaminant removal processes are required is also a critical component of the process design. Properly addressing each potential contaminant is important both due to downstream product specifications as well as protecting the cryogenic processing equipment in the Refinery Off-Gas Recovery unit. Finally, each location will have a unique set of requirements for downstream processing of the recovered olefins and NGL's, which will depend upon the type of Refinery units that are within the complex as well as potential customers across the fence.

In this paper, LPP will provide an overview of how cryogenic processing can be used to recover light olefins, hydrogen, and NGL's from Refinery Off-Gas and will also detail some challenges and solutions in designing a Refinery Off-Gas Recovery unit.

443633 A Solution for the Increasing Amine Cycle in Refinery Operations

Abstract:

Tramp amines are a growing issue for the refining industry. They impact desalter performance, increase the potential for overhead corrosion, decrease neutralizer effectiveness and increase loadings on the wastewater treatment plant. Levels as low as 5 ppm of certain tramp amines can have a dramatic impact on salt points and associated corrosion in the atmospheric section.

Desalting problems begin with an increase in the aqueous phase pH driven by the tramp amines. The higher pH results in increased amine partitioning to the oil phase. It also stabilizes the emulsion at the oil/water interface and extends the rag layer. Water carryover increases to the crude which results in increased chlorides in the crude tower overhead. Oil under carry is also increased which impacts the wastewater treatment plant. Increased corrosion of overheads is due to more than the higher chlorides from water carryover. Excess amine carryover buffers the overhead and makes it more difficult to control and maintain desired pH targets and thereby negating the effectiveness of neutralizers. Also the excess amines combined with the reduction of operational flexibility creates very high salt points. Salt deposition problems can result when the Salting Point exceeds the tower top temperature.

Lowering levels of tramp amines is very difficult because there are multiple sources of amines. In addition to H_2S scavengers and corrosion inhibitors in the incoming crude oil, amines are introduced through the sour water stripper, amine scrubber and from corrosion inhibitors used in steam production and refinery corrosion management. And with increased water recycling, amines and ammonia levels in the stripped sour water will tend to increase over time. This process can be described as the tramp amine cycle.

Conclusion

This paper will present a detailed overview of the cost of the amine cycle in the refinery, the solutions to break the amine cycle and provide supporting test and case study data.

16AIChE
Spring Meeting, Houston, TX

What has happened to the Enthusiasm for Computer Control in the Chemical and Petroleum Industries?

Charles Cutler, PhD-ChE
Cutler Technology Corporation
157 Westcourt Lane
San Antonio, Texas
Charles.Cutler@Cutler-Tech.com
Don Golden, MS-EE
CTC Consultant
Don.Golden@Cutler-Tech.com
James Cutler, BS-Math
CTC Consultant
James.Cutler@Cutler-Tech.com

Paper 147a

Prepared for Presentation at
American Institute of Chemical Engineers
2016 Spring Meeting
11th Global Congress on Process Safety
Houston, Texas
April 10-14, 2016

UNPUBLISHED

Keywords: Real Time Optimization, RTO, Model Predictive Control, MPC, Multivariable Control, Non-linear Optimization

Abstract

The entire Process Optimization and Control Activity are in decline. The enthusiasm for the activity relative to 20 years ago has almost vanished. The evidence of this is all around us. The Real Time Optimization and Control session of the 2014 AICHE meeting in New Orleans was almost unattended. 20 years ago it was a standing room only event. Further, 20 years ago there were many publications that described the profitability and success of Real Time Optimization (RTO) and Model Predictive Control (MPC). Today there are essentially none! What has happened? A number of problems have surfaced that the writers believe has led to the loss in interest in the Process Industries. Some of these require technological solutions, while others require management solutions. It is worth understanding these problems considering the money that is being left on the table. 20 years ago it was not uncommon to see control projects on a major process unit producing 4 to 8 million dollars per year.

Introduction

The principle author has been closely associated with computer control in the process industries. In 1963 in Houston, he was a member of the first Shell Oil computer control project to put a control computer to optimize and control a fluid catalytic cracking unit. While in Houston, he attended the University of Houston where he eventually wrote his dissertation "Dynamic Matrix Control an Optimal Multivariable Controller with Constraints". Later in Shell's New Orleans refinery and petrochemical complex, he was the project leader for computer control of a crude unit, a cat cracker, and two olefin plants. Also in New Orleans, he was a process manager and the technical manager for the complex. He was moved back to Houston where he managed the Instrumentation and Control department in Shell's Westhollow Research Lab for two years, This assignment was followed by nine years as the senior engineering manager for Instrumentation, Control, and Optimization for Shell USA. In 1984, he founded Dynamic Matrix Control Corporation. He and his associates developed the first commercial software to do Model Predictive Control and Real Time Optimization using open equation modeling. Both software packages were very successful and proved to be very profitable for clients. In 1996 he sold his company to Aspen Technology and after spending two years as a senior consultant to Aspen, he retired. In retirement he figured out how to remove the PID dynamics from a Finite Impulse Response (FIR) model and obtained a patent on it. In 2002 he started a new company to take advantage of the patent. He has observed the rise and fall of interest in computer control and optimization. He has also observed the rise and fall of the profit associated with this interest. The paper addresses what appears to be the problems with the decline in interest and profitability of computer control.

Summary

Control Software is viewed as a commodity and it has not improved significantly in recent years. The reality is that this perception is not true. Unfortunately, most of the people supporting Optimization and Control Systems do not know how to use the software they have. Further, these people do not understand that management needs to routinely see a return on their investment. Today, over 90 percent of the time, the control engineers supporting the optimization and control activity make the controllers go where operations feels comfortable. Current economics are not considered. At best the goals set by operations are based on economic calculations in the not too distant past; however, experience has shown that these calculations were made in most cases years ago and not in prior weeks or months.

One major control company advocates not using real economics in the MPC's objective function. It has a heuristic scheme that depends upon the knowledge of engineers to set the LP costs. The rationalization for the heuristic scheme is that it approximates what engineers do today, which is true; however, this practice is a major factor in the poor performance of todays' MPCs. It assumes that the engineers and operators can specify the manipulated variables in the order of their economic value. For a controller with 20 MVs and 49 CVs, there are 10^{30} possible combinations of active constraints. This is a number of possibilities that is incomprehensible to human beings.

From CRC Standard Mathematical Tables 11Th Edition pp342:
If M is the number of combinations of n things taken p at a time
$M = n!/(p!(n-p)!) = 1.250*10^{30}$ Where n = 20 MVs + 48 CVs = 68 p = 20 MVs
$n! = 2.480*10^{96}$ $p! = 2.432*10^{18}$ $(n-p)! = 8.159*10^{47}$

Even for the rare case where the economics are put into the controller, most engineers do not understand that the variables in a plant scheduling Linear Program (LP) are not in the same context as the variables in the multivariable controller, i.e. The independent variables in the controller are not exactly the same as the independent variables in the LP model. To put the correct economics into a controller, requires the control engineer to be familiar with the technology in the LP and the structure of the controller's model.

Most people do not understand what optimization really means. This includes engineers, managers, and operators. For example, the optimum economic operation of a process will use up all of its degrees of freedom at some set of constraints for a linear programming system. For a controller, the number of degrees of freedom is equal to the number of manipulated variables. A while back, C. Cutler taught an advance control course to the senior control engineers of a major oil and chemical company. Only two of the twenty-five attendees understood the relationship between the number of manipulated variables, the degrees of freedom, and the number of active constraints.

In the last 10 years, we have not found a client who routinely updates the economics in their controllers. This is one reason there are no publications. Without current economics in the optimizers, it is not possible to talk about their profitability. Further, management cannot be shown the profitability of the optimizer. Another pattern of behavior has evolved over the last 15 to 20 years that affects the profitability picture. Control engineers have gravitated into building small controllers with 3 to 6 manipulated variables that are easy to understand. They are basically controllers that only deal with a limited set of simple objectives. A profitable controller for a typical process unit will have between 25 and 50 manipulated variables and between 50 and 100 dependent controlled variables. The scope of the controller needs to be large enough to capture the essential benefits from the process.

Mismanagement of computer control by most companies is common. One of the underlying problems was and is managements' strategy for developing people to become managers. The strategy is to expose high potential engineers to as much of the company's business as possible in the shortest time. It takes 2 to 3 years for an outstanding engineer to understand the technology involved in realizing high profits from RTO and MPC. The short exposure for computer control provided in the management development programs, leaves the manager poorly equipped to make wise decisions with regard to real time computer control. The level of ignorance of the potential for computer control is sustained by the development system for managers.

An optimization and control system should be the repository of a company's best technology and operating experience. The best technical people in a company should keep technology in the control computers up to date. In general, the research and technical departments in companies are satisfied with producing reports that get placed in file cabinets and on book shelves and little net value is added to a company's bottom line. Putting new technology into a process control system is where the rubber meets the road. New technology is proven or rejected by the simple observation of whether the process went to the predictions the technology made when the computer moves the process.

A career in process optimization and control is no longer attracting the best engineers because it is viewed as a dead end street from an advancement point of view. To be a good control engineer requires the engineer have outstanding process engineering skills, familiarity with computer programming, and knowledge of the state of the art software. Inadequate training of support engineers is common. The best training that we have observed in the last 50 years is one that uses the apprenticeship approach to the training. High quality class room training, followed by case studies, which are followed by working from one to two years on projects with an experienced control engineer.

Why is the preceding analysis relevant? If management is not continuous reminded of the value to his company of the optimization and control technology, he will stop supporting it and it will eventually die from lack of care. We believe many companies are on the downhill side of giving enough support and attention to optimization and control; therefore, it is just a matter of time before they turn off the computers and reassign the people. In general, management does not understand the requirements of optimization and control. The very best technical people in a company should be the heart of maintenance support for the optimization and control. The optimization and control systems are the best way they have to apply their technology and experience. Maintenance support should not depend on inexperienced engineers and poor performers.

Refer to the following paper for details on the calculations for LP cost for a MPC controller:

Gulf Publishing, Aug 25, 1998, no. 172 p.2,

"LP integrates economics into dynamic matrix control" by

R. C. Sorenson (Chevron Research) and Charles Cutler

General Objectives and Benefits of an Advanced Process Control System for a Refinery

By

Neda Felorzabihi, David H. Jones and Javier Vazquez-Esparragoza

KBR Technology

Paper ID# 443419

Prepared for Presentation at American Institute of Chemical Engineers

2016 Spring Meeting

11th Global Congress on Process Safety

Houston, Texas

April 10-14, 2016

1 ABSTRACT

The objective of the paper is to present the design basis for the Advanced Process Control (APC) applications in a typical oil refinery, and a survey of APC potential benefits for applicable units in the refinery. It is also intended to provide a common understanding of the design requirements between all the project team members participating in an APC project.

2 SCOPE OVERVIEW

Advanced Process Control is a critical component of "best-in-class" operation, enabling significant improvements in asset utilization and energy consumption for most processes. Specifically, APC is any process control system application that has functions beyond those commonly associated with basic or regulatory controls.

Advanced Process Control could be defined as multivariable, model-predictive control (MVC) consisting of a suite of applications that use either linear or non-linear dynamic models to design, implement and maintain MVC strategies in order to effectively improve the plant stability, increase plant safety and maximize plant profitability.

In this article, the general functionality is presented for an APC system design developed for a FEED (Front End Engineering Design) phase of a 400,000 barrels per day (BPD) grassroots Refinery Complex.

The refinery was separated into several units for APC application as listed below:

- CDU, crude distillation unit
- VDU, vacuum distillation unit
- Naphtha stabilizer
- NHT, naphtha hydrotreater
- CCR, Continuous catalytic reformer unit
- Isomerization unit
- Hydrocracker unit
- Distillate Hydrotreater
- Other units

Figure 1 shows a picture of a section of a Refinery.

Figure 1. An overview of a refinery

As an example of general control philosophy, the following items are included in this article:

- APC General Control Objectives
- APC General Control Strategy
- Typical Benefits per plant and global

The estimated potential cost benefits from the APC applications are included in the scope of work, and the expected cost for a refinery wide APC is presented. Proper implementation and maintenance would be necessary to achieve and maintain the benefits.

3 APC DESIGN BASIS AND PHILOSOPHY

3.1 APC Philosophy

Some of the philosophies and design principles that need to be followed for a specific refinery are listed below. The APC design philosophy document will provide a common understanding between all the stake holders and other project teams including the refinery owner, the engineering contractor, and the APC contractor.

- All APC application decisions and estimates should be based upon the total cost of ownership with emphasis on lifecycle management of the system.
- The design of APC is value driven. The APC application design and specification should be driven by business needs and value generation.
- The APC application system architecture should be designed such that interfaces are economically justified.
- APC lifecycle management processes and program management would be designed in order to support value - sustaining objectives.
- The high-level business objectives guiding the design for APC are to provide significant additional value above basic process control in the following areas:
    - Improved process agility
    - Reduced time to achieve stable operations after upset or operational mode changes.
    - Increased throughput
    - Improved product quality

- Improved product yields
- More efficient energy utilization
- Improved operational stability
- Longer plant uptime by tighter and smarter controls

- The design of APC should be be consistent with existing specifications, standard, business infrastructure and best practices.
- The design of APC should minimize technical, commercial and project schedule risk and maximize the potential for success.
- APC is to be based on proven technology.
- APC would account for and be based on organizational alignment by adopting agreed business processes and workflows integrated with roles and responsibilities of the established organizational structure of the Refinery in the use and maintenance of APC applications.
- APC implementation would secure continuity of engaged resources.

Figure 2 shows a typical APC project implementation timeline. In general, for a grassroots plant the implementation starts after 4 to 6 months, when plant is stable and running smooth operations.

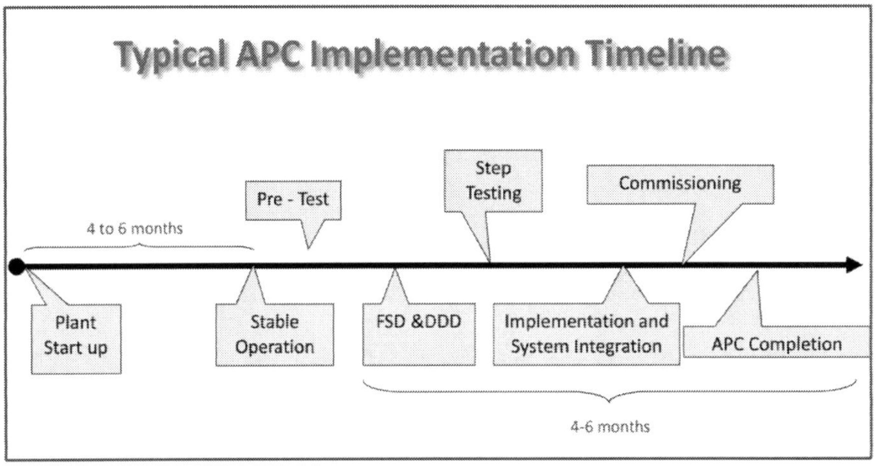

Figure 2. APC Implementation Timeline

- APC software used should be capable of generating APC Key Performance Indicators (KPIs) which will accurately reflect the effectiveness of the APC system, and not just uptime. The following table show some of the KPIs that can be implemented on the APC system:

Table 1: **Key Process Indicators**

KPI	**Description**
Controller ON/OFF	To be used to indicate if MVC is turned on and being used t process unit.
Service Factor	Historical trend of the ON/OFF status for a controller.
Utilization factor	To measure how effectively the MVC is being utilized during th in-service.
Constraint analysis	Charts showing the number, distribution and identification constrained Control Variables (CV's) and Manipulated Variab over a time period.
CV's/MV's statistics	This KPI will include standard deviations for CV's/MV's, CV's/I errors and CV's/MV's limit errors.
MVC model quality	Verification of model quality by measuring unbiased predi prediction errors.
CVs/MVs limit change frequency	Track number of times that operator limits are changed over a period. Excessive changes indicate trouble.

3.2 APC Architecture

The overall Refinery Automation System Architecture should comply with the four level architecture model of ISA-95 that is composed of business level Enterprise Systems (Level 4), Operations Management (Level 3), Process Control (Level 2), and Field Instrumentation (Level 1, 0).

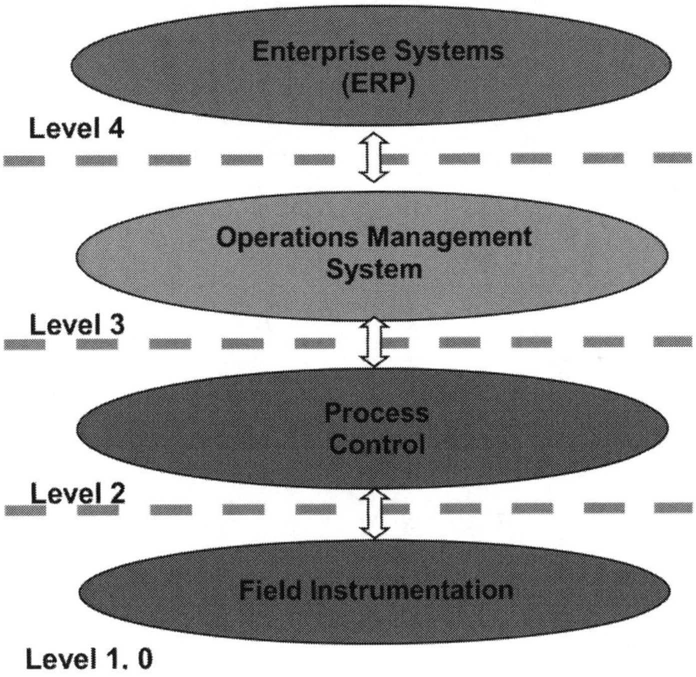

Figure 3. ISA S95 Automation

Conceptually, the APC applications may reside between the Operations Management System and the Process Control levels. For the purpose of APC front end engineering design (FEED) for a grassroots Refinery Complex, the following assumptions were made:

- Computer servers will be used for the APC applications
- There will be a minimum of one APC application server for each operational area where APC is applied.
- All of the APC application servers would be housed in a Central Control Room.
- Other network configurations such as Virtual Machines may be considered when implementing APC.

Figure 4 presents an example of a typical architecture for a site with three units.

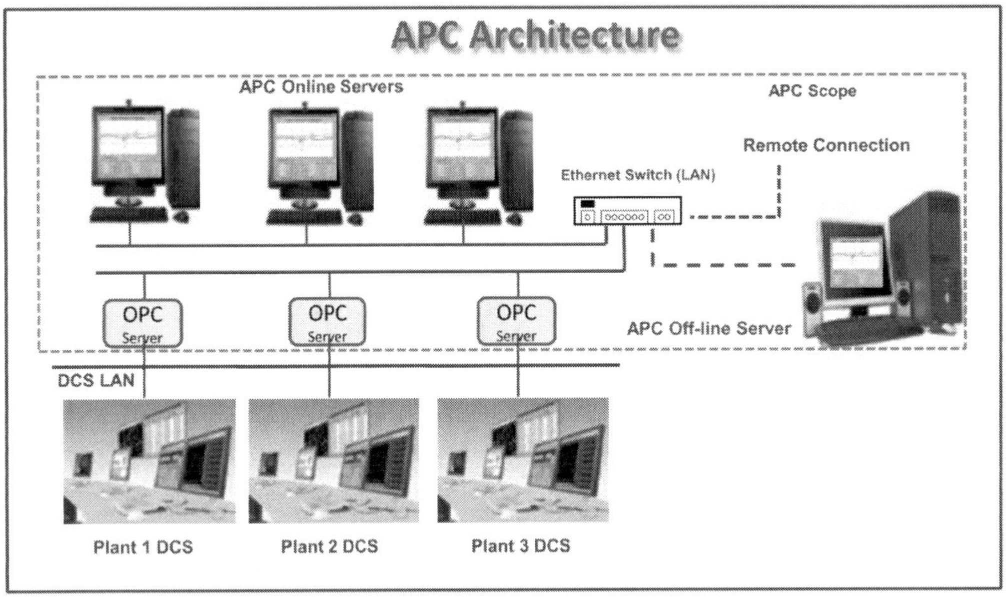

Figure 4. Typical APC architecture for a refinery with multiple units.

3.3 APC Design

3.3.1 Selection of Controller Type

Linear controllers have proven satisfactory in most continuous refinery processes and specifically for those processes in the scope of the refinery. Linear controllers are the preferred solution for processes that are linear or moderately non-linear over the normal operating region of the process.

3.3.2 Predictive Functionality

The controller would have the capability to take past information from the process and use it to predict future behaviour of all controlled variables. Process feedback should be taken into account to reconcile any difference between the predicted values and actual measured values.

3.3.3 Steady-State Optimization Functionality

The controller should have the capability to calculate an optimal local steady-state solution to the control problem. Typically, either a linear program (LP) or a quadratic program (QP) is used to come up with the manipulated and controlled variables' steady-state solution based on the current process values, predictions, constraint limits and economic factors.

3.3.4 Move Calculation Functionality

Given the steady-state targets obtained from the local LP optimizer, the controller technology must have the means to implement calculated dynamic moves to achieve the desired values.

3.3.5 Data Requirement

All the necessary data and information which will quantify the actual performance of the process relative to its operating targets and the control objectives is to be identified. The following types of data should be collected in sufficient quantity to be statistically significant and consistent:

- Operating targets.
- Actual values achieved.
- Process operating constraints and limits, reasons and values.
- Availability of process (stream days) and reasons for outages or restricted operation.
- Measurement availability.
- Discrete events.
- Key performance indicators.

- Future plans and timing for process modifications and their potential impact.
- Production operating target changes which may be planned.
- Future operational flexibility requirements.
- Economic driver data and long term plan.

3.3.6 Data Validation

The APC application(s) are to include methodology for data validation of field inputs.

During data collection, the validity of the data should be evaluated. Criteria to evaluate validity are, for example:

- Do flows add up to satisfy material balance?
- Are energy balances correct?
- Any output resulting from the data should be analyzed carefully to make sure that it is reflective of actual conditions.
- Has an unmeasured disturbance occurred during the data collection? If such a disturbance occurs during a test run, that data must be discarded.

Data validation criteria are to include "bad" status, high limits, low limits, rate of change limits, switch/flag points.

The validation of directly measured variables such as flow, temperature, and pressure are generally performed in the DCS, while inferred variables and chromatographic analyses are validated in the advanced control programs.

The APC strategy would be designed to manually and automatically turn off all or part of the application and return to the DCS regulatory control if key variables become "bad" or "uncertain'.

Depending of the nature of a problem, automatic or manual application shutdown should return to the regulatory control system without disturbing the process or requiring additional intervention by the operator. All application shutdowns, shedding or degradation should be alarmed or logged, depending on the severity or criticality of the event.

Messaging and alarming on the status of the advanced control and optimization operations would be provided on the appropriate DCS consoles.

3.3.7 Controller Model Development

The data used to develop a controller model would span the normal operating region of the process.

Preliminary model development would be done through step tests. The size and number of repetitions would be selected and monitored to ensure that the quantity and quality of collected data are sufficient to facilitate good model development.

Controller models developed for APC applications would be based on the first principles of the process and fundamental cause and effect relationships.

Models would use the simplest model form that achieves the desired accuracy, since a simpler model is easier to maintain and more robust.

Inferential variables would be developed and implemented on-line before step tests, wherever possible,

3.3.8 Controller Matrix Size Considerations

The APC provider as an expert would consider the best practices available for APC rather than what is known as common practice. Controllers would be sized such that they capture the business drivers of the process using the smallest matrix necessary. Only meaningful responses would be included in the controller matrix.

The design of individual APC controllers with more than 30 MVs would be avoided. Also, for large units such as CDU. VDU and a feed preheating furnace, separate APC controllers would be considered.

3.3.9 Controller Configuration

Controller frequency would be selected based on the process settling time, process dynamics, and the frequency and severity of disturbances.

Consideration would be given to the presentation of data and operation of controllers, especially for very large controllers. Logical groupings of common process variables would be considered in order to enhance operator/engineer effectiveness.

3.3.10 Calculated Variable Principles

The status of calculated variable inputs must be propagated to the output. For example, in the calculation of an average temperature, the status of the calculated average temperature point would be questionable if any of the inputs to the calculation were questionable.

3.3.11 Controller Tuning

Initial APC controller parameters for commissioning and field tuning could be established from APC simulations if available.

3.3.12 APC Validation and Training

An Operator Training Simulator (OTS) would be used for APC validation and training, if available

3.3.13 Inter-Unit Dependencies and Constraints

Due to potentially competing demands for the different products, it may be necessary to communicate operating targets and capacity constraints across process units as they change. For example, a unit may run up against a capacity constraint that limits production in an upstream unit. APC design would include capability to handle such dependencies.

3.3.14 Future Considerations

The APC applications would be designed such that additional CVs and MVs can be added in the future. The hardware would have sufficient capacity to add additional controllers as they become necessary.

3.4 Inferential Variables

3.4.1 Inferential Variable Development

Use of continuous online analyzers would be considered wherever possible. A combination of an inferential variable and an online analyzer provides the most reliable control. If both are available, the inferential would be used directly as the key control variable. This practice overcomes the deadtime, delays, slowness, routine calibrations, and troubleshooting associated with analyzers.

Special care would be taken in selecting the process data used in developing the inferential models. Process data should be properly aligned with the lab analysis results and/or analyzer reading.

The data used to develop an inferential model would span the normal operating region of the process.

3.4.2 Inferential Variable Application and Maintenance

Inferential variables would have a periodic validation mechanism to maintain accuracy as process conditions change.

Best practices for inferential calculation validation would be identified, e.g. against lab data or against analyzers.

All signals associated with the inferential would be validated: input, feedback, and output signals. Analyzer feedback would incorporate additional validation including but not limited to freeze and spike detection.

3.5 Process Control System Principles

3.5.1 Interface with the Basic Distributed Control System

The APC applications and the DCS must be configured such that transfer from DCS control to APC control and vice-versa are bumpless.

3.5.2 Operator Interface

Plant operators would control the APC applications from screens on the DCS. For all the control applications that involve constraints (variables to be kept within a high and low limit), the application and the operator interface would be designed to allow the operator to easily activate or deactivate the constraints through a field in the DCS interface screen. Also, the operator interface would show when a particular constraint is being considered as being active by the control algorithm. The operators would be responsible for turning APC applications on and off, and would have the ability to remove individual manipulated variables from service.

3.5.3 Engineer Interface

The APC system would include an Engineering Workstation that would serve as the interface of the Refinery's engineers to the APC system. The engineers would be able to make changes to maintain and monitor the system from this station. Among other things, this station would allow the engineers to modify the applications configuration and tuning, access historical data, analyze operating data to develop and validate the dynamic models used by the multivariable predictive control applications, etc.

3.5.4 Regulatory Control Configuration and Tuning

The following principles would guide the configuration of regulatory controls to accommodate APC:

- Disturbances would be rejected at the lowest possible control level.
- DCS controllers receiving setpoints from APC would be tuned for an appropriate level of responsiveness and performance.

3.6 APC Software

An APC proven software would be used for the purpose of coordination of lower level controllers and optimization of the process.

A performance monitoring tool (CPM) or equivalent would be provided for monitoring and diagnosis of APC. The performance monitoring tool would identify, improve and sustain APC control performance.

3.7 Hardware Requirements

APC would be implemented in a hardware platform that is supported as a standard offering by a computer hardware manufacturer or supplier.

4 APC FUNCTIONAL SCOPE

4.1 General

The functional scope of APC for the refinery is to provide comprehensive solutions that are consistently applied across the entire complex in the areas of:

- implementation strategy and deliverables
- software platform
- hardware platform
- interface with the basic process control system
- operator and engineering user interfaces

Following is a simplified representation of the APC components that would be provided and their potential architectural hierarchy. The diagram below represents the conceptual system for one plant area. It is envisaged that each process area would have similar APC architecture at Level 2, connected to plant wide OMS architecture for the communication of operating targets, constraint priorities, economic values that may be required for operation of the APC systems.

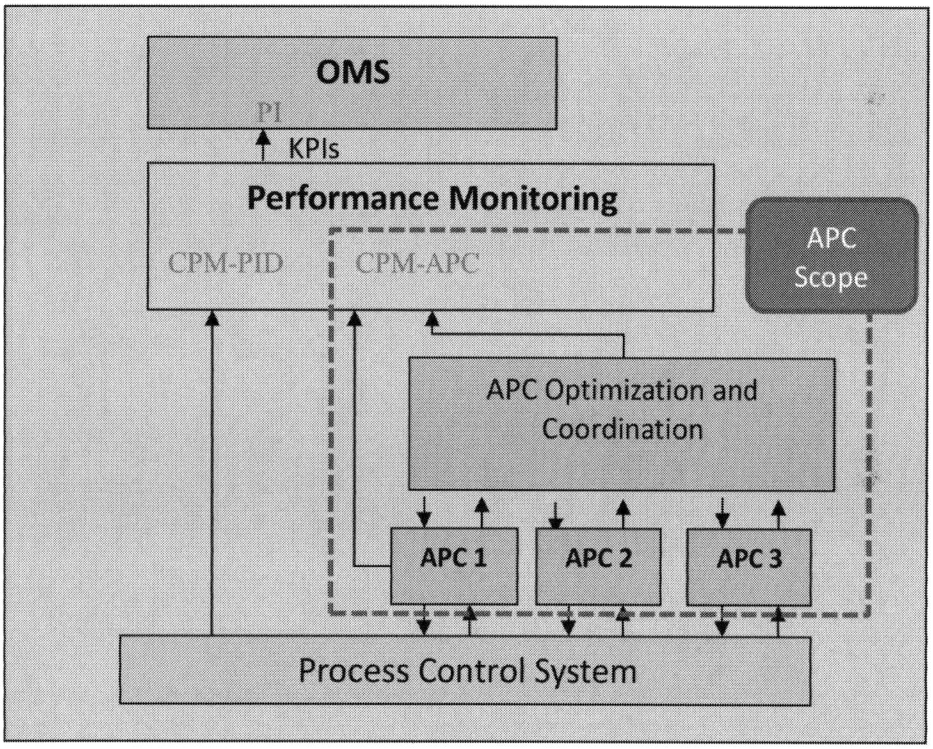

Figure 5. APC Conceptual Architecture & Hierarchy

. The table below presents a summary of the typical contents which would be delivered as part of the Functional Design Specification documents (FDS).

These documents consist of:

- APC General Specification
- APC Project Specification for each APC area

Table 2. APC General FDS Table of Contents

Section	Description

Section	Description
1	INTRODUCTION
1.1	Scope Overview
1.2	Project Overview
1.3	Conflicts and Deviations
1.4	References
1.5	Codes and Standards
1.5.1	Industry & Other Standards
1.6	Definitions
1.7	Language
1.8	Acronyms& Abbreviations
1.9	Instructions
1.10	Responsibilities
2	FUNCTIONAL REQUIREMENTS
2.1	General
2.1.1	Selection of Controller Type
2.1.2	Predictive Functionality
2.1.3	Steady-State Optimization Functionality
2.1.4	Move Calculation Functionality
2.1.5	Controller Model Development
2.1.6	Controller Configuration
2.1.7	Data Requirement
2.1.8	Data Validation
2.1.9	Inferential Property Principles
2.1.10	Basic Process Control System Principle
2.2	Operator Interface
2.3	Engineer Interface
2.4	Performance Monitoring
2.4.1	Key Performance Indicator (KPI)
2.5	Communications with Other Systems
2.6	Future Considerations
3	APC ARCHITECTURE
6	BENEFITS
74	HARDWARE SYSTEM REQUIREMENTS
5	SOFTWARE REQUIREMENTS
6	PROJECT EXECUTION
7	DOCUMENTATION
8	TRAINING
9	LIFECYCLE MANAGEMENT

4.2 APC Functional Specification Example- CDU/VDU

For example, in the Crude Distillation Unit (CDU)/Vacuum Distillation Unit (VDU), crude oil is processed in order to produce various products for further processing in downstream units.

The CDU is designed to fractionate crude oil into four main products, which are naphtha-minus, kerosene, diesel, and aromatic residue, at near atmospheric pressure. Besides the above mentioned products, brine and sour water are produced.

In the VDU unit, atmospheric residue coming from the CDU unit is separated into four main streams, viz. Vacuum Diesel Oil (VDO), Light Vacuum Gas Oil (LVGO), Heavy Vacuum Gas Oil (HVGO) and Vacuum Residue (VR). Besides the above mentioned products, VDU produces vacuum waste gas, sour water, slop oil and slop cut as by products

This Specification together with the APC General Functional Specification is to define the minimum requirements for Multi-variable model predictive Control (MVC) for the specific Refinery Crude Distillation Unit (CDU)/High Vacuum Distillation Unit (VDU) and the APC hardware /software platform. Together with referenced Client Standards, Specifications, and Project documents, it also defines the overall scope of work for implementation of APC at the Refinery complex.

4.2.1 APC Control Strategy

The Advanced Process Control package in association with on-line inferentials and cut point technology to maximize the benefits from CDU/VDU Process by:

- Dynamic estimation of the True Boiling Point (TBP) curve of the crude using heat and mass balance around the crude unit
- Side stream cut point calculation (using estimated TBP)
- Product quality estimation using on-line analyzer, inferred properties or a combination
- Adjustment of pumparounds to ensure adequate heat removal, maximum heat recovery and desired product separation, subject to loading constraints

Traditional Advanced Process Control will be commissioned after plant start-up and all other advanced applications would be executed after performance test and establishing a process performance baseline by the Refinery.

4.2.2 APC Preliminary Functional Design

The following table summarizes the CDU preliminary design. The APC Control Strategy for CDU/VDU units is separated into 6 sets of controllers (sub-controllers) which are linked and controlled by a supervisory controller. These sub-controllers are:

- Crude Feed and Preheat Section Sub-controller
- Crude Furnace Sub-controller
- Crude Column Sub-controller
- Naphtha Stabilizer Sub-controller
- Vacuum Furnace Sub-controller
- Vacuum Column Sub-controller

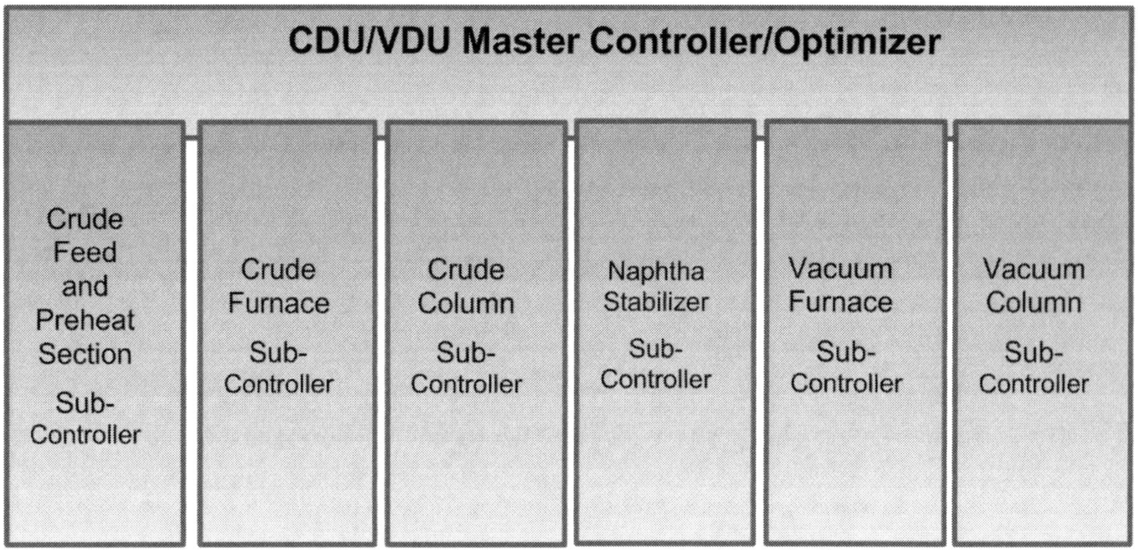

Figure 6. **Conceptual APC Diagram for CDU/VDU Application**

The control strategies in the sections below do not limit the scope of the applications. The actual CVs and MVs will be determined at the time of the APC implementation. The strategies described are a preliminary APC functional design for the refinery CDU/VDU.

4.2.2.1 Crude Feed and Preheat Section Sub-controller

The following proposed APC control strategy is for the Crude Feed and Preheat Section:

- The APC control strategy would raise the crude feed rate to achieve a target feed rate until it reaches a plant capacity constraint. This control strategy would tie into the product quality controls. It would give stable responses to feed rate quality and state changes.
- The APC control strategy for the preheat section would maximize the combined outlet temperature of the heat exchangers from two parallel trains while maintaining pressure drops. The heat recovery by the preheat section would be maximized without violating constraints on the preheat section such as: Desalter temperature, Furnace loads, etc.

4.2.2.2 Crude Furnace Sub-controller

The following proposed APC control strategy is for the Atmospheric Crude Furnace:

- The APC control strategy would address the following operating objectives:
  - Maintain the furnace coil outlet temperature at its desired temperature and minimize deviation from that value.
  - Operate at low excess air for higher furnace efficiency and operate at only slightly negative draft pressure.
  - When necessary overrule the first two objectives to keep the heater always within safe operating limits.
- The APC control strategy would maintain the COT (Coil Outlet Temperature) at its desired target by manipulating the fuel gas pressure or vacuum residue pressure or fuel oil flow rate to the burners. It should have a constraint monitoring and override feature. The following constraints should be monitored.
  - Bridge - wall temperature (BWT)
  - Tube skin temperature
  - Draft stack pressure
  - Stack oxygen concentration
  - Fuel burner pressure
- The APC control strategy for the Crude Furnace would be constructed as a constraint monitor, reading draft pressure and stack oxygen concentrations as two constraints and manipulating the stack damper to keep both constraints within safe limits.

- The APC control strategy for the Crude Furnace pass balancing would manipulate feed flows among passes to keep pass outlet temperatures reasonably close together while keeping the tube skin temperatures and pass flows (as two constraints) within safe limits.
- The APC control strategy for the Crude Furnace would control the low pressure steam (LPS) outlet pressure from Crude Furnace convection zone coils by manipulating the flow rate of LPS from VDU LCR LPS Crude Column Sub-controller

4.2.2.3 Crude Column Sub-controller

The following proposed APC control strategy is for the Crude Column:

- The APC control strategy for the Crude Column would maximize production of valuable products (e.g. kerosene, diesel) while maintaining all product specifications and would stabilize crude flow to the Vacuum Column by:
    - Controlling the column top temperature by manipulating column top pumparounds and column top pressure (CDU Overhead Accumulator top pressure).
    - Controlling superheated stripping LPS to Diesel ratio by manipulating the flow rate of LPS and diesel from Crude/Diesel Exchangers.
    - Controlling the product quality control as side cut distillation end point, flash point, viscosity etc. using MVC technology combined with inferential based quality models.
- The APC control strategy for the Crude Column would calculate the heat duties using temperature and flow measurements and maintain the heat duties by manipulating the pumparounds and/or return temperatures.
- The APC control strategy for the Crude Column would address several objectives on pumparound controls:
    - Stabilize pumparound such that the column is not disturbed upon changes in individual heat exchanger duties.
    - Respond quickly to disturbances on the column which can occur upon crude switching or furnace upsets.
    - Adjust internal reflux in the column such that a reasonable economic balance is achieved between recovery and product separation and tray loading does not exceed the flooding limit.
- The APC control strategy for the Crude Column would adjust the furnace COT controller set point and column top pressure to control the column bottom level and column overflash at its desired target.

4.2.2.4 Naphtha Stabilizer Sub-controller

The following proposed APC control strategy is for the Naphtha Stabilization Unit:

- The APC control strategy for the Naphtha Stabilization unit would maximize the production of valuable product e.g. stabilized naphtha,

4.2.2.5 Vacuum Furnace Sub-controller

The following proposed APC control strategy is for the Vacuum Furnace:

- The APC control strategy would address the following operating objectives:
    - Maintain the furnace coil outlet temperature at its desired temperature and minimize deviation from that value.
    - Operate at low excess air for higher furnace efficiency and operate at only slightly negative draft pressure.
    - When necessary overrule the first two objectives to keep the heater always within safe operating limits

4.2.2.6 Vacuum Column Sub-controller

The following proposed APC control strategy is for the Vacuum Column:

- The APC control strategy for the Vacuum Column would maximize production of valuable products (e.g. LVGO, HVGO) while maintaining all product specifications by:
    - Controlling the product quality control as sidecut distillation end point, flash point, viscosity, freeze point etc. using MVC technology combined with inferential based quality models.

- Controlling the Vacuum Column bottom temperature by manipulating vacuum residue quenched flow rate to the column bottom. .
- Controlling superheated stripping LPS to Diesel ratio by manipulating the flow rate of LPS and diesel from Crude/Diesel Exchangers.

- The APC control strategy for the Vacuum Column would calculate the heat duties using temperature and flow measurements and maintain the heat duties by manipulating the pumparounds and/or return temperatures.
- The APC control strategy for the Vacuum Column would address several objectives on pumparound controls:
  - Stabilize pumparound such that the column is not disturbed upon changes in individual heat exchanger duties.
  - Respond quickly to disturbances on the column which can occur upon column pressure or vacuum furnace upsets.
  - Adjust internal reflux in the column such that a reasonable economic balance is achieved between recovery and product separation and tray loading does not exceed the flooding limit.

4.2.3 Inferential Modules

The APC system design will include an Analyzers and Inferential Model Module. The Analyzers and Inferential Model Module includes the inferential models to estimate stream compositions which are used by the APC system but are not analyzed on-line or are analyzed on-line at a lower frequency than is required.

The module shall include one or more DCS interface screens to allow the operator to specify, for each inferred property, whether the inferential model should be updated and auto-corrected from laboratory data or from an on-line analyzer. The screen interface also provides the necessary fields for the operator to enter and/or see the latest laboratory analysis results.

4.2.4 Multivariable Controller Supporting Calculations

As part of the APC system the Contractor shall specify a Plant Performance Calculations module. The Plant KPI Calculations module shall provide valuable information about the operation of the plant. The calculation results shall be accessible from the Engineer Workstation personal computer and from the DCS screens.

5 ESTIMATED BENEFITS

5.1 Typical Benefits

Honeywell[1] reports the following typical benefits of APC for refining applications. Another reference[2] reports similar benefits for different units by implementation of APC.

Table 3. Typical APC Benefits - Honeywell

Unit	*Benefit [cents/bb*	*Benefit [$/yr]*
Crude Distillation (150 KBPD)	5-13	2.7MM – 7.0MM
Hydrocracking (70 KBPD)	13-30	3.3 MM – 7.5 MM
Catalytic Cracking (50 KBPD)	13-30	2.4 MM - 5.4 MM
Reforming (50 KBPD)	10-26	1.8 MM - 4.7 MM
Alkylation (30 KBPD)	10-26	1.1 MM - 2.8 MM
Isomerisation (30 KBPD)	3-17	.3 MM - 1.8 MM

KBR's analysis validates the above range of benefits for the indicated units. Based on the above benefits, it is clear that even assuming 5 cents/bbl benefit from the crude tower, the total benefit per year for the refinery is likely to be about $6.5MM. This amount of benefit would yield a pay out of about 1year for the entire APC implementation, even without considering benefits from any other unit.

5.2 Crude Distillation

5.2.1 Crude Unit/Vacuum Unit (CDU/VDU)

Using the range of benefits for Crude Distillation for a 400 KBPD unit, there would be between $6.6 million and $17.1 million in savings per year. This corresponds to about a 2-4% increase in throughput. Typical APC applications that are allowed to increase feed, lead to a 3-5% increase in feed. There seems to be no doubt that APC for the CDU/VDU unit is justified.

---

1 https://www.honeywellprocess.com/library/marketing/presentations/EmailAssets/AS-Egypt-Roadshow-APC-BusinessBenefits.pdf

2 Source Hart's Fuel Technology and Management, Nov/Dec 1996

5.3 Naphtha Complex

5.3.1 NHT Unit

A 1% increase in throughput for the NHT unit would result in an increase in profit of about $3 million per year. Units that are constrained in production rate by limited feed cannot claim APC benefits for production increases. In this case, however, there may be justification since sometimes the Naphtha can be used as a marginal feed to the unit.

5.3.2 CCR Reformer

The CCR Unit is more likely to be constrained by feed from other units. Any additional production due to increased Naphtha is likely going to be claimed by other applications. For this application it may be better to assume that the product slate can be improved. If there is a 2% decrease in the heavy reformate and that material is produced as light reformate, the increase in profit would be on the order of $600,000/year. A 2% reduction in fuel use would produce an additional $260,000/year. A 2% reduction in hydrogen would create a savings of around $230,000/year. This sums to $1,090,000/year. This is in line with the Honeywell number of $1.8M/year.

5.3.3 Isomerization

Reducing hydrogen by 2% would reduce costs by about $200,000/year. Energy savings of 2% result in about $60,000/year. The low end of the range from Honeywell is about $300,000/year. If all of the light naphtha and light reformate produced can be processed by this unit without APC, then the savings expected are closer to the low end of the range.

5.4 Hydrocracker

Using the cents/bbl numbers from Honeywell, the expected benefits from APC for the HCK unit is between $4.5M and $10M per year. This level of benefit can be achieved with a 0.5% to 1% increase in production.

5.5 Distillate Hydrotreater

In addition to the feed from the CDU/VDU unit, high sulphur diesel is also processed in this unit. If this flow is increased such that the DHT unit increases its throughput by 1%, the increase in profit would be around $5M/year.

5.6 Other Units

5.6.1 Sulphur Recovery

Axens reports that they have applied APC to Sulphur recovery. The benefits sited include more stable operation, reduction in energy usage and reduction of sulphur in emissions. The marketing material did not indicate a dollar value to the application.

5.6.2 Amine/Water Stripping

The recovery of acid gases leads to the generation of sulphur. Many of the process streams are sour and there is a large volume of material processed. While it is difficult to assess the economic benefit of APC, this area should be examined after plant start-up to see if it is a bottleneck to production or if better control will result in more sulphur being recovered.

5.6.3 Utilities

Most utilities units do not have APC on them. Cooling water, instrument air, fuel oil system, water treatment and fire water systems are low cost utilities and would be unlikely to benefit from APC. Boilers and power generation have been known to have APC applications in certain circumstances. At this site the power generation is controlled by an independent organization and would then be out of scope for APC.

Steam generation for the site would be expected to cost around $10 million/year. Some of the APC applications for process units are expected to reduce that somewhat. Some additional savings could come from better management of steam production and reduction in the amount of steam that is vented. If this resulted in a 2% reduction in steam production costs there would be a savings of $200,000/year.

5.6.4 Water and Effluent Treatment

Treating of streams that enter or leave the process is usually done with a package unit. Few of the valves are controlled by the DCS. There are quality constraints, but these are usually handled in a batch fashion. The cost of over treating a stream is usually low compared to the value of product streams. While there may be some benefit by reducing chemical usage costs, adding APC would be difficult and unlikely to have significant benefit.

6 CONCLUSIONS

The APC applications included in the project scope of work would be expected to have more than $20 million/year in benefits. Including additional areas may increase this by $300 thousand to $500 thousand. The expected total installed cost for a refinery wide APC is estimated to be around $5 million. Proper implementation and maintenance would be necessary to achieve and maintain these benefits.

Paper 147c : Steam System Control and Optimization

Doug White
Principal Consultant
Emerson Process Management
Houston, Texas

Speaker

Doug White

- **Principal Consultant**
- **PlantWeb Solutions Group**
- **Emerson Process Management**

Background: Many years experience designing, justifying, installing and commissioning advanced real time computer/ automation applications in the process industries.

EMERSON. Process Management

Typical Steam System Automation Problems

- Site was planning a major expansion which would have added additional boilers and users on common steam headers – how to handle boiler and other major equipment trips? Where should master controls be located and how should they be configured?
- Site(s) had excessive boiler trips due to process upsets – i.e. process equipment trips; process cycles generating major fuel gas composition changes
- Site was attempting to maximize on site power generation but header pressure instabilities limited generation

Found lack of generalized design procedure for these types of complex control problems

Outline

- Process Industries Energy and Environmental Background
- Large Site Steam System Overview
- Steam System Controls
- Procedural Automation Background
- Case Study
- Summary

Process Plant Emission Regulations - US

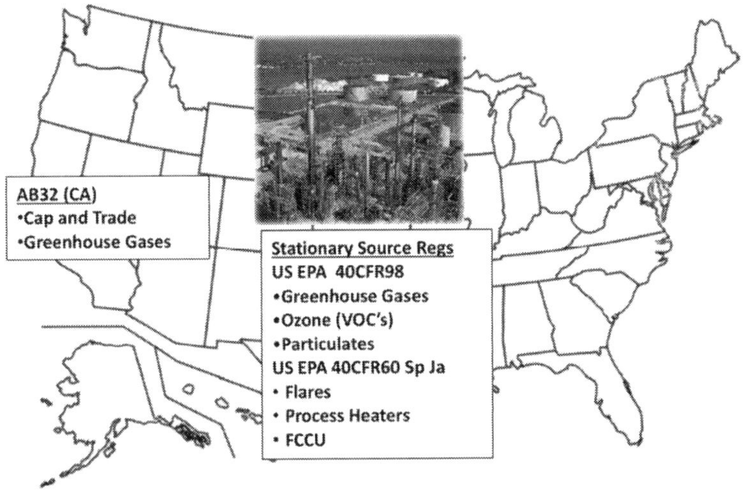

3/28/2016 16AIChE Spring: Paper 147c 5

Typical Complex Site - Oil Refining Energy Flow

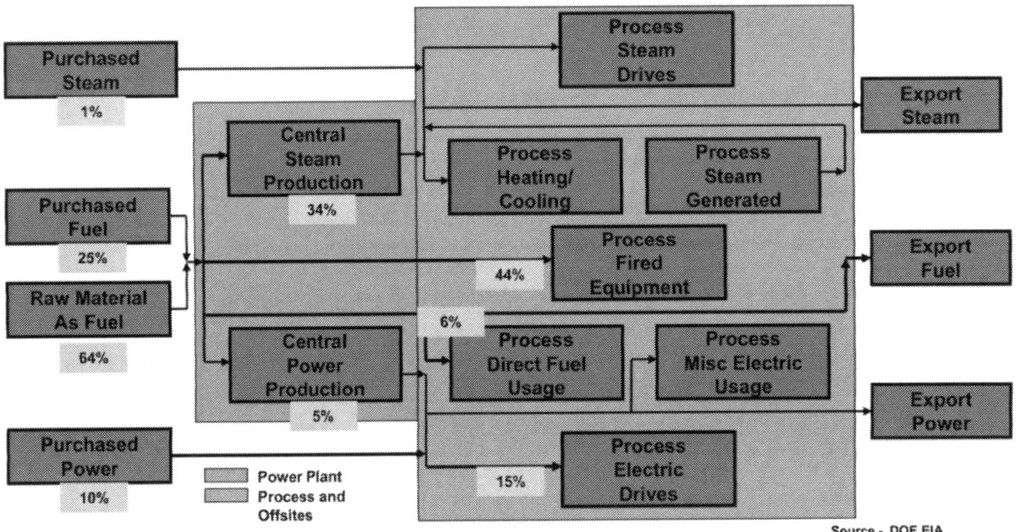

%-Equivalent BTU basis (Including losses) on total input

Source - DOE EIA

3/28/2016 16AIChE Spring: Paper 147c 6

Typical Refinery/ Chemical Complex Steam Systems

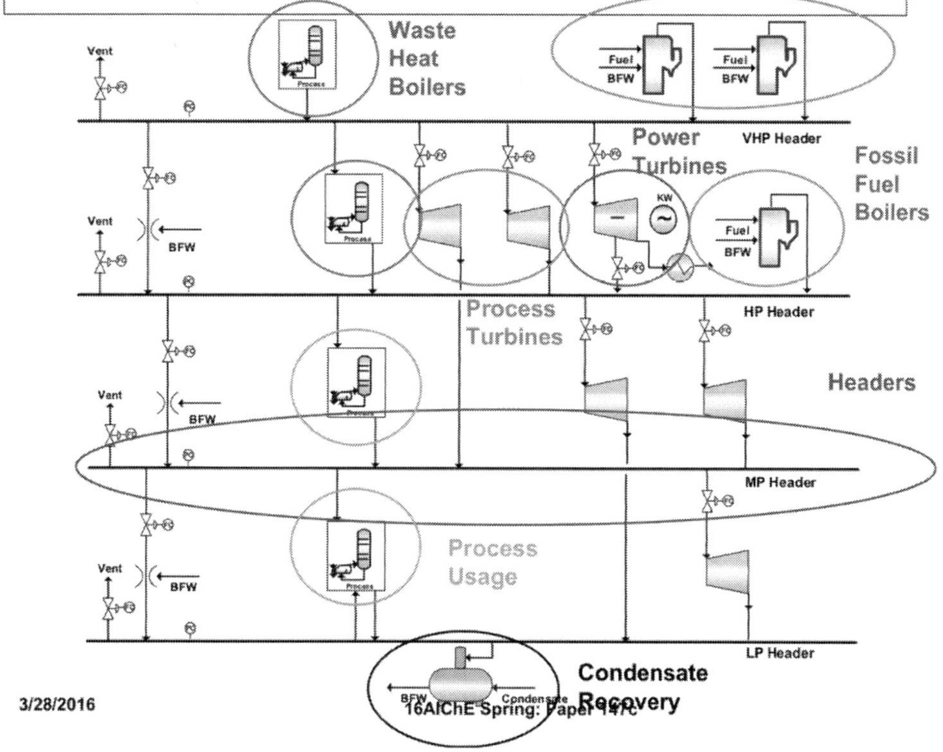

3/28/2016 16AIChE Spring: Paper 147c 7

Condensate Recovery and BFW

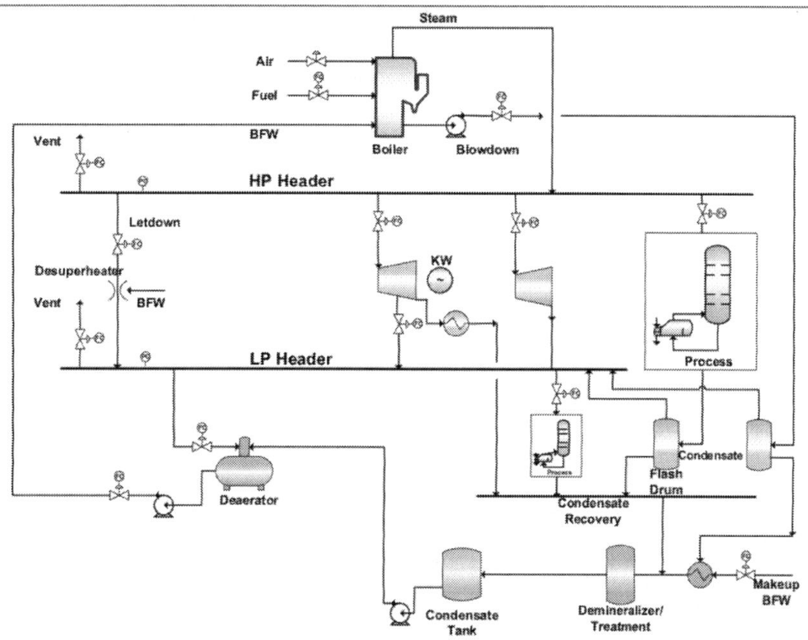

3/28/2016 16AIChE Spring: Paper 147c 8

Interaction of Steam, Electricity, Hydrogen, Fuel Systems

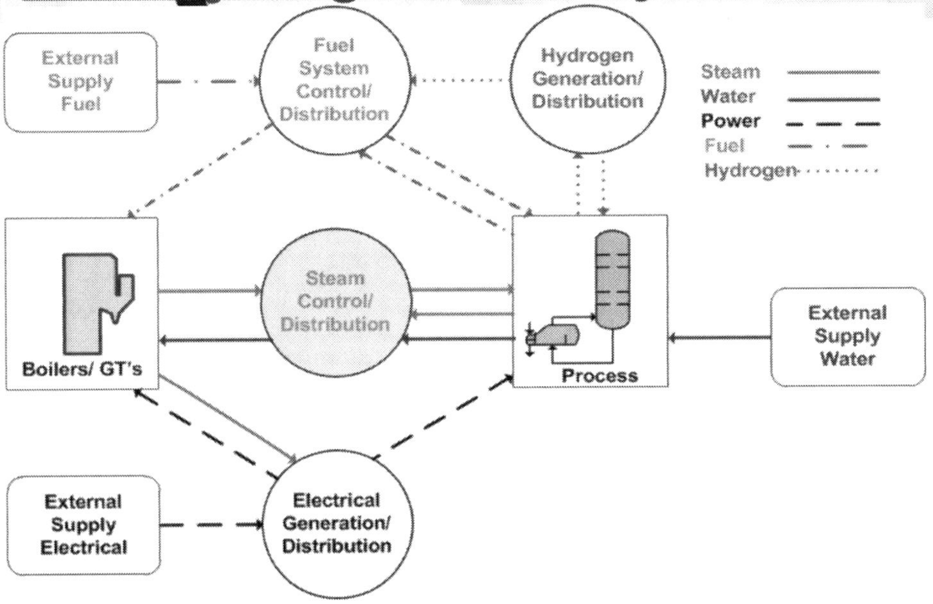

3/28/2016 16AIChE Spring: Paper 147c 9

Why is Steam System Management and Control Important?

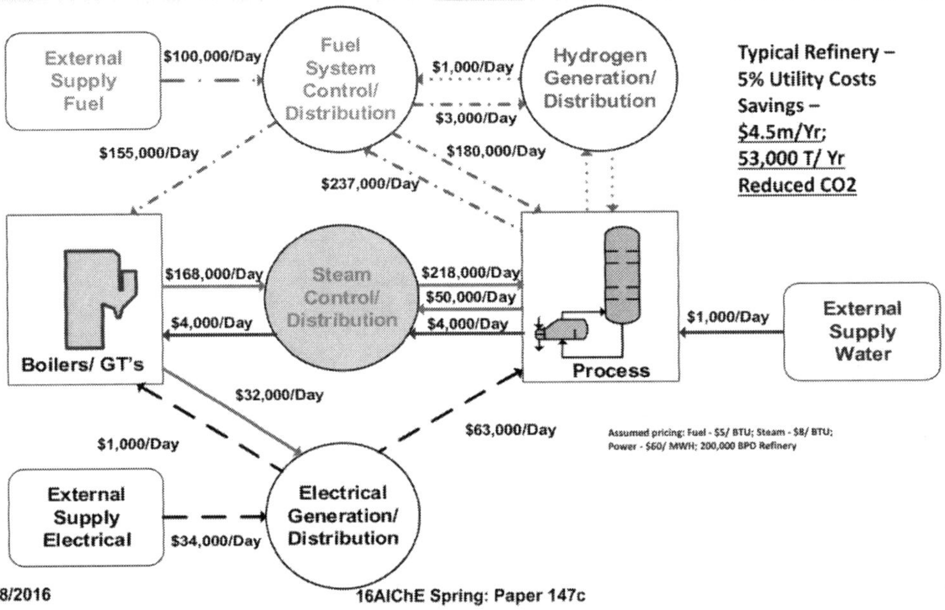

Typical Refinery – 5% Utility Costs Savings –
$4.5m/Yr;
53,000 T/ Yr Reduced CO2

Assumed pricing: Fuel - $5/ BTU; Steam - $8/ BTU; Power - $60/ MWH; 200,000 BPD Refinery

3/28/2016 16AIChE Spring: Paper 147c 10

Why is Steam System Control Challenging?

- Constantly Changing Demand at Different Steam Pressures
- Producers and Consumers are widely distributed geographically around the refinery
  - Multiple groups have control over their local steam equipment – uncoordinated
- Key measurements are often missing
- Wide range of dynamic responses for manipulated variable changes
- In older plants, many additions and modifications to steam systems and controls over the years – sometimes with limited documentation

Control Modes

Steam System Control Modes					
Priority	Mode	Objective	Typical Issue	Typical Response	Modern Control Approach
1.	Outage Control	Insure Safe Operation	Boiler Trip; Compressor Trip; Unit upset	Load Shedding; Meet flare steam requirements Maintain minimum flows to main turbines	May require operator action for seam curtailment
2.	Upset Control	Minimize Upset Operation Duration	Steam header pressure outside normal control range	Steam Letdown Control; Venting	Special Multivariable Control
3.	Normal Control	Respond to Normal Disturbances	Meet varying steam demand requirements at each pressure levels while compensating for supply variations	Control boilers' steam rates; Control Turbines' Extraction Steam Steam Letdown Control Vary power generation	Multivariable Control
4.	Optimization	Optimize Utility Supply	Economics	Maximize internal power generation Minimize letdown Manage Import/ Export Steam Steam/ Electric Turbine Selection	Steady State Optimization Model

Manipulated Variable Priorities

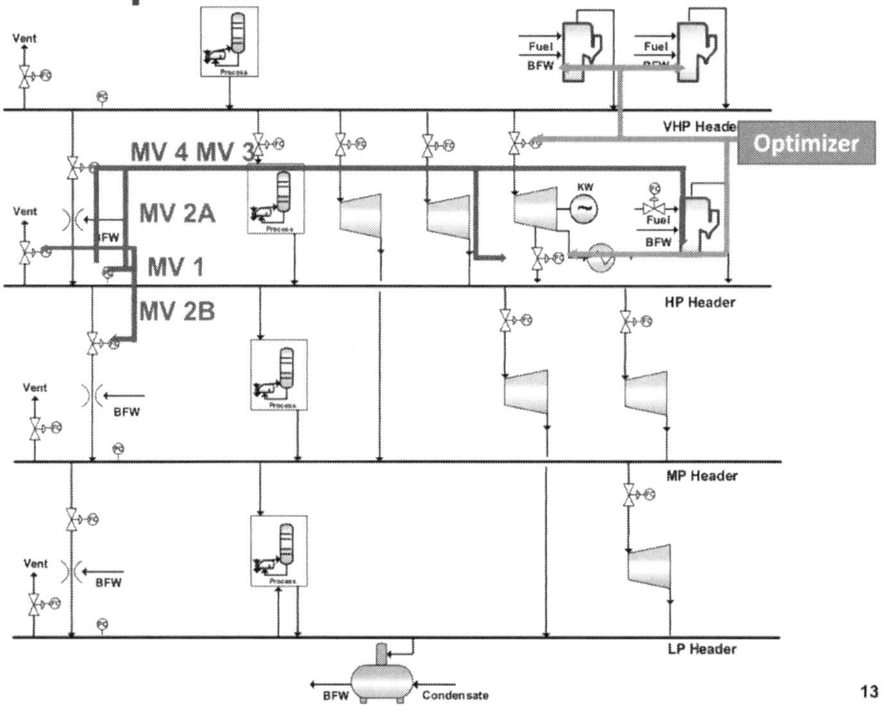

3/28/2016 13

Modes of Control

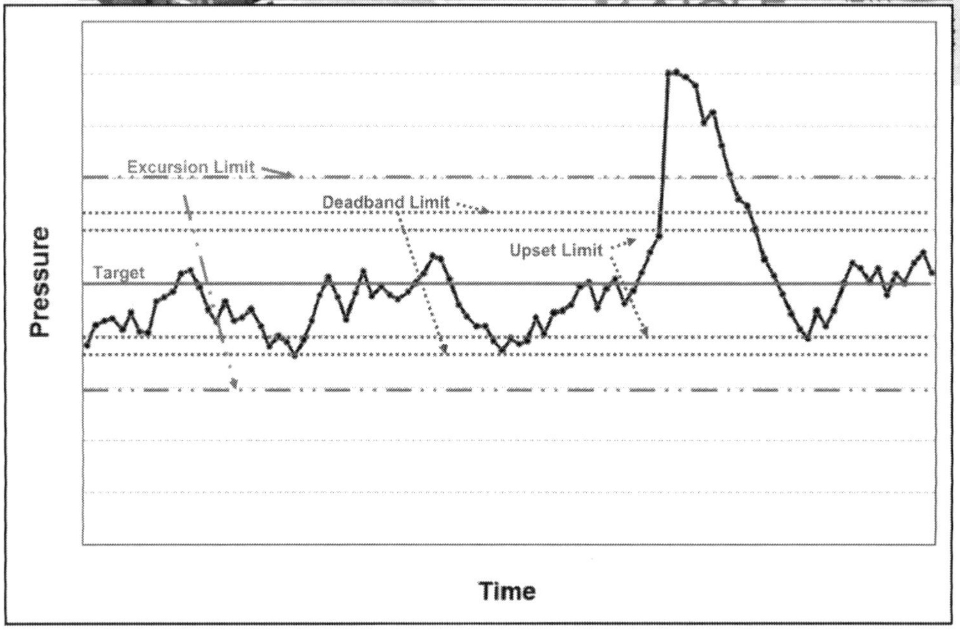

3/28/2016 16AIChE Spring: Paper 147c 14

What is Procedural Automation?

- Procedures are used to accomplish a specific complex process-oriented activity by executing a predetermined set of tasks in a set order to a set performance standard. Operational procedures often deal with startup, shutdown, product grade change, blending and response to abnormal conditions among other situations.
- Procedural automation is the methodology and technology to implement full or partial automation of these operational procedure.
- Also related is state-based (NOT state-space) control since the control action depends on the state of the system (off-line, startup, running, shutdown, etc.)

Historical Procedure Codification- ISA S88 (1995)

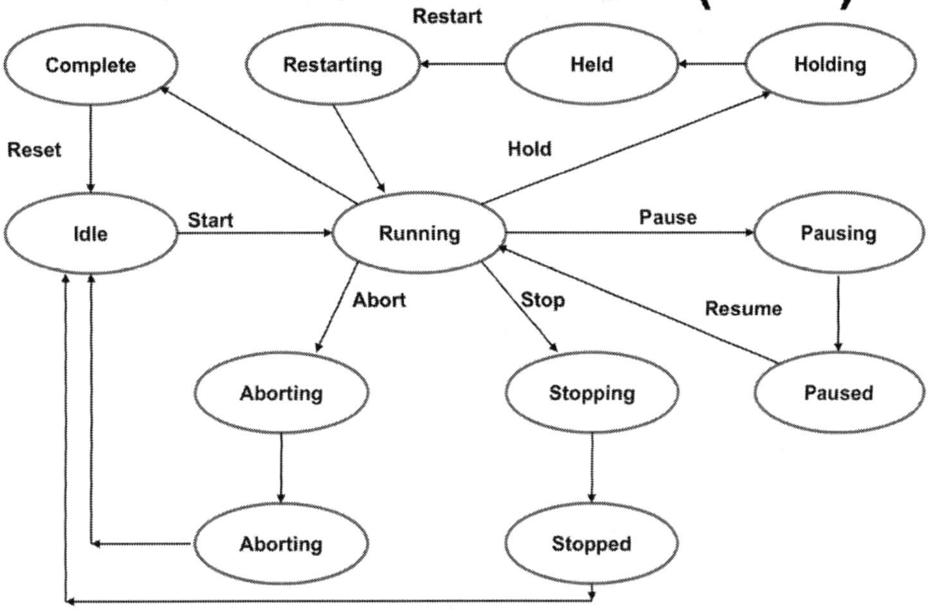

Plant Procedures defined as ordered sets of unit procedures, operations and phases.

Historical Procedure Codification -IEC61131-3 Sequence Function Charts (1993)

- Initially a PLC programming language but has been extended to DCS'
- Sequences refer to user actions and computer logic that initiate, interrupt, or terminate transactions. Sequence control governs the transition from one transaction to the next

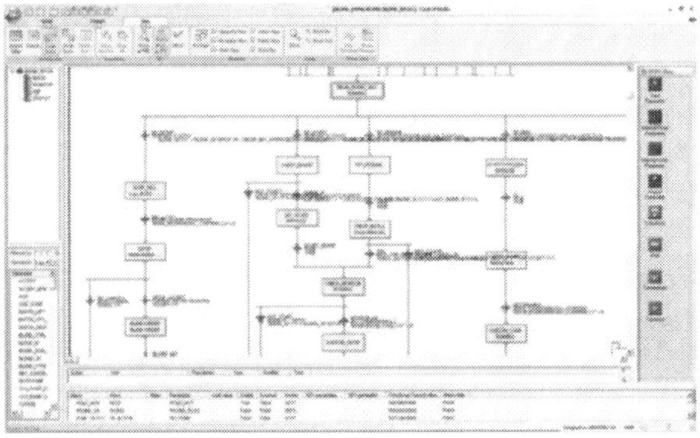

Historical Procedure Codification- ISA 106 (2010)

- ISA 106 – Procedure Automation For Continuous Operations Definitions: *Procedure – a specification of a sequence of actions or activities with a defined beginning and end that is intended to accomplish a specific objective. Procedure automation – Implementation of a procedure on a programmable mechanical, electric, or electronic system.*
  - Document and standardize operational SOP's - particularly startup, shutdowns, grade changes, abnormal situation responses – can be manual, prompted or automatic
  - Released Volume 1 (Aug 2013)

  Models
    - Physical Model (ISA88/95)
    - Procedure Requirements Model
    - Procedure Implementation Model

Unified Modeling Language – UML; System Modeling Language – SysML

- SysML is graphical modeling language focused on defining general system engineering problems that is tool and methodology independent
- SysML is based on UML with exclusions and extensions
- First released in 2007, it has been gaining popularity for control and automation applications for the aerospace, automotive, and defense industries

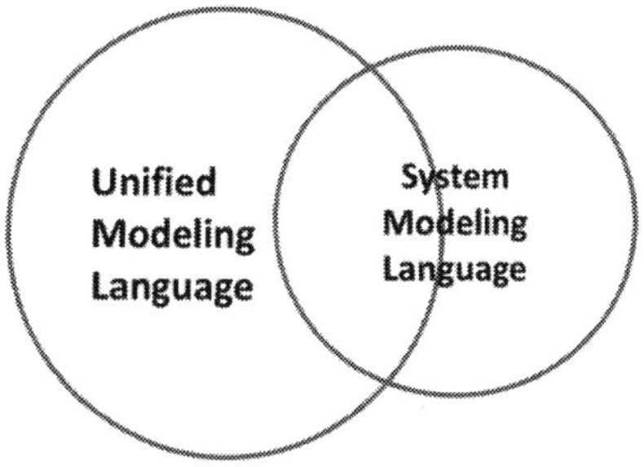

SysML Structure

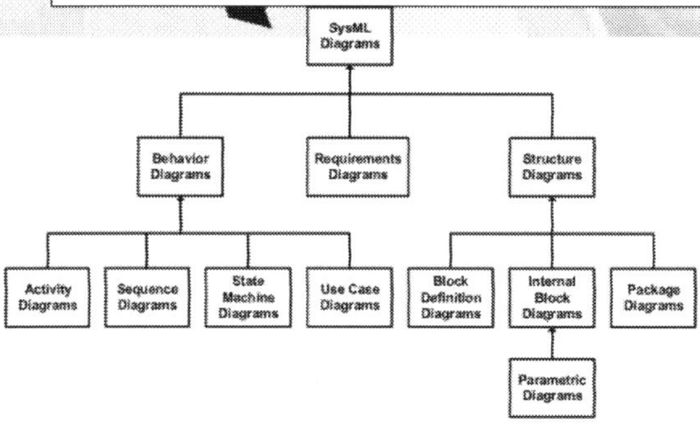

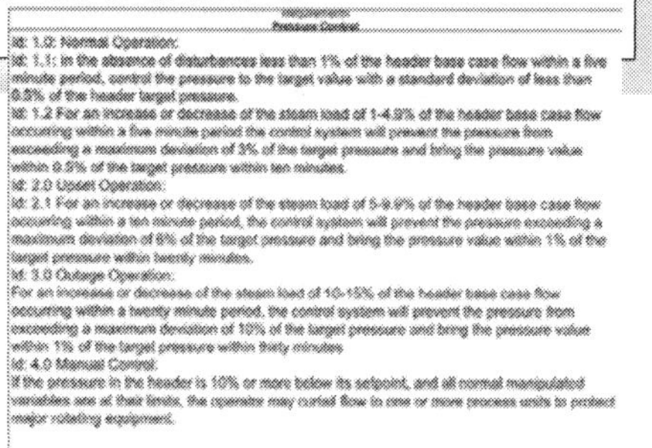

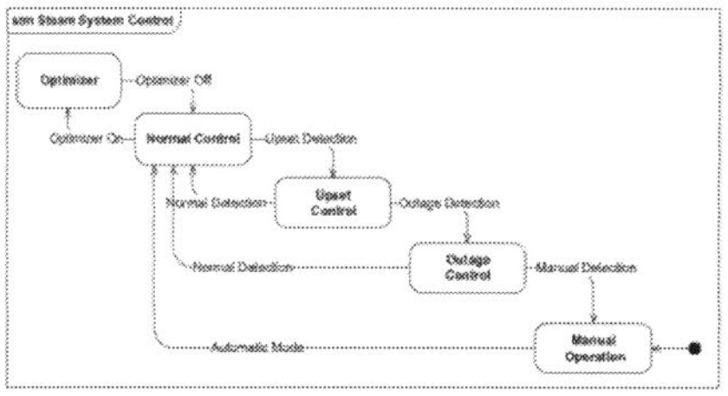

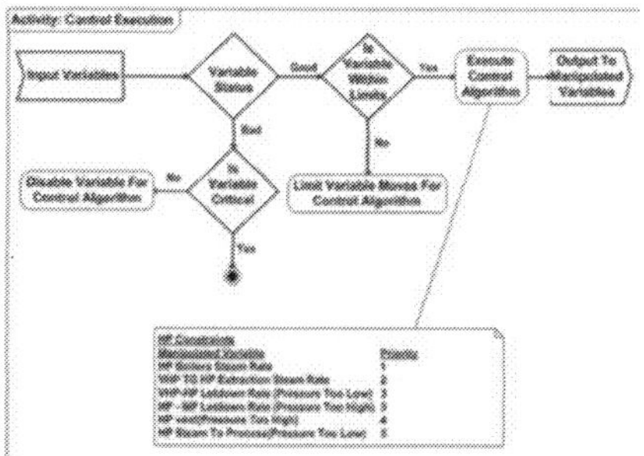

Case Study - Objectives

- **Design control structure to meet process steam and power demand:**
  - **Maintain Header Pressures with Minimal Deviation**
  - **Automatically Respond to Most Demand Changes without Operator Intervention**
    - **Process steam curtailment decisions will still be made and implemented manually**
  - **Minimize venting and letdowns**
  - **Optimize power production when feasible**

Original Control – HP Header

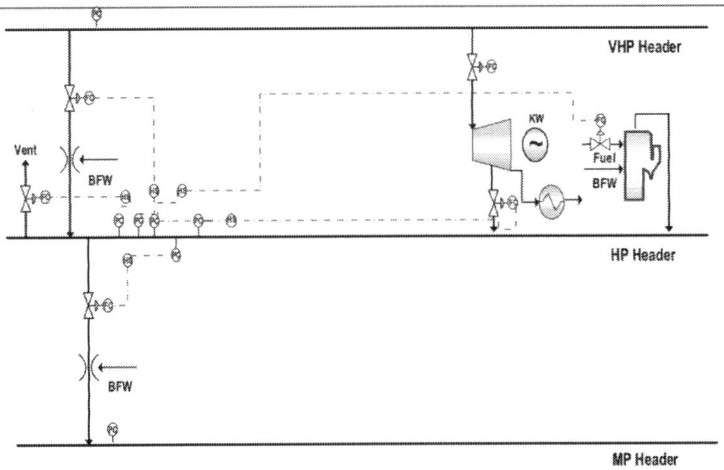

Problems: Excessive Letdown Steam; Frequent Manual Adjustment; Boiler Trips; Reduced Production of Cheaper Internal Power

Steam System Procedures (SOP's)

- Loss of One Boiler
- Cracked Gas Compressor Trip
- Process Steam Curtailment

Requirements Analysis

«requirement»
Pressure Control

Id: 1.0: Normal Operation:
Id: 1.1: In the absence of disturbances less than 1% of the header base case flow within a five minute period, control the pressure to the target value with a standard deviation of less than 0.5% of the header target pressure.
Id: 1.2 For an increase or decrease of the steam load of 1-4.9% of the header base case flow occurring within a five minute period the control system will prevent the pressure from exceeding a maximum deviation of 3% of the target pressure and bring the pressure value within 0.5% of the target pressure within ten minutes.
Id: 2.0 Upset Operation:
Id: 2.1 For an increase or decrease of the steam load of 5-9.9% of the header base case flow occurring within a ten minute period, the control system will prevent the pressure exceeding a maximum deviation of 6% of the target pressure and bring the pressure value within 1% of the target pressure within twenty minutes.
Id: 3.0 Outage Operation:
For an increase or decrease of the steam load of 10-15% of the header base case flow occurring within a twenty minute period, the control system will prevent the pressure from exceeding a maximum deviation of 10% of the target pressure and bring the pressure value within 1% of the target pressure within thirty minutes
Id: 4.0 Manual Control:
If the pressure in the header is 10% or more below its setpoint, and all normal manipulated variables are at their limits, the operator may curtail flow to one or more process units to protect major rotating equipment.

Mode Diagrams

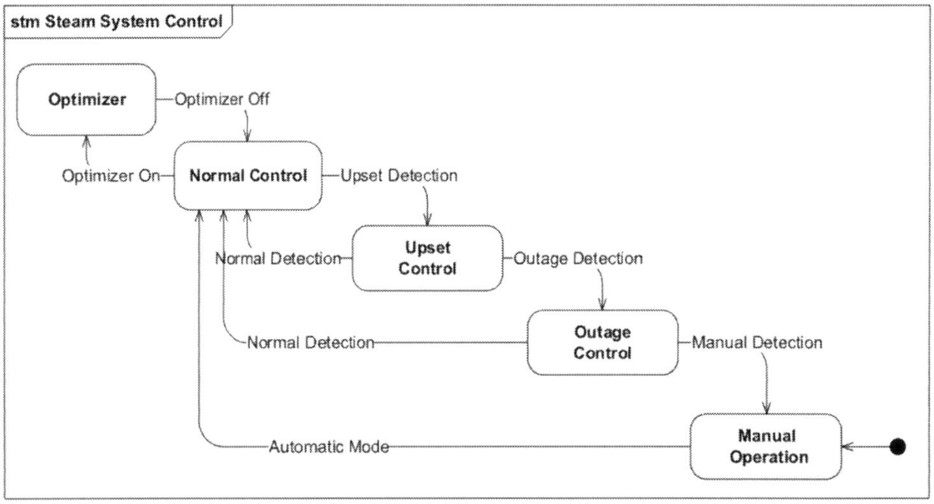

Activity Diagram

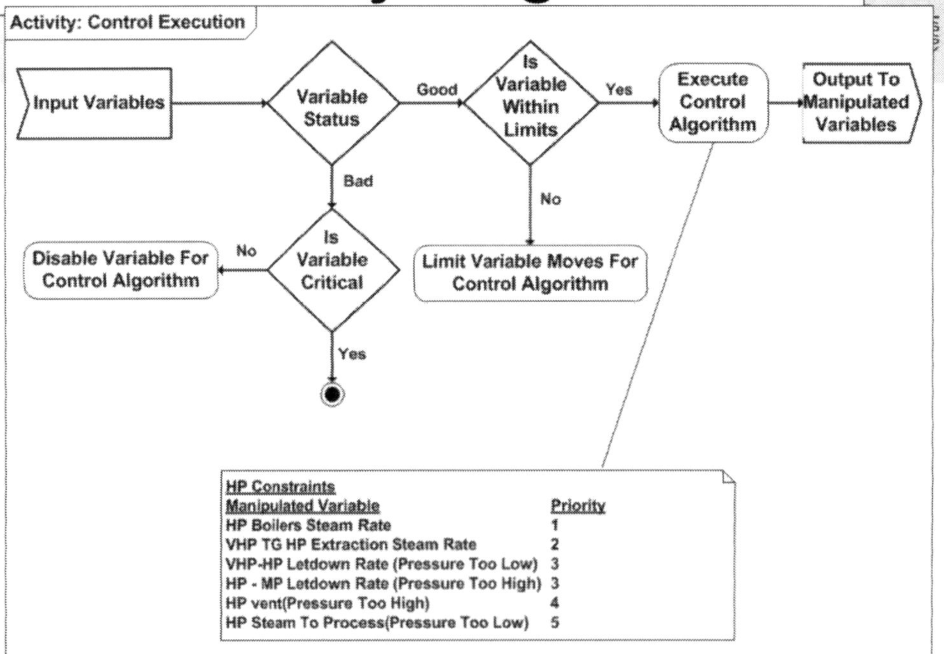

Conventional Implementation

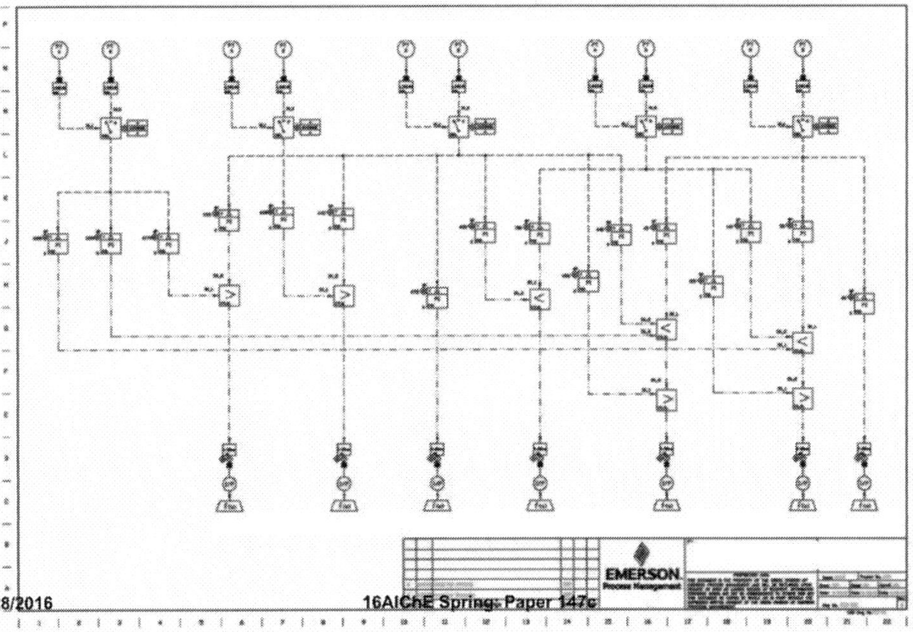

Implementation – Multivariable Control

Controlled Variables

- VHP Pressure
- HP Pressure
- MP Pressure
- LP Pressure

Disturbance Variables

- WHB To VHP
- WHB To HP

Constraint Variables

- Air Rates To VHP Boilers
- Air Rates To HP Boilers
- Electricity Import

Multivariable Controller

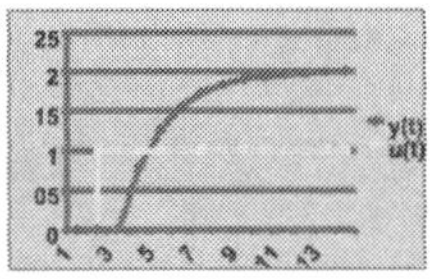

Manipulated Variables

- VHP Boilers Steam Rate
- VHP TG Steam Rate
- VHP-HP Letdown Rate
- VHP Vent Rate
- HP Boilers Steam Rate
- VHP TG HP Extraction Steam Rate
- HP - MP Letdown Rate
- HP Vent Rate
- VHP TG Steam Rate
- MP - LP Letdown Rate
- MP vent
- LP Steam Rate to Deaerator

What do we gain by considering this as procedural automation?

- **Structured problem definition – leads to clarity about requirements, performance and expected actions – graphical documentation helps understanding**
- **Can support testing system design by dynamic simulation (There are some software products which can be interfaced directly – still require some custom modeling)**

Financial Benefits

- **Typical Benefits**
  - **2 – 5% Increase in on-site power generation (reduction in purchased power)**
  - **Less "wasted" steam – 3 – 7 % Reduction in fuel costs for steam generation**
  - **Reduced incidence of boiler trips**
  - **Much more stable pressure control**

Benefits can be $4 million per year for a 200000 BPD refinery

Summary

- Process Industry steam systems provide complex management and control challenges involving differing types of control dependent on the plant conditions
- Analysis of these types of complex control problems and design of the appropriate control structure is aided by the use of structured control design

Thank you for attending. Questions?

444662 Simplifying Hydrogen Production

Abstract:

Steam reforming based hydrogen plants involve complex and energy intensive, multi-stage operations to convert hydrocarbons into high purity hydrogen. In a complex refinery, these plants are relied on to make hydrogen for hydroprocessing units in which the hydrogen is becoming of greater value to the overall operation. The barriers to achieve this reliability are increasing for the operator. For feed purification, high level sulfur specs on natural gas pipelines, varying sulfur speciation and levels in natural gas sources, and use of high-hydrogen refinery off-gases with heavier tails can result in condition changes, spikes, and transients in operation that challenge this reliability. For the steam methane reformer (SMR), heat transfer limitations across the reformer tubes, carbon formation and poisoning, rate changes in response to hydrogen demand, and increase in pressure drop in response to shutdowns and operation also challenge this reliability.

New purification and reforming technologies that deal with and remove these barriers for the hydrogen plant operator will be presented; these technologies can be implemented in existing plants with little to no capital expenditure. Experiences, support work and case studies will demonstrate how these barriers to reliability are being removed. For new plant designs, these technologies can miminize cost by removing vessels or decreasing the size of the plant, providing process simplification and intensification to hydrogen production as ever increasing demand for hydrogen and pressure on costs in the refinery make these technologies an attractive option.

445327 Hy-(drogen) Hopes in Clean Energy Landscape

Abstract:

Manifesting the challenge of future energy demand against global strife and mandates for curbing GHG induced climate change; hydrogen is poised to be a strong contender in the future energy spectrum. Strategic scenarios and roadmaps have been stipulated for addressing the future energy needs against the backdrop of strong environmental focus and clean energy targets. Its appreciable leverage and share is put on the power and transport sectors with gas, syngas and eventually hydrogen pathways rendering a progressive impetus in the longer run.

Hydrogen's role as a unique carrier of ultra clean energy is imminent despite any barriers or roadblocks put by skeptics debating its claimed viability. Based on its unparallel versatility and flexibility in culminating the different contemplated energy sources, whether fossil, renewable or nuclear into 'one clean energy platform', the hydrogen landscape is destined to be in the offing.

Deliberations over fossil-derived syngas with carcon capture and storage or sequestration (against current IGCC and NGCC) and gradual deployment of hydrogen from fuel substitution to stationary and mobile fuel cells is expected to take off exponentially in the coming decades. The availability of abundant low-cost (unconventional) gas is presently boosting as a cost-effective low-carbon feedstock for generating large scale syngas and hydrogen while well proven state-of-the-art technology.

With renewable energy as well as nuclear power expected to gain more grounds in future defying current setbacks, concerted roadmaps and future trajectories are being forecasted on syngas and hydrogen as dominant clean energy routes in the longer term. The clean energy drivers will be well-to-wheel efficiency in conjunction with mandatory reduction in the CO2 footprint expanding to use of renewable power to generate "green hydrogen". The economies of scale may also shift from large centralized power facilities to decentralized and distributed power generation, even digressing to household level.

The talk will provide an overview on the imminent role of hydrogen in the realms of future clean energy landscape.

444264 Modelling and Integration of Diesel Hydrotreater Kinetics with Refinery Hydrogen Networks

Abstract:

The current global trends to refine heavier crude oils and the stricter environmental regulations have enhanced the importance of hydroprocessing. Hydrogen, the essential feed of hydroprocesses, has a direct impact on the process performance and the final economic profits. Therefore, kinetic modelling of hydroprocessing units would provide detailed insights into the process as well as would enable opportunities to perform process optimisation and better hydrogen network management.

Based on an industrial configuration, a diesel hydrotreater model has been developed using industrial operating data. Bulk properties of feedstock measured through laboratory analysis are transformed into 18 molecular lumps based on the database that contains both bulk properties and molecular information of pre-collected samples. A kinetic model with molecular lumps is used for the reaction considering the operating conditions, including LHSV, makeup H2, recycle H2, reactor inlet temperature, H2 quench, and catalyst life. The model also predicts the light hydrocarbon generation that is critical for accurately modeling H2 recycle composition (on process level) and H2 network optimisation (on site level). The difference in properties between product and feed, such as distillation profile, cetane number and flash point, is predicted from the change in molecular lumps. A multi-period process optimisation model has also been developed. The proposed methodology changes the optimisation objective from the traditional methodology of reducing operating cost to improving the overall economic performance.

The developed kinetic hydroprocessing model is then applied to overall H2 network optimisation. This integration enables accurate evaluation of several changes to the diesel hydrotreating unit, such as: increased throughput to make full use of hydrogen surplus, increased proportion of cheaper feedstocks, variable H2 purity in the makeup for maximum H2 cascade utilization, etc. A case study is carried out to show the benefits of the developed methodology.

16AIChE
Spring Meeting, Houston, TX

Use of hiTRAN Tube Side Enhancement in Revamps to Improve Operability and Efficiency

Peter Drögemüller, Cal Gavin LTD, Minerva Mill Technology Centre, Alcester, UK
Peter.Droegemueller@calgavin.com

Peter Ellerby, Cal Gavin LTD, Minerva Mill Technology Centre, Alcester, UK
Peter.Ellerby@calgavin.com

Prepared for Presentation at
American Institute of Chemical Engineers
2016 Spring Meeting
11th Global Congress on Process Safety
Houston, Texas
April 10-14, 2016

UNPUBLISHED

Use of hiTRAN Tube Side Enhancement in Revamps to Improve Operability and Efficiency

Peter Drögemüller, Cal Gavin LTD, Alcester, UK

Peter Ellerby, Cal Gavin LTD, Alcester, UK

Peter.Droegemueller@calgavin.com

Keywords: hiTRAN, revamp, maldistribution, viscous flows, heat transfer

Abstract

Over the past decades various innovations in the design of heat exchangers have been major driving forces in the improvement of thermal process efficiency. hiTRAN thermal systems is an example of one of those technologies. The enhancement technique is applied on the tube side of tubular heat exchangers and can play a vital role in boosting the performance of this equipment. It is of special interest in revamp situations since it can be installed in situ during operational shut down. The technology can be used most efficiently in tubular single and two-phase heat exchangers were poor tube side heat transfer is the limiting factor. Poor performance can be found in exchangers operating under laminar and transitional flow conditions. Here hiTRAN technology can offer up to 8 times higher tube side heat transfer for the **same** pressure drop compared to the empty tube design. This increased performance can be used in different ways. Often increased throughput is the highest priority. But reduced or increased outlet temperatures or lower temperature driving forces between the fluids can be of equal value for the plant operator to apply this technology. Apart from the clear economic benefits, the use of this technology offers also operational benefits. Examples of this are suppression of waxing problems in aircooler operation or the suppression of flow instabilities in thermosiphon reboilers. The paper explains the basic concept how and when to apply this technology in single and two phase applications. Real case studies are presented, which show in detail the benefits when using hiTRAN technology in retrofit scenarios.

1 Introduction

hiTRAN wire Matrix technology has been available for more than 35years to improve tube side heat transfer in tubular heat exchangers. Its main application area is the use of this technology in new equipment design for exchangers with laminar tube side flow where considerable size reductions are achievable due to the increased tube side performance. Increased focus is now given for the use of this technology in exchanger revamps. Aside from the obvious potential of increased duty and therefore increased exchanger efficiency, applying this technology can also often provide improved operability as will be shown for specific operation conditions.

2 Principles of hiTRAN technology

2.1 Hydrodynamic in hiTRAN enhanced flow

hiTRAN wire matrix technology alters the flow field inside the tube. This can be demonstrated with dye stream injection technique. Coloured ink with similar properties to the bulk flow is injected into the flow near to the tube wall as shown in **Figure 1**. In the area of plain empty tube laminar flow (A) the dye is not mixed with the bulk flow and remains almost stagnant at the tube wall. Once the dye hits the first loop of the insert wire the coloured fluid is deflected towards the bulk flow (B) indicating a change in flow direction on a macro scale.

***Figure 1**: Dye stream experiments.*

This radial fluid mixing shortens the fluid residence time near to the wall, reducing the time the fluid is exposed to high or low wall temperatures. This benefits applications where fouling and product degradation are caused by excessive wall temperatures [1], [2]. It also improves considerably the rate of mass and heat transfer from the tube wall and between the fluid layers as explained in this paper.

2.2 Tube side heat transfer enhancement in hiTRAN enhanced flow

Tube side heat transfer measurements carried out at Cal Gavin Ltd, the manufacturer of hiTRAN Wire Matrix Elements, show significant differences to the empty tube behaviour. Typical curves are shown in **Figure 2** for hiTRAN Elements for two different Matrix geometries and compared with empty tube heat transfer rates. Ideally Elements are designed such that the entire allowable frictional tube side pressure drop for the exchanger design is used. This is done by varying geometrical parameters such as matrix density and wire sizes. Those parameters can be varied continuously within a wide range and therefore the induced heat transfer and pressure drop varies accordingly. The dotted lines in **Figure 2** show some of various additional hiTRAN geometries possible. Heat transfer and pressure drop correlations for all Element variations are implemented in hiTRAN.SP a free software tool which integrates with HTRI and Aspen software [3]. From the graph it is noted that the heat transfer increases continuously at a constant rate with

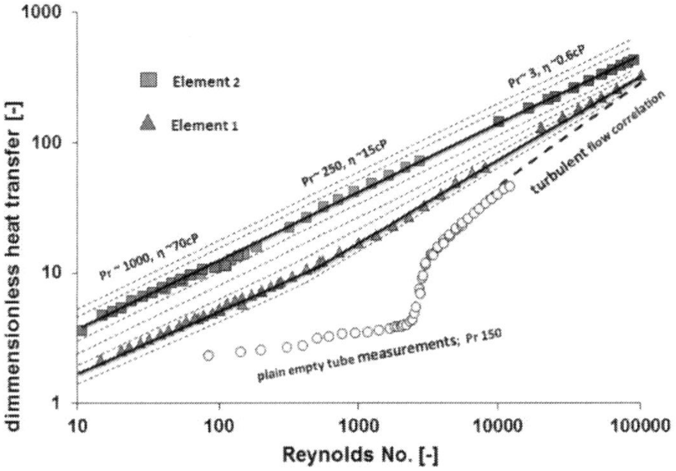

***Figure 2**: Heat transfer performance range of hiTRAN wire matrix insert compared with plain empty tube data.*

increased flow velocity and Reynolds number respectively. For the Reynolds numbers investigated, there is no appreciable difference between laminar and turbulent flow conditions. Uncertainty of empty tube performance in the transition region is therefore removed. Compared to empty tube conditions an increase of heat transfer coefficient of up to 16 times is possible. It can be seen that the enhancement levels achieved are greatest in laminar and transitional flow in the Reynolds range from 50 to 10000. As a rule of thumb for the majority of industrial applications this equates to a viscosity range from about 5cP to 200cP. Unlike laminar flow, heat transfer in turbulent flow is a strong function of flow velocity therefore the same levels of heat transfer enhancement under those flow conditions can often be achieved with higher flow velocities in a multi pass design without the need for inserts. For this reason hiTRAN applications in turbulent flow are typically found where a multi pass design is not possible, e.g. in situations of temperature cross. Under these conditions hiTRAN offers up to 5 times empty tube heat transfer. At the lower end hiTRAN has been shown to offer benefit in terms of heat transfer for Reynolds numbers as low as 1.The use of hiTRAN increases the associated pressure drop compared to the pressure drop for a plain empty tube at the **same** flow velocity. When used for revamp purposes this poses challenges but also opportunities as explained in the next chapters.

3 Pressure drop considerations

As discussed above applying hiTRAN can increase the tube side heat transfer several fold but installing any passive enhancement device inside tubes will also increase the tube side friction factor f_t and therefore the frictional pressure drop for the **same** flow velocity. When used for revamp purposes this poses challenges which can be overcome by minor modifications. But increased frictional pressure drop can also offer opportunities in terms of operability as explained later. In revamp scenarios the heat transfer surface area and also the allowable tube side pressure drop is in general given and fixed.

3.1 Modification of pass arrangement

Figure 3: Pass partitions in header.

In situations where after revamp the allowable tube side pressure drop would be exceeded, installing tube internals requires a reduction in flow velocity in order to stay within the allocated tube side pressure drop. In a multi-pass exchanger this is often easily achieved by the removal of pass partitions (illustrated in **Figure 3**). In general mechanical removal of pass partitions in the header does not alter the mechanical integrity of shell&tube heat exchangers in any way, nor does it change the ASME code. As a result the flow pass length is reduced and also the flow velocity, resulting in a reduction in tube side bundle pressure drop.

For an **entire** tube bundle the tube side pressure drop as function of pass partitions $\boldsymbol{N_p}$ can be expressed as follows:

$$\Delta p = \boldsymbol{f_t} \cdot \frac{32 \cdot L_t \cdot m_b^2}{\pi^2 \cdot \rho \cdot d_t^5 \cdot N_t^2} \cdot \boldsymbol{N_p^3} \qquad \textbf{(1)}$$

From equation (1) it can be deduced that the tube side pressure drop after hiTRAN revamp can be kept equal or lower compared to the plain empty tube pressure drop if the number of passes can be reduced according equation (2).

$$\left[\frac{N_{p\,(plain)}}{N_{p(hiTRAN)}}\right]^3 \cdot f_{plain} \geq f_{hiTRAN} \qquad \textbf{(2)}$$

An 8 pass plain empty tube exchanger, when reduced to a 2 pass hiTRAN exchanger after revamp, can be designed with a 64 times higher friction factor at the reduced velocity and stay within the same pressure drop compared to the empty tube design. In order to make the revamp thermally more efficient the heat transfer for this reduced pass arrangement (velocity) needs to be higher than the empty tube arrangement before pass reduction. When using hiTRAN in laminar or transitional flow this is generally the case. It is demonstrated with a revamp example in **Figure 4**, where the tube side heat transfer and pressure drop is shown for a heat exchanger operating under laminar flow conditions. The tube side fluid is a viscous oil (η = 18cP). For the fixed bundle size, tube side heat transfer and pressure drop change as a function of flow velocity. The markers represent operation with different tube passes, from one to eight and the corresponding tube side velocities from 0.2m/sec (one pass exchanger) to 1.5m/sec (eight pass exchanger). Under laminar flow conditions an increase in flow velocity only adds a limited amount of heat transfer but the pressure drop increases with the square of velocity as shown in the graph. In this example the original exchanger is designed as an eight pass exchanger (Re ~ 1700) in order to maximise heat transfer (160W/m^{2}K) within the allowable pressure drop of 1 bar. The constraint for the revamp with hiTRAN wire matrix technology is usually the allowable tube side pressure drop. It is evident from the remarks earlier, that installation in the 8-pass heat exchanger would yield a high heat transfer but an excessive tube side pressure drop. In order operate within the allowable tube side pressure drop the revamped exchanger must have a lower tube side velocity. This can be achieved by the removal of pass partitions. From **Figure 4** it can be seen that a two pass hiTRAN design yields the same pressure drop with much improved tube side heat transfer coefficient of about 1400W/m^{2}K. A one pass hiTRAN design would still increase the heat transfer coefficient more than 5fold with only 1/5 of the empty tube eight pass pressure drop.

Engineers often have to design with a minimum flow velocity in order to suppress fouling. The minimum velocity specified relates to the wall shear stress generated in the empty tube arrangement. Measurements show that hiTRAN alters the near wall velocity profile and yield higher wall shear stresses compared to the plain empty tube design [4], [5]. For that reason wall shear equivalent flow velocities are much lower in a hiTRAN enhanced design.

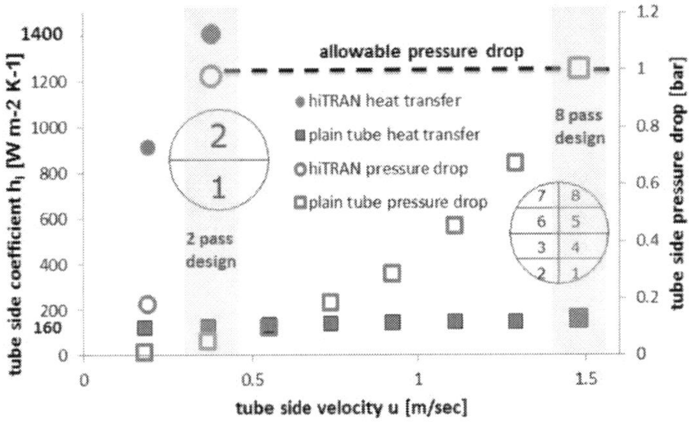

***Figure 4**: Heat transfer and pressure drop as function of tube velocity (pass arrangement) for a typical hiTRAN design.*

3.2 Switch of shell arrangement

In case a shell in series configuration is targeted for revamp, the methodology in order to keep the pressure drop after hiTRAN installation within the allowable limit is similar to a single bundle revamp. In a first approach a change of pass arrangement for each shell as discussed in section 3.1 should be considered. Another option is a switch from shells in series to shells in parallel arrangement. The rearrangement results in a shortened flow path and an enlarged cross-sectional entrance area, with lower flow velocity and corresponding lower pressure drop. The spare pressure drop can then be utilised in order to equip the exchangers with hiTRAN enhancement technology. The additional benefit of a parallel arrangement can be seen in the possibility to use designated spare exchangers for stand by purposes either for spare capacity or chemical cleaning purposes.

Multi-pass shells in series configurations are in many cases used in order to compensate for a temperature cross between the process streams. Those units are common in situations with small temperature approach between the two process streams. Often hiTRAN offers a better option by operating a single pass arrangement, even at low flow velocities the tube side enhancement will generate high heat transfer coefficients. This option is especially of interest in new exchanger design.

3.3 Part installation of hiTRAN

Modifying the pass arrangement may not be practical or economic in certain cases, such as where the required change is from an even to uneven arrangement. In such scenarios, if spare pressure drop is available, there is the option of installing only certain passes with inserts. In general passes with the greatest potential for enhancement are selected. In situations where the starting position is a single pass exchanger and the spare pressure drop prevents an installation over the whole tube length, enhancement can be achieved by installing the inserts only in part of the tube length. The installation length is dictated by the spare pressure drop available.

Due to process requirements, single pass exchangers can often be found in two phase applications. Those applications are often also very sensitive to changes in pressure along the flow path. For that reason part installation is common in thermosiphon reboilers or down flow condensers for example.

4 Revamp considerations

The main driving forces to undertake heat exchanger revamps on existing units can be divided as follows:

- Benefits in terms of increased overall heat transfer coefficient
- Improved fluid distribution
- Benefits in terms of more stable operation and longer runtimes

hiTRAN enhancement only affects the thermal tube side performance, whereas for an exchanger revamp the impact on overall performance has to be determined. The heat duty of a heat exchanger can be approximated in the form:

$$Q = U \cdot \boldsymbol{A} \cdot \text{EMTD} \tag{3}$$

The increase in overall coefficient U determines the efficiency of the revamp. Since this coefficient is also a function of the wall and fouling resistance and the tube outside coefficient the impact of the revamp depends very much on those variables:

$$\frac{1}{U} = \frac{1}{h_i} + \frac{1}{h_o} + foulingresistance + wallresistance \tag{4}$$

In the revamp example from **Figure 4** the tube side heat transfer increased almost 9 fold from 160W/m2K to 1400W/m^{2}K at identical pressure drop of 1 bar. The impact on the overall coefficient and therefore on the effectiveness of the revamp depends very much on the other heat transfer resistances in equation (4).

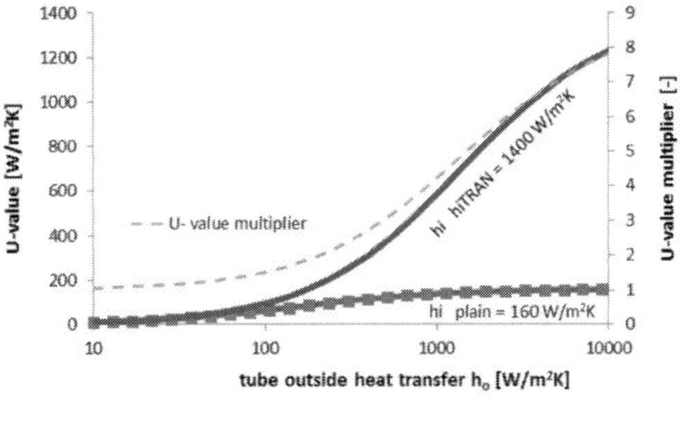

***Figure 5**: Change of U-value and U-value multiplier with increases tube outside coefficient h_o*

In graph **Figure 5** this is demonstrated by taking into account different tube outside coefficients h_o. It is evident that with higher tube outside coefficient the revamp becomes more tube side controlled and therefore more efficient. For a heavily tube side controlled scenario the overall multiplier approaches the increase in tube side coefficient only. hiTRAN revamps are most efficient in case of heavily tube side controlled operation.

4.1 Benefits in terms of increased overall heat transfer coefficient

Equation (3) can be used in order to discuss the different revamp scenarios which focus on the increase in tube side heat transfer and consequently in the increase of overall coefficient. An example for the retrofit of a viscous oil heater demonstrates the different possible revamp scenarios. In the base case before revamp the oil is heated from 120° to 163°C with condensing steam (20bar) on the shell side. Due to the viscosity of the viscous oil between 7cP and 17cP the flow in the 8pass AES heat exchanger (236 tubes 6m x 26.4mm x 2.11mm) ranges from laminar to transitional resulting in a low tube side heat transfer coefficient. The constraints for the different revamp scenarios are a constant heat transfer area A and also the maximum allowable pressure drop of 1bar. The base conditions are summarized in the **Table 1**. Where the exchanger operates already prior to revamp, as in this case, at the tube side pressure drop limit of 1bar the number of tube passes has to be reduced, as outlined in section 3.1

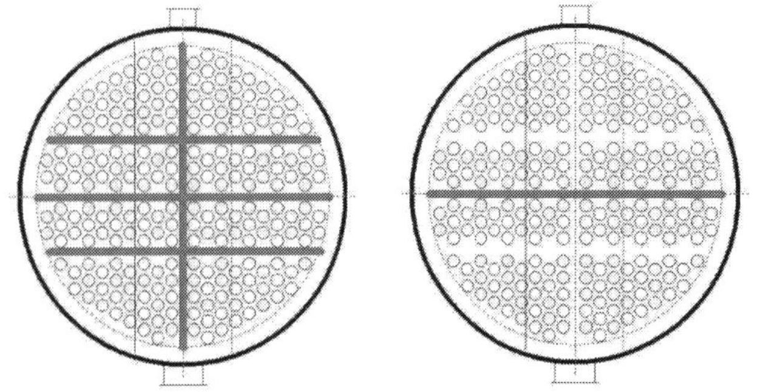

Figure 6: Pass arrangement reduced from 8 to 2 passes shown in HTRI [6] for the example case.

This option is only feasible if the exchanger is designed as a multi-pass unit. To keep the piping arrangement for the exchanger unchanged in general only an even to even modification is considered. After removal of pass partition plates the flow velocity is reduced accordingly. The different possible revamp scenarios are calculated with HTRI Xchanger suite software [6] comprising hiTRAN.SP plug-in [3] and later summarized in **Table 1**.

It should be noted that the impact on the shell side performance must be taken into account, increased tube side duties may result in change shell side outlet temperatures which affect the EMTD. In this case, the use of condensing steam means the overall EMTD is largely unaffected

4.1.1 Increased throughput

The purpose of a revamps is often to increase the plant throughput; here heat exchangers can be the bottleneck since they are designed for a specified mass flow. In case of tube side controlled operation as shown in the example above, the increase in tube side heat transfer translates to a much improved overall coefficient U. In this case with condensing steam on the shell side the temperature levels for both streams are kept constant and the potential increase in mass flow will be proportional to the increase in overall heat transfer.

$$\dot{m}_{after\ Revamp} = \frac{U_{after\ Revamp}}{U_{before\ Revamp}} \cdot \dot{m}_{before\ Revamp} \qquad (5)$$

Since both the increased mass flow and also the use of enhancement inside the tubes will add pressure drop, the number of passes has to be reduced in order to stay within the allowable pressure drop as explained before. In this particular case a reduction from 8 to 2 passes will allow, according to equation (2), an increase in friction factor of 64 times for the modified arrangement, without exceeding the pressure drop. The mass flow can be increased from 12kg/sec to 32kg/sec within the allowable pressure drop of 1bar. The increased heat transfer from hiTRAN allows even further heating of the fluid than the required 151°C. This suggests a considerable further increase in flow rate is possible, which could be achieved by modification to a 1 pass arrangement but requiring the piping and headers be modified. As a result of this the flow rate can be increased by the factor of 7 to about 85kg/sec within the allowable pressure drop.

4.1.2 Increased / reduced outlet temperatures

In other revamp situations the mass flow through the exchanger is kept unchanged and the aim is to achieve a higher temperature change of the process stream. Examples are product air coolers where the outlet temperature for storage can be reduced. In feed / effluent exchangers the increased heat transfer equates to higher heat recovery rates and can be used to reduce the load on the fired heater. Again the level of improvement possible depends on the tube side flow condition and how much the overall coefficient is affected by the change in tube side performance. In our example the pass arrangement of the exchanger again has to be modified in order to operate within the allowable pressure drop. As a result of the much improved tube side heat transfer coefficient the outlet temperature increases from 151°C to 211°C as shown in **Table 1**. The outlet temperature pinches with the steam temperature of 212° at this level. Highlighting that often for such revamp options temperature pinches must be taken into account. An additional benefit here is that there is additional capacity to allow for fouling margins.

4.1.3 Reduced temperature driving force

The aim for a particular revamp can also be to run an exchanger for the same duty with a reduced temperature driving force. Examples are where increasing ambient air temperatures beyond the design values lead to an increasing number of underperforming air coolers. In those situations an increased tube side coefficient and consequently higher overall heat transfer can compensate for a loss in effective mean temperature difference as evident from equation (3). A different scenario is demonstrated with the viscous steam heater example. The increased tube side coefficient can here be used in order to operate the exchanger with much reduced condensing steam pressure. In this case the steam pressure can be reduced from 20bar to 5.2bar. This often means lower quality steam can be used. Alternatively due to the lower temperature level, the process can be switched from a steam heated to a heat transfer fluid heated process with associated benefits. Apart from the cost reduction a reduced temperature driving force will lead to much reduced wall temperatures. For liquids sensitive to temperature driven chemical reaction fouling the drop in wall temperatures can have a dramatic effect on the fouling behaviour.

***Table 1**: Summary of the different revamp scenarios discussed, red values do highlight main achievements.*

	Base case	**4.1.1** increased throughput 2-pass	**4.1.1** increased throughput 1-pass	**4.1.2** Increased outlet temp	**4.1.3** Reduced steam pressure
Shell side					
Condensation pressure [bar]	20	20	20	20	5.2
Condensation temperature [°C]	212	212	212	212	153
Tube side					
No of passes	8	2	1	2	2
Flow rate [kg/sec]	12	32	85	12	12
Temperature out [°C]	151	180	151	211	151
Calc. pressure drop [bar]	1	1	1	0.85	1
Velocity / shear equiv. [m/sec]	1.3 / 1.3	0.9 / 1.4	1.15 / 2.0	0.33 / 0.9	0.33 / 1.1
Reynolds No. [-]	920/2650	650/2800	860/2200	250/1700	240/650
Heat transfer [W/m^2K]	**172**	**1016**	**1354**	**1405**	**1001**
Duty [MW]	1.3	5.4	9.1	2.8	1.3

4.2 Reducing the adverse effects of uneven fluid distribution.

Design calculation of the thermal performance of tubular heat exchanger does in general assume an even fluid distribution within a tube bundle but also within a single tube. For certain operating conditions this is not necessarily the case and can lead to underperforming units. hiTRAN enhancement technology can be useful to solve those issues.

4.2.1 Bundle fluid maldistribution

Severe maldistribution of the tube side flow within a bundle can result in an underperforming heat exchanger. Low flow regions within the bundle are also more prone to fouling, resulting in an even worse performance. Geometrical aspects like large bundle diameter, single pass configurations, small nozzle sizes and axial nozzle location, in conjunction with low frictional tube side pressure drop contribute to an uneven fluid distribution. In a revamp the geometrical conditions are difficult to alter, whereas the low frictional pressure offers the possibility to use hiTRAN technology. As discussed earlier the matrix density and also installation length can be adapted to the level of pressure drop required to remedy poor distribution. With this flexibility it is even possible to fine-tune the frictional pressure drop requirements over the cross-section of a bundle by varying Element geometries. Over the years CFD simulation has been proven a powerful tool to evaluate and predict fluid distribution in bundles [7]. In addition to the improved distribution the Elements also contribute to a much improved tube side coefficient as discussed earlier.

4.2.2 Tube fluid maldistribution

In laminar flow the heat transfer is often dominated by mixed convection. In mixed convection, a secondary flow profile caused by density differences is superimposed on the forced velocity profile in the flow direction [8], [9], [10]. Under these conditions natural convection superimposed on the main flow causes a rise of less dense fluid towards the upper tube region, whereas the more dense fluid accumulates at the bottom of the tube. The direction of movement within the tube depends on whether the tube is heated or cooled. Computational fluid dynamic (CFD) can be used in order to simulate such behaviour. This is shown in the cross sectional flow pattern on the left (a) in Figure 7.

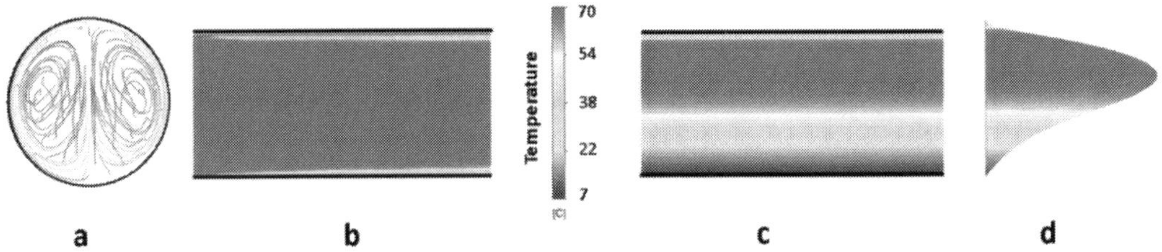

a b c d

***Figure 7**: CFD simulation of flow in 2.5m x 25.4mm horizontal tube at Reynolds 250, inlet 70°, wall 7°C and dynamic viscosity 12cP (b: inlet; c: outlet; d: velocity profile at outlet)*

As a result of the natural convection effects the temperature over the cross section of the tube can become stratified (c), with the highest temperature at the top and the lowest towards the bottom of the tube. This flow stratification has an impact on the predictability of heat transfer since, as shown in **Figure 7c**, the temperature of the fluid at tube bottom area approaches the temperature of the cooling fluid outside of the tube (7°). This in effect equates to a loss in heat transfer area, since here the temperatures pinches. Due to the complex flow patterns, the extent of this pinch area is difficult to predict and can severely limit heat transfer performance [11]. For this reason heat transfer in mixed convection flow is difficult to correlate. It is also evident that the stagnant flow areas can promote fouling due to the absence of sufficient fluid movement. Again CFD simulation was employed in order to investigate the impact of fluid mixing on flow stratification with hiTRAN. The simulation results presented in **Figure 8** show a flow pattern over the cross section of the tube which is substantially different to the plain empty tube.

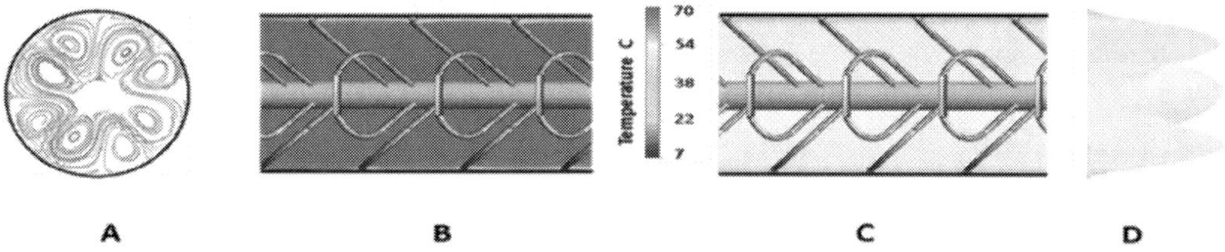

Figure 8*: hiTRAN CFD simulation of flow in 2.5m x 25.4mm horizontal tube at Reynolds 250, inlet 70°, wall 7°C and dynamic viscosity 12cP (B: inlet; C: outlet; D: velocity profile at outlet)*

Due to the mixing action of the Element, temperature differences between adjacent fluid layers are much smaller. For that reason the driving force for natural convection, which causes flow stratification is diminished. Due to the considerably higher heat transfer the measured / simulated mixed outlet temperature is much lower with 49.9°C compared to 61.8°C for the empty tube case. No stagnant velocity zones towards the bottom of the tube are present when operating with hiTRAN.

This also has implications on the residence time distribution which is much narrower than in case of empty tube flow. This is beneficial for applications which are sensitive to long residence times at cooled or heated surfaces and might suffer degradation under such conditions.

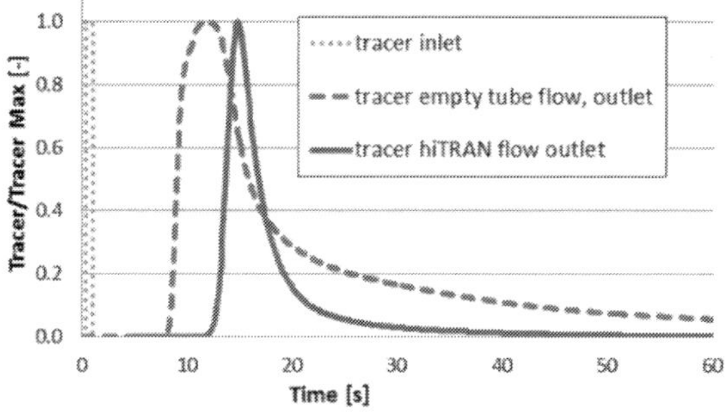

Figure 9*: Calculated residence time distribution in plain empty tube and tube equipped with hiTRAN at Re = 250.*

4.2.3 Thermodynamic non equilibrium in stratified two phase flow

Another form of fluid mal distribution within the tubes can be encountered in two phase flow heat transfer. In gravity controlled two phase flow regimes the liquid phase will accumulate towards the bottom of the horizontal tube.

Figure 10*: Wavy stratified two phase flow at mass flux of 50kg/m^{2}sec and gas mass fraction of 0.18 in transparent tube; left plain empty, right hiTRAN enhanced.*

In **Figure 10** the flow conditions in a 20mm transparent tube are shown with an air water mixture. The two phases are separated and flow with different velocities of 3.3m/sec (Re~11000) for the gas and < 0.1m/sec (Re ~1000) for the water. There is little interaction between the water and gas phases. Those situations can be found often towards the end of a condensation process in a horizontal tube. In multicomponent condensation the gas and vapour phase have to be cooled in order to maintain the condensation process but due to the low Reynolds numbers the cooling potential is limited. In addition due to the limited interaction between gas and liquid phases non-condensable components will accumulate at the interface and limit the mass transfer in turn limiting the condensation process. In general heat exchanger design programmes assume equilibrium conditions for their calculations and in such situations temperatures can be very different for the phases causing insecurity when designing for those scenarios. As shown in **Figure 10** hiTRAN technology improves the interfacial mixing and mass transfer. In addition the heat transfer and therefore the cooling duty will increase considerably in the liquid and vapour phases. Since those situations are often encountered towards the exit of condensers, hiTRAN may be installed as a partial length to limit the pressure drop penalty which can be important for such applications.

4.3 Operational benefits in terms of more stable operation

Often hiTRAN tube side enhancement is employed in order to improve the operation of shell and tube heat exchangers. This is valid for laminar and also for turbulent flow.

4.3.1 Operation in transition region from laminar to turbulent flow

As seen in **Figure 2** after onset of turbulence in an empty tube at about Reynolds 2300 the heat transfer characteristic changes significantly. Under laminar flow conditions heat transfer increases only little with increasing Reynolds number. In transitional flow there is a very strong dependency on Reynolds number. This can be expressed for the different flow conditions as follows:

$$Nu \approx Re^{exponent} \quad \textbf{(6)}$$

***Table 2**: Exponent in equation (6) for different flow conditions.*

Flow	**exponent**
Laminar flow	~ 0.2
Transitional flow	**~ 4**
Turbulent flow	~ 0.8
hiTRAN all flows	~0.5 – ~0.7

For a Reynolds number change of 25% this equates in laminar flow to ~5% change in heat transfer, in turbulent flow to about 20% change but in transitional flow region the heat transfer would change by more than 150%. Should an exchanger be designed to operate in this region this must be taken into account as such nonlinear behaviour may affect the operation of the exchanger under part load or in start-up or shut-down. Sometimes a minor change or uncertainty in viscosity can shift the operation into this region. It is notable from **Figure 2** when designing with hiTRAN there is no transition to turbulent flow and therefore no sudden change in heat transfer performance. The heat transfer performance will scale similar to turbulent flow exchangers over a range from single figure Reynolds to high turbulent flow.

4.3.2 Suppression of flow instabilities in thermosiphon reboilers

Unsteady flow in reboiler applications can cause operational and performance problems. The causes for these instabilities are complex. Instabilities in two-phase flows are sensitive to pressure changes in the system. Low system pressures tend to show a higher tendency towards instabilities. Under thermosiphon operation conditions "density wave oscillations" are the most common form of fluid fluctuations. According to Boure [12] certain types of flow instabilities encountered in reboilers can be avoided by installing a throttle valve upstream of the exchanger. According to Arneth [13] it is common practice to restrict the flow with a suitable orifice or gate valve in the inlet piping of the reboiler to manage fluctuating flow behaviour. In a more recent review paper Nayak and Vijayan [14] report always positive effects in forced circulation reboiler in case of additional pressure drop in the subcooled single-phase region. In general this is also the case for thermosiphon reboiler applications. When designing a reboiler in such a way that hiTRAN is installed only in the sub-cooled region additional pressure drop is induced into this region. This has similar effects as a flow restriction in the inlet of the exchanger with the additional benefit that the additional pressure drop is converted into flow turbulence with higher tube side heat transfer in this region. Hammerschmidt [15] investigated flow instabilities in a single tube thermosiphon reboiler. Under empty tube vacuum conditions the water flow collapsed, when operating with liquid heads below 80% of the tube length, whereas with hiTRAN a stable recirculation was maintained with even lower liquid heads. When operating with the more viscous water/glycerol mixtures under vacuum conditions the changes in flow stability were even more pronounced. In the empty tube experiments for liquid levels of 100% tube length and below severe fluctuations were measured. In contrast the tube with hiTRAN installed over the whole tube length showed, for all liquid levels investigated, higher recirculation flow with only minor fluctuations. As reported previously this also has a beneficial impact on the measured heat transfer in such scenarios.

4.3.3 Operation in transitional and film boiling flow regime

Vaporizers which operate under dry wall conditions can be difficult to predict since the mechanism for the onset of transitional and film boiling are poorly understood. The extent of the so called transitional and film boiling region has to be known since dry wall heat transfer rates can be as low as 1/30th of the wet wall heat transfer coefficient [16]. As shown in the case study below (**Table 3**) hiTRAN technology can be used to debottleneck poor performing vaporizers where the cause is dry wall condition. The use of hiTRAN affects film boiling mechanisms in two ways.

- Experimental research shows increased critical heat flux values for the onset of film boiling when compared with empty tube conditions [17]. The shift to higher heat flux can be explained with lower wall superheat due to higher convective boiling rates and a suppression effect on the generation of nucleation sites [18].

- In cases where film boiling is sustained even in the presence of hiTRAN, it can be expected that the heat transfer rates in the vapour blanket are increased. The mechanism of enhancement is similar compared to single phase flow where the matrix generates near wall turbulence in the vapour blanket.

Feedback from industrial applications does confirm these assumptions. This is illustrated in a real case study, a BEU 2 pass TEMA type shell and tube heat exchanger (702 tubes, 4m length) with ethylene evaporating on the tube side. The exchanger originally performed below specification. Evaluating the process conditions, convective film boiling over large sections of the exchanger was suspected. After retrofitting the exchanger with hiTRAN elements the increase in performance indicated suppression of film boiling with much higher heat transfer. In **Table 3** the conditions before and after retrofit are shown.

Table 3*: Revamp of ethylene vaporizer with hiTRAN.*

	Empty	**hiTRAN (after retrofit)**
Flow rate [kg/sec]	14.5	21.1
Temp. in/out [°C]	-100 / - 1 (sat.)	-100 / 30 (superheated)
Pressure in / out [bar]	40 / 39.93	40 / 39.74
Heat transfer [W/m^{2}K]	613	2390
Heat duty [kW]	261	618

4.3.4 Reduction of mist flow in vaporizers

Droplet carry-over can lead to operational problems with rotating process equipment. Mist flow in vaporizers is where remaining liquid is dispersed as small droplets in the vapour stream. The saturation temperature of the droplet increases with decreasing radius and makes it therefore more difficult to evaporate small droplets. In addition the velocity-slip between the vapour and the droplet decreases with smaller droplet size. A smaller velocity difference in turn reduces the interfacial heat transfer. This means that for small droplets the vaporization is mainly controlled by the temperature difference between droplet and vapour. Since hiTRAN increases the single phase heat transfer in gas flows several fold the superheat can be increased substantial for a given tube length compared to an empty tube. This superheat relative to the droplets is beneficial in vapourising the remaining liquid.
Additionally hiTRAN can have a demister effect agglomerating droplets and therefore increasing the relative velocity between vapour and droplets, increasing therefore the heat transfer at the interface. In revamp applications it is often advised to install the wire matrix only over a short flow length where droplet carry over is most likely. This will help to reduce the additional pressure drop after the revamp as explained in section 3.3.

4.3.5 Longer operation time under certain fouling conditions

Fouling characteristics are influenced largely by the properties of the thermal and hydrodynamic boundary layers. Heat transfer enhancement devices such as hiTRAN Matrix Elements have been used successfully as a tool to influence both wall temperatures and wall shear stress in revamp situations. Research into the change of velocity profile in hiTRAN enhanced flow shows a much steeper near wall velocity gradient compared to empty tube behaviour [4], [5]. This equates to higher wall shear stress as indicated by equation (7):

$$\tau_W = \eta \cdot \frac{\partial v}{\partial y} \tag{7}$$

Wall temperatures are affected by the much increased tube side heat transfer coefficient. Neglecting wall resistance and any fouling layer, the wall temperature can be expressed as follows:

$$T_W = T_o - \left(\frac{h_i}{h_i + h_o}\right) \cdot (T_o - T_i) \tag{8}$$

For an increased tube side heat transfer coefficient the wall temperature approaches the tube side bulk temperature.

- **Chemical reaction fouling**

Crude oil fouling can be characterised as chemical reaction fouling. The fouling behaviour can be modelled reasonably well with the Ebert and Panchal Model [19] and its variants. Common to these types of model is a deposition and a suppression term. The deposition term is strongly influenced by the film (wall) temperature of the fluid. The suppression term describes how the influence of wall shear forces counteracts the build-up of foulant. Those models indicate that crude oil fouling is sensitive to wall temperature level and wall shear forces. Research also show that for particular crude a threshold temperature and wall shear force for which fouling can be supressed can be found [20].
hiTRAN is of special interest where the wall temperature can be reduced considerably as shown for the steam heated crude oil heater in example show in section 4.1.3. Another area of successful applications in

crude oil services are revamps in exchangers with very low flow velocities and therefore also very low wall shear forces. These revamps are undertaken without modification of pass partitions and can lead to a considerable increase in wall shear force and tube side heat transfer

- **Waxing inside tubes**

In viscous cooling applications the viscosity gradient of the fluid to be cooled can be very steep. Particularly in air cooled applications, where the temperature outside the tubes is a function of ambient conditions, the wall temperatures can fall below the pour point temperature resulting in waxing inside the tubes. The stagnant fluid will form an insulating layer which significantly impacts the heat transfer and hydraulic performance of the cooler. hiTRAN technology can be used to mitigate these problems. As shown in equation (8) due to the increased tube side coefficient the temperature gradient from the bulk flow to the wall is reduced considerably which can help to keep the temperature level above the pour point temperature. **Figure 7** and **Figure 8** in section 4.2.2 also indicate the much improved back mixing of viscous fluid near to the wall to the bulk flow. Revamping such coolers may require pass modification as explained in section 3.1.

5 Conclusion

This paper gives detailed information about the potential benefits when using hiTRAN enhancement technology. Those benefits are divided into operational process improvements, improvements in terms of increased exchanger duty and improvements which affect a better fluid distribution in the equipment. The constraints for using this technology are clearly outlined and explained in detail. It can be seen that a substantial duty increase is possible where the tube side heat transfer is the controlling heat transfer resistance in the unit. It is also evident that minor changes to the pass partitions in the header opens up considerable possibilities for revamp in existing units. The aim is to encourage the plant operator and thermal design engineer to understand the opportunities to revamp with this technology.

6 Symbols used, Units, Abbreviations

6.1 Abbreviations

CFD	Computational Fluid Dynamic
EMTD	Effective Mean Temperature Difference [°C]

6.1 Symbols

A	[m^2]	area
d	[m]	diameter
h	[W/m^2K]	heat transfer
L	[m]	Length
v	[m/sec]	velocity
$\dot{m}$	[kg/sec]	mass flow
N	[-]	number
Δp	[Pa]	pressure drop

6.2 Subscripts

p	Passes
b	bundle
t	tube
i	internal, inside
o	external, outside
W	Wall

6.3 Dimensionless numbers

f	fanning friction factor
Nu	Nusselt number

Q	[W]	duty	Re	Reynolds number
T	[°C]	temperature		
U	[W/m²K]	Overall heat transfer		
y	[m]	Wall coordinate		

6.1.1 Greek Symbols

α	[W m^{-1} K^{-1}]	heat transfer coefficient
τ	[Pa]	shear stress
ρ	[kg/m^3]	density
η	[Pas]	dynamic viscosity
∂		differential

7 References

[1] P Drögemüller, "Heat Exchanger equipped with hiTRAN Wire Matrix Elements," in *Heat Exchanger Fouling*, H. Müller-Steinhagen and H.U. Zettler, Eds.: PP Publico, 2011, pp. 180 - 188.

[2] J.M. Ritchie, P. Droegemueller, and M.J.H. Simmons, "hiTRAN Wire Matrix Inserts in Fouling Applications," in *ECI Symposium Series, Heat Exchanger Fouling and Cleaning VII*, Tomar, Portugal, Volume RP5, 2007, pp. 410 - 418.

[3] P. Droegemueller and C. Mann. (2009) hiTRAN.SP Software. [Online]. http://www.calgavin.com/software-downloads/

[4] A. Smeethe, P. Droegemuller, W. Bujalski, and J. Wood, "Fluid dynamics in a tube equipped with wire matrix inserts," in *CHISA 16th INTERNATIONAL CONGRESS OF CHEMICAL AND PROCESS ENGINEERING*, Prague, 2004.

[5] P. Drögemüller, W. Osley, and D. Phillipp, "Increased Shear, Reduced Wall Temperatures, Use of hiTRAN Wire Matrix Inserts in Systems subject to fouling," in *Proceedings of Onternational Conference on Heat Exchanger Fouling and Cleaning*, Budapest, Hungary, 2013.

[6] Heat Transfer Research Inc. (2015) HTRI Xchanger suite 7.2. Software. [Online]. www.HTRI.net

[7] P. Ellerby. (1999, September) http://www.chemicalonline.com. [Online]. http://www.chemicalonline.com/doc/computer-simulation-facilitates-ninefold-impr-0001

[8] D.R. Oliver, "The effect of natural convection on viscous-flow heat transfer in horizontal tubes," *Chem. Eng. Sci.*, vol. 17, pp. 335 - 350, 1962.

[9] A. J. Ghajar and Lap-Mou Tam, "Flow Regime Map for a Horizontal Pipe with Uniform Wall Heat Flux and Three Inlet Configurations," *Exp. Thermal and Fluid Science*, vol. 10, pp. 287 - 297, 1995.

[10] B. Metais and E. R. G. Eckert, "Forced, Mixed and Free Convection Regimes," *Trans. ASME J. Heat Transfer*, vol. 10, pp. 295 - 296, 1964.

[11] J.D. Dooley, "CFD Study of Diabatic Single-Phase Laminar Liquid Flow in Horizontal Tubes," HTRI, HTRI Research Report S-ST-1-13, 2010.

[12] J.A. Boure, A.E. Bergles, and L.S. Tong, "REVIEW OF TWO-PHASE FLOW INSTABILITY," *NUCLEAR ENGINEERING AND DESIGN*, vol. 25, pp. 165 - 192, 1973.

[13] S. Arneth and J. Stichlmair, "Characteristics of thermosiphon reboilers," *Int. J. Therm. Sci.*, vol. 40, pp. 385 - 391, 2001.

[14] A. K. Nayak and P.K. Vijayan, "Flow Instabilities in Boiling Two-Phase Natural Circulation Systems: A Review," *Science and Technology of Nuclear Installations*, vol. 2008, p. 15, 2008; Article ID 573192.

[15] J. Hammerschmidt and S. Scholl, "Experimentelle Untersuchungen zum Einsatz von Drahtgestrickeinbauten und berippten Rohren in Naturumlaufverdampfern," *Chemie Ingenieur Technik*, vol. 84, no. 11, pp. 1939 - 1947, November 2012.

[16] R. Thome, "Post Dryout Heat Transfer," in *Wolverine Heat Transfer Engineering Data book III.*: Wolverine Tubes Inc., 2004 - 2010, ch. 10.

[17] F.E. Mergerlin, R.W. Murphy, and A.E. Bergles, "Augmentation of heat transfer by use of mesh and brush inserts.," *J. Heat Transfer*, vol. 96, pp. 145 - 151, 1974.

[18] P. Drögemüller, "The Use of hiTRAN Wire Matrix Elements to Improve the Thermal Efficiency of Tubular Heat Exchangers in Single and Two-Phase Flow," *Chemie Ingenieur Technik*, vol. 87, no. 3, pp. 188 - 202, March 2015.

[19] W. Ebert and C.B. Panchal, "Analysis of Exxon crude oil slip stream coling data," in *Proc. Mitigation of Fouling in Industrial Heat Exchanger Equipment*, San Luis Obispo, USA, 1997, pp. 451 - 460.

[20] Y., Crittenden, B. Mengyan, "Fouling thresholds in bare tubes and tubes fitted with inserts," *Applied Energy*, vol. 89, no. 1, pp. 67 - 73, 2012.

446669 Surviving the Down Times. Which Projects Will Make It

Abstract:

With oil prices reaching near historical low in terms of constant dollars, and oil producing countries having budget deficits, what will be the future of the mega projects world wide and in the oil producing countries? Will we see more fuel expansion projects? Renewables? Solar? Will infrastructure and housing projects have the highest priority? Will it be a good time to invest while construction market slowed down? This is what we will discuss in this paper giving the high uncertainty in the oil market and the efforts of opec countries to decrease the dependency on oil.

Process for Reconfiguring Refinery Assets for Renewable Fuels

James T. Wexler, Honeywell UOP, Des Plaines, IL, USA
José R. Rojas, Honeywell UOP, Houston, TX, USA

Abstract

The processing of "renewable feedstocks" is becoming more popular around the world, especially due to concerns about increasing energy demand and greenhouse gas emissions as they relate to climate change concerns, and long-term views about petroleum resources and prices.

The economic situation for renewable fuels is challenging, from competing with low value crude oil through policies and mandates of its use. In such an economic environment, there is pressure on refiners who have underperforming assets and there is often a need to reduce capital costs in order to have more favorable renewable project economics.

Reconfiguring existing equipment assets in a refinery can address the problem. Existing hydrotreating, hydrocracking, reforming and isomerization units and other refinery assets could potentially be converted in order to produce renewable fuels, particularly green diesel and green jet fuel. In order to achieve the best product quality, highest yields and optimal processing capability of such a reconfiguration, it is recommended to install two-stage processing (first stage deoxygenation followed by the second stage isomerization). Optimizing equipment reuse and reducing equipment replacement or modifications could achieve capital cost and operating cost objectives.

The Ecofining™ process and the UOP Renewable Jet Fuel Process are versatile two-stage processing solutions for producing green diesel or green jet fuel from a range of sustainable feedstocks such as inedible vegetable oils and waste fats and greases. The Ecofining process was developed jointly between Honeywell UOP and Eni S.p.A.

Today these Honeywell UOP technologies have been implemented at three commercial facilities, including two sites where existing refining assets were repurposed for the production of renewable fuels. The Eni Biorefinery in Venice, Italy, produces green diesel fuel and the AltAir Fuels facility in Paramount, California, USA, produces green diesel and green jet fuel.

Green jet fuel viability has been demonstrated via partnering with multiple military and commercial aviation leaders. In aircraft, Honeywell Green Jet Fuel™ can replace as much as 50 percent of the petroleum jet fuel used in flight, without any changes to the aircraft technology, while meeting key ASTM jet fuel specifications. AltAir Fuels successfully demonstrated commercial production of Honeywell Green Jet Fuel beginning in February 2016 by reconfiguring Alon's Paramount, California petroleum refinery to a renewable fuel production facility, using Honeywell's UOP Renewable Jet Fuel Process. This marked the world's first revamp of a fossil fuel refinery dedicated to produce both renewable diesel and renewable jet fuel.

Keywords: renewables, renewable fuels, Ecofining, Green Diesel, Green Jet Fuel, revamps.

Introduction

Oil refiners have been challenged in recent years to meet stricter standards for fuel quality such as reducing the sulfur content of diesel fuel and reducing the aromatics content in gasoline. They have retooled their refineries to meet these challenges and optimized their operations to remain competitive. In many cases, this has resulted in consolidating some of their hydroprocessing capabilities and the idling of some less competitive process units or other refining assets (including the shutdown of entire refineries sometimes).

Across the world, people are concerned with energy security, agricultural and rural development, and the environment. A common denominator for positively affecting these areas is renewable fuels because they provide substantial reductions in greenhouse gas emissions from sustainable raw materials. Fuels and chemicals from renewable sources have the potential to support growing energy needs while addressing concerns regarding climate change and greenhouse gas emissions.

Many governments have established policies, mandates or incentives to reduce their emissions by increasing the use of renewable fuels. In the U.S., the federally mandated Renewable Fuel Standard, known as RFS2, calls for substantial increases in the volume of advanced renewable transportation fuels over the next several years. In Europe, the Renewable Energy Directive and the Fuel Quality Directive require that 10% of transportation fuel come from renewable sources by 2020. These policies provide strong incentives that have been growing the supply of renewable fuels and many other countries of the world have policies and plans in place to drive further growth of renewable fuels. Ecofining and the UOP Renewable Jet Fuel Process, Honeywell UOP's innovative renewable fuel technologies, allow the energy industry to favorably contribute to improving the environment by lowering the greenhouse gas emissions compared with petroleum-based fuels, and to producing drop-in fuels that are compatible with existing engines and infrastructure for the delivery of transportation fuels. These technologies and their products are compatible with existing oil refineries and substantial costs savings can be achieved by co-locating these process units with existing refineries (either within or adjacent to the refinery).

In order to decrease the amount of capital required to implement a project existing refinery assets (single unit or multiple units and other associated facilities) could be reconfigured. Idle, underperforming or underutilized units such as hydrotreating, hydrocracking, reforming or isomerization units among others could be considered for potential conversion into renewable fuel units to produce green diesel and/or green jet fuel.

The advantages from revamping existing assets to produce renewables are: 1. Produce high-quality, high-reliability drop-in renewable fuels that meet or exceed product specifications. 2. Expand the fuel pool and enhance its characteristics by blending renewable fuel with lower quality fuels. 3. Reuse existing assets and infrastructure to reduce overall capital investment required to produce renewable fuels. 4. Meet renewable fuel obligations.

When a refiner is looking for options to produce renewable fuels, the first step should be to analyze its existing assets and determine the feasibility of repurposing. The thermal instability and high reactivity of the feedstocks should be considered, as well as the handling of CO_2 acid gas generation. The renewables technology flowscheme requires a first stage for hydrodeoxygenation operating at relatively medium to high pressure similar to hydrocracking/hydrotreating reactors, and a second stage for hydroisomerization operating at lower pressure than the first stage and that is similar to reforming or isomerization units. Existing fired heaters and heat exchangers can be kept in service or re-optimized depending on the heat integration. All other pieces of equipment could be evaluated for re-use. Probably the most important pieces of equipment are the compressors, either recycle or makeup gas compressors could be the limiting factor for the unit capacity. The next important parameter is metallurgy of the existing equipment as it may not be suitable for the new processing conditions. And last but not least will be the integration between units to have the desired two stage configuration.

The second step is to do a high level analysis on how to approach a revamp under capital cost constraints for Inside Battery Limits (ISBL), and while optimizing use of Outside Battery Limits (OSBL) systems. The analysis includes a look at the feedstock and chemistry involved during transformation to fuels, repurposing of existing refinery assets including a potential flow scheme, and a capital cost constraint overview.

Renewable Sourced Feedstocks and Renewable Fuels Chemistry

Biological oils and fats can be converted into diesel, naphtha and jet fuels using many different processes, such as hydrodeoxygenation and hydroisomerization. Renewable sources (vegetable oils and waste fats and greases) have chemical structures known as triglycerides, which are the main constituent of the fatty components on vegetables and animals. A triglyceride is a chemical compound formed from one molecule of glycerol and three fatty acids. Animal triglycerides generally have more saturated fatty acyl groups than plant triglycerides. Due to the typical length of the hydrocarbon branches, triglycerides have a chemical structure similar to diesel and jet fuel (12 to 22 carbon atoms). The hydrocarbon branches in a triglyceride molecule are less branched than a typical crude oil. Typical triglyceride feeds for renewable fuel production either originate from vegetable (e.g., soy, palm, corn, jatropha, pennycress, camelina, inedible corn oil, etc.), animal sources (beef tallow, white grease, and poultry fat), waste greases (used cooking oil/yellow grease) and algal and microbial oils. Feedstocks such as soy, corn, jatropha, pennycress, camelina, and waste greases have more unsaturated acyl groups and feedstocks such as palm oil or beef tallow have more saturated acyl groups.

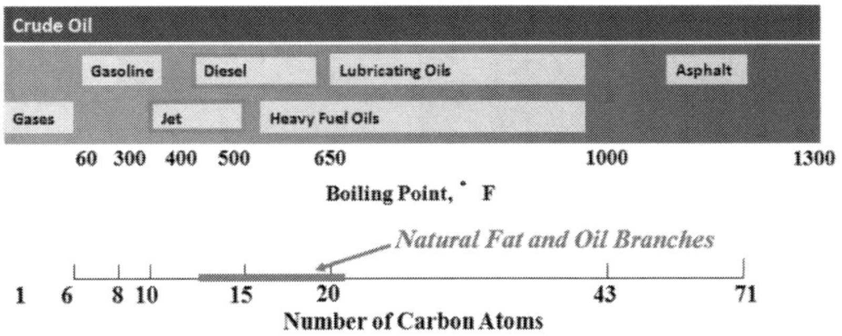

Figure 1. Crude Oil derivatives compared to Natural Fat and Oil

General chemistry for hydrotreatment of biological oils and fats includes olefin saturation, hydrodeoxygenation, decarboxylation, decarbonylation, water gas shift, and hydroisomerization. The general chemical formulas for each reaction are listed below.

Olefin Saturation:

$$C_nH_{2n} + H_2 \xrightarrow{Catalyst} C_nH_{2n+2}$$

Hydrodeoxygenation:

$$C_nCOOH \xrightarrow[\;]{\substack{+3H_2 \\ Catalyst}} C_{n+1} + 2H_2O$$

Decarboxylation:

$$C_nCOOH \xrightarrow{Catalyst} C_n + CO_2$$

Decarbonylation:

$$C_nCOOH + H_2 \xrightarrow{Catalyst} C_n + CO + H_2O$$

Water Gas Shift:

$$CO + H_2O \xleftrightarrow{Catalyst} CO_2 + H_2$$

Hydroisomerization

$$nC_{n+1} + nC_n \xrightarrow{\substack{H_2 \\ Catalyst}} iC_{n+1} + iC_n$$

The following formulas represent the chemical reactions taking place in the treating reactors for producing renewable fuels via the Ecofining and the UOP Renewable Jet Fuel Process. Triglycerides and free fatty acids (FFA) are deoxygenated to produce a straight chain paraffin waxy diesel.

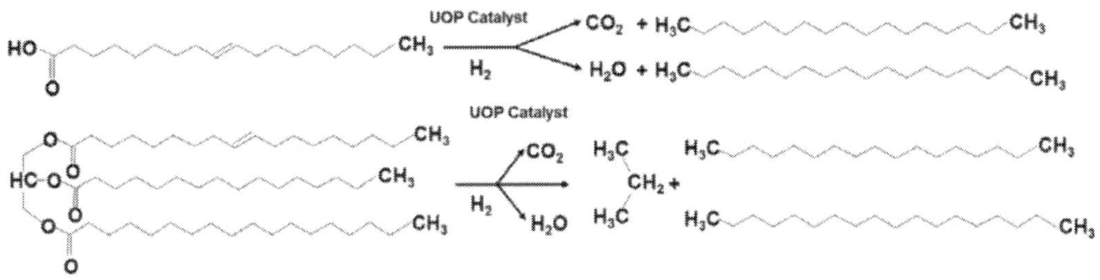

The following formula represents the chemical reactions taking place in the isomerization reactors for producing renewable fuels. Straight chain paraffins are isomerized to produce green diesel, green jet fuel, and green naphtha.

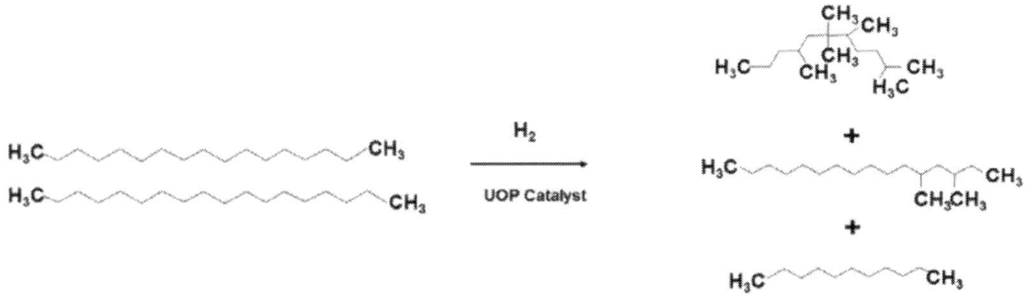

After review of the process chemistry, there are four key processing differences between first stage deoxygenation of natural oils and fats and petroleum gas oil desulfurization:

1. The heat release from deoxygenation of natural oils is significantly higher than heat release from desulfurization.
2. Natural oils and fats have fouling tendencies at high temperatures. Therefore, deoxygenation requires a heat sink to moderate the temperature rise, provide enhanced catalyst stability, and to limit feed side fouling.
3. Hydrogen consumption is much higher for the new service, and similarly the hydrogen recycle gas requirements are higher.
4. Water is produced along with CO_2 from the deoxygenation reactions which leads to metallurgical concerns in the unit.

Green diesel and green jet fuel are chemically identical to components of diesel and jet fuel made from petroleum but with some improved properties that allow these products to be a drop-in replacement or high value blender as shown in Table 1 and Table 2 below.

Table 1. Comparison of petroleum diesel and green diesel

	Petroleum ULSD	Green Diesel
Oxygen, %	0	0
Cetane	40-55	75-90
Energy Density, MJ/kg	43	44
Sulfur, ppm	< 10	< 2
Cold Flow	Baseline	Excellent
Oxidative Stability	Baseline	Excellent

Advantages of green diesel are:

- Lower density: Green diesel is lighter than the average density required of diesel, so when you add a molecule of green diesel, a heavier molecule of a diesel component can be added to compensate. For a refiner, this blending benefit could allow the addition of hydrotreated light cycle oil to enter the diesel pool.
- High blend limits: Green diesel can be used at 100% with no blending required in fleet operations or locations without blending restrictions.

Table 2. Comparison of petroleum jet fuel and green jet fuel

	Jet A-1 Spec	50/50 Blend	Green Jet
Flash Point, °C	38 min	38 min	38 min
Freeze Point, °C	-47 max	-47 max	-47 max
Energy Density, MJ/kg	42.8	43.6	43.9
Thermal Stability	Baseline	Excellent	Excellent
Aromatics	8 min, 25 max	8.5	Nil

Advantages of green jet fuel are:

- Green jet fuel can be blended in a 50/50 ratio with petroleum-based jet fuel and it offers significant advantages over petroleum jet fuel.
- Higher energy density in flight, which allows aircraft to fly farther on less fuel.

Repurpose of Existing Refinery Assets

The process flow scheme for producing renewables is comparable to the one required for processing crude oil derivatives with the main differences as explained above. Therefore, the repurpose of an existing unit or units should be focused on addressing these main aspects and reducing the number of changes to make it economically feasible.

The reconfiguration of existing equipment assets in a refinery unit or multiple refinery units can substantially reduce the amount of capital required and provide favorable project economics for producing renewable fuels. Due to the processing requirements of the renewable feedstocks, the redesign of processing units does involve technical challenges that must be addressed in order to successfully reutilize the existing petroleum refinery units for producing renewable fuels. The main technical challenges that exist in reconfiguring petroleum refinery units include the following: the high acidity and contaminant levels of the renewable feedstock, the high reactivity and thermal instability of the renewable feedstock, the carbon dioxide acid gas generation, and lastly the requirement for two-stage processing (first stage deoxygenation followed by second stage isomerization).

Reconfiguration of existing refinery units in order to produce renewable fuels, particularly green diesel or green jet fuel, may utilize two units and/or individual pieces of equipment added to one or more existing refinery units. Existing hydrotreating (hydrodesulfurization), hydrocracking (mild hydrocracking or full conversion), fixed bed reforming, and isomerization units and other refinery assets could potentially be used in the conversion to a renewable fuel producing facility. An example of reconfiguration of existing refinery assets into an Ecofining™ Process Unit or UOP Renewable Jet Fuel Unit is included in Figures 2 and 3.

The first stage deoxygenation step of the Ecofining process and UOP Renewable Jet Fuel Process is completed in a sour environment with sulfide base metal catalyst. An existing hydrotreating or hydrocracking unit is preferred for reutilization as the first stage deoxygenation section. The goal for reutilization of an existing hydrotreating or hydrocracking unit into this service would be to maintain the existing feed pumps, recycle gas compressor, combined feed fired heater, reactor(s), and product stripper column. Other equipment and piping shall be considered for re-use if considered adequate.

The potential revamp process flowscheme for the first stage deoxygenation section is as follows and is illustrated in Figure 2. Feed is pumped from a feed surge drum towards reactor section and mixed with recycle gas before it is pre-heated with the effluent reactor stream and sent to a combined fire heater to increase the temperature as required by the treating reactor inlet condition. The combined feed flows through the reactor catalyst beds where the contaminants will be removed and the composition will be transformed. Because the reaction is exothermic, it is necessary to have quench addition between each catalyst bed. Reactor effluent is cooled against the cold combined feed, and if possible cooled further with any other process stream and eventually is cooled down further in the air cooled products condenser and sent to a high pressure separator. Liquid from the separator is sent as feed to stripper section. Gas from separator is sent to an amine unit to have contaminants removed and then mixed with makeup hydrogen. The combined gas stream is sent to recycle gas compressor where it is compressed and mixed with the fresh feed. Stripping of light ends (propane rich gas and naphtha) from the intermediate paraffinic waxy diesel is achieved with a steam stripped column. The waxy intermediate diesel product is the fresh feed for the second stage isomerization section.

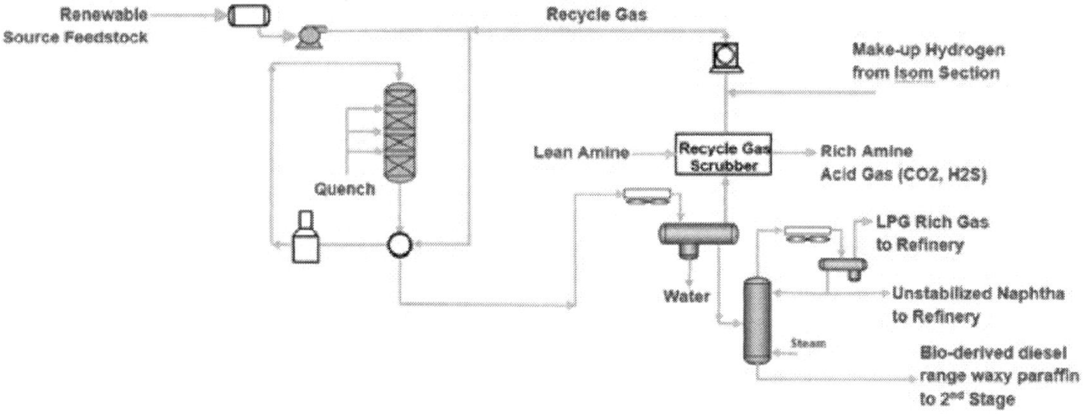

Figure 2. Potential process flow scheme for deoxygenation (treating) section

Several items must be reviewed in order to re-utilize an existing hydrotreating or hydrocracking unit as the first stage deoxygenation section. The renewable source feedstock will have some level of free fatty acids (FFA) and other contaminants. Feed system metallurgy must be reviewed depending on the feedstock composition and contaminant content. Natural fats and oils contain high levels of mineral contaminants, primarily phosphorous and alkali metals such as sodium, calcium, magnesium, and potassium. These contaminants require demetallization catalyst to absorb the metals. Demetallization catalyst is utilized in the first catalyst beds prior to the main treating catalyst. Due to the fouling tendency of the feedstock and the high heat of reaction, the unit must be designed with provisions to prevent fouling and maintain temperature control. The heat release will be much higher for treating of a renewable source feed versus a petroleum feed. Temperature limits in the reactor must be reviewed, and quench rates must be set accordingly in order to properly manage the heat release. The hydrogen consumption will be higher for the new service as well, and therefore the makeup gas compressor and recycle gas compressor need to be evaluated for handling of the higher expected rates. The overall feed rate to the unit may be set by a compressor limitation. Treating of the renewable source feedstock will lead to production of carbon dioxide. This acid gas must be removed in a recycle gas scrubber to avoid build-up of CO_2 and CO (formed via a water gas shift reaction). Carbonic acid corrosion is possible in the high pressure separator due the presence of CO_2 and aqueous phase water. Appropriate metallurgy must be specified in this section, and the existing equipment may not be adequate. Lastly, it is expected that unit will operate at a similar pressure to the previous petroleum operation. This pressure may set a limit on certain contaminants in the feedstock, such as nitrogen.

The second stage isomerization step of the Ecofining process and UOP Renewable Jet Fuel Process is completed in a sweet environment with a noble metal catalyst. An existing hydrotreating, hydrocracking, reforming, or isomerization unit can be reutilized as the second stage isomerization section. Since this stage is completed in a sweet environment, there are fewer concerns with metallurgy in order to re-use equipment and piping. The goal for reutilization of an existing unit or individual pieces of equipment added to a portion of an existing unit into this service would be to maintain the existing feed pumps, makeup gas compressor, feed fired heater, reactor(s), and product fractionator column. The reactor section equipment may be utilized from one refinery unit (e.g., fixed bed reforming unit) and the product fractionation section equipment may be utilized from another refinery unit (e.g., naphtha or raffinate splitter). Other equipment and piping shall be considered for re-use if considered adequate.

The potential revamp process flowscheme description for the second stage isomerization section is as follows and is illustrated in Figure 3. The feed to isomerization section is routed from the first stage into a feed surge drum. From the feed drum, the waxy diesel feed is mixed with recycle gas and make-up gas hydrogen and pre-heated via exchange with any process stream available and sent to the isomerization heater to increase the temperature to the isomerization reactor inlet requirement. The stream from isomerization heater is a two phase stream which passes through the isomerization reactor beds where it reacts to modify the hydrocarbon straight chain into a branched chain. This enhances the cold flow properties of the green diesel.

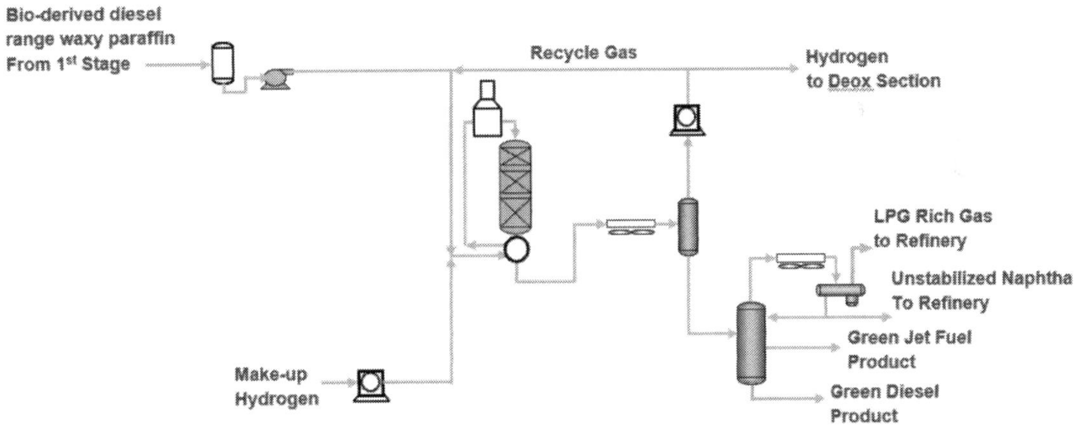

Figure 3. Potential process flow scheme for isomerization section

The isomerization reaction has a low heat of reaction and for that reason does not require reactor quench. However, in a jet mode operation that requires more hydrocracking, a quench addition system and recycle oil may be considered. Effluent from isomerization reactor is cooled with any process stream available (for example, waxy diesel feed or fractionator feed) and is further cooled in the air cooled products condenser before entering a high pressure separator. Vapor from the separator is sent to an existing recycle gas compressor and then split into an isomerization section recycle gas stream and a deoxygenation section makeup gas stream. Liquid from the separator is pressured to a fractionator column for final product separation. The reboiled fractionator column separates light ends (LPG rich vapor and naphtha) from green diesel. An optional green jet fuel side draw may be included.

Since the hydrogen consumption for the first stage section will be higher for the new service and the makeup hydrogen is cascaded from the second stage isomerization section, the makeup gas compressor and recycle gas compressor (now isomerization off gas compressor) need to be evaluated for handling of the higher expected rates. The isomerization section is a mild operation and has significantly lower hydrogen consumption than typical petroleum unit operations. It is expected that the second stage isomerization section will operate at a similar pressure to the previous petroleum operation.

Capital Cost Constraints

Each refiner will have its own capital cost constraints that include both inside battery limits (ISBL) and outside battery limits (OSBL) of the renewable fuel producing unit. A successful reconfiguration project for renewable fuels relies on good infrastructure present in the OSBL portions of the unit. For example, available feed and product tankage, acid gas handling systems, fuel gas systems, utility systems, and most importantly availability of hydrogen. If some of these systems are available, then project economics will be further enhanced.

For the ISBL design of the unit, the goal is to set a fresh feed rate in the treating (deoxygenation) section that will allow re-utilization of the charge pumps, recycle gas compressor, and reactors. The fresh feed rate will typically be limited by charge pump flow, recycle gas compressor flow, or catalyst volume availability. Ability to utilize these pieces of equipment will significantly reduce the capital cost investment for the refiner. At the fresh feed rate selected, other major equipment such as the combined feed fired heaters, reactor feed and effluent exchangers, high pressure separators, and fractionation columns would be evaluated for re-use.

Careful reviews have to be made for each piece of equipment in order to reduce the replacements and upgrades. The modifications are highly impacted by the amount of contaminants expected in the fresh feed. Operating conditions are also affected by these contaminants and the design conditions should be reviewed to stay below limits. Therefore, the contaminant limits should be set to allow for the re-utilization project to be economically feasible. This may mean lower limits on contaminants in the feedstock, such as free fatty acid, metals, phosphorous, nitrogen, and chloride.

As metallurgy is a concern in several areas of the deoxygenation section due to high level of feed contaminants, a thorough analysis should be done to maintain the mechanical integrity of the equipment considered for re-use, while considering the cost impact and long term equipment reliability for the new operating conditions. Equipment that could be affected in the deoxygenation section are the reactor effluent / fresh feed heat exchangers, the heater tubes, and associated reactor feed and effluent piping. Equipment metallurgy may need to be upgraded or equipment may need to be replaced depending on the required metallurgy for handling of the free fatty acid in the feedstock. The deoxygenation reactors may require replacement due to metallurgical concerns, so the availability of a suitable reactor may be the key to successful reutilization project. Similarly, the recycle gas treatment system may require modifications to handle CO_2 acid gas removal such as additional amine flow or the use of a different amine that is more selective for CO_2 removal. The high pressure separator section should be reviewed for metallurgy adequacy as it has to be able to handle potential corrosion due to carbonic acid (wet CO_2) and chlorides. Typically new and modified equipment is required in this section. Pumps are typically reused with few or no modifications. Fractionation section equipment is not highly affected and is typically re-used with few or no modifications.

Metallurgical modifications are typically limited for the isomerization section, as it is a cleaner service. Depending on the severity required in this section (e.g., increased jet fuel production), some modifications may be necessary for the reactor (number of catalyst beds, nozzles for interbed quench), but not at a high cost impact.

In a recent revamp study performed by Honeywell UOP, it was considered to repurpose existing refinery units with the objective to confirm achievable capacity while considering optimal equipment re-purpose for reduced capital cost. After all the constraints explained above were analyzed and an optimal configuration was determined, the score card report for the ISBL unit showed that 45% of the equipment was adequate, 25% of the equipment needed modifications and 30% of the equipment was new equipment, either as a new service or replacement of an existing service. Successful revamps of refinery facilities using Honeywell UOP renewable fuel technologies have been implemented for about one-third of the overall (ISBL & OSBL) capital costs of grass-roots projects.

Conclusions

The implementation of renewable fuel projects must consider many aspects to determine the best approach for each particular project. In some cases it can be best to install grass roots facilities to produce drop-in renewable fuels for improved process flexibility at a desired location. However, co-locating a facility adjacent to or within an existing refinery can provide substantial cost savings and should be considered as one of the options.

Refinery reconfiguration presents an opportunity for project developers and refiners seeking creative ways to enhance profitability and to meet government regulations for renewable transportation fuel production, while utilizing existing assets and reducing capital investment. If a refiner has idle, underperforming assets, or underutilized units such as hydrotreating, hydrocracking, reforming or isomerization units or associated facilities, it may be a good opportunity to seek options to produce renewable fuels. This could offer an advantageous opportunity for a refiner to optimize the use of its assets and provide overall value added to the final product chain, as renewable fuels help to reduce emissions and are mandated in many countries to be included in final products. Therefore, it is highly recommended to consider the reconfiguring of the units to produce green diesel and/or green jet fuel using the unique experience that Honeywell UOP has for such projects.

Although renewable fuels are similar to petroleum derived fuels, renewable source feedstocks in contrast present different type of contaminants and processing requirements than petroleum feedstocks. This imposes interesting technical challenges when trying to reuse existing refinery assets. Honeywell UOP has proven it has the expertise in renewable fuel technology development, and has commercially demonstrated reutilization of refinery assets for renewable fuel production at the Eni Biorefinery in Venice, Italy and AltAir Fuels facility in Paramount, California. These successful reconfiguration projects produce renewable fuels that reduce greenhouse gas emissions, help to reduce dependence on imported oil, and most importantly deliver it safely, reliably and economically.

Acknowledgements

The authors would like to express sincere thanks to the following Honeywell UOP personnel who participated in the work described above: James Andersen, Donald Eizenga, Daniel Ellig, Ralph Davis, Gabriela Gonzalez Lippke, Ralph Norton, Chad Dominic Cavan, Anthony Barnette, Prateek Gupta, and Afrin Ahmed.

444563 Dynamic Simulation for APC Projects: A Case Study on a Reformate Splitter with Side Draw

Abstract:

What happens when one tries to control the same impurity – here benzene content – in the top, the bottom and the side draw of a Reformate Splitter fractionation column? If there is no benzene in the bottom of the fractionator (e.g. because column temperature profile is high), what impact does this have on the APC models? This presentation addresses some of the challenges of trying to handle non-linear process responses with APC linear models, and how steady-state and dynamic simulation helped APC engineers gain the necessary process knowledge to make the right decisions.

Assessing Benefits of Advanced Process Control using Dynamic Simulation Model

Bhaskar Iyer, Ravi Nath, Ken Allsford, Mike Yen
Honeywell Process Solutions
Houston, TX

2016 AIChE Spring Meeting
Houston, TX
April 13, 2016

Honeywell

1

Outline

- Introduction
- APC Benefits Assessment
- Integration of APC and Dynamic Simulation
- Benefits of APC using Dynamic Simulation
- Conclusion

Honeywell

Introduction

- Advanced Process Control (APC)
  - Multivariable control & optimization of an Industrial Process
  - Helps process control for
    - Difficult dynamics (Long dead time, Inverse response etc.)
    - Strong interactions and multiple constraints
    - Difficult control problems due to complicated line-up or sizing
    - Unmeasured disturbances that cannot be eliminated
- Dynamic Simulation Model
  - Model process startup / shutdown and upsets
  - Common use – Operator Training Simulators (OTS)
  - OTS is usually integrated with Distributed Control System (DCS) simulation

Main Benefits

- APC Benefits
  - Ability to operate the process closer to constraints
  - Ability to solve difficult control problems
  - Provides framework for optimization in real time
    - Higher throughput, higher profit
- Dynamic Simulation Benefits
  - Tool to perform engineering calculations of industrial processes
  - Tool to understand the dynamic behavior of processes
- OTS Benefits
  - Operator training and competency tool
  - Operational assurance: validation of procedures, plant and control system design, control system configuration

APC Benefits Assessment

- Cost-benefit analysis is usually required for APC Projects
  - Traditional techniques
    - Plant History Data
    - Estimate standard deviation reduction and mean-shift
- Some plants look for ways of continuous improvement in operation using APC
  - Reproducibility, broaden operability range, safety
- This work discusses an alternative when OTS already exists
  - Create a prototype APC on the OTS and switch it ON
  - Serves as a proof of concept
  - Early involvement of operations staff helps in getting the buy-in from all concerned parties
  - Provides a platform for demonstration of APC benefits

Honeywell

Dynamic Simulation Model

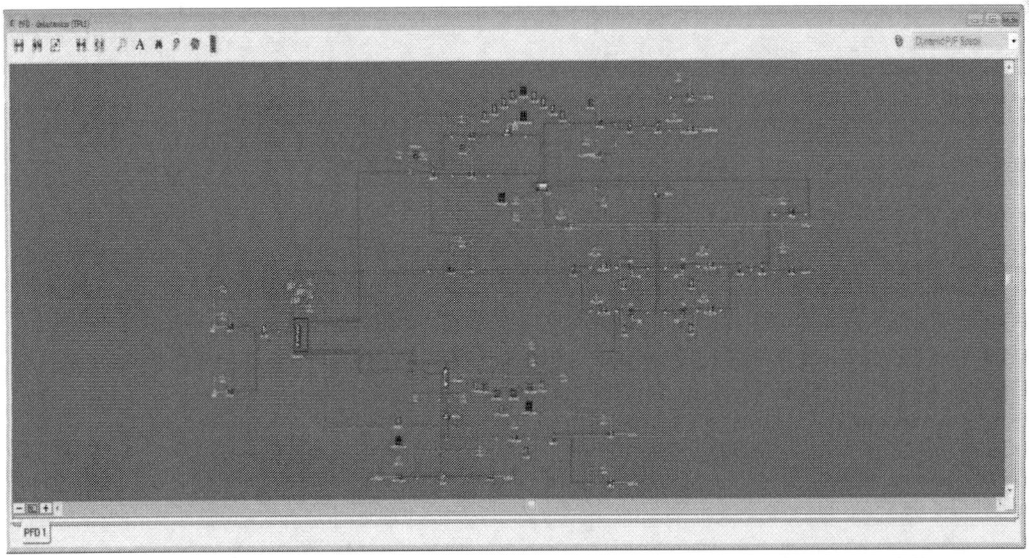

Honeywell

OTS Link with DCS

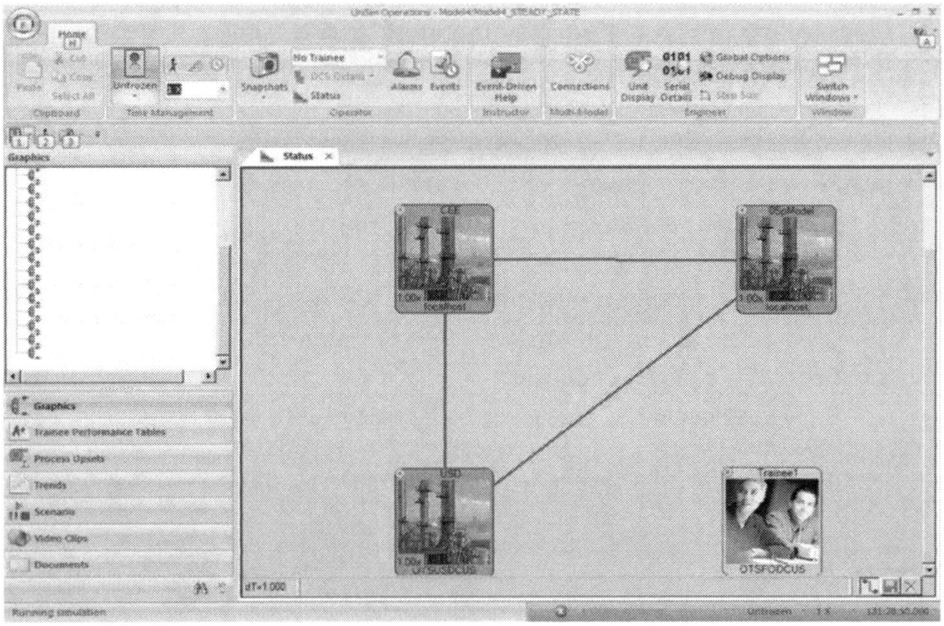

Honeywell

OTS Instructor / Operator HMI

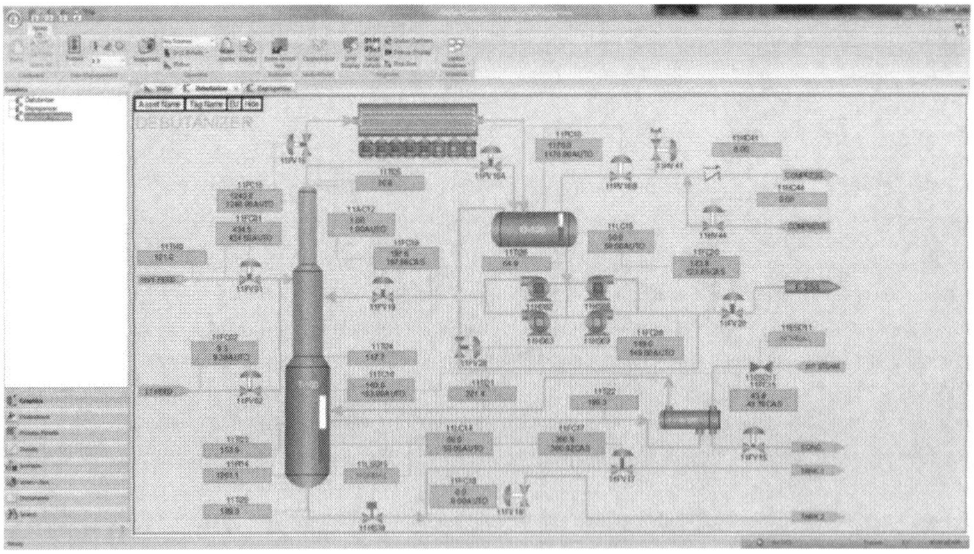

Honeywell

APC

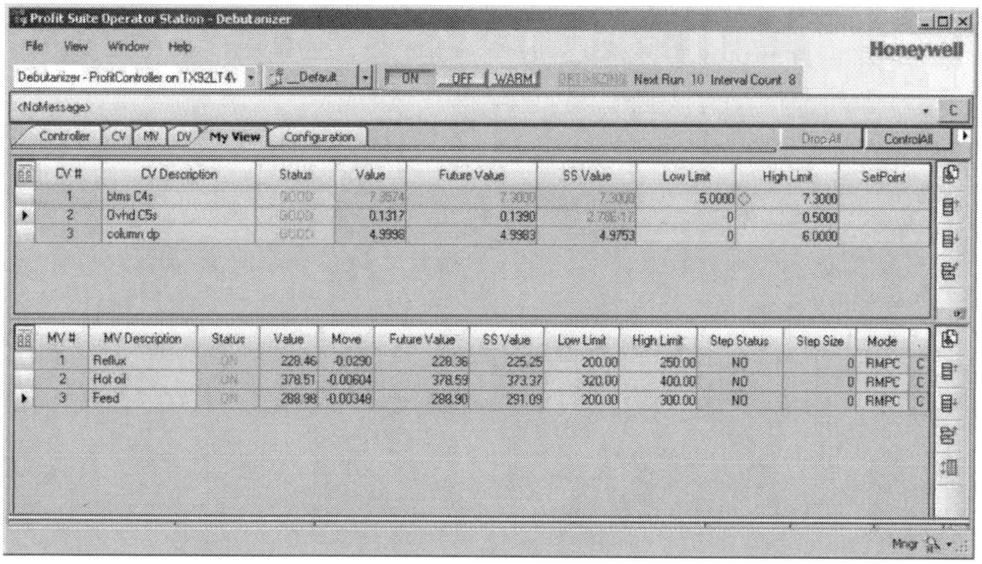

Honeywell

Architecture of APC on OTS & DCS Simulation

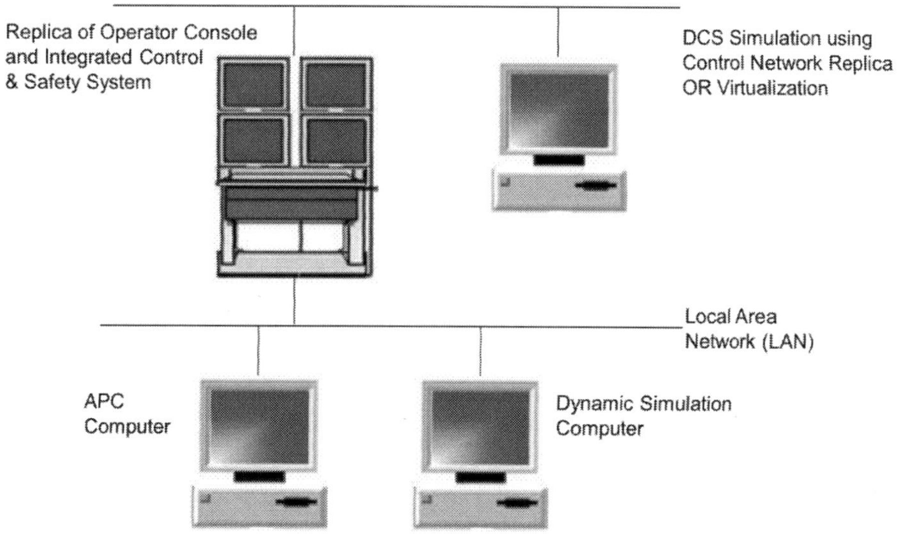

Honeywell

10

APC Project Steps

- Benefits estimation
- Regulatory and APC design
- Step test and model identification
- Controller building, acceptance test
- Implementation and commissioning
- Operator training
- Performance monitoring

APC on OTS offers benefits at each step of the project

11

APC Design

- Creating the APC design involves
  - Getting the plant objectives
    - Disturbance rejection, optimization etc.
  - Selecting the technology
    - Linear/nonlinear, gain based
  - Implementing
    - Base level and advanced regulatory control
    - PID Tuning
  - An iterative process involving
    - Step testing
    - Model identification
    - Control strategy

Honeywell

12

Step Testing Software

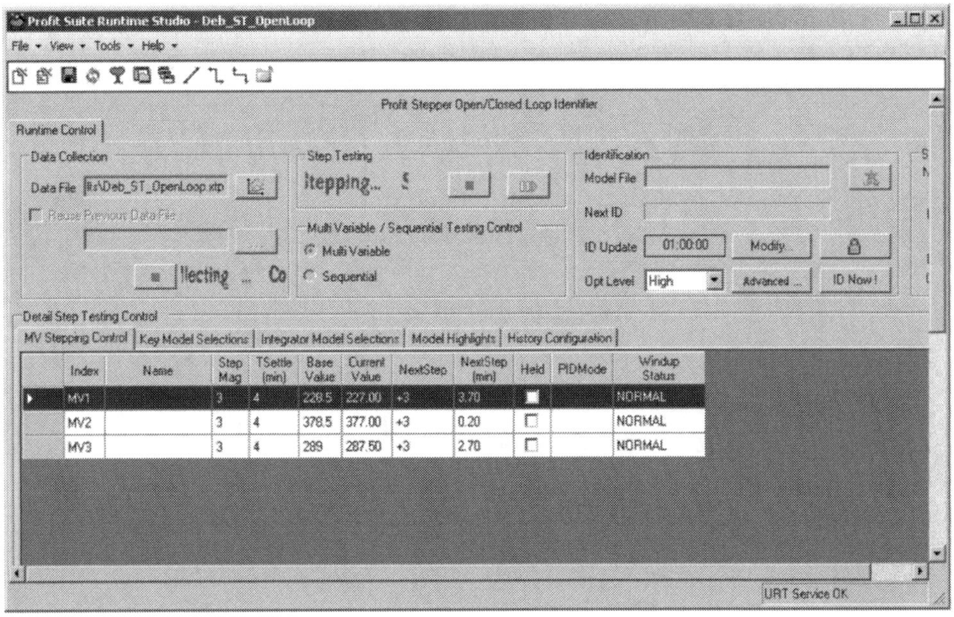

Honeywell

13

Open/Closed Loop Sequential Step Testing

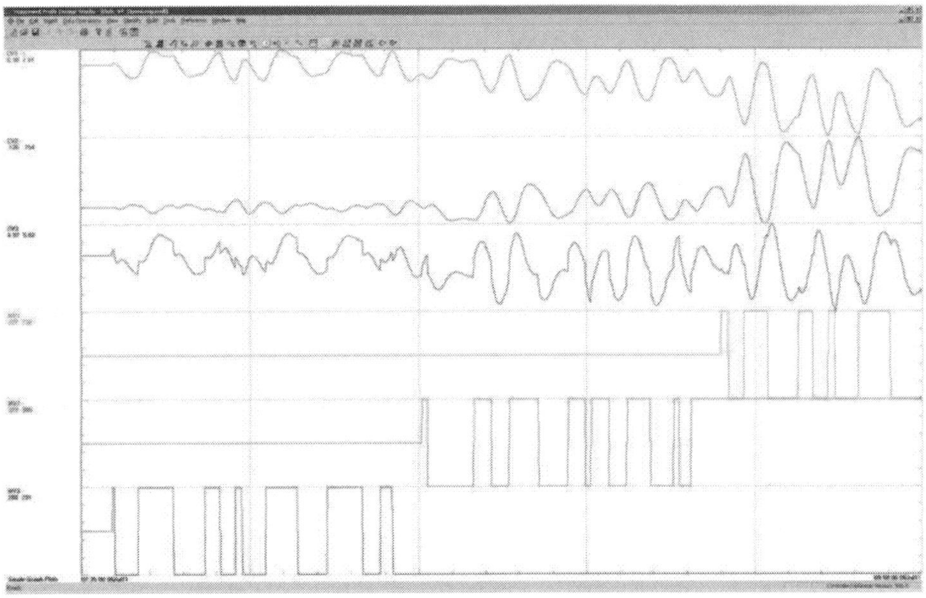

Honeywell

14

Open/Closed Loop Multivariable Step Testing

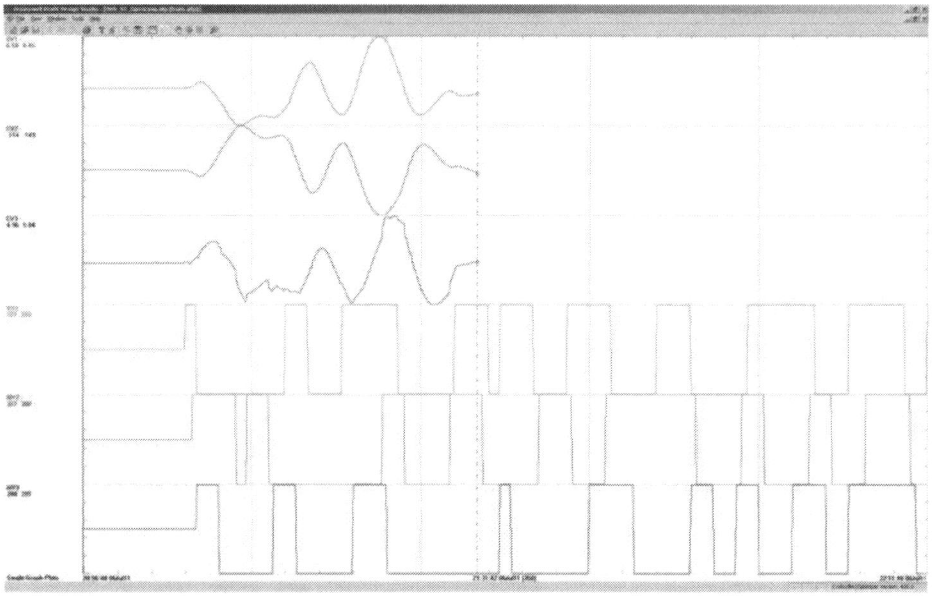

Honeywell

15

Simultaneous Model Identification

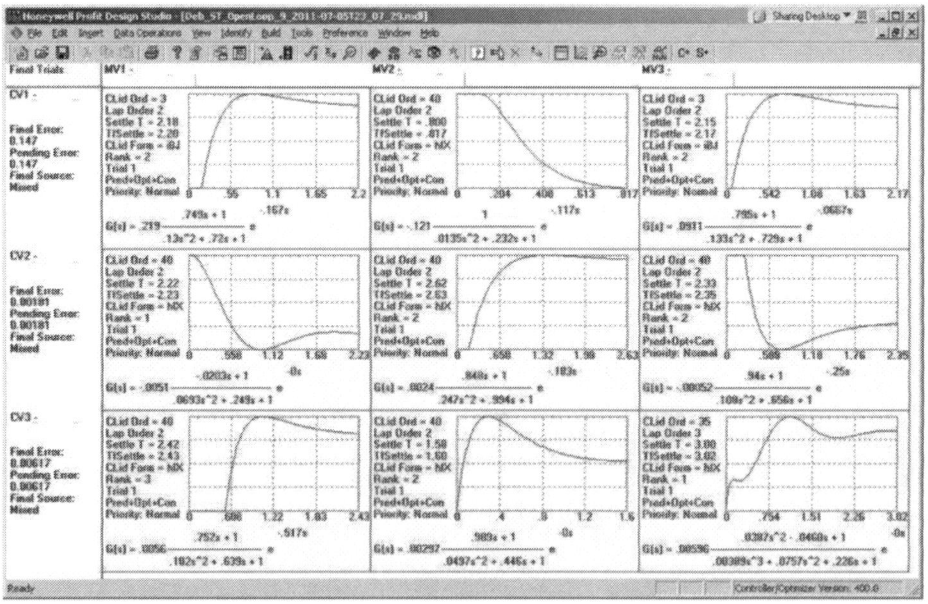

Honeywell

Benefits of APC on OTS for APC Design

16

- APC on OTS provides a test bed for
  - Designing the APC and base level DCS control strategies
  - Designing operator HMI, APC trends and integrity checks
  - Cross check results with the operations
  - Ability to simulate various disturbance scenarios
  - Reduced effort and APC project time-span
  - Facility for
    - Speed-up
    - Saving test cases

Honeywell

During APC Implementation

17

- Standard Workflow
  - Documentation for Management of change
  - Build APC and other control configuration on live system

- Benefits of APC on OTS
  - Documentation
    - Detailed design, acceptance test, commissioning procedure
    - Automate document creation
  - Framework for operator guidance and training
  - Offline control configuration and testing

Honeywell

During APC Performance Monitoring

- Standard Workflow
  - Key performance indicators (KPI) for APC reports
  - Change it based on feedback
  - Plant specific monitoring of APC benefits
- APC on OTS provides test bed:
  - Design custom performance monitoring reports
  - Make sure it meets the requirements

18

Honeywell

Conclusion

- APC on OTS is being successfully applied in various plants across the globe
  - Ammonia plants, gas plants, mineral extraction, refining, petrochemical and many more
- It gives a good idea of the benefits accruable with the use of APC
- It helps in identifying opportunities for improvement in control
- It offers many advantages in the design and implementation of APC

19

Honeywell

443952 Automated Data Screening and Selection for Model Identification and Adaptation Makes Adaptive MPC and Sustained Performance an Easy Project

Abstract:

--ABSTRACT

Multivariable Predictive Control (MPC) is the most widely used advanced process control technology in process industries, with more than 5,000 worldwide applications currently in service. A common and challenging problem is that MPC control performance degrades with time due to inevitable changes in the underlying subject process, such as equipment modifications, changes in operating strategy, instrumentation degradation, etc. Such degradation of control performance results in loss of benefits. Among all causes of control performance degradation, the process model's predictive quality is the primary factor in most cases. To sustain good control performance, the model needs be periodically audited and updated.

To address above challenging problem, a series of innovative technologies were emerging and put into APC practices with significant benefits. Such as automated non-invasive closed-loop step-testing, online model quality auditing, online closed-loop model identification and adaptation, and automated data screening and selection techniques. In this presentation, we will introduce the automated data screening, selection and repairing technology, which changed the traditional way to do an APC project, sped up the APC project tremendously and more important, made the Online Adaptive MPC a reality. This piece of technology can also serve many other data-related modeling and online applications, such as statistical analysis, root-cause analysis, process model calibration, data analytics, Big Data, etc.

By using this technique, all available information associated with process data, such as data sample status, High/Low operating limits, individual variable attributes in MPC controller (MV, DV or CV) and PID control loop associations are utilized to screen data, identify bad sections and repair them with interpolations or model predictions if possible. Internally built MISO (Multi-input, Single-output) predictive models are also employed to assist detection of spikes and unknown disturbances. Based on various criteria, "bad data" for specific applications are automatically identified and marked for removal/skip. Special algorithms for data point interconnection and adjustment techniques are provided to guarantee smooth/continuous replacement values. As a results, the usage of plant test data is maximized in model identification and adaptation. Examples will show how many different type of bad data slices can be automatically identified and how can this auto-data-slicing technique help improve model's gain accuracy as well as RGA stability in a collinear case.

441802 Improve Run Length of Crude Heaters Processing Shale Crudes

Abstract:

Refiners have now started processing shale crudes in their refineries. One of the major issues faced by the refineries is the run length of the crude heaters. Shale crude is light but it presents a lot of challenges in processing. One of the refineries in Texas started processing shale crudes in their unit and the run length of the heater came down from 5 years to less than 3 months. The refinery started injecting anti-foulants in the feed . They had limited benefit but the continuous cost of injecting anti foulants was in millions of dollar per year. The refinery approached Furnace Improvements to look at the crude heater operation and recommend improvements in the crude heater to extend the run length. FIS reviewed the crude heater design, analyzed the operating data and carried out CFD modeling of the existing heater. FIS recommended modifications on the fire side to improve heat transfer. FIS carried out CFD modeling of the fire side modifications and showed improvements in the film temperature and tube metal temperatures to the client. FIS implemented the modifications on the fire side and the run length has gone up by 6 times to about one and a half year without the expense of costly anti-foulants. We will present the challenges faced and improvements that were made to the heater in this presentation.

442002 Distillate Fractions from Hydrotreated Bio-Oils: Comprehensive Analyses

Abstract:

Producing an infrastructure-ready fuel from biomass requires a better understanding of the chemical constituents of the gasoline, diesel and jet fuel fractions of the biofuel product. Determining the heteroatom concentration in a specific fraction, especially O and S, impacts the acceptability of a distillate fraction within specific fuel specifications. These reactive and corrosive components can cause complications during downstream processing of the upgraded bio-oils. We have developed a suite of standardized methods to analyze and quantify the bio-oil components. We refined the analysis of distillates from pyrolysis oil by controlling the source and processing of hydroprocessed bio-oils with two oxygen contents. The upgraded bio-oils were subsequently distilled into five fractions. Analytical techniques such as elemental analysis, density, viscosity, simulated distillation (ASTM D2887), ^{13}C NMR, PIANO and HPLC measurements were used and results will be presented. Specific fractions were found to satisfy boiling point requirements. However, functional group characterization is needed to determine the fraction's actual utility, for example as a co-processing stream in the refinery or as a fuel blend stock. The methods developed, lessons learned, and application of these methods to different oils will be discussed.

442854 The Change Accelerator: Integrating Operability and Reliability into Your Culture

Wednesday, April 13, 2016: 10:15 AM
343A (Hilton Americas - Houston)

Mike Robin, Applied Knowledge LLC, Houston, TX, **Jim Mynaugh**, Applied Knowledge LLC, Lipan, TX and **Janet Hammill**, J L Hammill Consulting, Houston, TX

Abstract Outline

The call to action:

Investing in systems and structures to improve reliability will not be effective without engaging the organization in the change process. The diagram below reveals what is needed to successfully bring reliability to operations.

The Change Accelerator Model:

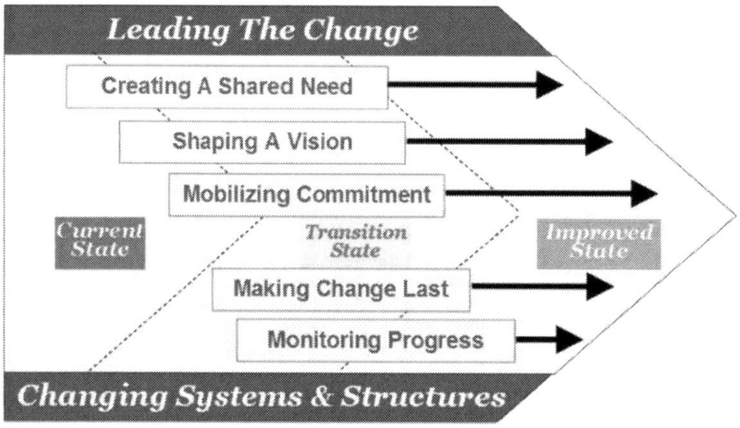

Topic, Tools and Examples:

The examples provided will describe how the change accelerator model was successfully used in a number of global plants with culturally different organizations.

Topic	Tool	Example
Leading change begins at the top	Explain leadership alignment, communication plan	Increased capacity in a Ni refinery in Australia
Operation and reliability improvement requires a shared vision with metric	Explain cascading metrics	Uptime improvement and injury reduction in Australia, First pass yield in USA
Mobilizing commitment	Explain the GRPI model and Leadership alignment	Change process and leadership alignment in the Ni refinery, Stakeholder analysis in Saudi Arabia
Making reliability last	Explain Poka Yoke and control charts with mitigation plans and cold eyes review	Defect reduction for a batch process in China

Benefits and expected outcomes:

Visible metric mix linking shop floor performance to business results

Sustained operations and higher reliability

Simpler, proven management process

More highly engaged workforce

Key points to remember:

1. Engaging leaders and the organization is easy with proven tools and techniques.

2. Our experience allows your efforts to be easier on your people and more effective and sustainable.

3. This experience is valuable wherever you are in your implementation efforts.

Benfield ™ Revamp Process for Lowering Operating Cost and Improving System Throughput

Executive Summary

AIChE Spring Meeting

Houston, TX

April 13, 2016

The UOP Benfield ™ Process is a thermally regenerated cyclical solvent process using an activated and inhibited hot potassium carbonate (HPC) solution to remove CO_2, H_2S and other acid gas components.

The basic chemistry of scrubbing acid gas by HPC solution is the exothermic acid-base reaction. Activator promotes the absorption of carbon dioxide by HPC solution by increasing mass transfer rates. Activated HPC is the basis of the Benfield ™ Process. The first activator was DEA, and this is still in use in many Benfield™ units worldwide. A newer activator, ACT-1™, has been commercialized for several years and shows considerable improvement over DEA, including improved CO_2 mass transfer, more favourable vapour-liquid equilibrium, and lower foaming tendency. Conversion to ACT-1™ from DEA activation can be done "on-the fly" without a complete replacement of the solvent inventory.

Benfield ™ Process has three basic flow schemes: single stage absorber, two stage absorber and two stage regenerator. The more sophisticated flow scheme offers more design and operationally flexibility and deeper acid gas removal capability. Rich solution flash drum can be added to any flow scheme to improve the acid gas purity.

Existing plants can be revamped for either capacity increase and/or energy savings through the use of UOP proprietary LoHeat™ technology. Typically this process uses an additional vessel, a lean or semi-lean solution flash tank, to reduce solution pressure. The multi-stage reduction in pressure removes CO_2 from the solution, improving its absorption ability in the CO_2 Absorber. Steam is also produced by the pressure decrease, which can then be compressed and injected into the Regenerator Column in place of reboiler steam. There is no change in the quantity of stripping steam needed to regenerate the carbonate solution. Rather, the reboiler duty is reduced and replaced with internally generated steam from the LoHeat™ flash tank.

For HPC systems that have the ability to process increased feed gas rate from upstream sources, UOP high performance packing can be used to de-bottleneck the absorber and regenerator towers to allow higher vapor rates. Higher performance packing reduces required tower diameter, packed bed height and pressure drop.

In summary, Benfield ™ Process is one of the most widely used solvent based acid gas removal processes available today. Over the last 50 years, UOP has licensed more than 700 Benfield ™ units. UOP offers the opportunity to upgrade these installed and operating units for operating cost reduction and capacity increases through the revamp options discussed above. Recent Benfield ™ revamp project is presented as an example of how the upgrading options are implemented. Challenges in the Benfield ™ units Revamp includes constraints from existing equipment, required battery limit stream conditions, unit water balance and available plot space and external steam.

444390 Template-Free Synthesis of Hierarchical Zeolites for Petrochemical and Refining Processes

Abstract:

Zeolites are wildly used catalysts in various refinery and petrochemical processes because of their inherently strong acidity and uniform micropore structure. Unfortunately, the microporous structure of traditional zeolites often suffer from serious diffusion problems for bulky reactants/products, leading to low utilization of their active sites and poor catalytic activity, Introducing mesopores into zeolites, with a combination of micropores and mesopores, could overcome such diffusion limitation while retain the size and shape selectivity. Here, we report a low-cost and environmental friendly synthetic synthesis route for mesoporous zeolites without the use of template. The produced mesoporous zeolites exhibit remarkably higher catalytic activity than their conventional counterparts. This could be very useful in various upgrading processes of crude oil.

443608 A Deep Dive into Iron Contamination on Fluid Catalytic Cracking Catalysts

Abstract:

Fluid catalytic cracking (FCC) utilizes a solid, fluidized catalyst, mainly made up of silica and alumina. There are many components in an FCC catalyst including amorphous matrix material, zeolite Y crystal, metals trapping ingredients, binders, and other additives. The FCC catalyst is prone to deactivation and unwanted effects by a number of contaminants that come into the FCC via the crude oil. Contaminants of concern include nickel, vanadium, sodium, and iron. Nickel acts as a dehydrogenation catalyst and leads to higher hydrogen and coke during FCC unit operation. Vanadium directly attacks the zeolite active sites of the FCC catalyst. Sodium neutralizes the acid sites of the FCC catalyst rendering its activity lower. Iron has many different facets to its contamination profile and they will be discussed in this presentation.

Iron comes in with the crude oil. It is found in especially high concentrations in tight oils, so has become more of a concern in the United States in recent years. Iron participates in a number of mechanisms in an FCC unit including mild dehydrogenation, CO promotion in the regenerator, and acts as in inverse SOx additive, resulting in higher SOx emissions from the FCC unit. More importantly, iron can block the surface pores of the catalyst, which prevents FCC feed from entering the catalyst particles. This will directly result in lower conversion and higher slurry yields coming off the FCC. This latter phenomenon has been studied in detail and will be presented here. The effect of various technologies and manufacturing techniques will be discussed. Case studies involving refinery operations will be presented, some of which showing very extreme effects on the FCC catalyst. The source of the problem and the mitigation steps to curb iron poisoning of FCC catalyst will be presented in this lecture.

443410 Gasoline Sulfur Reduction in FCC Using Catalyst Additives

Abstract:

Tier 3 standards are coming into effect in 2017 and are aimed at reducing sulfur in gasoline, ultimately to reduce the impacts of motor vehicles on air quality and public health. Tier 3 modifications can be made in a refinery by capital investments, via either pre or post hydrotreating of the crude oil or naphtha cut, respectively. For refiners looking to avoid or delay capital expenditure, FCC catalyst additives offer an alternative route. FCC catalyst additives are also used during hydotreater outages.

Reducing sulfur in the naphtha cut off an FCC is an important consideration since 90% of the naphtha sulfur in a given refinery typically comes from FCC naphtha. The amount of reducible sulfur depends on whether or not the FCC feed is hydrotreated. With hydrotreated feed, the sulfur in the gasoline cut is typically lower, but the sulfur that remains is more difficult to remove (aromatic sulfur). On the other hand, unhydrotreated feed contains higher sulfur but is easier to remove. Saturates, mercaptans, thiophenes, and alkyl thiophenes see good or reasonable conversion to H2S or coke with the use of an additive. Benzothiophenes, although present in lower quantities than other species, have much lower conversion rates and are harder to remove. A sulfur reduction from benzothiophene of 15% or less is common with an additive.

Overall, when looking at the entire sulfur species in FCC naphtha, an average of about 20-40% reduction across the entire boiling point range can be achieved with gasoline sulfur reduction additives. Typical loading is about 15-25% of the base catalyst and can be synergistically combined with the base catalyst formulation to minimize activity loss via dilution effects. This presentation will discuss recent testing with gasoline sulfur reduction additives and FCC unit examples.

Tuning PID Controllers on Integrating Process with Complex Dynamics

James Beall
Principal Process Control Consultant
Emerson Process Management

16AIChE
Spring Meeting, Houston, TX

EMERSON.
Process Management

Agenda

- Background
- Tuning Method
  - Identify the process type as Integrating
    - Integrator Plus Lag
    - Integrator Plus Lead
  - Measure Process Dynamics
  - Choose "Speed of Response", Lambda
  - Calculate Tuning
- Case Studies
  - Integrator Plus Lag
  - Integrator Plus Lead
- Questions

16AIChE
Spring Meeting, Houston, TX

EMERSON.
Process Management

"PID = Proportional, Integral, Derivative"

- **P = proportional** - output contribution is "proportional" to error (Set Point – PV)
- **I = Integral** - output contribution is from the "integration" of the error
- **D = Derivative** - output contribution is based on the rate of change of the error. Actually, often based on the rate of change of the PV.

16AIChE
Spring Meeting, Houston, TX

EMERSON.
Process Management

Caution!!! – Units for Tuning Constants

- Controllers have different units for their PID tuning constants. For this presentation:
  - **Proportional** action will be expressed in **Gain**
  - **Integral** action will be expressed as **Reset Time**, **Tr,** in **seconds**
  - **Derivative** expressed as **Rate** in units of **seconds (not used in this presentation)**

16AIChE
Spring Meeting, Houston, TX

EMERSON.
Process Management

Caution!!! – PID "Form"

- Controllers can have a different "**Form**" of the PID algorithm. In some cases this can significantly impact the response of the same tuning parameters
  - Forms include
    - Standard (aka Ideal, Non-interactive)
    - Series (aka Classical, interactive)
    - Parallel
  - The tuning calculations in this presentation are based on the "Series" Form. Conversion to "Standard" Form tuning is provided in Appendix A

16AIChE
Spring Meeting, Houston, TX

EMERSON
Process Management

Tuning Procedure

- Step 1: Identify the basic process type
  - Self-regulating
  - Integrating
    - Integrator plus lag
    - Integrator plus lead
- Step 2: Measure process dynamics
  - Must use %Process Variable (%PV) and %OUT
  - %ΔPV = ΔPV_Eng_Units*(100%/Span_PV_Eng_Units)
  - %ΔOUT= ΔOUT_Eng_Units*(100%/Span_OUT_Units)
- Step 3: Choose desired closed loop response time, "Lambda"
- Step 4: Calculate tuning constants

16AIChE
Spring Meeting, Houston, TX

EMERSON
Process Management

Step 1 – Identify Self Regulating or Integrating

Self Regulating Process

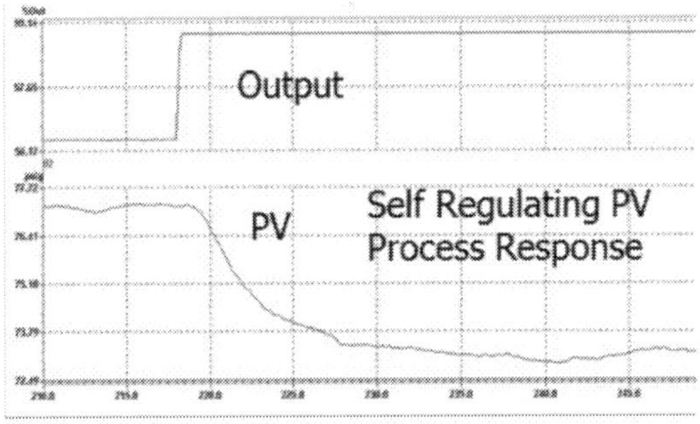

Integrating Process

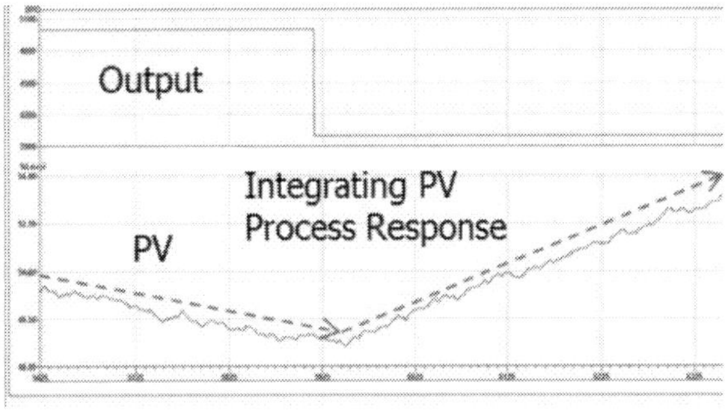

16 AIChE
Spring Meeting, Houston, TX

EMERSON
Process Management

Step 1 – Integrating, Complex

Integrator

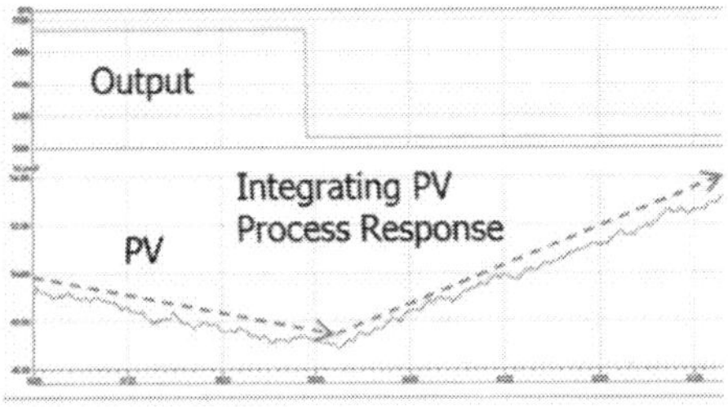

Integrator + Lag

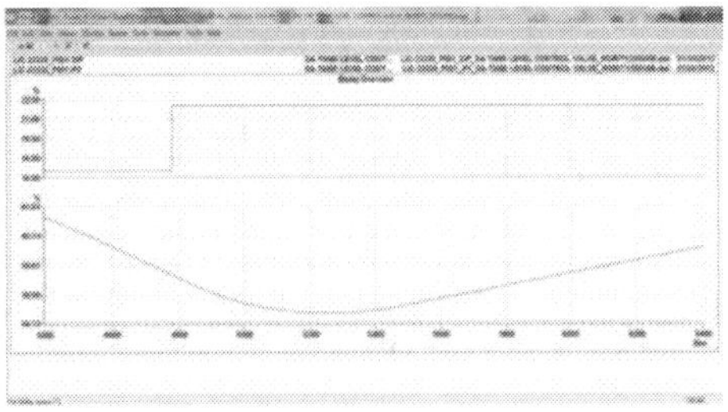

Integrator + Lead

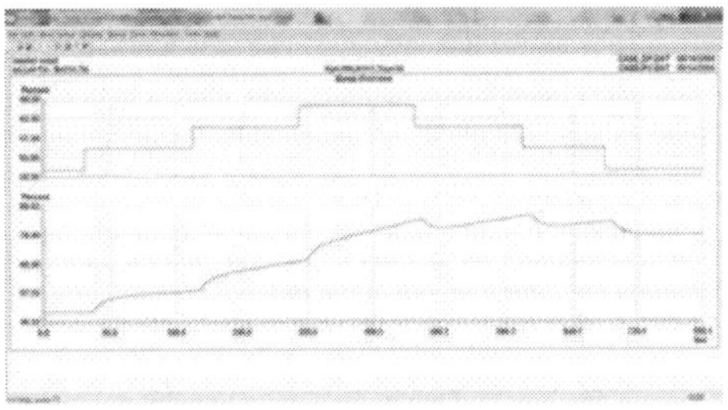

16 AIChE
Spring Meeting, Houston, TX

EMERSON
Process Management

Integrating Process - Open Loop Response

- Most levels
- Some temperatures
- Some Pressures (e.g. large pressure drop across valve)

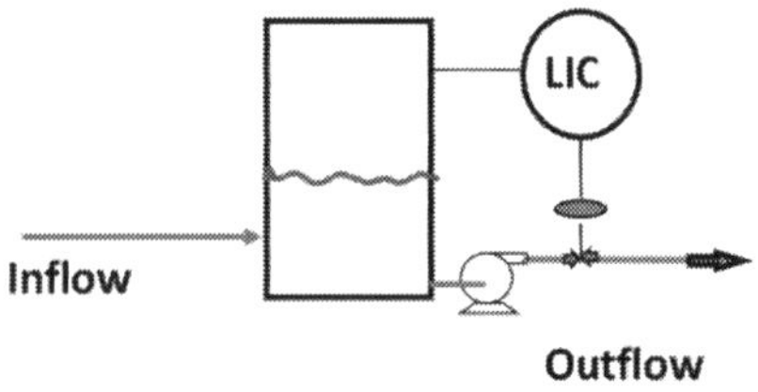

16AIChE
Spring Meeting, Houston, TX

EMERSON.
Process Management

Step 2 - Measure the Process Dynamics

- There is no "final value of the PV" so you can't measure $\frac{\Delta\%\text{PV}}{\Delta\%\text{OUT}}$ for Process Gain!
- So, you use the "change in slope of %PV"
- Integrating Process Dynamics are:
  - Process Gain, Kp = $\frac{\%\text{PV Slope Final} - \%\text{PV Slope Initial}}{\Delta\%\text{OUT}}$
  - Dead Time, Td in seconds

16AIChE
Spring Meeting, Houston, TX

EMERSON.
Process Management

Step 2a - Measure the Process Dynamics-Kp

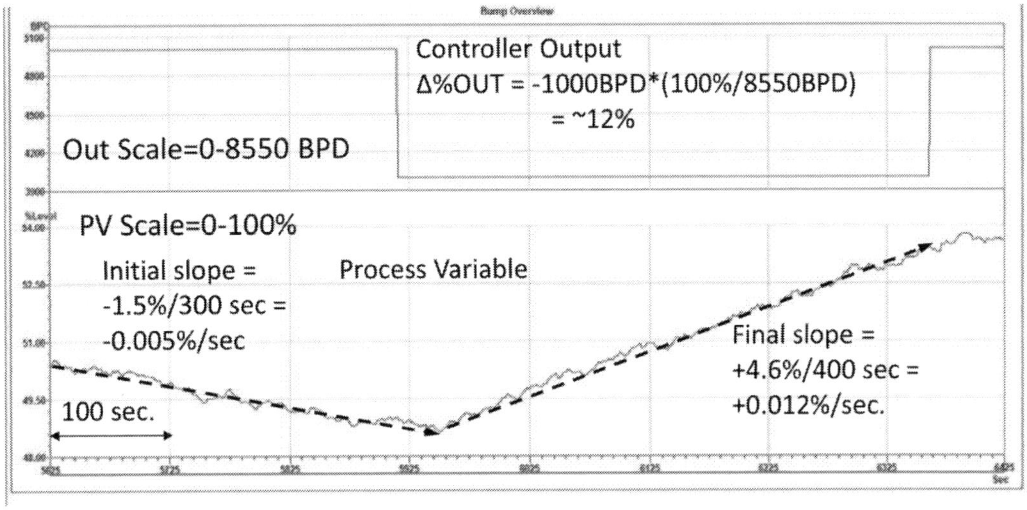

Kp = (final slope - initial slope)/ Δ%OUT

$$Kp = \frac{0.012\%PV/sec - (-0.005\%PV/sec)}{-12\%OUT} = -0.0014\%PV/sec/\%OUT$$

16AIChE Spring Meeting, Houston, TX

EMERSON Process Management

Step 2b - Measure the Process Dynamics-Td

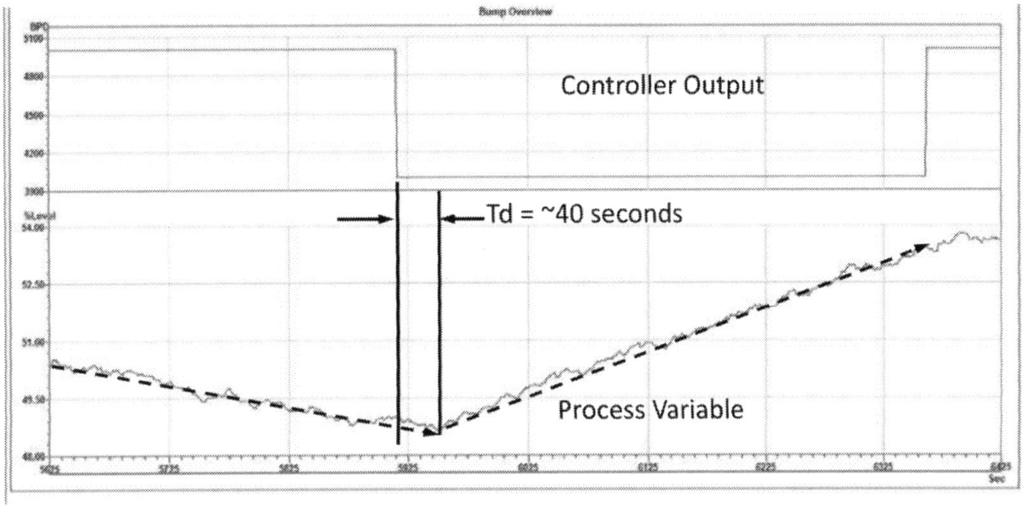

Process Model for this Integrating Process is:

- Process Gain = Kp = -0.0014%PV/sec/%OUT
- Dead Time = Td = 40 seconds

16AIChE Spring Meeting, Houston, TX

EMERSON Process Management

Step 2c - Measure the Lag

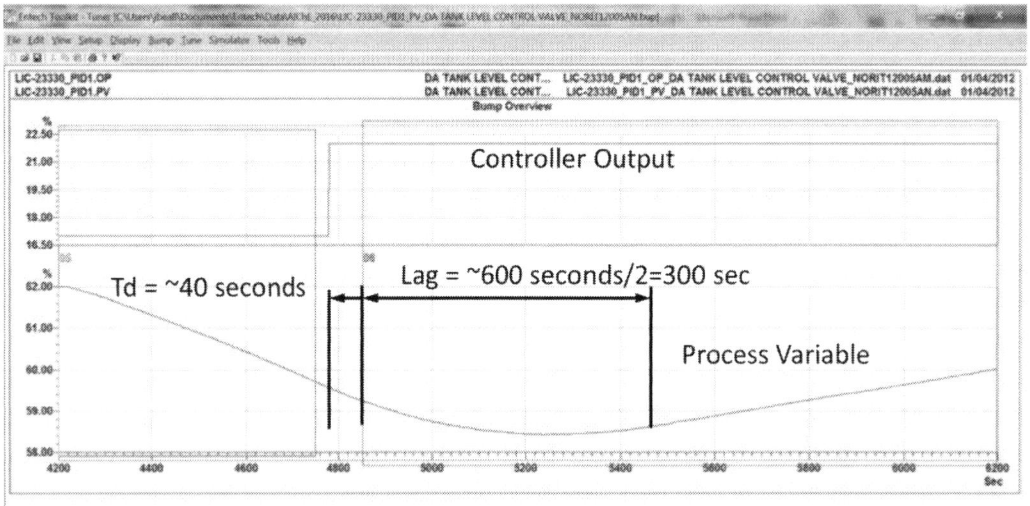

Process Model for this Integrating Process is:

- Process Gain = Kp = -0.0014%PV/sec/%OUT
- Dead Time = Td = 40 seconds
- Lag = 289 seconds

16AIChE Spring Meeting, Houston, TX

EMERSON Process Management

Step 2c - Measure the Lead

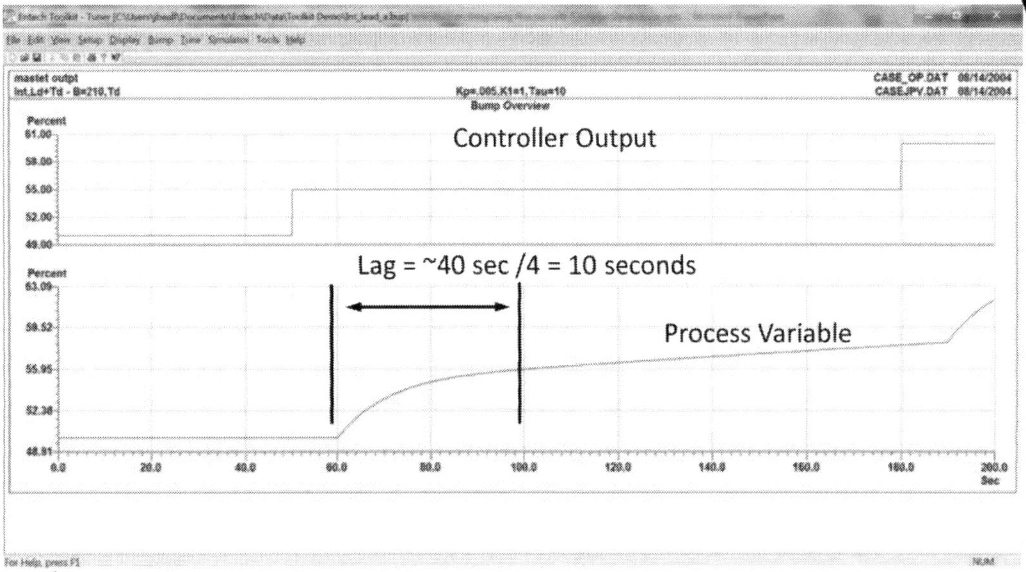

16AIChE Spring Meeting, Houston, TX

EMERSON Process Management

Step 3 – Choose the Lambda

Step Set Point Change Response

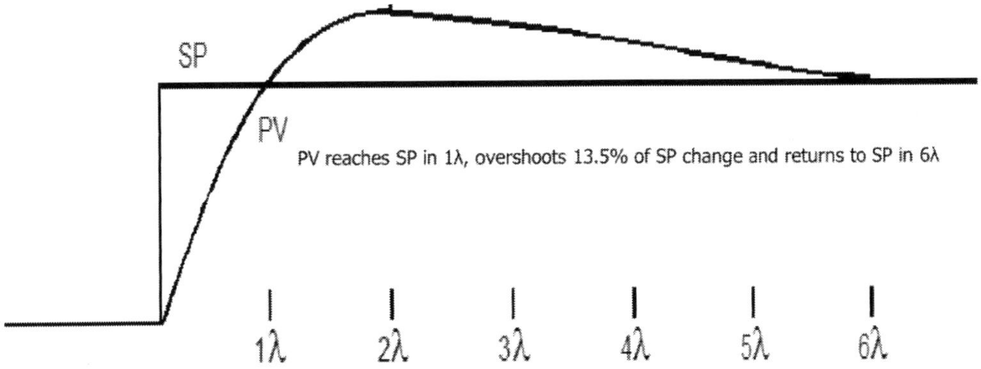

16AIChE
Spring Meeting, Houston, TX

EMERSON
Process Management

Step 3 – Choose the Lambda

Step Load Change Response

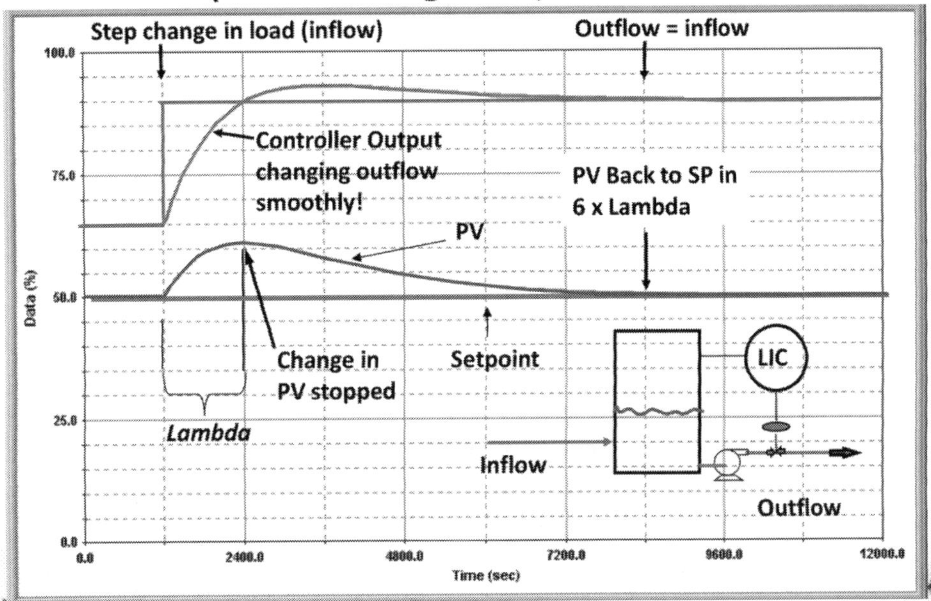

16AIChE
Spring Meeting, Houston, TX

EMERSON
Process Management

Step 4a – Integrator: Calculate the Tuning Constants

- For an Integrator plus dead time process:
  - Reset Time, Tr = $2\lambda + Td$
  - Controller Gain, Kc = $\frac{\text{Tr}}{|\text{Kp}|\ (\lambda+\text{Td})^2}$
  - Rate = 0 seconds
- Note, Reset Time is not constant. It varies with Lambda as does the Controller Gain

16 AIChE Spring Meeting, Houston, TX

EMERSON. Process Management

Step 4a – Integrator: Calculate the Tuning Constants

- Using Lambda = 600 seconds (more info later!)
- From example, Kp=-0.0014/sec., Td=40 sec
- Reset Time, Tr = $2\lambda + Td$ = 2(600)+40= 1240 sec
- Controller Gain, Kc = $\frac{\text{Tr}}{|\text{Kp}|\ (\lambda+\text{Td})^2}$

  $= \frac{1240 \text{ seconds}}{0.0014/\text{sec} * (600 + 40 \text{ sec})^2}$

  = 2.2
- Rate = 0 seconds
- Be sure to use the same time units for Kp, λ & Td

16 AIChE Spring Meeting, Houston, TX

EMERSON. Process Management

Step 4 a– Integrator: Simulated SP Response
λ = 600 sec

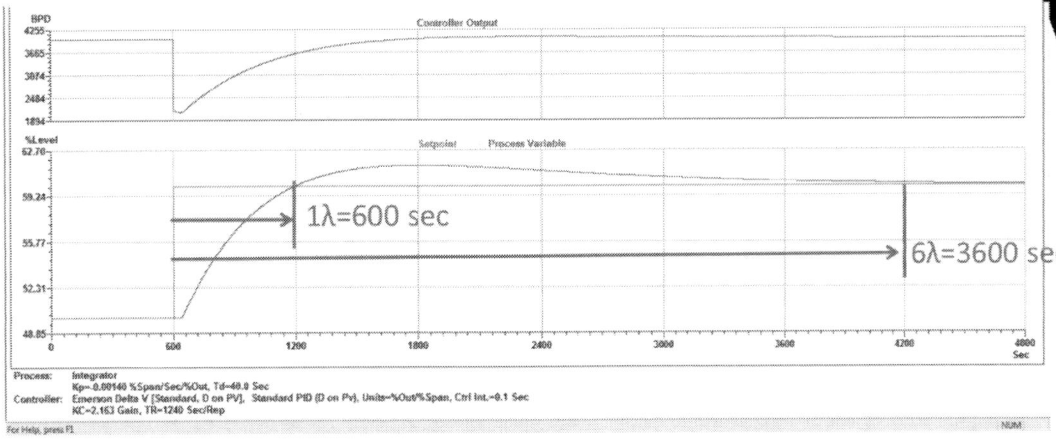

λ=600 sec., Kc = 2.2, Tr=1240 sec, Rate=0 sec

16AIChE
Spring Meeting, Houston, TX

EMERSON.
Process Management

Step 4a – Integrator: Simulated Load Response
λ = 600 sec

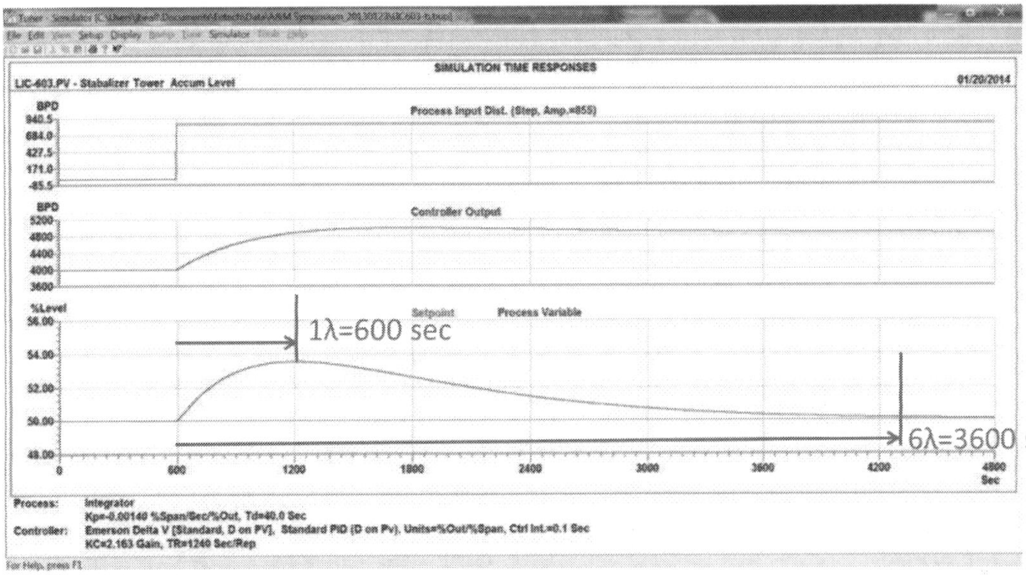

λ=600 sec., Kc = 2.2, Tr=1240 sec, Rate=0 sec

16AIChE
Spring Meeting, Houston, TX

EMERSON.
Process Management

Lambda for Integrating Processes

For the previous example of Kp = -0.0014%PV/sec/%OUT and Td = 40 seconds:

Lambda	Kc	Tr
600 sec	2.2	1240 sec
300 sec	4	640 sec
200 sec	5.5	440 sec
120 sec	7.8	280 sec

Note that for a lower Controller Gain, Kc, a "slower" or longer reset time is calculated. "Tune by feel" often results in too "fast" (short) reset time which will cause oscillation of an integrating loop.

16AIChE
Spring Meeting, Houston, TX

EMERSON.
Process Management

Step 4b– Integrator + Lag: Calculate the Tuning Constants

- For an Integrator plus dead time process:
  - Reset Time, Tr = $2\lambda + Td$
  - Controller Gain, Kc = $\frac{\text{Tr}}{|\text{Kp}|\ (\lambda + \text{Td})^2}$
  - Rate = Lag (seconds)
- Note, Reset Time is not constant. It varies with Lambda as does the Controller Gain

16AIChE
Spring Meeting, Houston, TX

EMERSON.
Process Management

Step 4b – Integrator: Calculate the Tuning Constants

- Using Lambda = 600 seconds (more info later!)
- From example, Kp=-0.0014/sec., Td=40 sec., Lag = 289 sec.
- Reset Time, Tr = $2\lambda + Td$ = 2(600)+40= 1240 sec
- Controller Gain, Kc = $\dfrac{\text{Tr}}{|\text{Kp}|\,(\lambda+\text{Td})^2}$

$$= \frac{1240 \text{ seconds}}{0.0014/\text{sec}*(600 + 40 \text{ sec})^2}$$

$$= 2.2$$

- Rate = 289 seconds
- Be sure to use the same time units for Kp, λ & Td

16AIChE
Spring Meeting, Houston, TX

EMERSON.
Process Management

Step 4 a– Integrator + Lag: Simulated SP Response
λ = 600 sec

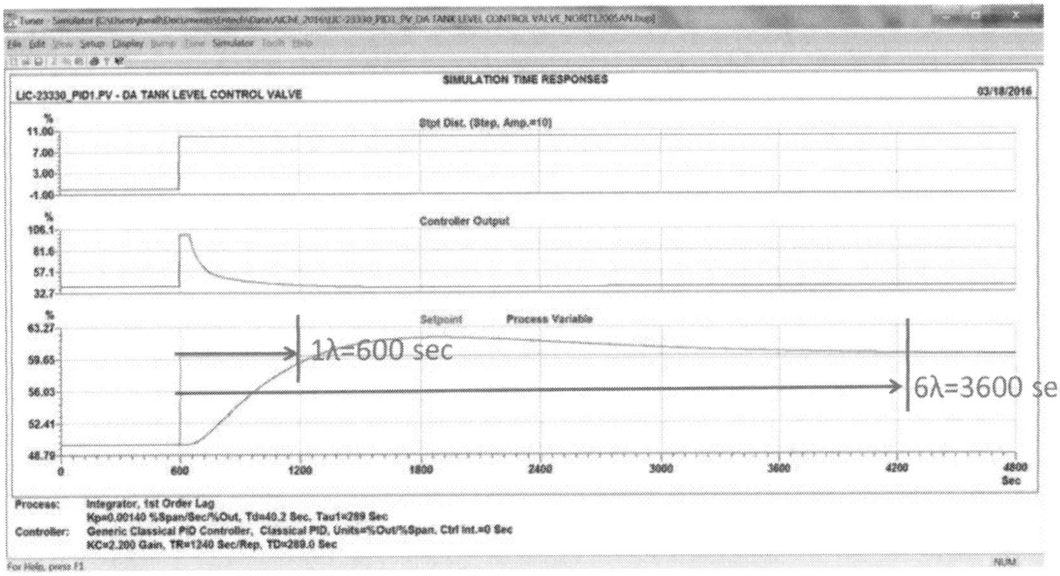

λ=600 sec., Kc = 2.2, Tr=1240 sec, Rate=289 sec

16AIChE
Spring Meeting, Houston, TX

EMERSON.
Process Management

Step 4a – Integrator + Lag: Simulated Load Response
λ = 600 sec

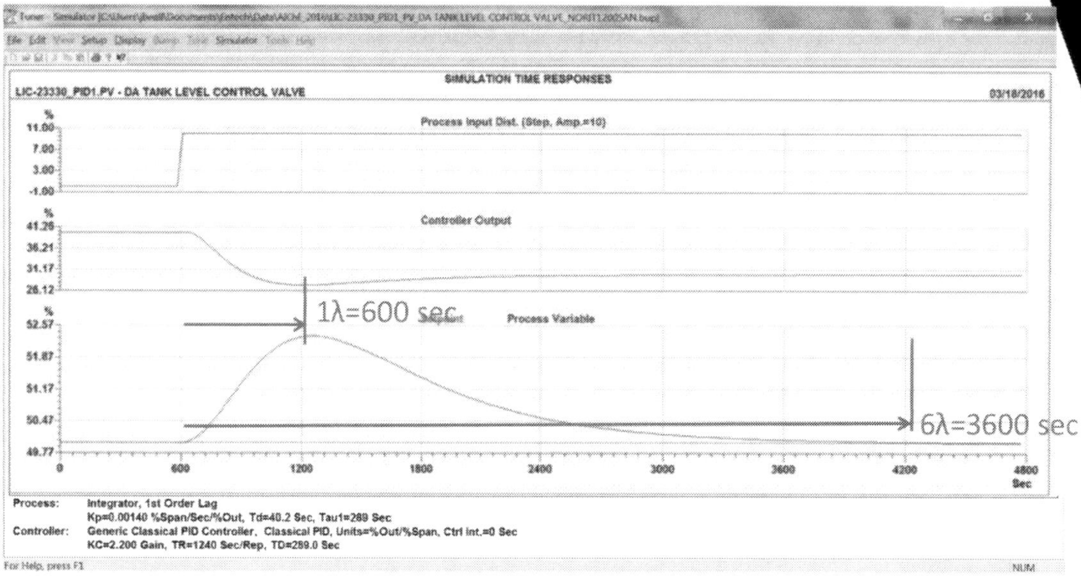

λ=600 sec., Kc = 2.2, Tr=1240 sec, Rate=289 sec

16AIChE
Spring Meeting, Houston, TX

EMERSON.
Process Management

Choosing Lambda

- For "tight" load regulation, choose a smaller Lambda – use caution if Lambda < 3*Dead time. Watch out for excessively high controller gain (though it might be appropriate)
- For cases you want the PV to absorb variability and reduce the variability of the manipulated variable (controller output) – choose a "slow" Lambda for "averaging" control

16AIChE
Spring Meeting, Houston, TX

EMERSON.
Process Management

Choosing Lambda

- Required Lambda~$\frac{2*APD}{Kp*MLD}$ where
  - APD = Allowable %PV Deviation from SP
  - MLD = Maximum Load Disturbance in units of % of controller output span
- Can be used for "tight" or "loose" control

16AIChE
Spring Meeting, Houston, TX

EMERSON.
Process Management

Case 1 – "Tight" Level Control

- Need tight level control
- Level was held close to SP but product outflow was varying dramatically
- Non-uniform distribution of additive in polymer product threating produc quality/sales
- The level controller had been re-tuned but it did not seem to help

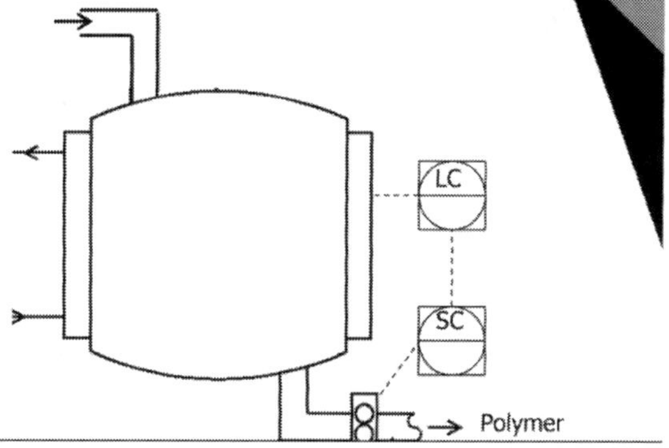

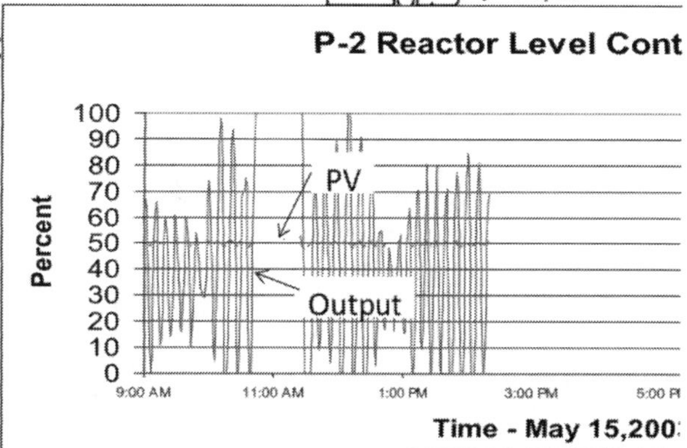

16AIChE
Spring Meeting, Houston, TX

Case 1 – "Tight" Level Control

- Measure process dynamics
- ADP = 5%
- MLD = 25%
- Lambda = ~$\frac{2*APD}{Kp*MLD}$

$$= \frac{2*5\%PV}{0.00029\%PV/sec\%out * 25\%}$$

= ~1400 seconds

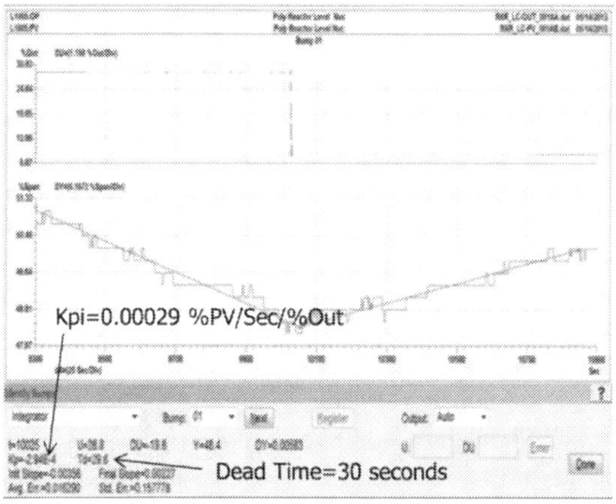

16 AIChE Spring Meeting, Houston, TX

EMERSON. Process Management

Case 1 – "Tight" Level Control

Calculate Tuning for Lambda = 1400 seconds

- Reset Time, Tr = $2\lambda + Td$ = 2(1400)+30= **2830 seconds**
- Controller Gain, Kc = $\frac{\text{Tr}}{|\text{Kp}|\ (\lambda+\text{Td})^2}$

$$= \frac{2830 \text{ seconds}}{0.00029/sec*(1400 + 30 \ sec)^2}$$

= **4.8**

- Rate = **0 seconds**

16 AIChE Spring Meeting, Houston, TX

EMERSON. Process Management

Case 1 – “Tight” Level Control

Test with simulation, ADP=5%, MLD=25%

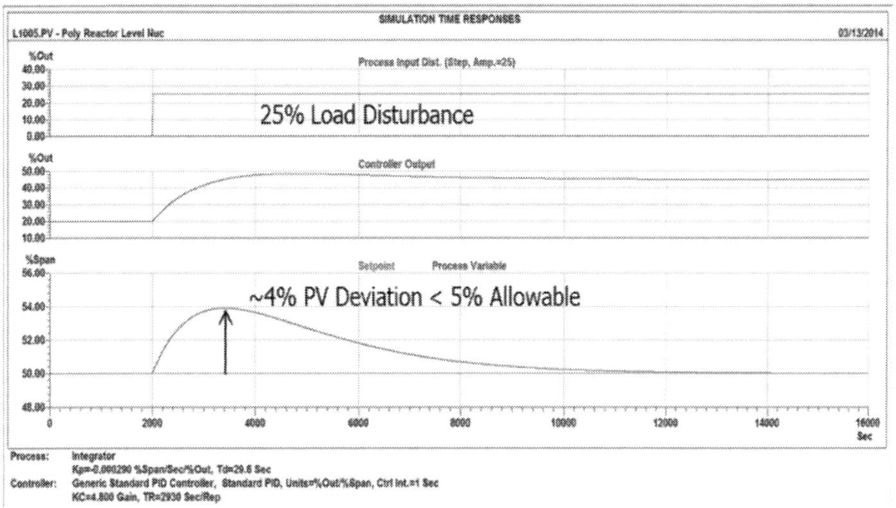

16AIChE Spring Meeting, Houston, TX

EMERSON. Process Management

Case 1 – “Tight” Level Control

Results on actual project

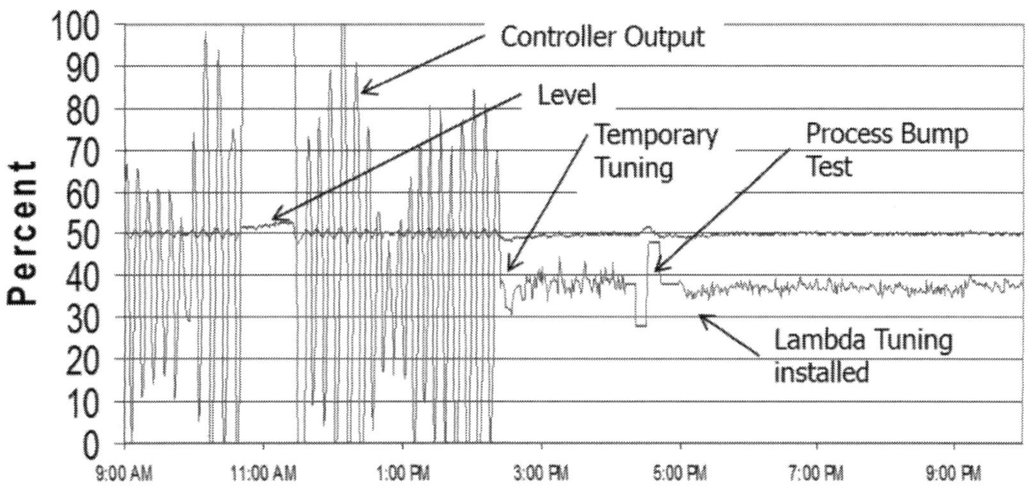

16AIChE Spring Meeting, Houston, TX

EMERSON. Process Management

Attenuate Variability with Control/Equipment

- "Capacity" in the process can be used to attenuate or absorb variability
- Primary source of process capacity is level control
- To utilize level control as a capacity tune the controller only "fast" enough to hold the PV within the allowable PV deviation (APD) for a maximum load disturbance (MLD)
- Called "Averaging" or "Surge" control

16AIChE
Spring Meeting, Houston, TX

EMERSON.
Process Management

Case 2 – "Averaging" Level Control

- Need to impact of inflow fluctuations
- Need to keep level within 30% of SP if 2 of 4 reactors shutdown (1 reactor = 20% load)

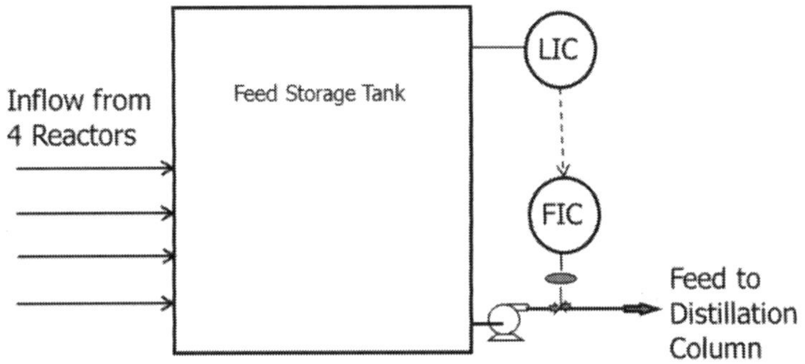

16AIChE
Spring Meeting, Houston, TX

EMERSON.
Process Management

Case 2 – "Averaging" Level Control

- Measure process dynamics
- ADP = 30%
- MLD = 40%
- Lambda = ~$\frac{2*APD}{Kp*MLD}$

  =2*30%PV / 0.000216%PV/sec%out * 40%

  = ~6900 seconds

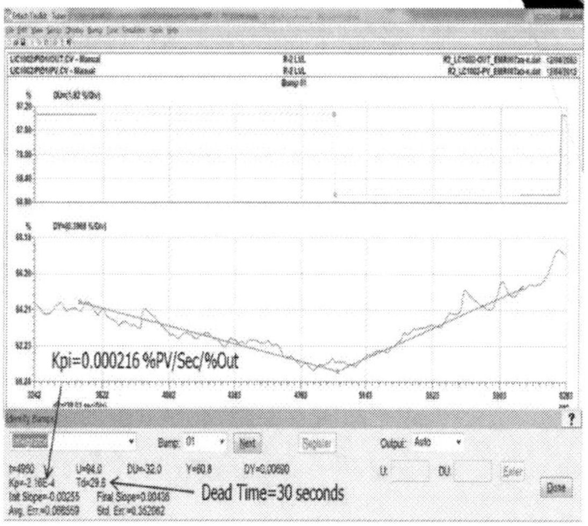

16AIChE Spring Meeting, Houston, TX

EMERSON. Process Management

Case 2 – "Averaging" Level Control

Calculate Tuning for Lambda = 6900 seconds

- Reset Time, Tr = $2\lambda + Td$ = 2(6900)+30= **13,830 seconds**
- Controller Gain, Kc = $\frac{\mathrm{Tr}}{|\mathrm{Kp}|\ (\lambda+\mathrm{Td})^2}$

  = $\frac{13{,}830 \text{ seconds}}{0.000216/\text{sec}*(6900 + 30 \text{ sec})^2}$

  = 1.33

- Rate = **0 seconds**

16AIChE Spring Meeting, Houston, TX

EMERSON. Process Management

Case 2 – "Averaging" Level Control

Test with simulation, ADP=30%, MLD=40%

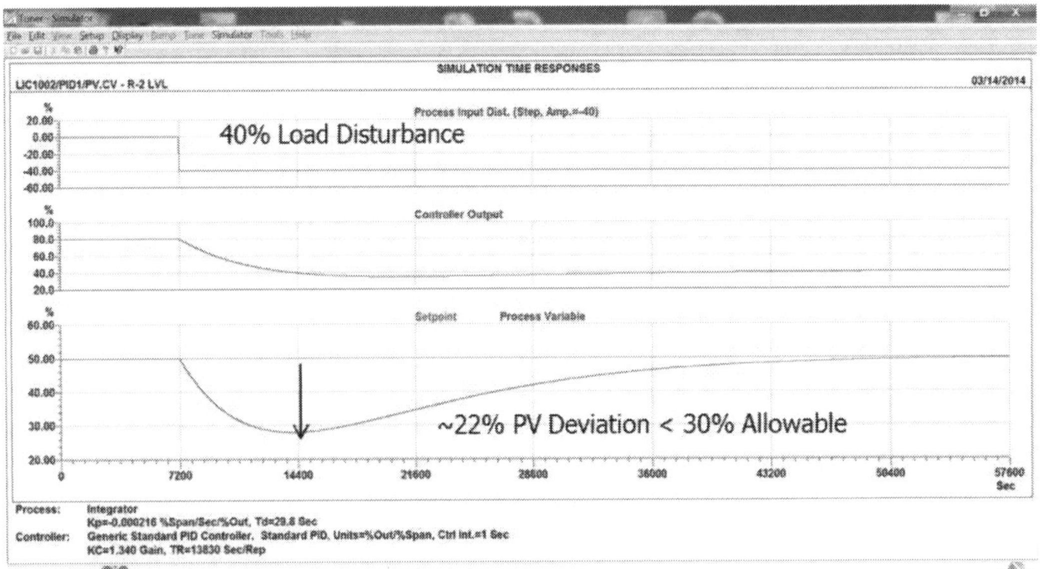

16AIChE Spring Meeting, Houston, TX

EMERSON. Process Management

Case 2 – "Tight" Level Control

Attenuation of Lambda vs. Ziegler-Nichols Tuning

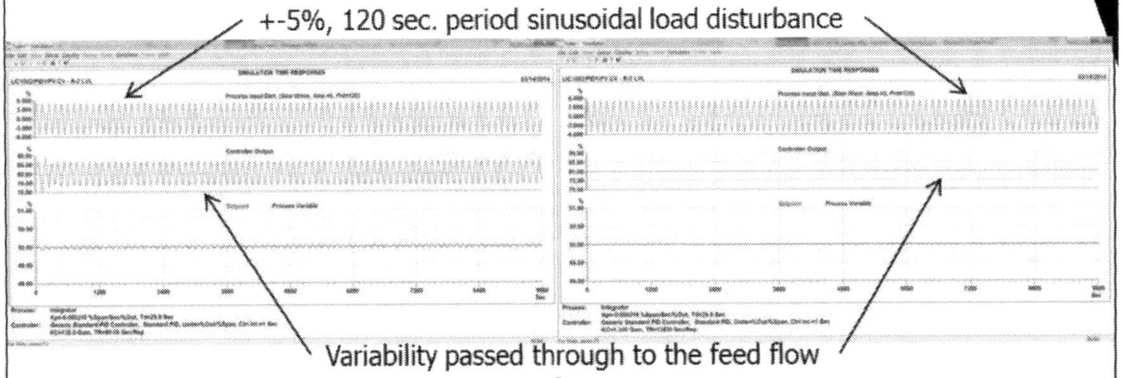

Zeigler-Nichols Reaction Rate Tuning | Lambda Tuning, Lambda = 6900 seconds

16AIChE Spring Meeting, Houston, TX

EMERSON. Process Management

Complex Integrators

- These "complex" integrator responses that can be handled with Lambda Tuning Methods (all include dead time)

- Integrator + lag
- Integrator + lead
- Integrator non-minimum phase

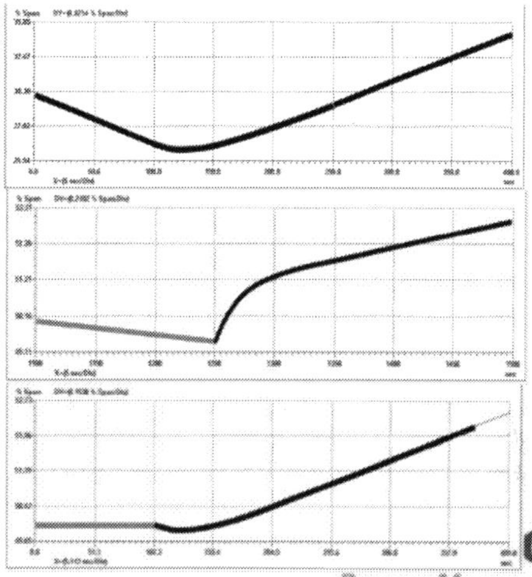

16AIChE
Spring Meeting, Houston, TX

ON
Process Management

Questions?

James.Beall@Emerson.com

903-235-7935

16AIChE
Spring Meeting, Houston, TX

EMERSON
Process Management

Modernizing Control Functionality

Easy Steps to Start Process Optimization

Laurie R. Ben
James F. Beall
Emerson Process Management

16 AIChE
Spring Meeting, Houston, TX

EMERSON.
Process Management

Agenda

- Introduction – Taking advantage of New Technology
- Examples – Best Practices & Lessons Learned
  - Split Range Duty
  - Smith Predictor
  - Gain Scheduling/Adaptive Control
  - Signal Characterization
- Summary

16 AIChE
Spring Meeting, Houston, TX

EMERSON.
Process Management

Introduction

- Modernizing Control System – Decision Done Check!
- 'Do I' or 'Don't I' take advantage of new technology
  - Implemented older cumbersome techniques?
  - Complex custom strategies that handle automation challenges?
  - Hard to implement & maintain?
- Standard features that are field proven, practical, and easy to implement
  - Easier to implement & maintain
  - More efficient and effective control strategies
  - Achieve improvements in process performance
  - Lays the foundation for optimization

16AIChE
Spring Meeting, Houston, TX

EMERSON.
Process Management

Agenda

- Introduction
- Examples – Best Practices & Lessons Learned
  - Split Range Duty
    - Control Challenge
    - Legacy Solution
    - Newer Technology Solution
    - Summary
  - Smith Predictor
  - Gain Scheduling/Adaptive Control
  - Signal Characterization
- Summary

16AIChE
Spring Meeting, Houston, TX

EMERSON.
Process Management

Split Range Duty – Control Challenge

- Use steam and hot oil as energy source for distillation column for temperature controller.
- Hot oil is “free energy” and should be maximized
- Control Scenarios
  - Maximize hot oil and supplement with steam
  - Base load steam (if bad meter) and supplement with hot oil
  - Base load hot oil flow (if bad meter) and supplement with steam
- Constraints
  - Minimum and maximum hot oil valve position and flow rate
  - Minimum steam flow rate

16 AIChE
Spring Meeting, Houston, TX

EMERSON.
Process Management

Split Range Duty – Legacy Solution

– Old legacy system implemented this feature using a velocity PID with incremental output and a “micro expert rules system”

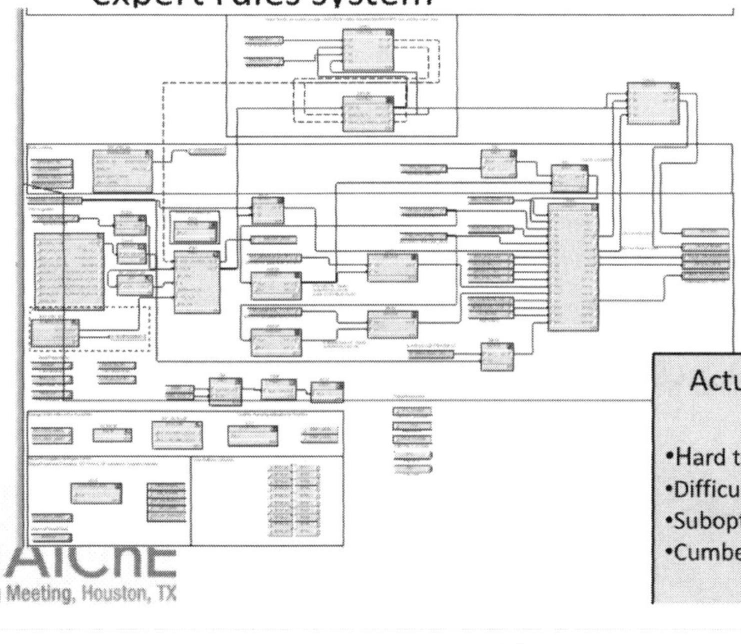

Actually Duplicated in New system, but still...
- Hard to maintain and modify
- Difficult to troubleshoot
- Suboptimal performance
- Cumbersome/Kludge/Ugly

16 AIChE
Spring Meeting, Houston, TX

Split Range Duty – New Technology Solution

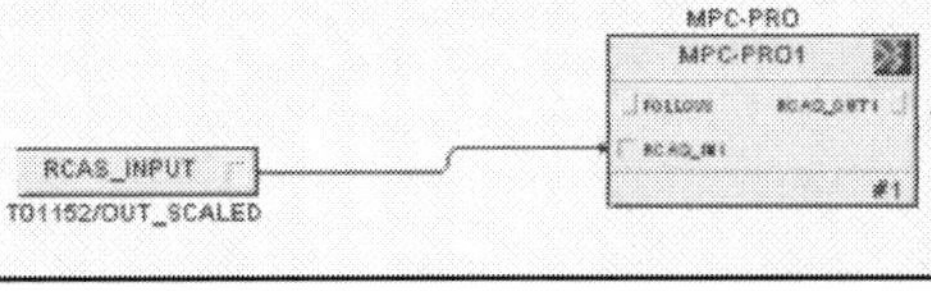

Embedded MPC Function Block provides all control functionality! EASY!!!

Simple calculations to calculate limits

16 AIChE Spring Meeting, Houston, TX

EMERSON. Process Management

Split Range Duty – New Technology Solution

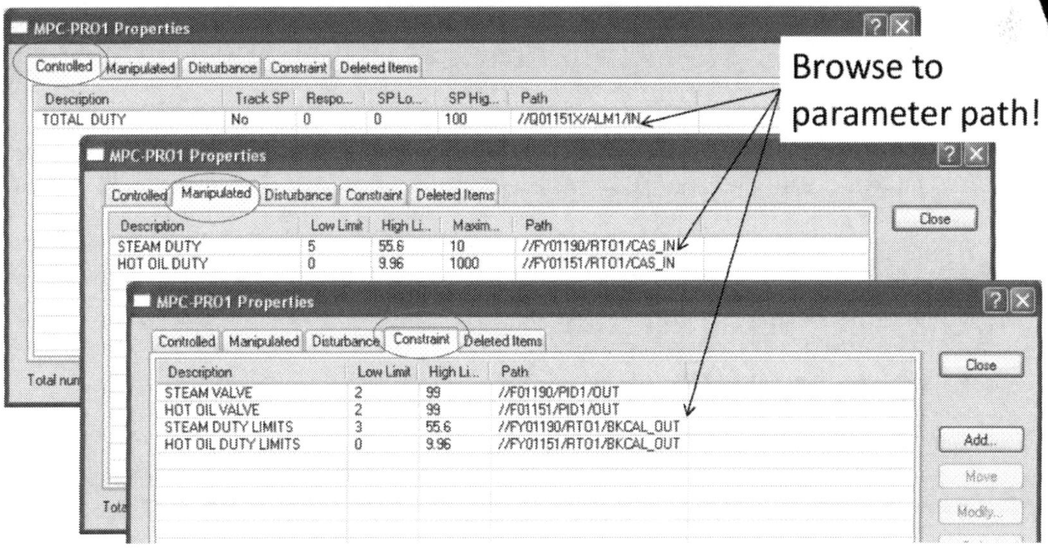

Browse to parameter path!

16 AIChE Spring Meeting, Houston, TX

EMERSON. Process Management

Split Range Duty – New Technology Solution

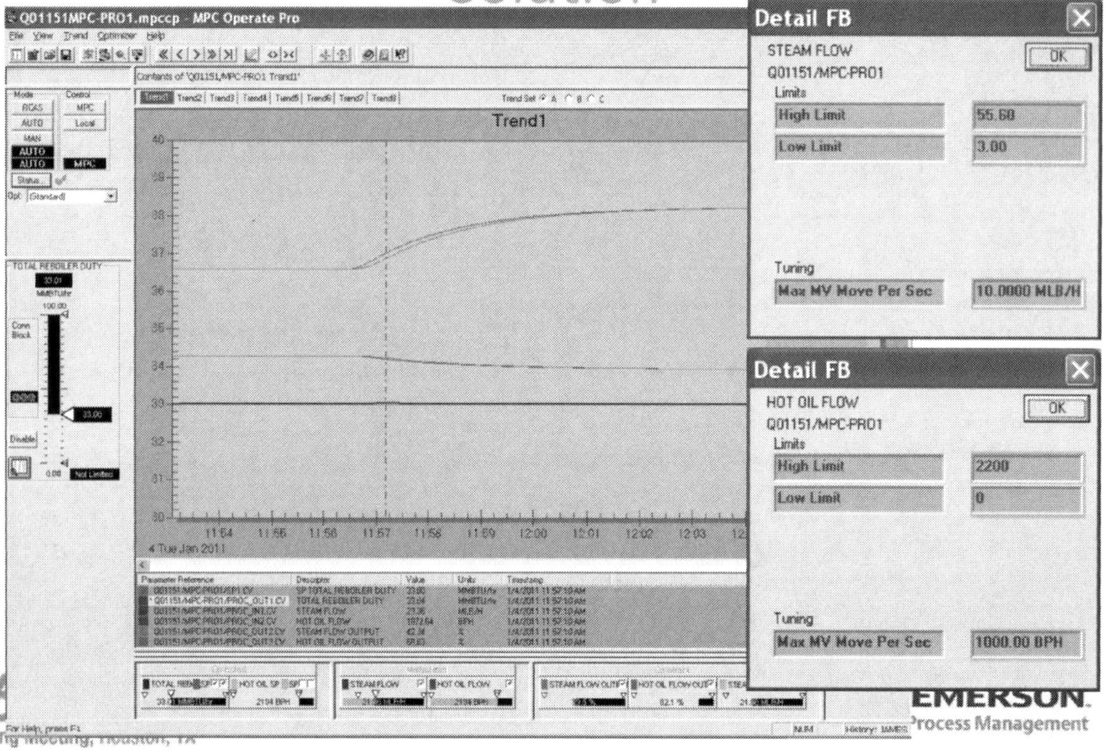

16 AIChE Spring Meeting, Houston, TX

EMERSON Process Management

Summary Split Range Duty Control

Embedded MPC (New Technology) vs Legacy Solution

- Better performing solution
  - Bumpless transfers and constraint control
- Easier to Maintain
  - Change limits and settings
  - Simpler implementation
  - Easy to understand, train others, document
- Easily expands to handle future constraints
- Reduce energy costs!

16 AIChE Spring Meeting, Houston, TX

EMERSON Process Management

Agenda

- Introduction
- Examples – Best Practices & Lessons Learned
  - Split Range Duty
  - Smith Predictor
    - Control Challenge
    - Legacy Solution
    - Newer Technology Solution
    - Summary
  - Gain Scheduling/Adaptive Control
  - Signal Characterization
- Summary

16AIChE
Spring Meeting, Houston, TX

EMERSON.
Process Management

Smith-Predictor PID – Control Challenge

- Control of processes that are "dead time dominant" (dead time > time constant)
- Typically custom configured
  - some legacy control systems have preconfigured options
- Algorithm uses a process model
  - gain, dead time, and time constant
- Challenge: as process dynamics change from model used, performance degrades rapidly

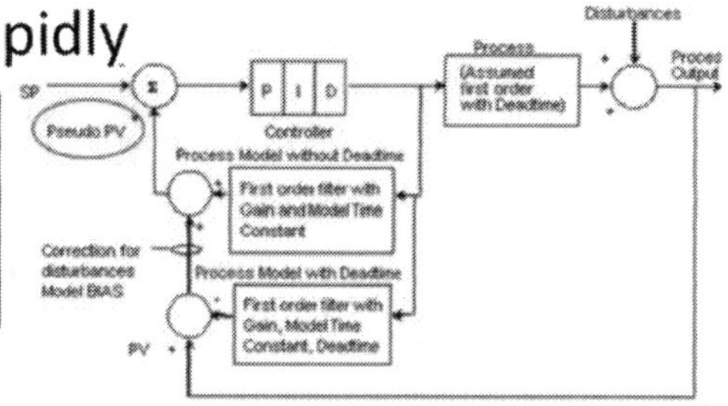

- High maintenance when process dynamics change
- Performance degrades rapidly as model mismatch increases
- Operations view distorted by Pseudo PV

16AIChE
Spring Meeting, Houston, TX

Smith-Predictor PID

Easy to implement the Smith-Predictor if standard function block exists, but still ...
- High Maintenance when process dynamics change
- Performance degrades rapidly as process dynamics change from configured model

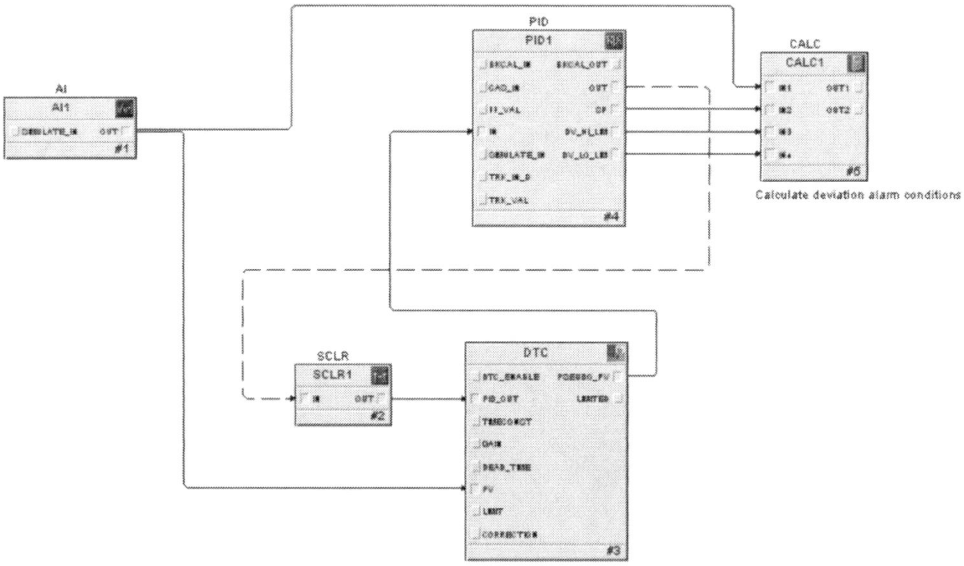

16
Spring

SON.
agement

Smith-Predictor PID – New Technology Solution

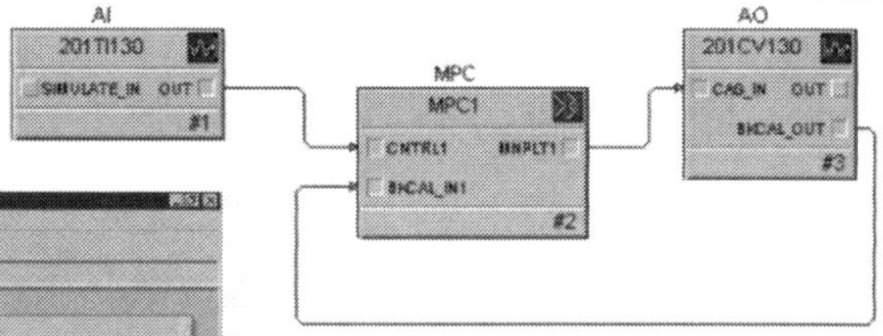

- Embedded MPC Soluti

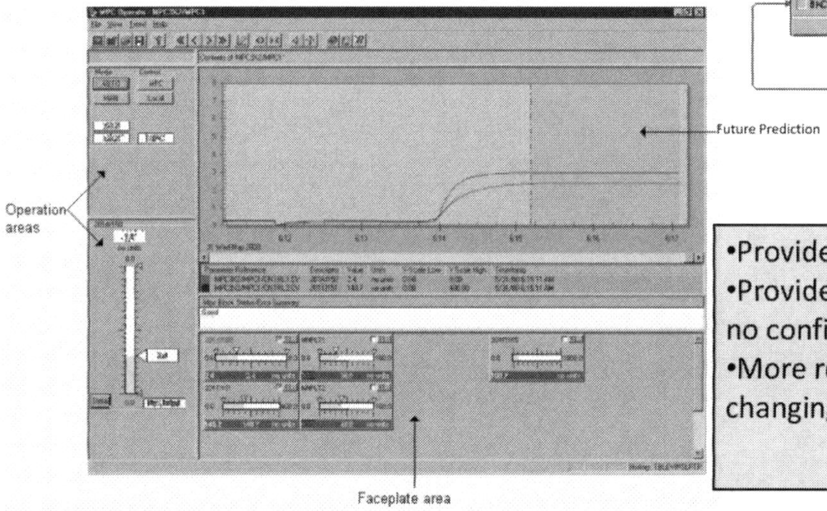

- Provides Future Prediction
- Provides MPC Faceplate – no configuration needed
- More robust under changing process dynamics

16AIChE
Spring Meeting, Houston, TX

EMERSON.
Process Management

Summary Smith-Predictor PID

Embedded MPC (New Technology) vs Legacy Solution

- More robust solution under a changing process model
- Provides visual display of the "future" PV
- Save money through better control performance

16AIChE
Spring Meeting, Houston, TX

EMERSON.
Process Management

Agenda

- Introduction
- Examples – Best Practices & Lessons Learned
  - Split Range Duty
  - Smith Predictor
  - Gain Scheduling/Adaptive Control
    - Control Challenge
    - Legacy Solution
    - New Technology Solution
    - Summary
  - Signal Characterization
- Summary

16AIChE
Spring Meeting, Houston, TX

EMERSON.
Process Management

Gain Scheduling – Control Challenge

- Legacy systems have limited capability to modify tuning based on process conditions
- Some Legacy systems had add-on application software
  - e.g. PROVOX "Gain Scheduler" system provided 3 sets of Gain, Reset, and Rate applied in 3 separate regions of a "state variable"
    - State variable = PV, OUT or "other"
    - Transition between tuning sets interpolated over "dead band"
- Beware: some suspend loop execution when tuning parameters change, so don't change the tuning every scan

16AIChE Spring Meeting, Houston, TX

EMERSON. Process Management

Gain Scheduling – Legacy Solution

- Custom code
- Multiple points, registers, displays
- Restricted (slower) update rate

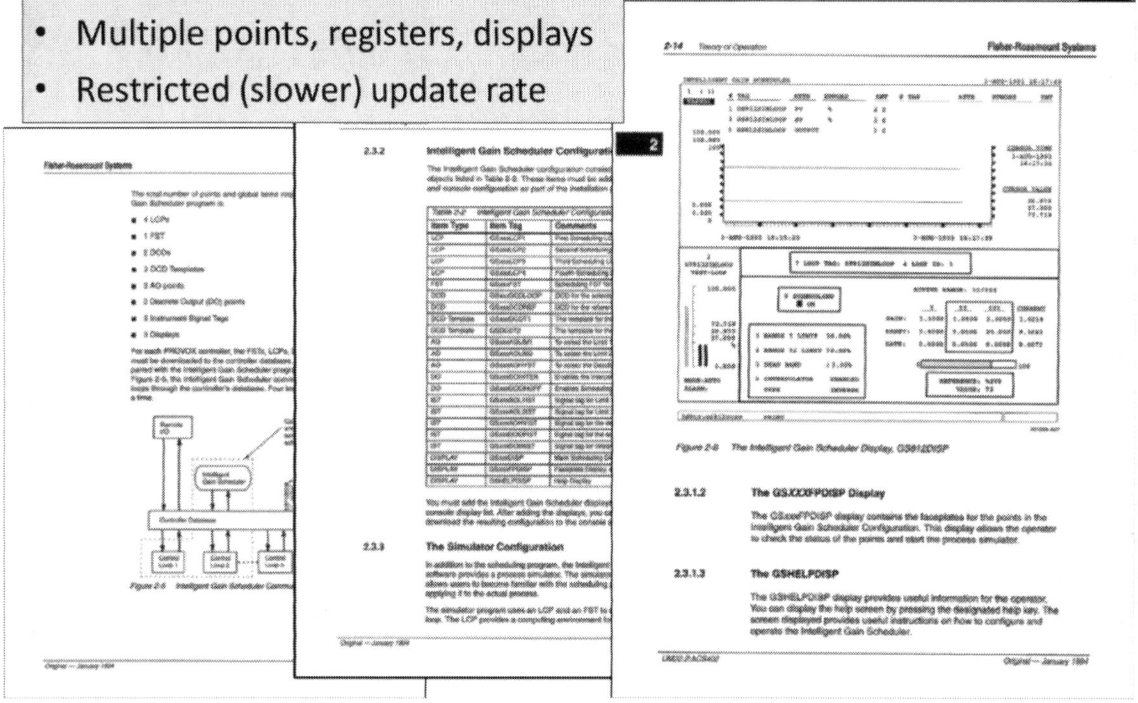

Gain Scheduling – New Technology Solution #1

Standard Modules

- Standard configuration
- Preconfigured faceplate & detailed display
- Easy to implement
- Tuning updates at loop execution

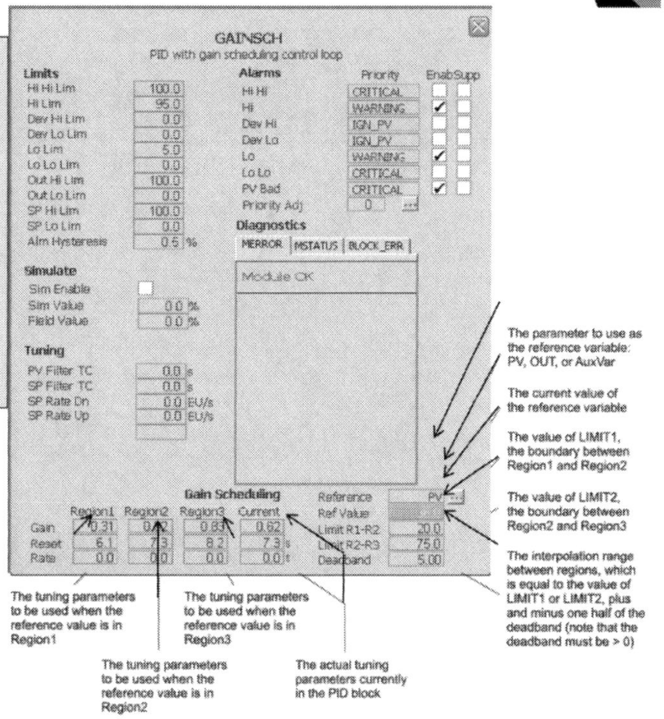

16AIChE
Spring Meeting, Houston, TX

Gain Scheduling – New Technology Solution #2

New Technology "Adaptive Control"

- "Learns" process models from normal operating moves
- Allows up to 5 tuning regions of the "state variable"
- Partial Adaptive Control
  - Use "approved model" for each region per loop selected tuning rule
- Optional Full Adaptive Control
  - Continuous "Learning" of process model in each region
  - Updates tuning based on 'learned' model while in each region per loop selected tuning rule
  - Set allowable deviation (%) from approved model in each region

16AIChE
Spring Meeting, Houston, TX

EMERSON.
Process Management

New Technology – Adaptive Control

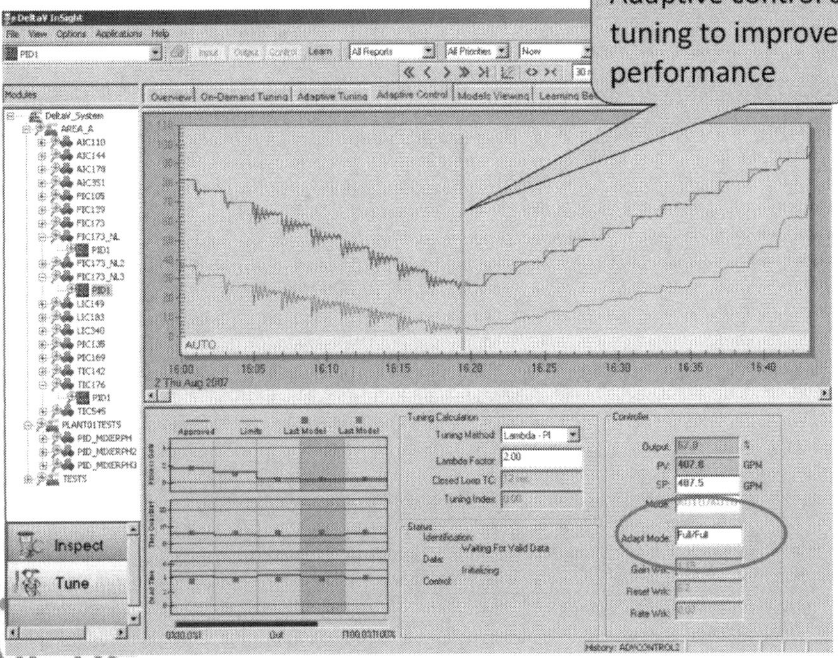

16AIChE Spring Meeting, Houston, TX

EMERSON. Process Management

Summary Gain Scheduling

New Technology Solution(s) vs. Legacy Solution

- Gain Scheduling easy to implement and understand!
  - Standard configuration
  - Preconfigured faceplate & detailed display
  - Tuning updates at loop execution
- Adaptive Control provides additional benefit and ease of implementation
  - Learns models during normal operation – NO STEP TESTs!
  - More Regions – Up to 5
  - Automatic tuning updates when process model changes
- Improved control performance saves money!

16AIChE Spring Meeting, Houston, TX

EMERSON. Process Management

Agenda

- Introduction
- Examples – Best Practices & Lessons Learned
  - Split Range Duty
  - Smith Predictor
  - Gain Scheduling/Adaptive Control
  - Controller Characterization
- Summary

16AIChE
Spring Meeting, Houston, TX

EMERSON.
Process Management

Summary

- Practical application of control features that deal with common challenges
- Highlighted best practices & lessons learned
  - Split Range Duty
  - Smith Predictor
  - Gain Scheduling/Adaptive Control
  - Signal Characterization
- Do look at new functionality available when modernizing
  - Allows jump start to process optimization
  - Leads to improved process performance
  - Easier to implement and maintain over the life investment

16AIChE
Spring Meeting, Houston, TX

EMERSON.
Process Management

Contacts & Questions?

Laurie.Ben@Emerson.com
512-834-7237

James.Beall@Emerson.com
903-235-7935

16AIChE
Spring Meeting, Houston, TX

EMERSON.
Process Management

Specifying Control Valve Performance to Achieve Process Performance Goals

James Beall

Principal Process Control Consultant
Emerson Process Management

16AIChE
Spring Meeting, Houston, TX

EMERSON.
Process Management

Introduction

- Importance of Valve Performance
- Standards and References
- Dead Band and Resolution
- Response Time
- Loop Gain
- Comments on Valve Troubleshooting
- Summary
- Speaker Info

16AIChE
Spring Meeting, Houston, TX

EMERSON.
Process Management

Control Loops with Excessive Variability

Loops

Why?

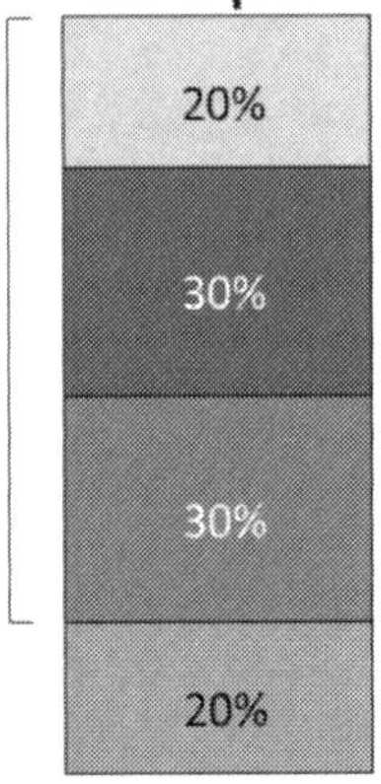

Design

Tuning

Control Valve Performance

Increases Variability

Source: EnTech - Results from audits of over 5000 loops

- **"The undesirable behavior of control valves is the biggest contributor to poor loop performance and the destabilization of product uniformity".**

W. L. Bialkowski, President
EnTech Control Engineering

16AIChE
Spring Meeting, Houston, TX

EMERSON.
Process Management

LNG Plant, Main Gas Feed Flow Loop

Cycle caused by nonlinearity in control valve and aggravated by high process gain – The loop had to stay in Manual- for a valve that cost $1million!

High Process Gain amplifies valve non-linearities!

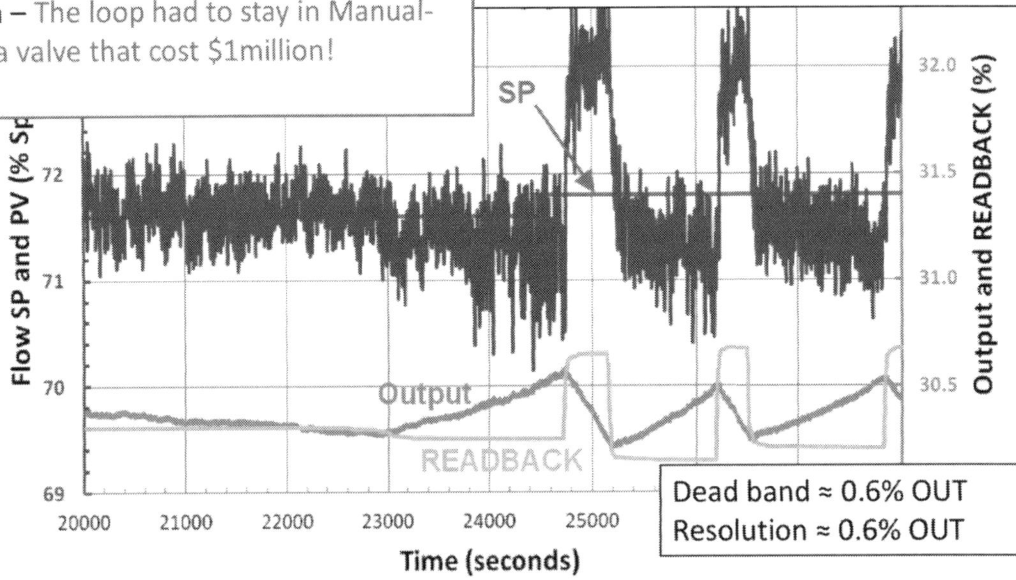

Dead band ≈ 0.6% OUT
Resolution ≈ 0.6% OUT

Vitamin plant—Steam Valve

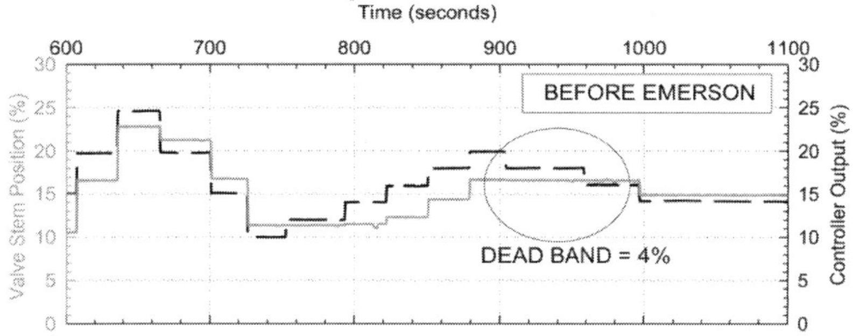

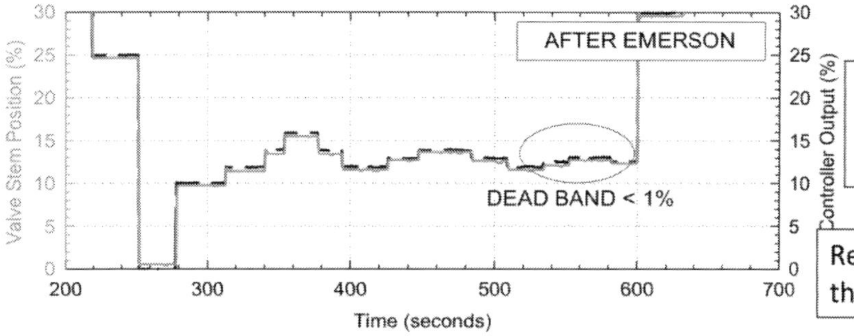

As-found positioner

Digital Valve Controller (DVC) installed same day

Reduced trim installed the following week

Batch time reduced 27% and steam consumption reduced

16AIChE
Spring Meeting, Houston, TX

EMERSON.
Process Management

Chemical Plant – Steam Valve

Flow cycles in automatic

Flow

PID Output

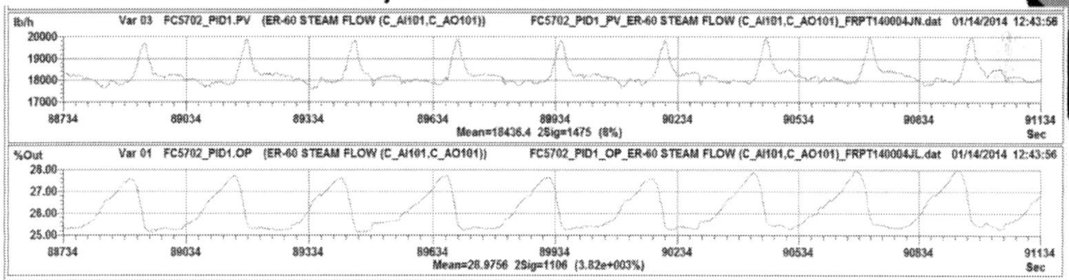

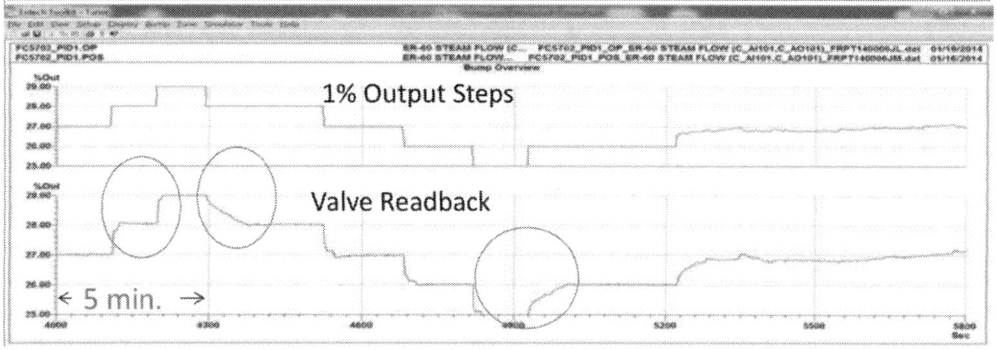

Step test: Valve had a 30 – 90 second Response Time! – Root cause of cycle!

16AIChE
Spring Meeting, Houston, TX

EMERSON.
Process Management

Control Valve Performance

- ANSI/ISA–S75.25 (Standard) - *Test Procedure for Control Valve Response Measurement from Step Inputs*
  - Defines metrics and test procedure
  - Does NOT provide guidance to specify valve performance requirements
- ANSI/ISA–TR75.25 (Technical Report) - *Control Valve Response Measurement from Step Inputs*
- *EnTech Control Valve Dynamic Specification V3.0*
  - V2.0 update to V3.0 to align with ANSI/ISA–75.25
  - DOES provide guidance to specify valve performance requirements

Definitions from ANSI/ISA-S75.25

- **Dead band: the range through which an input signal may be varied, with reversal of direction, without initiating an observable change in output signal** [ANSI/ISA-S51.1-1979 (R1993)]. In this standard and ISA-TR75.25.02-2000, it is defined in percent of input span. Note that in some other literature this definition is used for dead zone.

Definitions from ANSI/ISA-S75.25

- **Resolution: smallest step increment of input signal in one direction for which movement of the output is observed.** Resolution is expressed as percentage of input span. The term in this document means: the tendency of a control valve to move in finite steps in responding to step changes in input signal applied in the same direction. An example of this is when the valve is driven by a stepper motor or an AC motor or when the pneumatically actuated control valve sticks in place, having stopped moving after the previous step change.

16AIChE
Spring Meeting, Houston, TX

EMERSON.
Process Management

Resolution

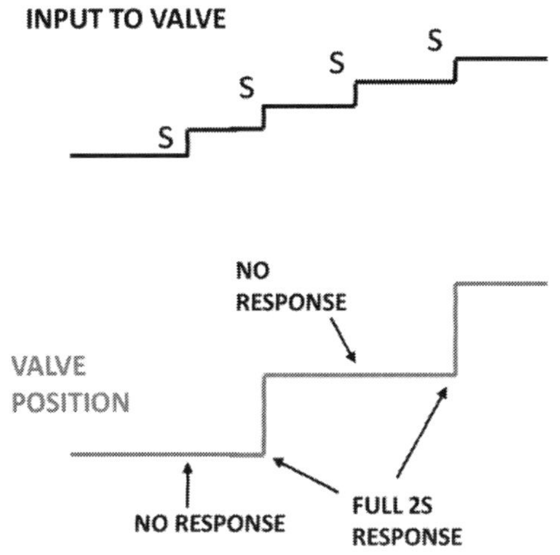

SMALLEST
INPUT CHANGE IN THE SAME
DIRECTION TO
GET A RESPONSE

S<RESOLUTION<2S

16AIChE
Spring Meeting, Houston, TX

EMERSON.
Process Management

Resolution

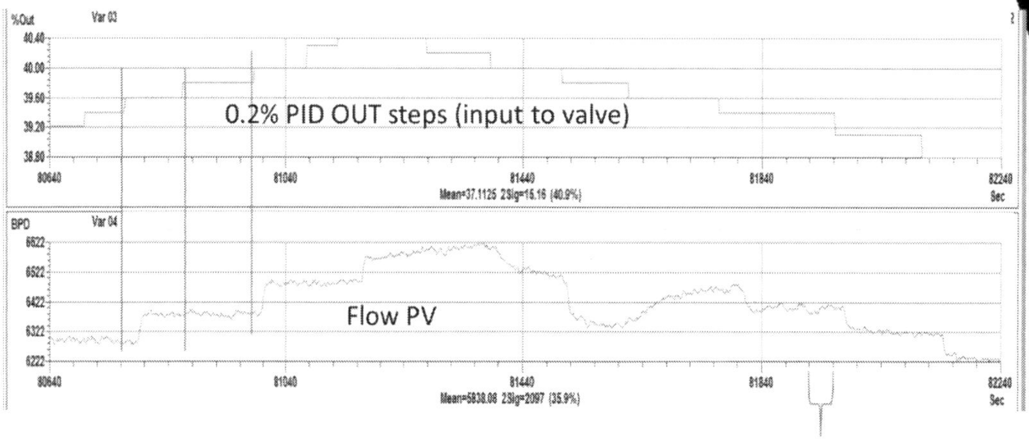

0.2%< Resolution < 0.4% 40 seconds

16AIChE Spring Meeting, Houston, TX

EMERSON. Process Management

Dead Band

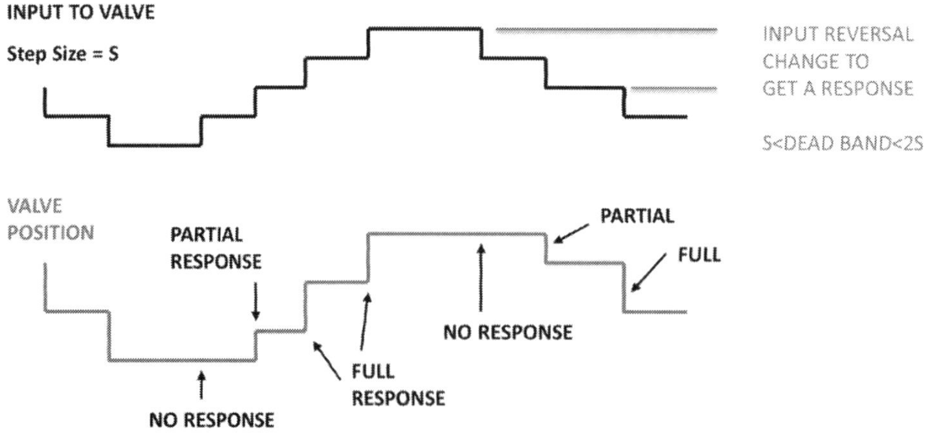

16AIChE Spring Meeting, Houston, TX

EMERSON. Process Management

Dead Band

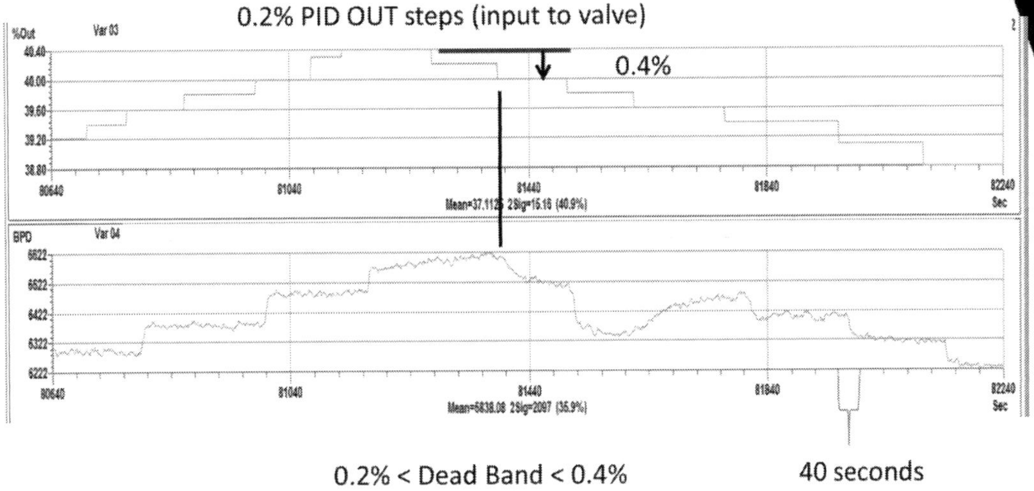

16AIChE
Spring Meeting, Houston, TX

EMERSON.
Process Management

Dead Band and Resolution Cause Limit Cycles

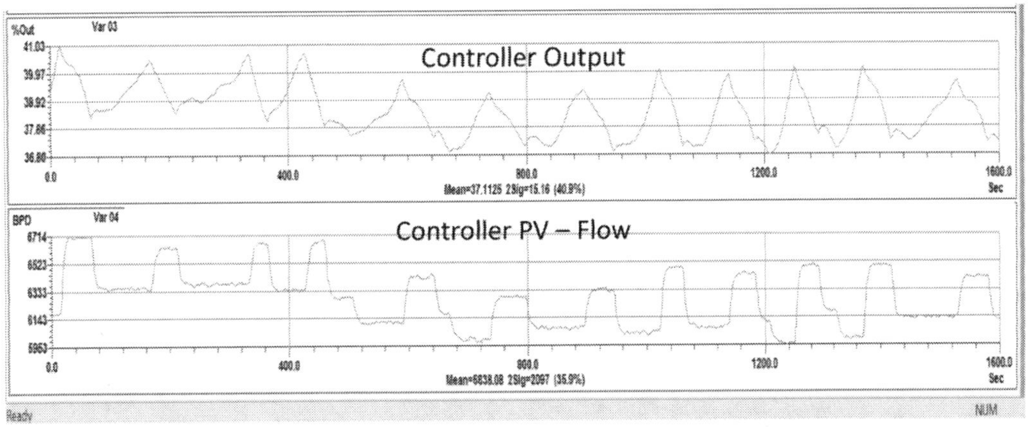

16AIChE
Spring Meeting, Houston, TX

EMERSON.
Process Management

Definitions from ANSI/ISA-S75.25

- **step response time (T86):** the interval of time between initiation of an input signal step change and the moment that the response of a dynamic reaches 86.5% of its full steady state value. The step response time includes the dead time before the dynamic response.

16AIChE
Spring Meeting, Houston, TX

EMERSON.
Process Management

Step Response Time (T_{86})

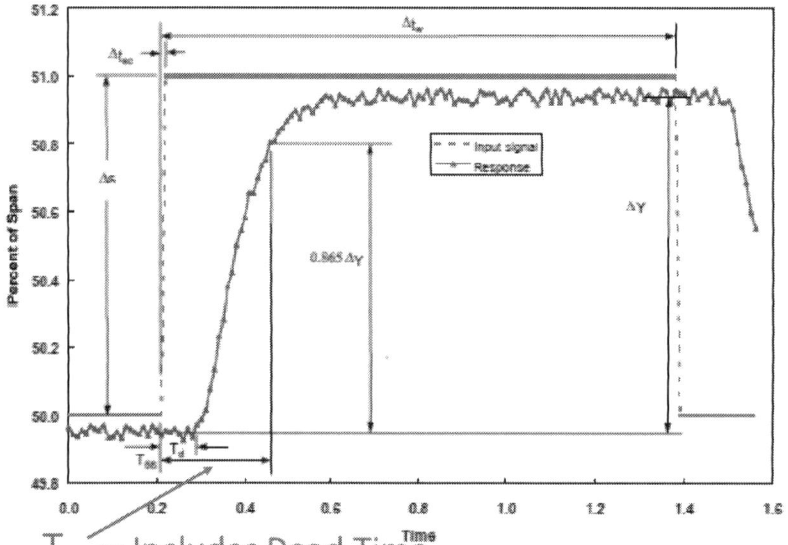

T_{86} – Includes Dead Time

Figure 2 — Typical step change and response without overshoot

16AIChE
Spring Meeting, Houston, TX

EMERSON.
Process Management

Step Response Time VS. Step Size

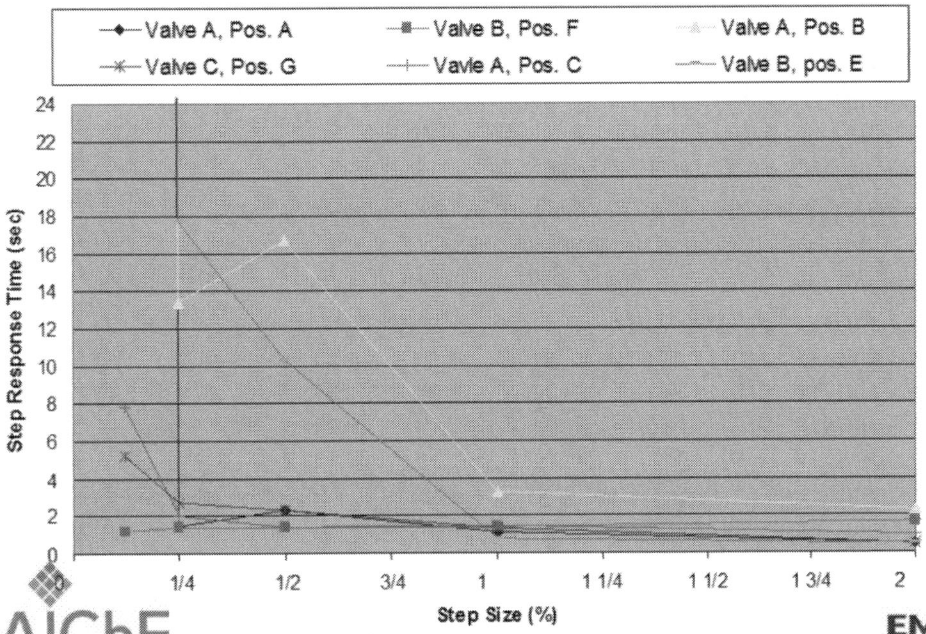

16AIChE
Spring Meeting, Houston, TX

EMERSON.
Process Management

From ANSI/ISA–TR75.25

- **Region 1** is defined as small input steps which result in no measurable movement of the closure member within the specified wait time.
  - Comment: This is in the region of dead band and resolution.

- **Region 2** is defined as input step changes which are large enough to result in some control valve response with each input signal change, but the response does not satisfy the requirements of the specified time and linearity.

16AIChE
Spring Meeting, Houston, TX

EMERSON.
Process Management

From ANSI/ISA–TR75.25

- **Region 3** is defined as step changes which are large enough to result in flow coefficient changes which satisfy both the specified maximum response time and the specified maximum linearity.
  - Comment: This is the range of "typical" control moves
- **Region 4** is defined as input steps larger than in region 3 where the specified magnitude response linearity is satisfied but the specified response time is exceeded.
  - Comment: This is typically caused by velocity limited operation.
- *EnTech Control Valve Dynamic Specification V3.0* is similar, using A, B, C, D instead of 1, 2, 3, 4.

16AIChE
Spring Meeting, Houston, TX

EMERSON.
Process Management

Step Response Time VS. Step Size

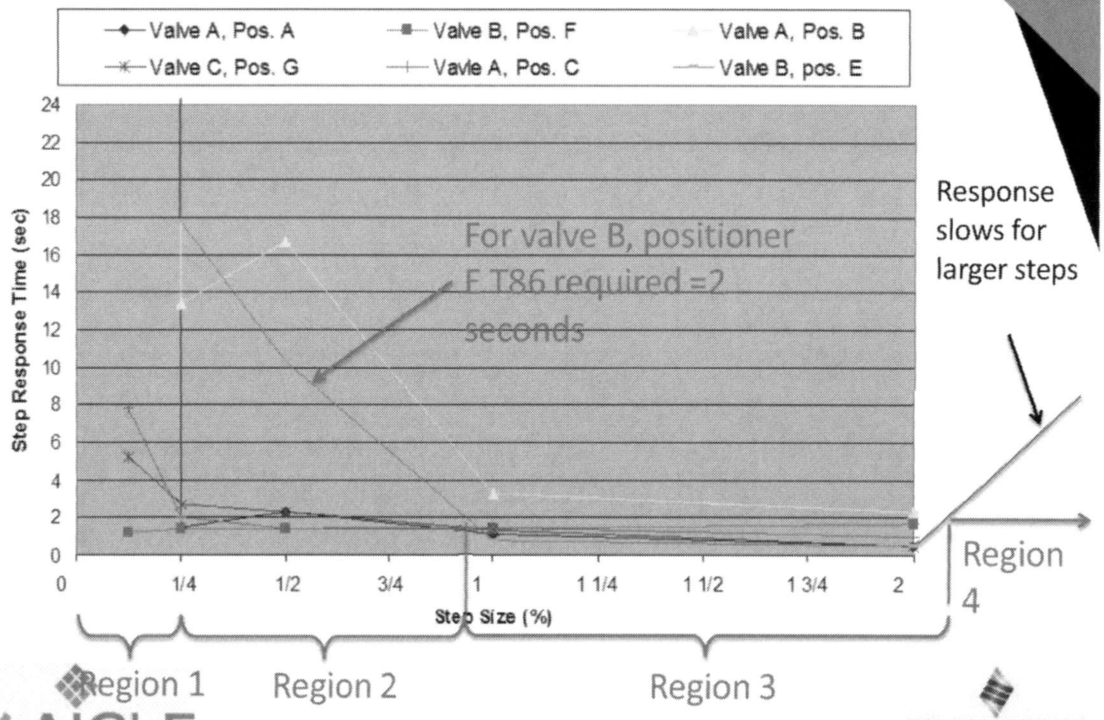

16AIChE
Spring Meeting, Houston, TX

EMERSON.
Process Management

Region 2 – Causes Limit Cycle

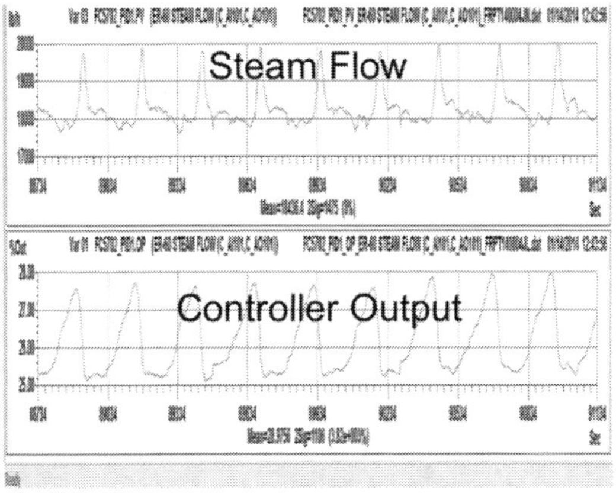

Limit cycle in Auto

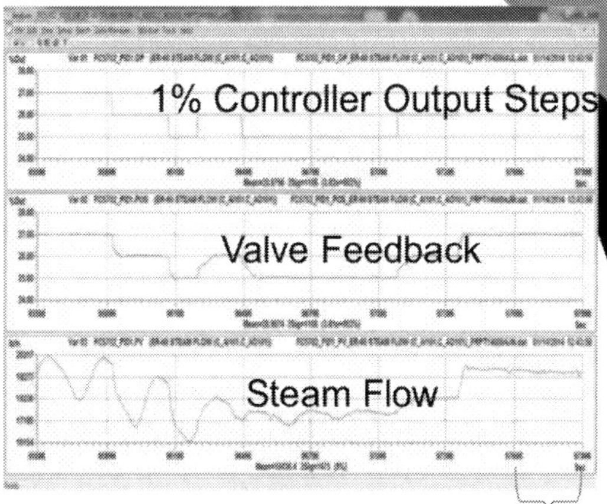

Step test in Manual

300 sec.

Slow response of control valve, due to slow response of valve positioner, causes cycling of PV. Solution – retune the digital positioner!

16AIChE
Spring Meeting, Houston, TX

EMERSON.
Process Management

Region 2 – Causes Limit Cycle

Region 2: 0.2 - 0.6%; T_{86} = 43 seconds; Region 3 T_{86} = 1 second

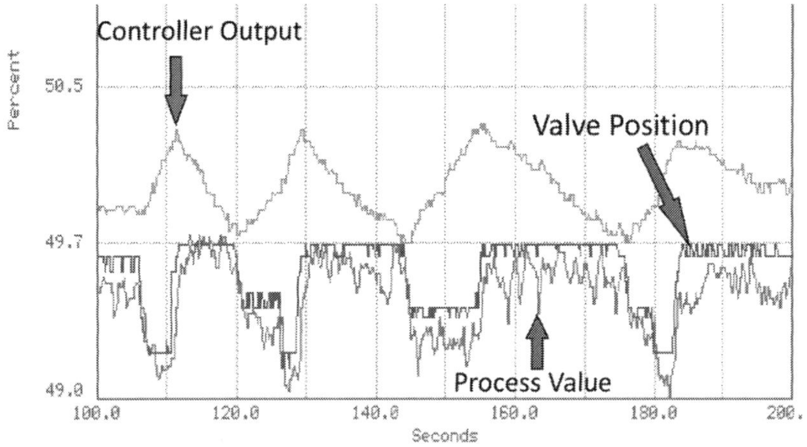

Slow response of control valve, due to slow response of valve positioner, causes cycling of PV. Solution – replace the pneumatic positioner with a high performance digital positioner!

16AIChE
Spring Meeting, Houston, TX

EMERSON.
Process Management

Summary: Step Response Time

- **Region 1 and 2** are bad for control performance
- **Region 3** performance provides the most linear operation of the valve and hence the best control performance
- **Region 4** performance (large step, slow or velocity limited) is typically inconsequential to control performance except in special cases like anti-surge control
- So, **minimize** the size of **Region 1 and 2**, **maximize** the size of **Region 3**

"Gain"

- **General:** change in *output* divided by the change in *input*

$$Gain = \frac{\Delta Output}{\Delta Input}$$

- **Process control:**

$$Process\ Gain = \frac{\Delta Controller\ PV\ (\%\ of\ span)}{\Delta Controller\ Output\ (\%\ of\ span)}$$

16AIChE Spring Meeting, Houston, TX

EMERSON. Process Management

Loop Gain

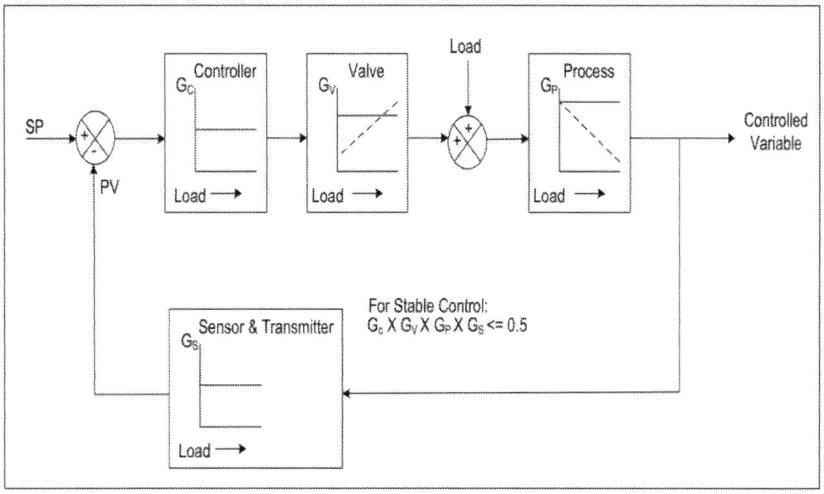

$G_C \times G_V \times G_P \times G_S \le 0.5$

Total Loop Gain = (Controller Gain) x (Valve Flow Gain) x (Process Gain) x (Transmitter Gain)

Valve flow gain magnitude and consistency are important!

16AIChE Spring Meeting, Houston, TX

EMERSON. Process Management

Control Valve Flow Characteristics

Control Valve Inherent Flow Characteristics :

The relationship of flow through the valve VS. valve position at constant process conditions and pressure drop

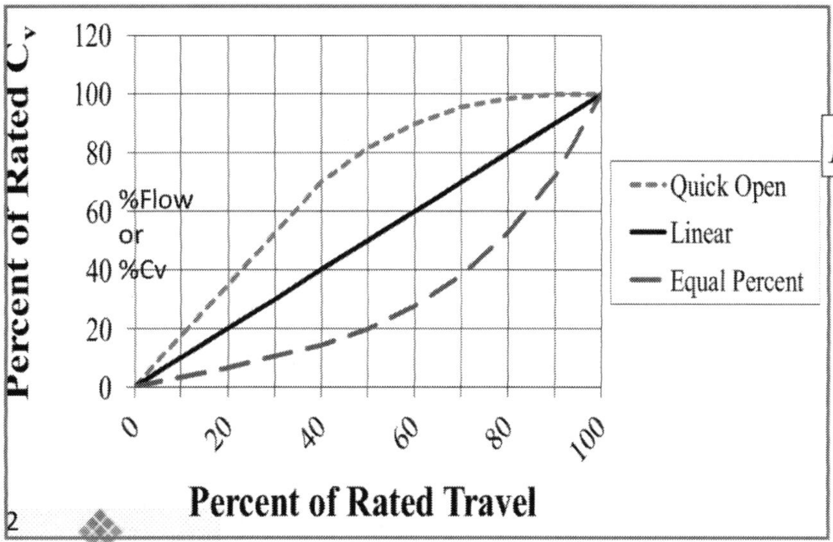

$$Flow = Cv\sqrt{\Delta P / SG}$$

C_V = valve flow coefficient (a function of valve position)
*Thus, at constant ΔP and SG, %Flow = % Rated Cv

26

16AIChE Spring Meeting, Houston, TX

EMERSON. Process Management

Total Loop Gain & Installed Valve Gain

- **Recall: Valve gain magnitude and consistency are important!**
- **Valve Flow Gain as Installed**

$$ValveFlowGain = \frac{\Delta Flowrate(\%)}{\Delta StemPosition(\%)}$$

$$Flowrate = Cv\sqrt{\Delta P / SG}$$

- **Thus, the Valve Flow Gain as installed, depends on inherent valve flow characteristics AND the actual ΔP across the valve**

16AIChE
Spring Meeting, Houston, TX

EMERSON.
Process Management

Valve Trim Selection Example:

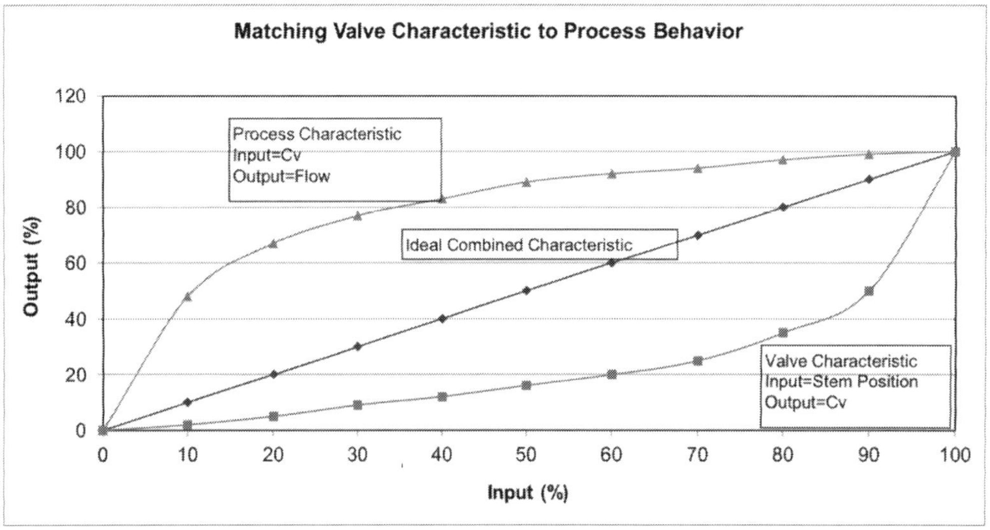

16AIChE
Spring Meeting, Houston, TX

EMERSON.
Process Management

Installed Gain – Goals and Selection

From a "% of flow span* perspective"…

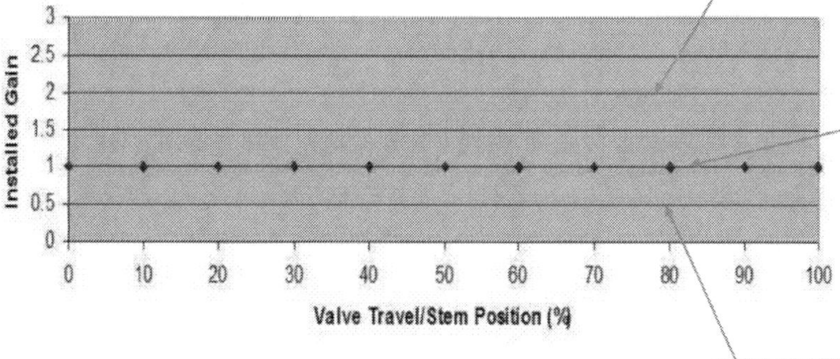

Gain ≤ 2.0 (to avoid amplification of valve non-linearities)

Gain as close to 1 as possible and as constant as possible

Gain ≥ 0.5 (to reach 100% PV)

* Could be span of flow controller, or maximum valve flow or maximum desired flow

16AIChE Spring Meeting, Houston, TX

EMERSON. Process Management

Installed Characteristics and Process Gain

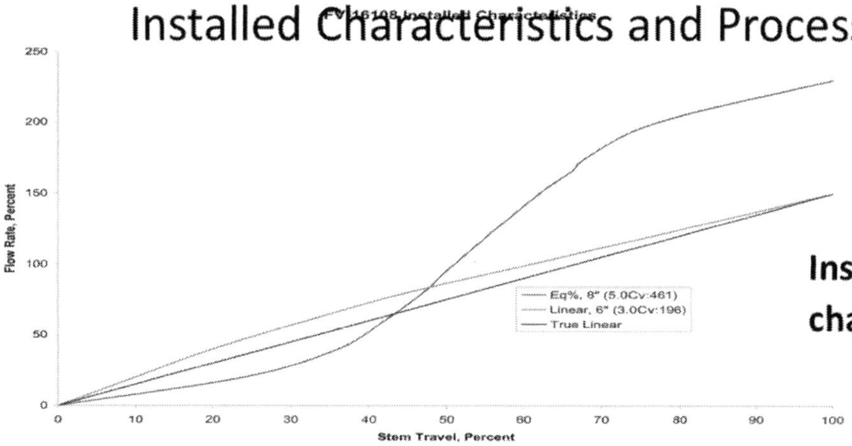

Installed characteristics

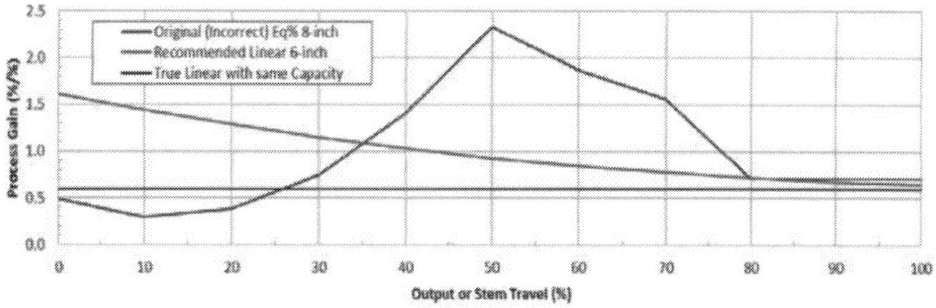

Installed gains

16AIChE Spring Meeting, Houston, TX

EMERSON. Process Management

Impact of Change in Valve Gain

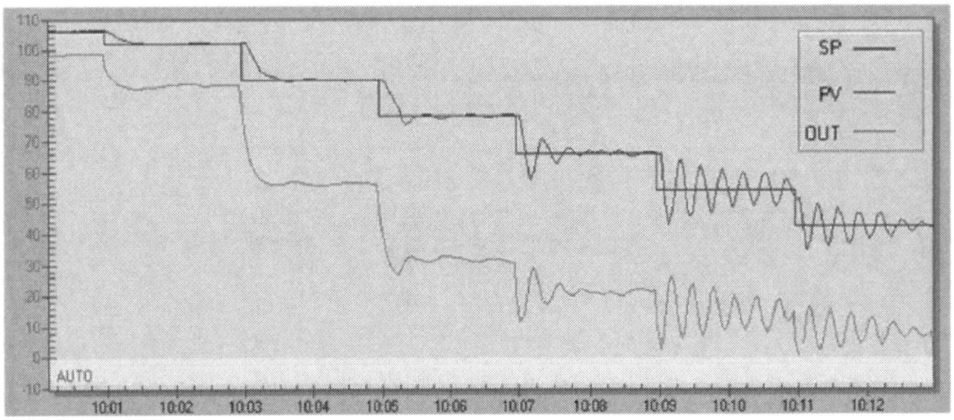

Process Gain increases as controller output decreases, causing loop instability.

Gain Characterizer

- A signal (gain) characterizer on the output of the controller, or in a smart valve positioner, can help linearize the total process and/or valve flow gain
  - Note that the use of this technique can cause confusion. The actual output to the valve, should be displayed to the operator.
  - An alternate method is to adjust the controller tuning based on the valve position (controller output)
- However, note that the characterizer DOES NOT reduce the PROCESS non-linearities due to valve non-linearities
  - If the characterizer REDUCES the process gain, it correspondingly INCREASES the effective valve non-linearity (e.g. dead band) resulting in the same effective PROCESS non-linearity!

Specifying Valve Performance

- Process or Valve Gain – close to 1 %PV/%OUTPUT and "constant as possible"
  - "Span" could be span of flow controller if applicable, or maximum valve flow or maximum specified flow (latter two choices assume 0 min. flow)
- Key Valve Tracking Performance Parameters
  - Valve Dead Band
    - Note Process Dead Band=Process Gain*(Valve Dead Band)
    - Thus, required Valve Dead Band = Allowable Process Dead Band / Process Gain
  - Resolution
    - Note Process Resolution = Process Gain*(Valve Resolution)
    - Thus, required Valve Resolution = Allowable Process Resolution / Process Gain

Specifying Valve Performance

- Response Time Parameters
  - Response Time (T_{86}) - < 1/3 of process time constant
  - Size of Region 1- (~=dead band/resolution) - bad, minimize
  - Size of Region 2 – bad, minimize
  - Size of Region 3 – wide as possible but over 10% typically not required
  - Size of Region 4 – not of great consequence to control
- *EnTech Control Valve Dynamic Specification V3.0* helps *SPECIFY* control valve performance

16 AIChE
Spring Meeting, Houston, TX
EMERSON.
Process Management

Summary

- Many operational problems can be prevented, and process control significantly improved, by proper selection of the control valve
  - Consider control loop stability and performance during valve procurement!
  - Valve gain, dead band, resolution, response time regions are clearly defined parameters
  - Severe control valve nonlinearities **cannot** be corrected by the process controller
- Applies across all process industries

16AIChE
Spring Meeting, Houston, TX

EMERSON.
Process Management

Presenter – James Beall

- Principal Process Control Consultant, Emerson Process Management
- 34 Years experience in process control
  - 15 Years Emerson Process Management
  - 19 Years Eastman Chemical Company
- Chairman ISA 75.25 Committee - Control Valve Performance Testing
- James.Beall@emerson.com 903-235-7935

EMERSON.
Process Management

16AIChE
Spring Meeting, Houston, TX

EMERSON.
Process Management

443948 Refinery Wastewater Treatment for 'zero Liquid Discharge'

Abstract:

Entire wastewater from a large refinery was selectively treated for oil removal, then for BOD/COD removal and comingled with freshwater cooling tower blowdown. After equalization of the streams divalent cation were reduced by physical chemical treatment followed by multi-pass multistage Reverse Osmosis membrane process. Reverse Osmosis brine was further treated with falling film thermal evaporator. More than 95% of the water was recovered.

443539 New Refinery Wastewater Treatment Plant at Petron Bataan Refinery in the Philippines

Abstract:

Paper Is about the Design and Performance of a New Waste Water Treatment Plant with a Capacity of 700 m3/h for the Petron Bataan Refinery in the Philippines.Six Different Influents from the RMP-2 Refinery Are Sent to the WWTP. Treatment Included: Deoling - Biological Treatment and Solids Separation - Filtration - Disinfection - VOC and Deodorization

442942 It's All about the Money, Reducing Operating Costs in Heavy Metals Wastewater

Abstract:

It's all about the Money

Acid Mine Drainage (AMD) is naturally occurring water that continues draining from mines after they are closed. This water usually has high concentrations of heavy metals and a low pH. This water will flow for centuries and needs to be treated before it mixes with surface water.

Many of the mines that caused this drainage are closed so there is no responsible party. Many states and the EPA have set aside trust funds to treat this waste water from some of these sites. Existing mines are required to set up treatment plants and establish trust funds for operation of the treatment system before mine closure.

This worked for many years until interest rates dropped. The amount of interest on the trust fund was insufficient to operate the plants. The plants had to dip into the principal to pay operating costs. This paper outlines what was done to reduce costs and the problem that the changes caused.

443256 Managing Water Treatment Risks in Refining

Abstract:

Against the backdrop of global water scarcity, the industrial sector is the fastest growing source of water consumption. If current trends persist, by 2030, industrial water use will rise by 65 percent. Petroleum refiners and plants rank among the top industrial water users, consuming huge volumes of fresh water for cooling, steam generation and process needs.

Few would argue against the industry's need to reduce water and energy consumption, but accomplishing it is a challenge. Refiners are operating in an environment with wide variability in converting feedstock into high-quality products and where profitability hinges on keeping key process units running, controlling operating costs, and effectively meeting regulatory requirements. They are under pressure to expand their operating envelope, while addressing growing environmental concerns.

In this presentation, water treatment technologies, commercially available and under development, will be explored that facilitate the responsible and economical use of water in an increasingly demanding future. This presentation will highlight:

- Crude blend and desalter management while utilizing opportunity, or discounted, crudes in refining.
- Managing wastewater plant variation and risks associated with processing these crudes.
- Use of alternative influent waters to minimize fresh water consumption.

445584 Consider Thermal and Hydraulic Impacts of Fouling in Crude Preheat Exchanger Design

Abstract:

Consider thermal and hydraulic impacts of fouling in crude preheat exchanger design

Krishnan S. Chunangad, Robert Chang, Lou Curcio, Richard Casebolt

ExxonMobil Research and Engineering, 22777 Springwoods Village Parkway, Spring, TX 77389, USA

Abstract

The overall energy efficiency of an oil refinery is highly affected by the performance of its crude preheat train which is often severely hindered by fouling, the deposition of unwanted material on thermal surfaces. Crude oils are complex fluids and many species such as salts, particles, wax, asphaltenes, coke and colloids can deposit on the tube surfaces thus impeding heat transfer and increasing pressure drops. As a result, the energy recovered in the preheat train is reduced, more fuel is burned at the furnace downstream of the pre-heat train, which in turn increases CO_2 emissions, throughput may be reduced, and units need to be periodically taken offline for cleaning. Annual losses associated to fouling can be quantified in the order of millions of dollars each year in individual refineries.

Traditional design methodologies deal with fouling in a very simplistic way by using fixed fouling factors. This neglects entirely any dynamics of the deposition process, as fixed fouling factors are not capable of capturing variations in fouling behavior as a function of process conditions. Moreover, information on the hydraulic impact of fouling is typically neglected. As a result, it is not possible to quantify the effects that a proposed design may have on the heat transfer and pressure drop performance through the exchanger run-length. To improve this aspect, a methodology that accounts for both thermal and hydraulic effects of fouling is needed.

This paper reports on the work where thermal data and hydraulic data as available, on crude preheat exchangers in seven different preheat trains, over several cleaning cycles, have been used to estimate the key fouling parameters in the Ebert-Panchal (1999) model using Hexxcell *Studio*™, a commercial software package for the thermo-hydraulic analysis, estimation and prediction of fouling in refinery heat exchangers. The data covers many crude blends with tubeside shear stresses ranging from 2 to 40 Pa at the hot end of the crude preheat train with multiple metallurgies. The simultaneous consideration of both temperatures and pressures allowed estimating the growth of fouling thermal resistance and hydraulic resistance (fouling layer thickness) with time and its effect on the exchangers' thermal and hydraulic performance. This information can be used to improve the design methodology for crude preheat exchangers to minimize fouling and maximize energy performance and train throughput.

In general, a good fit of the thermal-hydraulic plant data was achieved with the fouling model. The work also highlighted some known limitations of the Ebert-Panchal (1999) model. Both aspects will be briefly discussed in the presentation.

Keywords: heat exchange; crude oil; fouling; thermo-hydraulic performance; Ebert-Panchal; resistance;

444830 Cleaning Heat Exchangers: How, When, Where and How Much?

Abstract:

AIChE Spring Meeting 2016

ABSTRACT

Cleaning heat exchangers: how, when, where and how much?

Laura Lanchas-Fuentes, Emilio Díaz-Bejarano, Sandro Macchietto and Francesco Coletti*

Hexxcell Ltd., Imperial College Incubator, Bessemer Building Level 2, Imperial College London, SW7 2AZ (UK)

* Corresponding author: f.coletti@hexxcell.com

Energy recovery and production in oil refinery pre-heat trains are greatly affected by fouling, the progressive build-up of unwanted material on the heat exchanger surfaces. Even when good design and operating practices are followed and mitigation technologies are in place, very often crude oil fouling cannot be completely eliminated. As a result, it is necessary to carry out periodical cleaning of the heat exchangers.

Refineries typically use two methodologies to clean their heat exchangers: mechanical (e.g. hydro-blast) and chemical cleaning. Chemical cleaning methods represent a quicker and cheaper, though less effective, option than the more commonly used mechanical ones. They allow cleaning a heat exchanger in place, removing the need to open the shell and pull the bundle, thus significantly shortening the downtime. However, they typically return the exchanger to operations in a less clean condition. If the chemical used is ineffective, not only the benefits of a faster turnaround time can be lost, but the overall costs to the refinery may increase. To accurately assess the overalle economic benefits for the different options, it is therefore important to capture the trade-offs between the effectiveness of the cleaning action and the offline time required for it.

In this paper, the advantages and disadvantages between chemical and mechanical cleaning methods are discussed. Hexxcell *Studio*™, a thermo-hydraulic fouling analysis and prediction simulation software, is used to assess the impact of and techno-economic trade-offs between cleaning options. This cleaning analysis includes:

i) Establishing the effectiveness and depth of cleaning achieved by different chemical cleaning agents and the exchanger conditions left at the end (which depend on the deposit state prior to the cleaning and cleaning time). These will also affect fouling build-up in subsequent operation.

ii) Minimising the cleaning time for a specific chemical cleaning agent (which we term "condition-based cleaning")

iii) Calculating the detailed interaction effects that cleaning individual exchangers has on the fouling, heat duty and pressure drop of other exchangers in a network and overall

iv) Generating alternative cleaning schedules involving the selection of the appropriate type of chemical and mechanical cleaning, and assessing the cleaning schedule performance based on thermal and hydraulic considerations, costs of energy and refinery margins.

Results are presented using an industrially relevant case study.

443138 The Stratification of Crude Oil Residues at Elevated Temperatures Detected Via *in-Situ* ESR

Abstract:

The prevention or reduction of fouling caused by asphaltene deposition in pipelines, heat exchangers, and furnaces is one of the most serious challenges that oil industry faces nowadays. The temperature and pressure conditions involved in crude oil and resid processing may also trigger or exacerbate the formation of asphaltene deposits. In order to select the optimum temperature and/or pressure, as well as the appropriate feedstock composition that does not lead to asphaltene instability, it is important to understand and evaluate the behavior of the heavy components present in crude oils at elevated temperatures.

There are limited number of physical methods and techniques that can provide *in situ* information about the heavy components in crude oils in molecular scale and at elevated temperatures and pressures. Recently, we have shown that ESR using the VO^{2+} as a natural spin probe can provide quantitative information about the size distribution of vanadyl-containing fragments (asphaltenes and their aggregates) in heavy oil fractions. It was also shown that the precise simulation of the fine structure (FS) of ESR spectra of the VO^{2+}ions , measured in oils at temperatures when the partial averaging of the FS anisotropy is observed, allow us to separate the contributions of asphaltene molecules of different sizes that move with different characteristic rotational time.

The developed approach was used to register *in situ* the ESR spectra of atmospheric residues at temperatures of up to 420 °C. The simulated spectra fitted well with the experimental results obtained for all the evaluated residues at $T < 350$ °C using the smooth size distribution of vanadyl-containing fragments (VCF). Increase in temperature from 250 °C to 350 °C, decreased the viscosity of the fluid with concomitant decrease in the maximum size of the VCF; this response seems to be aligned with the disaggregation of the asphaltene agglomerates.

At 420 °C, the satisfactory agreement between experimental and simulated spectra could not be obtained using just plain size distribution of vanadyl-containing aggregates alone. The relatively good fitting between experimental and calculated spectra has been achieved using a two-step distribution function. Optimization of the parameters provided a satisfactory fitting of the simulated spectra with the experimental values. Similarly, we could obtain quantitative information about the size distribution of VCF present in the sample at elevated temperatures. The results also showed that some vanadyl-containing fragments remain mobile, while some "precipitate" or move to a sedentary state at high temperature.

Thus, the precise simulation of ESR spectra of vanadyl-containing fragments obtained *in situ* provides quantitative information on the dynamics and size changes of heavy components in crude oils at elevated temperatures.

AUTHOR INDEX

AUTHOR INDEX

AUTHOR INDEX